Operator Commutation Relations

Mathematics and Its Applications

Operator Commutation Relations

Commutation Relations for Operators, Semigroups, and Resolvents with Applications to Mathematical Physics and Representations of Lie Groups

Palle E. T. Jørgensen
Department of Mathematics,
University of Pennsylvania, Philadelphia, U.S.A.

and

Robert T. Moore
Department of Mathematics,
University of Washington, Seattle, U.S.A.

D. Reidel Publishing Company

A MEMBER OF THE KLUWER ACADEMIC PUBLISHERS GROUP

Dordrecht / Boston / Lancaster

Library of Congress Cataloging in Publication Data

CIP

Jørgensen, Palle E. T., 1947-
Operator commutation relations.

(Mathematics and its applications)
Bibliography: p.
Includes index.
1. Partial differential operators. 2. Commutation relations (Quantum mechanics) 3. Lie groups. 4. Representations of groups. I. Moore, Robert T., 1938- II. Title. III. Series: Mathematics and its applications (D. Reidel Publishing Company)
QA329.42.J67 1984 515.7'242 83-26957
ISBN-13: 978-94-009-6330-6 e-ISBN-13: 978-94-009-6328-3
DOI: 10.1007/978-94-009-6328-3

Published by D. Reidel Publishing Company,
P.O. Box 17, 3300 AA Dordrecht, Holland

Sold and distributed in the U.S.A. and Canada
by Kluwer Academic Publishers,
190 Old Derby Street, Hingham, MA 02043, U.S.A.

In all other countries, sold and distributed
by Kluwer Academic Publishers Group,
P.O. Box 322, 3300 AH Dordrecht, Holland

TABLE OF CONTENTS

PREFACE

In his Retiring Presidential address, delivered before the Annual Meeting of The American Mathematical Society on December, 1948, the late Professor Einar Hille spoke on his recent results on the Lie theory of semigroups of linear transformations, ..."So far only commutative operators have been considered and the product law ... is the simplest possible. The non-commutative case has resisted numerous attacks in the past and it is only a few months ago that any headway was made with this problem. I shall have the pleasure of outlining the new theory here; it is a blend of the classical theory of Lie groups with the recent theory of one-parameter semigroups." The list of references in the subsequent publication of Hille's address (Bull. Amer. Math.. Soc. 56 (1950)) includes pioneering papers of I.E. Segal, I.M. Gelfand, and K. Yosida. In the following three decades the subject grew tremendously in vitality, incorporating a number of different fields of mathematical analysis. Early papers of V. Bargmann, I.E. Segal, L. Gårding, Harish-Chandra, I.M. Singer, R. Langlands, B. Konstant, and E. Nelson developed the theoretical basis for later work in a variety of different applications: Mathematical physics, astronomy, partial differential equations, operator algebras, dynamical systems, geometry, and, most recently, stochastic filtering theory. As it turned out, of course, the Lie groups, rather than the semigroups, provided the focus of attention. However, the operator theoretic viewpoint is the same, and the influence of Hille's original ideas is present in the later applications. This book is devoted to some of the important developments since 1948.

Dedicating our book to the memory of Einar Hille, we find it quite appropriate to begin with a quote from Kipling: "And each man hears as the twilight nears, to the beat of his dying heart, the Devil drum on the darkened pane: 'You did it, but was it Art?'". This, in fact, was Hille's own beginning quote in his first edition of Functional Analysis and Semi-groups.

Although, on the surface, the different topics of the

present book are quite distinct, we shall show that a certain analysis of operator commutation relations (or rather a structure of identities for families of operators, and iterated commutators of the individual operators) leads to an unexpected unification, as well as to a variety of new results, in diverse areas of mathematics and applications.

As a first indication of what is involved, consider the familiar regularity problem for solutions to partial differential equations. Suppose A is a linear partial differential operator, and suppose that a fixed linear space of functions, or sections in a vector bundle, has been chosen such that the resolvent operator $(\lambda I-A)^{-1}$ provides a solution formula for the inhomogeneous equation $\lambda u - Au = f$ with given right-hand f in E. As is well known, the regularity problem for the solution u can then be stated in terms of domains for the first order vector fields $\partial/\partial x_i$, $i=1,\ldots,n$. One way to check regularity, i.e., to check whether the solution u is in the domain of $\partial/\partial x_i$, or in the domain of higher order polynomials in the $\partial/\partial x_i$'s, is to consider the formal expansion,

$$\sum_{k=0}^{\infty} (-1)^k(\lambda I-A)^{-(k+1)}(\text{ad } A)^k(\partial/\partial x_i)f.$$

Suppose the convergence questions which are implicit in the series can be resolved, then by simple associative algebra one would expect the formula to represent the x_i-derivative of the solution, viz., $\partial u/\partial x_i$ where

$$\partial u/\partial x_i = (\partial/\partial x_i)u = (\partial/\partial x_i)(\lambda I-A)^{-1}(f).$$

Hence, the regularity question has been reformulated in terms of a commutation-relation, and a domain problem.

Historically, our work on the subject originated with a different problem. Rather than the commuting family of first order operators $\partial/\partial x_i$, we took, as the starting point, a given Lie algebra $\mathfrak{L}$ of operators on a complete normed space E. What this amounts to is that the elements in $\mathfrak{L}$ are unbounded operators on E, but the different operators have a common dense invariant domain D in E. The Lie algebra structure on $\mathfrak{L}$ refers to the commutator bracket, i.e., elements A and B in $\mathfrak{L}$ are considered as operators, and the bracket is $[A,B] = AB - BA$. Then note that $[A,B]$ is again an operator with the same dense domain, viz., D. We consider $\mathcal{L}$ as a finite-dimensional real Lie algebra. By Ado's theorem, it is also the Lie algebra of some (analytic) Lie group G, and we shall choose G to be simply connected (for the purpose of the present introduction). The exponentiation problem was considered in a number of earlier papers, notably [Bg], [Sr 1], and [Nℓ 1]. This was the starting point of our work. The problem is to find sufficient conditions on the operators (alias elements in $\mathcal{L}$) which ensure that $\mathcal{L}$ exponentiates to a strongly continuous representation of the Lie group G by bounded operators on the space E.

Our approach to this problem goes through a system of formal commutation relations. The sufficient conditions on $\mathcal{L}$ are precisely those which provide rigorous convergence criteria for the iterated commutator expansions. Again, it emerges that domain considerations for the unbounded infinitesimal operators play a central role.

As a particular example, we could have a Lie algebra of partial differential operators. While the theory allows for operators of any order, the special case of first order operators has an intrinsic geometric significance. This case was considered in [Pℓ]. The solution by Palais is an interesting application of the theory of foliations. Palais' theorem (which provides a simple condition on a given Lie algebra of smooth vector fields on a manifold to integrate to an action of the corresponding Lie group) is taken up in Chapter 10 where we show that the integrability comes out quite easily as a direct application of our first theorem on exponentiability for operator Lie algebras. This means, in particular, that the proof only relies on operator theory, with no reference to the geometry of the underlying smooth manifold. Other applications of our first exponentiation theorem are taken up in Chapters 9 through 12. As one new result here we mention only our perturbation theory for representations of Lie groups.

In this preface we have chosen to limit the discussion to two topics, rather than a summary of all the items in the twelve chapters. In each of the different problems, the underlying analysis will be shown to rely, through a sequence of steps, on one of the three different commutation relations treated in the book. For a given pair of operators A and B, we shall consider, (i) the resolvent operators $(\lambda I-A)^{-1}$ for values of λ in the resolvent set, (ii) the functional calculus $\varphi(A)$ for a certain class of scalar functions φ; and, finally (iii) the strongly continuous semigroup $V(t,A) = \exp(t\bar{A})$ in case A has a closure as an operator on E, and the closure $\bar{A}$ is the infinitesimal generator of a semigroup in the sense of Hille and Yosida. The three types of commutation relations then take the following form,

(i) $$B(\lambda-A)^{-1} = \sum_{k=0}^{\infty} (-1)^k(\lambda-A)^{-(k+1)}(\mathrm{ad}\ A)^k(B)$$

(ii) $$B\ \varphi(A) = \sum_{k=0}^{\infty} \frac{(-1)^k}{k!}\ \varphi^{(k)}(A)(\mathrm{ad}\ A)^k(B)$$

and

(iii) $$B\ V(t,A) = V(t,A)\exp(-t\ \mathrm{ad}\ A)(A)$$

respectively. Here $\varphi^{(k)}$ denotes the kth derivative of the scalar function φ, and $\varphi^{(k)}(A)$ is the corresponding functional calculus operator. The expression $(\mathrm{ad}\ A)^k(B)$ is defined inductively through $(\mathrm{ad}\ A)^0(B) = B$, and $(\mathrm{ad}\ A)^{k+1}(B) = [A, (\mathrm{ad}\ A)^k(B)]$, and finally $\exp(-t\ \mathrm{ad}\ A)(B)$ is defined via the formal exponential power series

$$\sum_{k=0}^{\infty} \frac{(-t)^k}{k!} (\mathrm{ad}\ A)^k(B) .$$

The rigorous convergence questions for the respective formal expansions will differ from problem to problem, and the analysis may take entirely different forms even though the underlying set of formal commutation relations is the same throughout.

It turns out, in each case, that the solution to the particular problem amounts to the answers to two questions which are implicit in the statement of identities (i) through (iii). The first question, which is often the easier one, is the convergence issue for the infinite series on the right-hand side of the three identities. A particular setting where this question has an easy answer is the one where A and B both belong to a given finite-dimensional Lie algebra of operators. In this case, the expression $\exp(-t\ \mathrm{ad}\ A)(B)$ is then just a one-parameter orbit in the finite-dimensional adjoint representation.

The other question refers to the left-hand side of the three commutator formulas: Although D is a common invariant domain for the two unbounded operators, A and B, the three bounded operators $(\lambda-A)^{-1}$, $\varphi(A)$, and $V(t,A)$, generally do not leave D invariant. This means that a rigorous analysis of (i) through (iii) must include a discussion of the closure $\bar{B}$ of the operator B. The essential and deep analytic question then amounts to this: For elements f in D, can we show, in case (i), that $(\lambda-A)^{-1}f$ belongs to the domain of $\bar{B}$? Similarly, in cases (ii), and (iii), the question is whether $\varphi(A)(f)$, respectively $V(t,A)f$, falls in the domain of the closed operator $\bar{B}$. A special issue which must also be addressed in each case is, of course, the closability of the operator B.

The book consists of seven different parts, the parts are subdivided into chapters, and the chapters again into sections. Parts are numbered I through VII, chapters, 1 through 12, and, in each chapter, the sections are denoted A, B, etc. We have strived to make the seven parts independent: Each part starts with an introduction which describes the main ideas that will be used. It summarizes the prerequisites and establishes special notation and symbols.

Three parts depend particularly heavily on results from different places in the book. These parts, IV, VI, and VII, each conclude with appendices where the relevant results are recorded. Part IV, for example, on the exponentiability problem uses important theorems from Chapter 3, and Chapters 5 through 7. Therefore, in the appendix to Part IV, we have simply recorded the theorems in question. A few comments have been inserted to inform, and orient, the reader where the given theorem fits into the application to the proofs of the main results inside Part IV. A reader who wants to begin directly with Part IV should be able to do this after reading only the introduction and the appendix. He would not have to be chased through a complicated net of cross references, even though many cross-dependencies exist.

With a slight over-simplification we have charted the main cross-dependencies between the twelve chapters. The interpretation of the arrows in the Chapter-Dependency-Diagram below can be explained by the following example:

The proofs of the main results in Chapter 9 depend on theorems from Chapter 3, and Chapters 5 through 7. In some cases, this dependency is indirect: the theorems in question get combined and applied in Chapter 8 where the integrability theorem for smooth operator Lie algebras is proved. The exponentiation and perturbation theorems in Chapter 9 are then, in turn, applied in two different areas, commutation relations in mathematical physics (Chapter 10), and analytic continuation of $s\ell(2,\mathbb{R})$-modules (Chapter 12). Note that the results in Chapter 12 also rely on separate developments in other chapters of the book.

In some cases where the end of a proof or a remark is not obvious from the context we have used the symbol E.O.P. (resp., E.O.R.) to signal this fact.

CHAPTER DEPENDENCY DIAGRAM

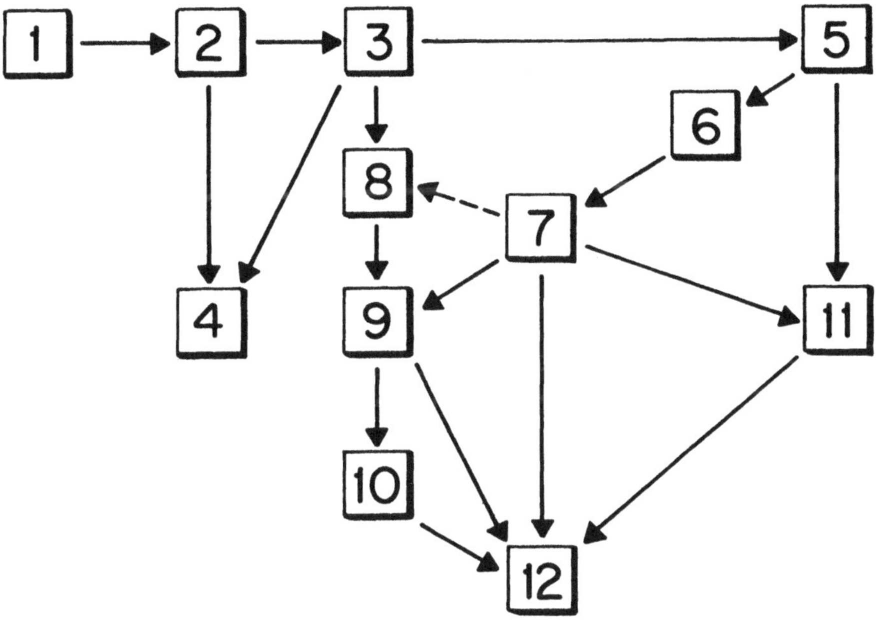

... thought Alice, and she went on. "Would you tell me, please, which way I ought to go from here?"
"That depends a good deal on where you want to get to", said the Cat.
"I don't much care where -" said Alice.
"Then it doesn't matter which way you go," said the Cat.
" - so long as I get *somewhere*," Alice added as an explanation.
"Oh, you're sure to do that," said the Cat, "if you only walk long enough."

Alice's Adventures ...
Chapter VI
LEWIS CARROLL

ACKNOWLEDGEMENTS

First and foremost, we wish to acknowledge the invaluable advice and encouragement supplied by Professors Edward Nelson and Niels Skovhus Poulsen during the initial student efforts of the co-authors which began the series of investigations leading ultimately to the body of results presented in this monograph.

We also wish to acknowledge direct and indirect debt to Professor Irving Segal. Our whole subject area has evolved through his pioneering efforts in this interface between physics, differential equations, operator theory, group representations, and functional analysis. Moreover, his guidance not only of the student efforts of Professors Nelson and Poulsen, but also of Goodman (and indirectly of several others acknowledged below) has in large part created the mathematical community from whom we have received most of our stimulation and encouragement. Our indirect debt to Professor Lars Gårding is similar in nature.

Both authors would like to thank the following colleagues for helpful conversations over the years concerning these and related matters:Professors V. Bargmann, M. Flato, G. Folland, J.M.G. Fell, R. Goodman, R. Kadison, R. Powers, S. Sakai, E.M. Stein, and A.S. Wightman. The first author has also had useful discussions with Professor C. Radin on certain technical aspects of the graph density condition for operator Lie algebras isomorphic to $s\ell(2,\mathbb{R})$.

The second author also expresses his appreciation to R.F.V. Anderson, W.G. Faris, R.E. Howe, O.E. Lanford, A. Cohen Murray and R.S. Palais for useful discussions on portions of this material.

Both authors wish to thank the various Universities and funding agencies which have offered facilities and support during the conduct of the research reported here. Specifically, the first author's thanks go to Aarhus University for student support and to Odense University for a research stipenium which has funded subsequent research and travel expenses, while the second author's thanks are similarly directed to Princeton University (for student support), to the National Science Foundation (for a decade of grant support, most recently NSF GP 7727 and amendments). Both authors thank Aarhus University, Princeton University, the University of Pennsylvania, the University of Washington and Stanford University for facilities, support and hospitality at one time or another during the gestation period of these ideas, and the second author's appreciation also extends to the University of California (Berkeley).

Finally, we wish to thank the publisher, and Professor M. Hazewinkel - the editor of Mathematics and its Applications - who have been most helpful in bringing this monograph to publication. Special thanks are due to Dr. D.J. Larner for efficiently organizing and managing the production of the final manuscript, and to mrs. Fenny Staal, who typed the entire final typescript of the book, for her fine and devoted work.

PART I

SOME MAIN RESULTS ON COMMUTATOR IDENTITIES

"The Other Professor is to recite a Tale of a Pig - I mean a Pig-Tale," he corrected himself. "It has Introductory Verses at the beginning, and at the end."
"It ca'n't have Introductory Verses at the *end*, can it?" said Sylvie.
"Wait till you hear it," said the Professor: "then you'll see. I'm not sure it hasn't some in the *middle*, as well."

Sylvie and Bruno Concluded
The Pig-Tale
LEWIS CARROLL

... I saw that the noncommutation was really the dominant characteristic of Heisenberg's theory ... So I was led to concentrate on the idea of noncommutation and to see how the ordinary dynamics which people had been using until then should be modified to include it. (Dirac, 1971, pp. 20-24.)

Chapter 1

INTRODUCTION AND SURVEY

This brief chapter provides a general overview of the kind of mathematics treated in our monograph. Section 1A indicates the general questions and topics addressed here, while Section 1B describes some of the prior work which bears upon these topics. Then Sections 1C through 1E supply precise formulations of the main theoretical results, and Sections 1F and 1G sketch the principal applications of these results.

1A. General Objectives of the Monograph

This monograph is the first of two projected publications by the co-authors concerning commutation theory, spectral analysis, and joint exponentiation and perturbation theory for several noncommuting unbounded operators in Banach (and general locally convex) spaces. Our approach emphasizes auxiliary core subspaces of C^∞ vectors, with the usual stronger C^∞-vector topology. Some of the main results can be viewed as simultaneous generalizations of two classical lines of development. First, we supply natural infinite-dimensional extensions of classical facts from the commutation theory and Lie theory of matrices (qua finite-dimensional operators). Second, we obtain abstract generalizations for functional-analytic versions of classical results in the commutation theory and integration theory which connects vector fields on manifolds with global smooth flows or Lie transformation groups. Here, vector fields and diffeomorphisms are viewed as operators on various Banach or locally convex function spaces. This distinction is reflected to some degree in two qualitatively different types of sufficient conditions (below) which ensure that the various formal commutation and exponentiation identities of interest in applications are rigorously correct. It has its historical roots in the two different formalisms of early quantum physics: the Heisenberg formalism, which viewed operators (or "observables") as $\infty \times \infty$ matrices; versus the Schrödinger formalism, where operator-observables were represented by partial differential operators.

The main theoretical results in commutation theory concern

pairs of closable linear endomorphisms A,B, acting on a dense subspace D of the representation Banach (or locally convex) space E; most require that the family of operators $\mathcal{O}_A(B) = \mathrm{span}\{(\mathrm{ad}\ A)^k(B): k = 0,1,\ldots\}$ be finite-dimensional, where $\mathrm{ad}\ A(B) = AB-BA$, etc. Our two types of sufficient conditions then rigorously justify the following sorts of formal commutation relations

$$B(\lambda-A)^{-1} = \Sigma\{(-1)^n(\lambda-A)^{-(n+1)}(\mathrm{ad}\ A)^n(B): 0 \leq n < \infty\} \quad (1)$$

$$\begin{aligned} B\exp(tA) &= \exp(tA)\exp(-t\ \mathrm{ad}\ A)(B) \\ &= \Sigma\{(-1)^n/n!\ t^n\exp(tA)(\mathrm{ad}\ A)^n(B): 0 \leq n < \infty\} \quad (2) \end{aligned}$$

$$B\varphi(A) = \Sigma\{(-1)^n/n!\ \varphi^{(n)}(A)(\mathrm{ad}\ A)^n(B): 0 \leq n < \infty\} \quad (3)$$

where $\varphi(A)$ denotes the image of A under a suitable operational calculus. Variants of (1)-(3) are obtained which exploit the Jordan canonical form of ad A on finite-dimensional $\mathcal{O}_A(B)$. Identity (2), in particular, is exploited in justifying a simple local construction of a Lie group representation from one-parameter operator groups by "canonical coordinates of the second kind". Extensions of (3) are used to show that certain spectral projections for A still may reduce non-commuting B, and to derive results on the spectral structure of A and B that are entailed by their commutation relations.

For the "Schrödinger formalism" situation, with differential operators acting on function or distribution spaces, the appropriate sufficient condition for (1)-(3) and their applications requires primarily that the C^∞-vector domain D be invariant under the C_0-semigroup-exponential $V(t,A) = \exp(t\bar{A})$. For the contrasting "Heisenberg formalism", with matrix operators on sequence spaces, the appropriate assumption imposes a "zero-deficiency" condition on A with respect to a stronger generalized Sobolev or C^1-norm topology. Both lead to useful recasting of problems on a familiar type of dense re-topologized Fréchet space of C^∞ vectors. Stability of the zero-deficiency condition under small perturbations is used to obtain an analytic perturbation theorem for Lie group representations related to Phillips' perturbation theorem for one-parameter semigroups.

These techniques are applied to several previously-studied problems:

(i) A new, pure C^∞-vector treatment is supplied for the O'Raifeartaigh-Jost-Segal "mass-splitting" principle of elementary particle physics.
(ii) A functional analytic proof of Palais' global integration theorem for Lie algebras of vector fields is given, and this is applied to the study of certain polynomial coefficient hypoelliptic operators as invariant operators on nilmanifolds.
(iii) Strong commutativity, the canonical (PQ-QP = c) commutation relations, dynamics of the quantum harmonic oscillator and other related problems are rigorously analyzed.
(iv) Bargmann's analysis of the infinitesimal "irreducible unitary representations" of $s\ell(2, \mathbb{R})$ as almost-diagonal matrices on $\ell^2(\mathbb{Z})$ is extended to a non-unitary theory on all ℓ^p, $1 \leq p \leq \infty$.
(v) On ℓ^2 we obtain a technically simple infinitesimal construction and analytic continuation for a class of quasi-simple unitary and non-unitary representations of the group $G = SL(2, \mathbb{R})$ via Phillips' perturbation theory. This leads to a new analytic embedding of the dual object $\hat{G}$ into bounded operators on Hilbert space.
(vi) For ℓ^p with $p \neq 2$ the situation is different. Rather than group representations we get representations of the convolution algebra of test functions on $SL(2, \mathbb{R})$ by algebras of bounded operators on ℓ^p, (smeared representations).
(vii) Ordinary group representations are obtained in the Fourier-transformed picture on $L^p(T)$, for $T = \{z \in \mathbb{C}: |z| = 1\}$ the unit circle and $p \neq 2$.

For those readers who desire a more precise picture of the main results of this monograph before reading the full development in Chapters 3-12, we supply abbreviated statements of the main definitions and results in Sections 1C-G below. Self-contained (and occasionally simplified) versions of the main theoretical theorems from Chapters 3-9 are previewed in Sections 1C through 1E, making explicit the two sufficient conditions suggested above. The applications sections 1F and 1G are less formal, stating some of the more definitive results and describing others. None of the material in Sections 1C-G is prerequesite to understanding the main development.

1B. Contact with Prior Literature

There is an extensive prior literature in operator theory and mathematical physics concerned with just the sorts of infinite-dimensional commutation-theoretic and Lie-algebraic matters treated below. Key early papers include those of Segal [Sg 1], Gårding [Gå], and Nelson-Stinespring [NS], which are concerned with density and core properties of C^∞-vector domains in the presence of a strongly-continuous Lie group representation. These are direct ancestors of our present work.

The fundamental paper of Nelson [Nℓ 1] in 1959 served, on the one hand, to introduce the study of general "infinitesimal" sufficient conditions for commutation and exponentiation theory that concerns us here, via the "essentially self-adjoint Laplacian" condition for exponentiating (alias integrating) Lie algebras of skew-symmetric operators. (The integrability problem was formulated first in the 1950 thesis [Sr 1] of I.M. Singer.) On the other hand, it initiated the emphasis upon analytic vectors, rather than C^∞-vectors, as the tool of preference for dealing with problems of the sort attacked here.

In most of the subsequent literature, as a result, one finds analytic vector methods playing a dominant role: Dixmier [Dx 2], Flato, Simon, and others [FSSS], [FS 1-2], [S], Goodman [Gd 1-5], Poulsen [Ps 2] and Powers [Pw 1-2]. (See also Gårding [Gå 2] and Chapter 8 of the second author's Memoir [Mr 5].) The most successful results of this type are confined to self- and skew-adjoint operators related to unitary groups in Hilbert spaces, with methods such as Cayley transforms, the spectral theorem, Stone's theorem, and more recently the powerful order theory and commutant theory of von Neumann algebras as indispensable tools (cf. [Pw 1-2]).

By contrast, the C^∞-vector approach which we adopt here has received less attention, despite its broader availability in the non-unitary Banach and locally convex setting (see [Wr 1], [Mr 5], Poulsen [Ps 1], and Jorgensen [Jo 2].) In order to achieve such non-unitary generality, it is of course necessary to replace Stone's generation theorem for unitary groups by the Hille-Yosida theorem (Hille-Phillips [HP], Kato [Kt 2] and Yosida [Yo 1]) and its modern locally convex or distribution-theoretic extensions (e.g., Komura [Ko], Chazarain [Cz]). The Stone-von Neumann operational calculus for bounded measurable functions supplied by the spectral theorem is correspondingly replaced by the holomorphic Dunford-Taylor calculus or the Bade-Phillips bilateral Laplace transform calculus (Dunford-Schwartz [DS 1]), and occasionally by Dunford's theory of general spectral operators [DS 3], or its generalizations due to Foias-Colojohara [FC] and others.

As indicated (in the context of group representations) by the comparative treatment of C^∞- and analytic vectors in the monograph of Warner [Wr 1, Chapter 4], there have in the past been sound technical reasons for preferring analytic vectors. However, we exhibit systematic techniques for surmounting some of the disadvantages of C^∞-vectors that are displayed by examples in [Wr 1]. It is for this reason that we have generally avoided the use of analytic-vector arguments even in the relatively few cases when these are available for the applications treated in the present monograph.

The main results on integrability of Lie algebras (Chapter 9) were obtained in joint work by the co-authors in 1974. These theorems in turn extend earlier independent work of the co-authors.

1C. The Main Results in Commutation Theory

Two different rigorous formulations for the formal commutation identities (1)-(3) (Section 1A) are developed in Chapters 3-6, and these are derived from the sufficient conditions mentioned in Section 1A. Here, we summarize typical cases of the main results from these chapters in a unified setting defined by the following assumptions.

(B) The linear operators A and B are closable endomorphisms of a dense domain D of "C^∞ vectors" in a Banach space E.

(F_c) The ad-orbit of B under A, $\mathcal{O}_A(B) = \text{span}\{(\text{ad } A)^k(B): 0 \leq k < \infty\}$ is finite-dimensional, and has a basis $B_1,\ldots,B_d$ of closable operators.

(E) The closure $\bar{A}$ of A generates a strongly continuous (C_0) semigroup $\{V(t,A): t \geq 0\}$ of bounded operators on E, which plays the rôle of "exp(tA)".

By the Hille-Yosida theorem (reviewed in Section 6A), condition (E) ensures that at least for all λ in a half-plane $\text{Re}(\lambda) > \omega$, the resolvent $(\lambda-\bar{A})^{-1} = R(\lambda,A)$ exists as a bounded operator, and various operational calculi ensure that at least for φ holomorphic in that half-plane, $\varphi(A)$ exists (see Sections 5D, 6D).

As is shown in Chapter 3, the simplest type of sufficient condition for commutation theory is the one most natural for A and B differential operators in some function space E (the "Schrödinger formalism"). There, we impose the additional domain invariance condition

(D) for all $t \geq 0$, $V(t,A)D \subset D$.

This permits identity (2) from Section 1A to be formulated quite simply as

$$B\, V(t,A)u = V(t,A)\, \exp(-t \text{ ad } A)(B)u$$
$$= \Sigma\{(-1)^n/n!\; t^n\, V(t,A)(\text{ad } A)^n(B)u\colon 0 \leq n < \infty\} \quad (4)$$

for all $u \in D$. (Here, $\exp(-t \text{ ad } A)$ is the group of endomorphisms of finite-dimensional $\mathcal{O}_A(B)$ generated by ad $A\colon \mathcal{O}_A(B) \to \mathcal{O}_A(B)$.)

All known results rigorously justifying (4) require an additional regularity assumption on B and $V(t,A)$ beyond condition (D). The following theorem is typical, and is probably the appropriate one for most applications.

Theorem 3.4'

Suppose that A and B satisfy (B), (F_c), (E) and (D) above. Suppose that the following regularity condition holds:

(R) $\bar{B}$ generates a C_o semigroup, and for all $u \in D$, and bounded linear functionals $f \in E^*$ the functions $G_u^f(t) = f(B\, V(t,A)u)$ are bounded on some $[0,\varepsilon)$. ($\varepsilon > 0$ may depend on u and f.)

Then (4) holds.

This result was obtained by the second author in 1963 [Mr 1,2], and independently by the first in [Jo 1], for the special case where A and B generate a finite-dimensional Lie algebra of endomorphisms of D. The extended versions given in Chapter 3 relax both condition (B): (E may be locally convex) and more importantly the closable basis condition in (F_c), necessitating a more technical proof due jointly to the co-authors. Another more recent joint result addresses the question of possible redundancy of condition (R): is (D) by itself sufficient in the presence of (B), (F_c) and (E)?

Proposition 3.6'

Suppose that (B), (F_c) and a strengthened form of (E) hold: $\bar{A}$ generates a C_o group $\{V(t,A): t \in \mathbb{R}\}$, and that $\bar{B}$ generates a C_o semigroup. Suppose further that D is a Fréchet space with respect to a topology stronger than the relative norm topology. Then all $G_u^f(t) = f(B\ V(t,A)u)$ are automatically bounded on compacta as in condition (R), and (4) holds.

Versions of the formal identities (1) and (3) from Section 1A can then be derived from (4), by methods implicit in Chapter 6. The details of formulation are discussed below, following our survey of Chapters 5 and 6.

The sufficient condition that has proven more suitable in the setting of matrix operators in sequence spaces ("Heisenberg formalism") requires the construction of an auxiliary "C^1-norm" of Sobolev type, which is also instrumental in the rigorous formulation of identities (1)-(3) when (4) fails. For $u \in D$, we put

$$\|u\|_1 = \max\{\|B_i u\|: 0 \le i \le d\} \tag{5}$$

where $B_0 = I$, the identity operator, and $\{B_1,\ldots,B_d\}$ supplies a closable basis for $\mathcal{O}_A(B)$. This norm is regarded as the graph-norm induced by $\mathcal{O}_A(B)$, and is independent of the choice of basis. The appropriate sufficient condition for commutation theory in Chapters 5 and 6 then reads:

(GD) for a given λ, the space $D_\lambda = (\lambda - A)D$ is $\|\cdot\|_1$-dense in D.

Closability of the B_i easily shows that the $\|\cdot\|_1$-completion D_1 of $(D,\|\cdot\|_1)$ is canonically injected into E with $D \subset D_1 \subset E$, and A gives rise to a natural closable operator in the Banach space $(D_1,\|\cdot\|_1)$ which we shall denote by A_1.

Condition (GD) can then be interpreted as a zero-deficiency condition: λ cannot be in the residual spectrum of the closure $\bar{A}_1$ of A in $(D_1, \|\cdot\|_1)$. In our results below, it is important that λ turns out to be in the resolvent set of $\bar{A}_1$, and its resolvent $R_1(\lambda,A) = (\lambda-\bar{A}_1)^{-1}$ is essential to the rigorous formulation of identity (1) from Section A as an identity in bounded operators from D_1 to E. We also need the fact that every $C \in \mathcal{O}_A(B)$ extends naturally to a bounded operator $C^{(1)}$ from $(D_1, \|\cdot\|_1)$ to $(E, \|\cdot\|)$.

The following result, which orginates with [Mr 1,2], indicates the flavor of the theory. (Assumptions (B) and (F_c) are in force; (E) is optional.)

Theorem 5.4'

Suppose that λ satisfies condition (GD), and that λ is in the resolvent set of A, with the property that the distance $\mathrm{dist}(\lambda, \sigma(\bar{A}))$ from λ to the spectrum of A is larger than the spectral radius $\nu(\mathrm{ad}\ A) = \max\{|\alpha| : \alpha \in \sigma(\mathrm{ad}\ A)\}$ of ad A on the complexification $\mathcal{O}_A^{\mathbb{C}}(B)$ of $\mathcal{O}_A(B)$. Then the following hold.

The resolvent $R(\lambda,A)$ of $\bar{A}$ on E leaves $D_1 \subset E$ invariant, and restricts there to the $\|\cdot\|_1$-bounded resolvent $R_1(\lambda,A)$ of $\bar{A}_1$.

The infinite series commutation relation.

$$B^{(1)}R_1(\lambda,A) = \Sigma\{(-1)^n R(\lambda,A)^{n+1}[(\mathrm{ad}\ A)^n(B)]^{(1)} : 0 \leq n < \infty\} \quad (6)$$

holds, as a norm-convergent expression in bounded operators from D_1 to E.

As stated in Chapter 5, Theorem 5.4 also contains a refinement of identity (6), and sharper results on the spectral behavior of $\bar{A}_1$ in D_1. This further information brings the spectral behavior of ad A on $\mathcal{O}_A^{\mathbb{C}}(B)$ more explicitly into play via the Jordan canonical form of ad A. That is, $\mathcal{O}_A^{\mathbb{C}}(B)$ has a basis $\{B_i\}$ of generalized eigen-elements, each associated with an eigenvalue $\alpha_i \in \sigma(\mathrm{ad}\ A)$ and a step (or ascent) s_i such that $(\mathrm{ad}\ A-\alpha_i)^{s_i+1}(B_i) = 0 \neq (\mathrm{ad}\ A-\alpha_i)^{s_i}(B_i)$.

If we suppose for simplicity that B itself is one of these generalized eigen-elements with eigenvalue α and step s, it turns out that whenever both λ and $\lambda - \alpha$ are in the resolvent set of A, λ is in the resolvent set of $\bar{A}_1$ and $R(\lambda,A)$ restricts to the $\|\cdot\|_1$-bounded resolvent $R_1(\lambda,A)$ of $\bar{A}_1$ as in (6), while the <u>finite sum commutation relation</u>

$$
\begin{aligned}
B^{(1)}R_1(\lambda,A) &= \Sigma\{(-1)^k R(\lambda+\alpha,A)^{k+1}(\mathrm{ad}\ A-\alpha)^k(B)\colon 0 \le k \le s\} \\
&= \Sigma\{(-1)^k R(\lambda,A-\alpha)^{k+1}(\mathrm{ad}\ A-\alpha)^k(B)\colon 0 \le k \le s\} \quad (6_\alpha)
\end{aligned}
$$

holds in $L(D_1,E)$. The reader is referred to Theorem 5.4 (claim (2)) for the general case. This refinement represents recent work of the second author; it is useful in the study of the way in which commutation relations restrict the spectral behavior of the operators involved. (Cf. Chapter 11 and its survey in Section F below.)

We note also a related "finite" or "truncated" commutation identity that was exploited by the first author [Jo 1] in place of (6).

$$
B^{(1)}R_1(\lambda,A) = R(\lambda,A)B^{(1)} - R(\lambda,A)[A,B]^{(1)}R_1(\lambda,A). \qquad (6_\tau)
$$

For many technical purposes, (6_τ) can provide an algebraically simpler substitute for (6_α) and its Jordan generalization, but it is not developed in detail in the present monograph.

The second rigorous form of the semigroup or exponential commutation identify (2) from Section A also uses the Sobolev Banach space D_1 and is formulated in terms of bounded operators in $L(D_1,E)$. Here, we <u>prove</u> semigroup-invariance of D_1, rather than <u>assuming</u> invariance of D as in Theorem 3.4'. (Assumptions (B), (F_c) and (E) are in force.)

<u>Theorem 6.1'</u>

Suppose that $\|V(t,A)\| \le Me^{\omega t}$, $t \ge 0$, and that (GD) holds for some λ with $\mathrm{Re}(\lambda) > \omega + \nu(\mathrm{ad}\ A)$, ($\nu(\mathrm{ad}\ A)$ the spectral radius of ad A as in Theorem 5.2')

(i) Then each of the semigroup elements $V(t,A)$ leaves D_1 invariant and restricts there to a bounded operator $V_1(t,A)$

such that $\{V_1(t,A): t \geq 0\}$ is a C_o semigroup on D_1 whose generator is $\bar{A}_1$.

(ii) In addition, the commutation relation

$$B^{(1)}V_1(t,A) = V(t,A)[\exp(-t \text{ ad } A)(B)]^{(1)} \tag{7}$$

holds in $L(D_1,E)$, for all $t \geq 0$.

This result was first obtained by the second author in [Mr 1,2] using identity (6) as a Lemma. The first author gave a more economical proof using (6_τ) in [Jo 1]. Identity (6_α) leads to an alternative form which is included in Theorem 6.1 as given in Chapter 6:

$$\begin{aligned} B^{(1)}V_1(t,A+\alpha) &= e^{\alpha t}V(t,A)[\exp(-t \text{ ad } A)(B)]^{(1)} \\ &= V(t,A+\alpha)[\exp(-t \text{ ad } A)(B)]^{(1)} \\ &= V(t,A)[\exp\{-t(\text{ad } A-\alpha)\}(B)]^{(1)} \end{aligned} \tag{7_α}$$

It is also of interest that the domain invariance-regularity conditions (D) and (R) can be used to obtain the conclusions of Theorems 5.2' and 6.1' from the hypotheses of Theorem 3.4'. This is carried out in Section 6E (Theorem 6.7).

Finally, we turn to the formal operational calculus commutation identity, which involves the Dunford-Taylor operational image

$$\varphi(A) = \varphi(\infty)I + (2\pi i)^{-1}\int_\Gamma \varphi(\lambda)R(\lambda,A)d\lambda \tag{8}$$

for φ holomorphic in a neighborhood U of $\sigma(A)$ and the point at ∞, and Γ the usual sort of contour about $\sigma(A)$ in U. (Again, assume (B), (F_c) and optionally (E).)

Theorem 5.8'

Suppose that U contains the closed $\nu(\text{ad } A)$-neighborhood $\{\lambda \in \mathbb{C}: \text{dist}(\lambda,\sigma(A)) \leq |\alpha| \text{ for all } \alpha \in \sigma(\text{ad } A)\}$ of the spectrum of A, and that (GD) holds for all $\lambda \in \Gamma$,

(a') Then $\varphi(A)$ leaves D_1 invariant, and restricts there to a $\|\cdot\|_1$-bounded operator $\varphi_1(A)$ (which is in fact the operational image $\varphi(\bar{A}_1)$).

(b') In addition, with $\varphi^{(k)}(\lambda) = d^k/d\lambda^k \varphi(\lambda)$, we have

$$B^{(1)} \varphi_1(A) = \Sigma\{(-1)^k/k!\varphi^{(k)}(A)[(\text{ad } A)^k(B)]^{(1)}: 0 \le k < \infty\} \quad (9)$$

Again, Jordan data permit the result as it appears in Chapter 5 to be sharpened in a variety of ways. Identity (6_α) corresponds to (and implies)

$$B^{(1)} \varphi_1(A) = \Sigma\{(-1)^k/k!\ \varphi^{(k)}(A-\alpha)[(\text{ad } A-\alpha)^k(B)]^{(1)}: 0 \le k \le s\} \quad (9_\alpha)$$

Variants of Theorem 5.8 can be derived directly from the kinds of semigroup commutation relations obtained in Theorems 3.4 and 6.1; these matters are discussed in Section 6E, with the methodology illustrated in Section 3D. This body of operational calculus results is recent joint work of the co-authors. Related ideas (technically quite different) can be found in the papers of Goodman [Gd 3] and Sternheimer [St 2].

The discussion of the results in Chapters 3-6 that is given above avoids one technical point which occupies a substantial portion of the development in those chapters: avoidance of the part of assumption (F_c) that requires $\mathcal{O}_A(B)$ to have a finite basis of <u>closable</u> operators. As we indicate in Section 6B, there are many practical instances in which closability of every operator in $\mathcal{O}_A(B)$ is readily available. But simple examples show that this is not automatic. Moreover, there are parts of the literature of unbounded operators, where our methods may prove appropriate, in which the issue of closability is distinctly non-trivial and is under active investigation. (The subject of unbounded derivations on C*- and W*-algebras is a case in point. Cf. [BR], [Jo 4,5], [GJ], [PS] and papers referenced there. As in our considerations above, a main problem in unbounded derivations is to infer closability from given algebraic assumptions which are present in applications. It has recently been shown in the references mentioned that the dynamics of infinite quantum systems is described quite well in terms of a closed unbounded derivation δ implemented by a <u>formal</u> Hamiltonian. This is so even in cases where δ is known not to generate a group of autmorphisms $e^{t\delta}$ of the time zero observable algebra.)

1D. The Main Results in Exponentiation Theory

Chapter 9 presents the three main results in "exponentiation theory": the construction of local and global representations of Lie groups from infinitesimal representations of their Lie algebras. Two of our theorems (9.1 and 9.2) are consequences of the two main results of semigroup commutation relations (Theorems 3.4 and 6.1); they carry qualitatively similar hypotheses. The third is rather different, treating the process of constructing new representations by "perturbing" the infinitesimal representations of known group representations. But it, too, makes implicit use of commutation theory, via stability of the enabling hypotheses under small perturbations.

In order to state these results, we require a precise formulation of the exponentiation problem in the C^{∞}-vector form that is treated in the present monograph.

Exponentiation Problem

Let E be a complex Banach (or locally convex) space, and let D be a dense linear subspace.

(a) An operator Lie algebra $\mathfrak{L}$ on D is a finite-dimensional linear subspace $\mathfrak{L}$ of the endomorphism algebra $\mathbf{A}(D)$, such that $\mathfrak{L}$ is closed under formation of commutators: $A, B \in \mathfrak{L}$ implies that $[A,B] = AB - BA = \mathrm{ad}\, A(B) \in \mathfrak{L}$.

(b) If G is a Lie group with Lie algebra $\mathfrak{g}$, an infinitesimal representation dV of G in E (or on D) is a homomorphism $dV: \mathfrak{g} \to \mathfrak{L}$ where $\mathfrak{L} = dV(\mathfrak{g})$ is an operator Lie algebra on D.

(c) A strongly continuous representation $V: G \to \mathrm{Aut}(E)$ is an exponential for dV if and only if D consists of C^{∞}-vectors for V (i.e., for all $u \in D$ the map $g \to V(g)u = \tilde{u}(g)$ is a C^{∞} vector-valued function on G) and for all $X \in \mathfrak{g}$, $u \in D$ we have

$$dV(X)u = \lim\{t^{-1}(V(\exp(tX)u-u): t \to 0\} \qquad (10)$$

(i.e., dV is the differential of V in the operator space L(D,E) with the strong-operator topology).

(d) Given an operator Lie algebra $\mathfrak{L} \subset \mathbf{A}(D)$, the exponentiation problem (EP) for $\mathfrak{L}$ is as follows: When does there exist an exponential $V : G \to \mathrm{Aut}(E)$ for the natural isomorphic infinitesimal representation $dV : \mathfrak{g} \to \mathfrak{L}$ of the underlying Lie algebra $\mathfrak{g}$ of $\mathfrak{L}$, where G is the unique connected, simply connecte Lie group whose Lie algebra is $\mathfrak{g}$.

It is necessary to call attention to a crucial distinction between the EP as discussed here and a conceptually similar problem concerning infinite-dimensional $\mathfrak{g}$-modules (and $\mathfrak{U}(\mathfrak{g})$-modules) as considered in the work of Harish-Chandra, cf. [Wr 1; Chapters 3, 5]. In that problem, one begins with what we would call an infinitesimal algebraic representation $dV : \mathfrak{g} \to \mathfrak{L} \subset \mathfrak{A}(D)$ of $\mathfrak{g}$ in the endomorphism algebra of a (frequently infinite-dimensional) vector space D <u>without topology</u>, and addresses the existence problem for a <u>pair</u> (V,E) where E is a completion of D with respect to some suitably chosen norm and V sends G into Aut(E). In our treatment, the Banach space E is prescribes <u>in advance</u>, and we must consider the existence of a V compatible with that pre-given structure, clearly a different and somewhat more difficult problem.

The following two results give intrinsic solutions to the exponentiation problem EP, formulated in terms of conditions imposed upon a Lie-generating subset S of the operator Lie algebra $\mathfrak{L}$. (Here, S generates $\mathfrak{L}$ if $\mathfrak{L}$ is the smallest Lie subalgebra of $\mathfrak{A}(D)$ containing S.) In the second, we use the same sort of Sobolev C^1-norm that was involved in the commutation results from Chapters 5 and 6 discussed in Section 1C above

$$\|u\|_1 = \max\{\|B_i u\| : 0 \le i \le d\} \tag{11}$$

where $B_0 = I$ and $\{B_1,\ldots,B_d\}$ is a basis for $\mathfrak{L}$ over $\mathbb{R}$. Also, we assume that a norm for the finite-dimensional space $\mathfrak{L}$ has been chosen, and $|\mathrm{ad}\ A|$ denotes the operator norm of ad A on $\mathfrak{L}$ for any $A \in S$.

<u>Theorem 9.1'</u>

Suppose that every $A \in S$, a generating set for $\mathfrak{L}$, is closable and satisfies the following conditions:

(D) $\bar{A}$ generates a C_o group $\{V(t,A): t \in \mathbb{R}\}$ which leaves the domain D invariant $(V(t,A)D \subset D$ for all $t \in \mathbb{R})$,
and

(R) for every $B \in S$, $u \in D$ and $f \in E^*$ the function $G_u^f(t) = f(B\,V(t,A)u)$ is bounded in a neighborhood of 0.
Then $\mathfrak{L}$ is exponentiable.

Theorem 9.2'

Suppose every $A \in S$ is closable and satisfies the following conditions:

(E) $\bar{A}$ generates a C_o group $\{V(t,A)\colon t \in \mathbb{R}\}$ such that $\|V(t,A)\| \leq Me^{\omega|t|}$ for suitable M, $\omega < \infty$, and for every $\lambda \in \mathbb{C}$ with $|\mathrm{Re}(\lambda)| > \omega + |\mathrm{ad}\,A|$, the graph-density condition (GD) is satisfied. ($D_\lambda = (\lambda - A)D$ is $\|\cdot\|_1$-dense in D.)

Then $\mathfrak{L}$ is exponentiable.

Special cases of both of these results were obtained by the second author in [Mr 1,2], with conditions imposed upon all elements $A \in \mathfrak{L}$. As proven in Chapter 9, Theorem 9.1 holds for general representations on locally convex spaces; the result in this form was obtained by the first author in [Jo 1]. We note that a 1962 result of Kato [Kt 1], relating the canonical commutation relations $PQ - QP = iI$ to the Weyl realtions $V(s,P)V(t,Q) = e^{sti}\,V(t,A)V(s,P)$, can be viewed as an independent special case of Theorem 9.2. This interpretations is developed in Chapter 11.

The third exponentiation result could be called extrinsic, since it relates exponentiability of operator Lie algebras $\mathfrak{L}_z$, in an "analytic family" to known exponentiability of a fixed base-point Lie algebra" $\mathfrak{L}_0$. These analytic families can be defined in terms of a basis $\{A_1,\ldots,A_d\}$ for $\mathfrak{L}_0 \subset \mathbf{A}(D)$ and a d-tuple $\{U_1(z),\ldots,U_d(z)\}$ of $\mathbf{A}(D)$-valued functions on an open domain $\Omega \subset \mathbb{C}^n$ such that each $U_i(z)$ extends to a bounded operator in $L(E)$ and gives rise to an $L(E)$-valued analytic function. Then

$$B_k(z) = A_k + U_k(z),\ 1 \leq k \leq d;\ \mathfrak{L}_z = \mathrm{span}_{\mathbb{R}}\{B_k(z)\colon 1 \leq k \leq d\} \qquad (12)$$

is assumed here to define an operator Lie algebra $\mathfrak{L}_z \subset \mathbf{A}(D)$ isomorphic to $\mathfrak{L}_0$. (Alternatively, if $\mathfrak{g}$ is the abstract Lie algebra isomorphic to $\mathfrak{L}_0$ and $dV_0: \mathfrak{g} \to \mathfrak{L}_0$ is the natural infinitesimal representation, with $(dV_0)^{-1}(A_k) = X_k$, we define $dV_z: \mathfrak{g} \to \mathfrak{L}_z$ by $dV_z(X_k) = B_k(z)$ and linear extension. Then dV_z is assumed to be an infinitesimal representation as well for all $z \in \Omega$.) The following result (essentially "half" of Theorem 9.3 as formulated in Chapter 9) gives conditions sufficient to infer exponentiability of $\mathfrak{L}_z$ from that of $\mathfrak{L}_0$, and describes parameter-dependence of the exponentials.

Theorem 9.3'

Supppose that $\mathfrak{L}_0 \subset \mathbf{A}(D)$ is an exponentiable operator Lie algebra whose exponential $V_0: G \to \mathrm{Aut}(E)$ leaves invariant the $\|\cdot\|_1$-completion D_1 of D (determined by the basis $\{A_1,\ldots,A_d\}$ for $\mathfrak{L}_0$). Let $\{\mathfrak{L}_z: z \in \Omega\}$ be an analytic family of perturbations of $\mathfrak{L}_0$ as described above. If $U_k(z)$ is $\|\cdot\|_1$-bounded on D for all $1 \leq k \leq d$ and $z \in \Omega$ (hence extends to a bounded operator $U_{k,1}(z) \in L(D_1)$) then every $\mathfrak{L}_z$ is exponentiable with exponential $V_z: G \to \mathrm{Aut}(E)$, and for all $x \in G$, the map $z \to V_z(x)$ is operator-norm-analytic into $L(E)$.

This result is a new generalization [Jo 3] by the first author of a jointly-discovered lemma used in an earlier unpublished treatment of the analytic families of representations for SL(2,$\mathbb{R}$) discussed in Chapter 12. It is based upon stability of the graph-density condition (GD) in Theorem 9.2 under perturbations of the type described. As given in Chapter 9, Theorem 9.3 also supplies an alternative sufficient condition (ii) which ensures stability under perturbation of conditions (D) and (R) from Theorem 9.1, on a domain D_∞ of C^∞-vectors discussed in Section 1E below. We refer the reader to Chapters 7, 8 and 9 for the details.

We close this section with a few remarks upon the way in which the commutation results discussed in Section 1G are applied in obtaining the three theorems stated above. Commutation theory comes into play in two rather different contexts: algebraic/Lie-theoretic and analytic/generator-theoretic. The analytic aspects can be viewed as special cases of general C^∞-vector results discussed in Section 1E below, and will be pointed out there. For the algebraic aspects we note that if $\bar{A}$ generates a group $\{V(t,A): t \in \mathbb{R}\}$ in the setting of Theorem 3.4; then identity (4) transforms easily into and $\mathbf{A}(D)$-identity

$$V(t,A)\ B\ V(-t,A) = \exp(t \text{ ad } A)(B). \tag{13}$$

Theorem 6.1 in that context yields an $L(D_1,E)$ identity of the same form,

$$V(t,A)\ B^{(1)}V_1(-t,A) = [\exp(t \text{ ad } A)(B)]^{(1)}. \tag{13_1}$$

When $\bar{B}$ generates a group $V(s,B)$, one then shows that so does $\exp(t \text{ ad } A)(B)^-$, and

$$V(t,A)\ V(s,B)\ V(-t,A) = V(s, \exp(t \text{ ad } A)(B)) \tag{14}$$

follows from either (13) or (13_1). This indicates that the operator groups generated by $A, B \in \mathcal{S}$ in Theorems 9.1 or 9.2 exhibit the same commutation-compatibility properties that characterize the one-parameter subgroups of the group G to be represented via the exponential for $\mathfrak{L}$: for all $X, Y \in \mathfrak{g}$

$$\exp(tX)\ \exp(sY)\ \exp(-tX) = \exp(s \exp(t \text{ ad } X)(Y)). \tag{14_G}$$

This kind of algebraic information is used to verify first that the pairs $A, B \in \mathfrak{L}$ for which (14) holds comprise not just a Lie-generating set, but a <u>spanning set</u> for $\mathfrak{L}$. Hence, a basic $\{B_1,\dots,B_d\}$ can be chosen for $\mathfrak{L}$ consisting of operators whose operator-groups satisfy correct commutation relations. If $\{X_1,\dots,X_d\}$ is the corresponding basis for $\mathfrak{g}$, all $x \in G$ close to e can be written uniquely ("canonical coordinates of the second kind") as

$$x = \exp(t_1X_1)\ \exp(t_2X_2) \cdots \exp(t_dX_d) \tag{15}$$

and we can define a local representation

$$V(x) = V(t_1,B_1)\ V(t_2,B_2) \cdots V(t_d,B_d) \tag{16}$$

The local representation property for x,y close to e

$$V(x)\ V(y) = V(xy) \tag{17}$$

can then be viewed as a consequence of the correct commutation properties of the defining operator groups, in combination with the analytic information from Section 1E below.

1E. Results on (semi) group-invariant C^∞-domains

The results developed in Chapters 7 and 8 provide a mechanism for unifying the two different commutation theories discussed in Section 1C: the domain-regularity results appropriate to "Schrödinger formalism" problems in differential operators theory and the graph-density results appropriate to "Heisenberg formalism" problems with matrix operators in sequence spaces. They accomplish a corresponding unification of the two intrinsic exponentiation theorems for operator Lie algebras discussed in Section 1D (Theorems 9.1' and 9.2'), while indicating one of the main mechanisms of proof for those theorems.

The main results concerning construction of semigroup-invariant C^∞-vector domains are obtained in Chapter 7, using assumptions of graph-density type as encountered in Theorems 5.4', 6.1' and 9.2' discussed above. Some concern semigroup-generators A and finite-dimensional ad-orbits $\mathcal{O}_A(B)$ while others concern operators A in a Lie-generating subset S of an operator Lie algebra $\mathfrak{L} \subset \mathbf{A}(D)$. We discuss the Lie algebra case here, since it connects in an interesting way with the phenomenon of non-exponentiable, "smeared-exponentiable" operator Lie algebras encountered in our applications to the representation theory of SL(2, $\mathbb{R}$) in Chapter 12.

The C^∞-domain constructions utilize the following variations on familiar notions from group representations. Let $\mathfrak{L}$ be a finite-dimensional real Lie algebra in $\mathbf{A}(D)$ for D a dense subspace of a Banach space E; let $\mathbf{B} = \{B_1,\ldots,B_d\}$ be a basis of closable operators, and let $E(\mathfrak{L})$ denote the associative subalgebra of $\mathbf{A}(D)$ generated by $\mathfrak{L}$ (i.e., the algebra of non-commutative polynomials in $B_1,\ldots,B_d$. Let $\bar{\mathbf{B}} = \{\bar{B}_1,\ldots,\bar{B}_d\}$ be the set of closures. Then a subspace $F \subset E$ is a <u>C^∞-domain for $\bar{\mathbf{B}}$</u> if $F \subset D(\bar{B}_k)$, and $\bar{B}_k F \subset F$ for $1 \le k \le d$. (In the Schrödinger formalism, each $u \in F$ is "$\bar{B}_k$-differentiable", and $\bar{B}_k u$ is back in F, hence is $\bar{B}_\ell$-differentiable for all ℓ,... .) The largest C^∞-domain (the union of all C^∞-domains) is denoted by $E_\infty(\bar{\mathbf{B}})$, the space of C^∞-vectors for $\bar{\mathfrak{B}}$, and clearly $D \subset E_\infty(\bar{\mathfrak{B}})$. Then $E_\infty(\bar{\mathfrak{B}})$ is equipped with a natural Fréchet space topology, most quickly described as the weakest locally convex topology τ_∞ such that every polynomial P in the $\{\bar{B}_k: 1 \le k \le d\}$ maps $E_\infty(\bar{\mathfrak{B}})$ continuously into $(E, \|\cdot\|)$. It turns out that every such P maps (E_∞, τ_∞) continuously into itself. This topology can be determined by the seminorms $p_{k(1),\ldots,k(n)}(u) = \|\bar{B}_{k(1)}\bar{B}_{k(2)}\cdots\bar{B}_{k(n)}u\|$ constructed from monomials in basis elements, or by various norms built from sums or maxima of these, but we prefer the following inductive construction of "C^1-norms" as in Section 1D above

$$\|u\|_{n+1} = \max\{\|\bar{B}_k u\|_n: 0 \le k \le d\} \tag{18}$$

where $B_0 = I$ and $\|\cdot\|_0 = \|\cdot\|$, the norm for E. Then $\|\cdot\|_1$ is the $\mathfrak{L}$-graph-norm from Section 1D, $\|\cdot\|_2$ is the $\mathfrak{L}$-graph-norm built from $\|\cdot\|_1$, etc. It turns out that $(E_\infty(\bar{\mathfrak{B}}), \tau_\infty)$ is then the projective limit of a chain $(E, \|\cdot\|) \supset (E_1, \|\cdot\|_1) \supset (E_2, \|\cdot\|_2)\ldots$ of Banach spaces. The objective of the construction in Chapter 7 is to obtain a closed subspace of $(E_\infty(\bar{\mathfrak{B}}), \tau_\infty)$ which is big enough to be invariant under the groups $\{V(t,A): t \in \mathbb{R}\}$ for A in a Lie generating set $\mathcal{S} \subset \mathfrak{L}$, but small enough that $\mathfrak{L}$ (and $\mathcal{E}(\mathfrak{L})$) extend in an isomorphic way. The following result is due primarily to the second author (Moore).

<u>Theorem 7.7'</u>

Suppose that for every $A \in \mathcal{S}$, $\bar{A}$ generates a C_0 group $\{V(t,A): t \in \mathbb{R}\}$ and that for all λ with large $|\mathrm{Re}(\lambda)|$,

$$D_\lambda = (\lambda - A)D$$

is $\|\cdot\|_1$-dense in D. Let $D_{\mathcal{S}}$ be the smallest subspace of E containing D and invariant under all $V(t,A)$ for $A \in \mathcal{S}$, $t \in \mathbb{R}$.

(a) Then $D_{\mathcal{S}} \subset E_\infty(\bar{\mathfrak{B}})$, and its closure $D_\infty(\mathfrak{L})$ in $(E_\infty(\bar{\mathfrak{B}}), \tau_\infty)$ is also $V(t,A)$ invariant for all $A \in \mathcal{S}$, $t \in \mathbb{R}$, each $V(t,A)$ restricts to a continuous automorphism $V_\infty(t,A)$ of $(D_\infty(\mathfrak{L}), \tau_\infty)$, and the group $\{V_\infty(t,A): t \in \mathbb{R}\}$ acts smoothly on $D_\infty(\mathfrak{L})$ (i.e., for all $u \in D_\infty(\mathcal{L})$, $V_\infty(t,A)u = \tilde{u}(t)$ is a C^∞-function with values in $(D_\infty(\mathcal{L}), \tau_\infty)$.)

(b-c) The space $D_\infty(\mathcal{L})$ is a C^∞-domain for $\bar{\mathcal{B}}$, and the space $\mathcal{L}_\infty$ spanned by the operators $(B_k)_\infty = \bar{B}_k|_{D_\infty(\mathcal{L})}$ is an operator Lie algebra in $\mathbf{A}(D_\infty(\mathcal{L}))$ which is isomorphic to $\mathcal{L}$ under the restriction map $A_\infty \to A = A_\infty|_D$.

(d) For all $C_\infty \in \mathcal{L}_\infty$, $A_\infty \in S_\infty$ (corresponding to $S \subset \mathcal{L}$), and $t \in \mathbb{R}$ the semigroup commutation relation

$$C_\infty V_\infty(t,A) = V_\infty(t,A)\exp(-t \text{ ad } A)(C_\infty) \tag{19}$$

holds as an identity in the algebra $L(D_\infty(\mathcal{L}))$ of $\tau_\infty - \tau_\infty$ - continuous linear maps.

We observe that identity (19) serves to replace its $L(D_1,E)$ version (7) as obtained in Theorem 6.1', by a simpler identity analogous to (4) in the invariant-domain commutation Theorem 3.4'. Similarly, the invariance and differentiability conclusions in (a) certainly ensure that $\mathcal{L}_\infty$ on $D_\infty(\mathcal{L})$ satisfies conditions (D) and (R) of Theorem 3.4' and 9.1. Hence, Theorem 7.7' places Theorem 9.1 and 9.2 in a common setting. (More is true, but not explicitly proved in our development: the hypotheses of Theorem 9.1' lead to the conclusions of Theorem 7.7' via arguments used in the proof of Theorem 3.4.)

This construction is then used in Chapters 8 and 9 to separate the algebraic/Lie theoretic aspects of the exponentiation problem (EP) from the analytic/unbounded operator-theoretic aspects as suggested at the end of Section 1D above. The algebra and Lie theory (with a bit of straightforward analysis) is packaged in the first result of Chapter 8 which is only implicitly concerned with C^∞-domains.

Theorem 8.1'

Let $\mathcal{L}_\infty$ be a finite-dimensional real Lie algebra of continuous endomorphisms of a locally convex space D. Suppose that every A in a Lie-generating set S_∞ for $\mathcal{L}_\infty$ is the infinitesimal generator of a smooth (C^∞) group $\{V_\infty(t,A): t \in \mathbb{R}\}$ of operators on D.

Then $\mathfrak{L}_\infty$ exponentiates ("integrates") to a smooth representation V_∞: G → Aut(D) of the appropriate simply connected, connected Lie group.

The main ideas for this result are due to the first author [Jo 1].

The conclusion of this result is applied to the construction of an exponential for the original operator Lie algebra $\mathfrak{L}$ in a Banach space E via the following result from Chapter 8.

Theorem 8.6'

Let D_∞ be dense in a Banach space E, and let $\mathfrak{L}_\infty \subset \mathbf{A}(D)$ be an operator Lie algebra such that D_∞ is a closed subspace of $E_\infty(\bar{\mathfrak{B}})$. Suppose further that for every A in a Lie-generating subset $S \subset \mathfrak{L}$, $\bar{A}$ generates a C_o group $\{V(t,A)\colon t \in \mathbb{R}\} \subset \mathrm{Aut}(E)$ that leaves D_∞ invariant.

Then $\mathfrak{L}_\infty$ "integrates" to a smooth representation $V_\infty\colon G \to \mathrm{Aut}(D_\infty)$ on D_∞ and exponentiates to a C_o representation V: G → Aut(E) on E whose restriction to D_∞ is V_∞. (In particular V(G) is a group of bounded operators on E.)

The interest of these results is enhanced by the fact that there exist operator Lie algebras $\mathfrak{L}$ with C^∞-domains D_∞ closed in $(E_\infty(\bar{\mathfrak{B}}),\tau_\infty)$ and with smooth integrals $V_\infty\colon G \to \mathrm{Aut}(D_\infty)$, but without classical exponentials V: G → Aut(E) as discussed in Section 1D and Theorem 8.6'. Some information concerning this matter is provided by the examples in Chapter 9, but the most interesting aspect is treated in Chapter 12, where certain smoothly integrable representations are shown to have exponentials on E "in the sense of distributions": smeared representations. That is, there exists an operator-valued distribution $V\colon \mathcal{D}(G) \to L(E)$, where $\mathcal{D}(G)$ is the Schwartz LF test function space of compactly supported C^∞ functions on G, satisfying

(a) V is a convolution algebra homomorphism

$$V(\varphi * \psi) = V(\varphi)V(\psi) \tag{20}$$

(b) the Gårding domain $D_0 = \mathrm{span}\{V(\varphi)u\colon \varphi \in \mathcal{D}(G),\ u \in E\}$ is dense in E, and

(c) $\bigcap\{\text{kernel } V(\varphi) : \varphi \in \mathcal{D}(G)\} = \{0\}$.

Every such operator-valued distribution is easily shown to give rise to a smooth representation V_∞: $G \to \mathrm{Aut}(D_0)$ with the property that

$$V(\varphi)u = \int_G \varphi(x)\, V_\infty(x)u\, dx \tag{21}$$

describes $V(\varphi)$ as the "integrated form" of V_∞. Such smeared representations can arise as generalized exponentials of non-exponentiable operator Lie algebras $\mathfrak{L}$ which <u>do</u> have smooth integrals on domains D_∞ whose closures in $E_\infty(\bar{\mathfrak{B}})$ agree with the closure of D_0 in $E_\infty(\bar{\mathfrak{B}})$. We refer the reader to Section 12C for the technicalities. Our examples there utilize the following general lemma due to the first author.

<u>Lemma 12.7'</u>

Let $\mathfrak{L} \subset \mathfrak{A}(D)$ be an operator Lie algebra in a reflexive Banach space E which exponentiates ("integrates") to a smooth representation on $E_\infty(\bar{\mathfrak{B}})$ for some basis $\mathfrak{B}$ of $\mathfrak{L}$ (as in Theorem 8.1', for example, with $D_\infty = E_\infty$). Suppose that there exists a closable element $\Pi \in E(\mathfrak{L})$ which <u>C^∞-dominates</u> $\mathfrak{L}$, in the sense that for every $\Gamma \in E(\mathfrak{L})$ there is an integer n and a constant C such that

$$\|\Gamma u\| \leq C\, \Sigma\, \{\|\Pi^k u\| : 0 \leq k \leq n\} \tag{22}$$

Suppose that $\bar{\Pi}$ has a nonempty resolvent set.
Then $\mathfrak{L}$ has a smeared exponential.

In Sections 12G and 12D, we exhibit examples where this result applies but the operator Lie algebras $\mathfrak{L}$ demonstrably fail to have exponentials as in Section 1D above. For many technical purposes, we believe that the smooth representations on C^∞-domains are the fundamental objects, and that the continuous exponentials supplied by Theorem 9.1' or 9.2', and the smeared ones as in Theorem 12.7', may eventually prove to be primarily of historical interest. If so, the "main results in exponentiation theory" in this monograph may turn out to be Theorem 7.7' and the smooth representation conclusions in Theorems 8.1 and 8.6.

1F. <u>Typical Applications of Commutation Theory</u>

Two areas of direct application for the commutation results in Chapters 3-6 are discussed in Chapters 4 and 10. One major theme in both chapters is the restricted spectral behavior of unbounded operators that is entailed by their commutation relations with other operators. Chapter 4, on the "O'Raifeartaigh mass-splitting phenomenon" in mathematical physics, is based upon domain invariance/regularity commutation theory as developed in Chapter 4 (Theorem 3.4', of Section 1C above). Chapter 11, on various low-dimensional commutation relations from mathematic physics and group representations, is based instead upon the graph-density commutation theory of Chapters 5 and 6, as in Theorems 5.4' and 6.1. For the most part, these applications are better suited to informal discussion rather than formally stated definitions and theorems as in Sections 1C through 1E above.

Our treatment in Chapter 4, which is close in spirit to that of Goodman [Gd 3], provides a number of qualitatively different uses of Theorem 3.4' (potentially of Theorem 6.1 as well).

The operators A and B of Section 2A involved in these applications arise in various natural ways from a strongly continuous group representation $V : G \to \mathrm{Aut}(E)$ of a connected Lie group G on a Banach space E. The space D consists of the C^∞ vectors $u \in D^\infty(V)$ for the representation: those such that the function $\mathfrak{U}(x) = V(x)u$ is C^∞ on G. The representation V defines in the usual way a Lie algebra representation $dV : \mathfrak{g} \to \mathfrak{L} \subset \mathfrak{A}(D)$ of the Lie algebra $\mathfrak{g}$ of G which extends to a representation (also denoted dV) sending the universal enveloping algebra $\mathfrak{U}(\mathfrak{g})$ (qua right-invariant differential operators on G) into the complex associative enveloping algebra $E_{\mathbb{C}}(\mathfrak{L})$ generated by $\mathfrak{L}$ in $\mathfrak{A}(D)$. (See [Gd 1] or [Ps 1] for background.)

The first application of Chapter 3 in Chapter 4 involves operators $B \in \mathfrak{L}$ and $A \in E_{\mathbb{C}}(\mathfrak{L})$. Initially, we study images $A = dV(T)$ of especially well-behaved hypoelliptic right-invariant differential operators $T \in \mathfrak{U}(\mathfrak{g})$. By the generalized Nelson-Stinespring methods introduced by the first author [Jo 2], it is indicated that $\bar{A}$ generates a semigroup $\{V(t,A) : t \in [0,\infty)\}$ which analytically continues into the open right half-plane $\{V(\zeta,A) : \mathrm{Re}(\zeta) > 0\}$, and that there exists on G a "generalized Gauss kernel" $p_\zeta(x)$ such that $V(\zeta,A)$ can be computed in terms of the group representation by integration:

$$V(\zeta,A)u = \int_G p_\zeta(x)\, V(x)u\, dx \tag{23}$$

Moreover, the right-invariant derivatives of $p_\zeta(x)$ can be shown to vanish so rapidly at ∞ that for any initial $u \in E$ (in particular $u \in D = D^\infty(V)$), $V(\zeta,A)u \in D^\infty(V)$. (Recall by the Gårding-Segal scheme that operators $B \in \mathfrak{L}$ act on vectors as in (23) by differentiating $p_\zeta(x)$ under the integral.) In this way, analysis on G is parlayed into invariance of D under $\{V(\zeta,A)\colon \mathrm{Re}(\zeta) > 0\}$ (hence in particular under $\{V(t,A)\colon t \in [0,\infty)\}$), in the process producing the sort of local boundedness demanded by Theorem 3.4'.

Algebraically, the special operators A considéred here are situated in $E(\mathfrak{L})$ so that $\mathcal{O}_A(B)$ is finite-dimensional for all $B \in \mathfrak{L}$. Theorem 3.4' can thus be used to deduce that for $\zeta = t \in [0,\infty)$, the commutation relation

$$B\,V(\zeta,A) = V(\zeta,A)\exp(-\zeta \text{ ad } A)(B) \tag{24}$$

holds. Analyticity of both sides of (24) for $\mathrm{Re}(\zeta) > 0$ and their agreement on $\zeta \in [0,\infty)$ then extends (24) to the whole half-plane.

The point of this extension is reminiscent of many arguments in mathematical physics (e.g., Euclidean versus relativistic field theory): the operators A have the property that $\{V(\zeta,A)\colon \mathrm{Re}(\zeta) > 0\}$ extends continuously to the <u>closed</u> right half-plane, with boundary value $\{V(it,A) = V(t,iA)\colon t \in \mathbb{R}\}$ a strongly-continuous <u>group</u>. The complex commutation relation (24) is then extended by limits to $\zeta = it$. (It is important that boundary group $V(it,A) = V(t,iA)$ cannot always be attacked by the kernel methods used for $\mathrm{Re}(\zeta) > 0$, and that one therefore does not know <u>a priori</u> that $V(t,iA)$ leaves D invariant. This invariance can quickly be otbained from (24) with $\zeta = it$, however.)

Next, more general operators $A \in E_{\mathbb{C}}(\mathfrak{L})$ are treated by resolving them as $A = i(A_+ - A_-)$ where A_+ and A_- are of the type discussed above (images of hypoelliptics under dV) and $[A_+,A_-] = 0$ on D. Theorem 3.4' is used a second time to conclude that $V(t,iA_+)$ and $V(t,iA_-)$ commute, whence

$$V(t,A) = V(t,iA_+)V(-t,iA_-) \tag{25}$$

holds, and we derive the commutation relation

$$B\,V(t,A) = V(t,A)\,\exp(-t \text{ ad } A)(B) \tag{26}$$

for this A and all $B \in \mathfrak{L}$ by (24) with $\zeta = it$, and the fact that ad A_+ commutes with ad A_- so that

$$\exp(-t\ \mathrm{ad}(iA_+ - iA_-)) = \exp(-t\ \mathrm{ad}\ iA_+)\ \exp(t\ \mathrm{ad}\ iA_-).$$

These last operators are the "generalized mass-operators" of interest in the O'Raifeartaigh phenomenon.

Yet another application of Theorem 3.4' is made. One computes via an operational calculus trick using (26) that certain spectral projections P for A leave D invariant and commute there will all $B \in \mathfrak{L}$. Reversing roles, P becomes "B in (4)" and $B \in \mathfrak{L}$ becomes "A in (4)" to obtain

$$P\ V(t,B) = V(t,B)\ P \tag{27}$$

which quickly shows that P commutes with the entire range V(G) of the group representation (that is, P reduces V).

In the prototype example [Sg 2, Gd 3], G is a large group of physical symmetries, containing a subgroup G_1 of internal ("particle") symmetries and the Poincaré group G_2 of relativistic kinematical symmetries. In fact G_2 is the semi-direct product L·H of the Lorentz group L with the four-dimensional Abelian ideal H of Minkowski space-time translations, where ad L acts on H according to the usual action of $L = SO(3,1)$ on $\mathbb{R}^4$ with the Lorentz metric $(x_0^2 - x_1^2 - x_2^2 - x_3^2)$. It turns out that (essentially because L acts transitively upon $\mathbb{R}^4$) the Lie algebra $\mathfrak{g}_2$ of G_2 maps the Lie algebra $\mathfrak{h}$ of H onto itself under the adjoint action: $[\mathfrak{g}_2,\mathfrak{h}] = \mathfrak{h}$, hence $[\mathfrak{g},\mathfrak{h}] \supset \mathfrak{h}$. O'Raifeartaigh [O'R] observed that this simple fact forces ad $\mathfrak{h}$ to act <u>nilpotently</u> on $\mathfrak{g}$. Moreover, he essentially showed that for an appropriate basis $B_0,\ldots,B_3$ in $\mathfrak{L} = dV(\mathfrak{h})$ the quadratic mass-energy $M = B_0^2 - B_1^2 - B_2^2 - B_3^2$ acts nilpotently on $\mathfrak{g}$ as well, hence has finite-dimensional ad-orbits. The work of Segal [Sg 2] and Goodman [Gd 3] shows that if Abelian $\mathfrak{h} \subset \mathfrak{g}$ with $[\mathfrak{g},\mathfrak{h}] \supset \mathfrak{h}$ for otherwise arbitrary $\mathfrak{g}$ and $\mathfrak{h}$, then $\mathfrak{h}$ is nilpotently embedded <u>and</u>, for any $M \in \mathfrak{U}_{\mathbb{C}}(\mathfrak{h})$ (the complex enveloping algebra of $\mathfrak{h}$, contained in $\mathfrak{U}_{\mathbb{C}}(\mathfrak{g})$), M has such finite-dimensional ad-orbits (cf. Chapter 4).

For the quadratic mass-energy treated in Theorem 4.2,

the operators $A_+ = dV(B_0^2)$ and $A_- = dV(B_1^2 + B_2^2 + B_3^2)$ are the two images of hypoelliptics that we discussed above, and the kernels for these can be written down explicitly as Gaussians in the appropriate variables in $\mathbb{R}^4$ - alias - H. The projections P are those associated with isolated eigenvalues of A = i dV(M) (eigenvalues that are not cluster points of others), or more generally with components of the spectrum $\sigma(A)$, and the conclusion is that if V is unitary on H and irreducible, then A has at most one eigenvalue (resp.: $\sigma(A)$ has at most one component). Physically, this means that all symmetry models of elementary particle physics of the type under discussion conflict with experimental data which are normally interpreted as exhibiting irreducible representations with several distinct mass eigenvalues (multiplets). This proof scheme (due to the first author) is modified, somewhat in the spirit of Goodman's work, to treat representations in Banach space and A = dV(T) for arbitrary $T \in \mathfrak{U}_{\mathbb{C}}(\mathfrak{h})$. The idea is to take $A_+ = dV(\Delta_H^n + T)$ and $A_- = dV(\Delta_H^n)$, where $\Delta_H = B_0{}^2 + B_1{}^2 + B_2{}^2 + B_3{}^2$ and n is chosen $\geq$ degree(T)/2 so that $\Delta_H^n + T$ is hypoelliptic. The algebraic calculations involved in treating these operators make use of certain "generalized Hermite polynomials" which we have not previously encountered in the literature; these permit, in addition, sharper estimates of degree of nilpotency than those given in [Gd 3].

In Chapter 11, the objective is to simplify, apply, and extend the commutation theory of Chapters 5 and 6 in the context of three familiar classical commutation relations which arise frequently in group representations and mathematical physics:

$$[A,B] = 0 \quad \text{(commutativity)}, \tag{28}$$

$$[A,B] = Z \text{ bounded}, \ [A,Z] = [B,Z] = 0 \ \text{(Heisenberg or canonical commutation relations)}, \tag{29}$$

$$[A,B] = \alpha B, \ \alpha \in \mathbb{C} \ \text{(solvable "ax+b" relations)}. \tag{30}$$

The cases (29) and (30) are treated by similar methods and details are omitted for (30). These three share the useful property that B "dominates" $\mathcal{O}_A(B)$, in the sense that for all

$C \in \mathcal{O}_A(B)$, there are constants a, $b < \infty$ such that for all $u \in D$,

$$\|Cu\| \leq a\|Bu\| + b\|u\|. \tag{31}$$

This means that $D_1 = D(\bar{B})$ and $\|\cdot\|_1$ is topologically equivalent to the B-graph norm $\|u\|_B = \|u\| + \|\bar{B}u\|$. These cases are covered by the following general result concerning natural conditions equivalent to the graph-density condition for A and B.

Theorem 11.2'

Let B dominate $\mathcal{O}_A(B)$, and let $\lambda \in \rho(\bar{A})$. Then the following are equivalent to the graph-density condition (GD) ($D_\lambda = (\lambda - A)D$ is $\|\cdot\|_B$-dense in D):

(a) (Kato's condition) For some, equivalently all, $\mu \in \rho(\bar{B})$,

$$D_{\mu\lambda} = (\mu - B)D_\lambda = (\mu - B)(\lambda - A)D$$

is dense in E.

(b) The space D_λ is a core for $\bar{B}$: $\bar{B}$ is the closure of B restricted to D_λ.

In Section 11B, the commutative case (28) is considered. We show that whenever the conditions of Theorem 11.2' are satisfied for any λ and μ in the respective resolvent sets, then A and all of its resolvents, operational images $\varphi(A)$, (semi-) group elements $V(t,A)$, and spectral projections P commute with all of the corresponding operators determined by B. In the case where A and B are both essentially self- (or skew-) adjoint, this yields a set of sufficient conditions for strong commutativity (commutation of spectral families). A counterexample in the spirit of Nelson's [Nℓ 1] is presented where essentially self-adjoint A and B commute "infinitesimally" as in (28), but have no common commuting projections. This example, which repairs a gap in one proposed by Powers [Pw 1], permits us to compute the space of bounded linear functionals which annihilate $D_{\mu\lambda}$ and thereby account for the failure of Kato's condition. We also compute the corresponding $\|\cdot\|_1$-bounded annihilators of D_λ which account for the faillure of (GD).

Section 11C carries out a similar program for the

Heisenberg relations (29), recovering and slightly generalizing Kato's original work [Kt 1] for the case where Z is scalar. Here, the equivalent conditions are sufficient for the semigroup Weyl relations

$$V(s,A)\ V(t,B) = \exp(st\ Z)\ V(t,B)\ V(s,A) \tag{32}$$

and for a number of other useful commutation relations connecting bounded operators determined by A and B. Two of the more interesting are the resolvent commutation relations

$$\begin{aligned}[R(\lambda,A),\ R(\mu,B)] &= -R(\lambda,A)\ R(\mu,B)\, Z\, R(\mu,B)\ R(\lambda,A) \\ &= -R(\mu,B)\ R(\lambda,A)\, Z\, R(\lambda,A)\ R(\mu,B)\end{aligned} \tag{33}$$

and the group-resolvent conjugation relation no. 3

$$V(s,A)\, R(\mu,B)\, V(-s,A) = R(\mu,\ B + sZ) = R(\mu - sZ,\ B), \tag{34}$$

where the last term applies only for $Z \in \mathbb{C}$ scalar. Following Kato [Kt 1], we discuss how to recover (29) from identities like (32) and (33). We also show how (34) forces the resolvent set and spectrum of B to be closed under translations in the direction of $Z \in \mathbb{C}$, for A,B essentially skew-adjoint and $Z \neq 0$ imaginary, (34) also forces uniform multiplicity in the spectrum of B. An example shows that this may fail if $\bar{A}$ does not generate a C_o group of operators. A variant by Reed and Simon [RS 1] on the Nelson-Powers counter-example is also discussed, where (29) holds but (32)-(34) fail; the failure of (GD) and of Kato's condition is described again in terms of annihilating functionals.

1G. Typical Applications of exponentiation theory

Like the applications of commutation theory, those for exponentiation theory are divided into a chapter on "Schrödinger formalism"-type applications using the domain-regularity assumptions in the function-space context (Chapter 10) and one using graph-density assumptions in the "Heisenberg matrix-formalism" to represent the Lie group $SL(2,\mathbb{R})$ on sequence spaces (Chapter 12).

Two direct applications of the domain-regularity exponentiation Theorem 9.1' are made in Chapter 10, first to Lie algebras of vector fields in continuous- and smooth-function spaces, and then to Lie algebras of first-order differential operators (with zeroth - order terms) in L^p spaces on Riemannian manifolds.

The general spirit of these applications is accurately represented by our treatment of the integration theory of vector field Lie algebras. Local results of this type date back to Lie himself, but we are concerned with a global version due first to Palais [Pℓ], which can be formulated in the following quasi-geometrical language paralleling our operator-theoretic formalism. First, a Lie transformation group or global Lie flow is a Lie group G together with a homomorphism $V_* : G \to \mathrm{Aut}(M)$ of G into the group of diffeomorphisms of a C^∞ manifold M, such that for each $x \in M$ the map $g \to V_*(g)x$ is C^∞ from G to M. The usual procedure of differentiation along G-orbits through a point $x \in M$ defines a homomorphism dV_* sending the Lie algebra $\mathfrak{g}$ of right-invariant vector fields on G onto a Lie algebra $\mathfrak{L} = dV_*(\mathfrak{g})$ of vector fields on M [Pℓ]. (These are the "infinitesimal transformations" on M, and the action V_* of G on M is usually called the integral of $\mathfrak{L}$. If $G = \mathbb{R}$ and $A = dV_*(d/dt)$ is the natural vector field such that $\mathbb{R}A = \mathfrak{L}$, we adopt the notation that $\{V_*(t,A): t \in \mathbb{R}\}$ denotes the one-parameter transformation group corresponding to A. If a vector field A has such an integral to $\{V_*(t,A)\}$ we often say A is complete (equivalently, integrable). All of this, of course, is simply the finite-dimensional non-linear version of the infinite-dimensional linear language of Chapters 8 and 9. In that language, the semi-classical theorem reads as follows.

Lie-Palais Integration Theorem

Suppose that every vector field $A \in \mathcal{S}$ in a Lie generating set for a finite-dimensional real Lie algebra $\mathfrak{L}$ of vector fields on M is complete (integrable).
Then $\mathfrak{L}$ itself is integrable to a Lie transformation group on M.

Our approach to this result is to "linearize" it by letting vector fields "differentiate" functions, and diffeomorphisms "push" them (by composition), in a suitable locally convex space D of functions. Our three variant results can be summarized within the following general framework.

Theorem-Format

Let $D = C_0^\infty(M)$, the compactly-supported smooth functions on M, and let a vector field Lie algebra $\mathfrak{L}$ as in the Lie-Palais theorem act upon D. Suppose that D is equipped with a locally convex topology such that the (multiplicative) functionals of point-evaluation are all continuous, and the operator groups $V(t,A)u = u \circ V_*(t,A)$ on D are all differentiable. Then $\mathfrak{L}$ exponentiates to a group representation $V: G \to \mathrm{Aut}(D)$ such that the contragredient representation $V^* : G \to \mathrm{Aut}(D^*)$ leaves the set "spec(D)" $= \{\delta_x : x \in M\}$ of point-evaluation functionals invariant in D^* and restricts there to (a copy of) an integral $V_* : G \to \mathrm{Aut}(M)$ for $\mathfrak{L}$ on M.

Specifically, in Theorem 10.1 we use the sup-norm topology (for which the completion of D is the complex Banach algebra $\mathfrak{U} = C_\infty(M)$ of smooth functions vanishing at infinity, and V extends to a representation in algebra automorphisms that transforms points in spec $(\mathfrak{U})$ = M. In Theorem 10.2, D carries the C^∞ topology (uniform convergence of derivativies on compacta) and has the Schwartz test-function Fréchet space $\mathcal{E}(M)$ as completion: the algebra of all C^∞ functions. In Corollary 10.3, D is equipped with the already-complete limit-Fréchet test-function topology to become $\mathcal{D}(M)$ in Schwartz' notation.

In all of these cases, the group representation claim is obtained essentially immediately from Theorem 9.1, in its full generality of locally convex spaces. (For all A, B $\in S$ the function B V(t,A) is obviously well-behaved: $B\,V(t,A)u = [(dV_*(t,A)B)u] \circ V_*(t,A)$ and the t-dependence of the image $dV_*(t,A)B$ of B under the differential of $V_*(t,A)$ is smooth.) In fact, the directness of these applications leads us to view our general exponentiation Theorem 9.1 as a "generalized functional-analytic Lie-Palais integration theorem."

The results just discussed require only the manifold structure of M (in fact, an artificial version of Theorem 10.1 makes sense for continuous transformation groups on locally compact spaces), while our other application of Theorem 9.1' requires almost the full structure of a Riemannian manifold. That structure gives rise not only to the existence of a metric on M, but also a natural Riemannian measure (defining L^p spaces on M for $1 \leq p < \infty$) and a notion of the divergence div(X) of a vector field X. (Local-coordinate descriptions of these notions are recalled in Chapter 10.)

Using these, a simplification of Theorem 10.6 can be formulated as follows.

Theorem 10.6'

Let $\{X_1,\ldots,X_n\}$ be complete vector fields on M; let $\{b_1,\ldots,b_n\}$ be C^∞ functions on M, and suppose that the Lie algebra generated by $\{X_j + b_j : 1 \le j \le n\}$ is finite-dimensional. Suppose further that for some $1 \le p < \infty$, the differences

$$p \operatorname{Re}(b_j) - \operatorname{div}(X_j), \quad 1 \le j \le n, \tag{35}$$

are bounded functions on M.
Then $\mathcal{L}$ exponentiates to a C_o group representation on the Banach space $L^p(M, dx)$.

The exponential $V : G \to \operatorname{Aut}(L^p)$ in this result turns out to be a "multiplier representation" of G on M: there exists a transformation group $\tau : G \times M \to M$ and a multiplier $\mu : G \times M \to \mathbb{C}$ such that

$$V(g)u\ (x) = \mu(g,x)\ u(\tau(g,x)) \tag{36}$$

describes V for $g \in G$, $x \in M$ in terms of the translation of u by τ and pointwise multiplication of u by μ.
The condition (35) on the vector fields X_j and "infinitesimal multipliers" b_j is used to obtain numerical range information (cf. Appendix F) on the $A_j = X_j + b_j$. This is then used to conclude that these generate one-parameter C_o groups on $L^p(M,dx)$ which leave the domain D of C^∞ functions with compact support invariant (Lemma 10.4). The rest is an application of Theorem 9.3'. (The full version of Theorem 10.7, proved in Chapter 10, allows pseudo-Riemannian manifolds and replaces the measure dx by m dx for m a non-vanishing C^∞ "density function", with technical changes in (35).

Secondary applications of Theorem 9.1, via Theorem 10.7' are given in Sections 10D and 10E. There we consider the semigroup-generation properties of "sub-Laplacians" $\Delta = \Sigma_1^n X_j^2$ on $L^p(M, dx)$, where $\{X_1,\ldots,X_n\}$ generates a Lie algebra $\mathcal{L}$ of (complete) vector-fields with bounded divergence. Here Δ is hypoelliptic on M, and turns out to be the image of a right-invariant hypoelliptic differential operator on G under the infinitesimal representation for the exponential $V : G \to \mathrm{Aut}(L^p)$ of $\mathcal{L}$ in L^p obtained by Theorem 10.7. The first author's work on generation properties of such hypoelliptic images [Jo 2] is applied to express the resolvents and (holomorphic) semigroup generated by Δ in L^p explicitly in terms of Green's and Gauss' kernels. In Section 10D, we discuss an operator used by Bony [By]

$$S_o = \left(\frac{\partial}{\partial x_0}\right)^2 + \left(x_0 \frac{\partial}{\partial x_1}\right)^2 + \ldots + \left(x_0^n \frac{\partial}{\partial x_n}\right)^2$$

on $\mathbb{R}^{n+1}$ as a sub-Laplacian of this type associated with the action of a suitable nilpotent Lie group on $\mathbb{R}^{n+1}$.

The applications discussed in Chapter 12 are more ambitious than those involved in the remaining applications, Chapters 4 and 10-11. Here, we build a new cross-section for the non-unitary dual objects of the matrix Lie group

$$G = SL(2,\mathbb{R}) = \left\{ \begin{pmatrix} a & b \\ c & d \end{pmatrix} : ac - bd = 1 \right\} \text{ and of its simply-}$$

connected covering group $\widetilde{G}$, acting on the Hilbert space $\ell^2(\mathbb{Z})$ of square-integrable functions on the integers.

This process begins with a base-point operator Lie algebra $\mathcal{L}_0$ defined on the subspace D of finitely-supported sequences by linear extension from the following matrix-operators acting on canonical basis elements e_n for $\ell^p(\mathbb{Z})$,

$$e_n(m) = \delta_{mn}, \quad A_0 e_n = in\, e_n,$$

$$A_1 e_n = \frac{n}{2}(e_{n+1} - e_{n-1}), \quad A_2 e_n = i\frac{n}{2}(e_{n+1} + e_{n-1}) \tag{37}$$

Here, the graph-density hypotheses of Theorem 9.2' are checked directly for $S = \{A_0, A_1\} \subset \mathcal{L}_0$ to obtain a non-unitary exponential for $\mathcal{L}_0$. This involves verifying that no sequences $\{a_n : n \in \mathbb{Z}\}$ in the algebraic dual of D that annihilate $D_\lambda = (\lambda - A_i)D$ define $\|\cdot\|_1$-bounded functionals orthogonal to D_λ, so that D_λ must be $\|\cdot\|_1$-dense. For A_0, this is a triviality, but for A_1 it involves solution of the following recurrence relation

$$a_{n+1} = n^{-1}(2\lambda a_n - a_{n-1}) \tag{38}$$

and demonstration that no nontrivial solution $\{a_n\}$ has $\{\frac{a_n}{n} : n \in \mathbb{Z}\}$ square-summable (Theorem 12.1). In fact, it is unbounded so (GD) is satisfied in any one of the $\ell^p(\mathbb{Z})$ spaces.

The next main step is to define an analytic family $\mathfrak{L}(q,\tau)$ of perturbations of $\mathfrak{L}_0$ which can be exponentiated using Theorem 9.3'. These operator Lie algebras are defined in terms of basis operators

$$B_0 e_n = i(n+\tau)e_n, \quad B_1 e_n = \tfrac{1}{2}(\gamma_n e_{n+1} - \gamma_{n-1}e_{n-1}),$$
$$B_2 e_n = \frac{i}{2}(\gamma_n e_{n+1} + \gamma_{n-1}e_{n-1}) \tag{39}$$

where the weight function γ_n is described by

$$\gamma_n(q,\tau) = \text{signum}(n)(q+(n+\tau)(n+\tau+1))^{\frac{1}{2}} \tag{40}$$

for $(q,\tau) \in \mathbb{C}^2$ with suitable conventions on determination of the square root. We verify as in Bargmann [Bg] that $\mathfrak{L}_0$ and all $\mathfrak{L}(q,\tau)$ are isomorphic to the Lie Algebra $s\ell(2,\mathbb{R})$, and that $Q = B_1{}^2 + B_2{}^2 - B_0{}^2 = q$, a scalar. It is then shown that the perturbations $U_k(q,\tau) = B_k - A_k$ $0 \le k \le 2$ are bounded on ℓ^2 and on $(D, \|\cdot\|_1)$. (In fact, for $\tau = 0$ they are Hilbert-Schmidt on ℓ^2 !) This yields in Theorem 12.15 an analytic family of representations $V_{q\tau} : G \to \text{Aut}(\ell^2)$ for the covering group $\widetilde{G}$ of $SL(2,\mathbb{R})$ with the curious property that for fixed τ and $g \in \widetilde{G}$, $V_{q\tau}(g) - V_{r\tau}(g)$ is a compact operator. (Thus dependence upon the "character" parameter q disappears in the Calkin quotient of $L(\ell^2)$ by the compact operators $K(\ell^2)$.) We have encountered neither this series of exponentials nor this qualitative perturbation behavior in the prior literature.

In Section 12G, we sort out the relationship of this analytic series to the dual objects for $\widetilde{G}$, $SL(2,\mathbb{R})$, and the Lorentz group for two space-dimensions. Briefly, for $-\frac{1}{2} \le \tau < \frac{1}{2}$ and $\frac{1}{4} \ne q > 0$, we pick up exactly one copy for each representation in the irreducible unitary continuous and complementary series for $\widetilde{G}$, in unitary form. For "most" other (q,τ) with $-\frac{1}{2} \le \text{Re}(\tau) < \frac{1}{2}$ the representations are non-unitary and irreducible. At exceptional points, $V_{q,\tau}$ direct-sum decomposes into appropriate unitary discrete series, finite-dimensional, and non-unitary discrete series summands. Integer changes in τ lead to unitarily equivalent representations. Some of these qualitative properties are technically more convenient than those of the cross-sections obtained by Harish-Chandra's "sub quotient" construction [Wr 1, Chapter 5].

We also examine the extent to which the results generalize to two natural families of Banach space completions of D closely related to $\ell^2(\mathbb{Z})$: $\ell^p(\mathbb{Z})$ and $L^p(T)$ for $p \neq 2$. For $\ell^p(\mathbb{Z})$ with $p \neq 2$, it turns out both that the basis operators A_k, $0 \leq k \leq 2$, for $\mathfrak{L}_0$ satisfy the graph-density condition, and that D is a dense set of analytic vectors for them. They also act as unbounded * - derivations on the Banach *-algebra $\ell^1(\mathbb{Z})$ with the convolution product. In addition the $\mathfrak{L}(q,\tau)$ exhibit boundedness (indeed, compactness) of the perturbations U_k. But $\bar{A}_1, \bar{A}_2$ and their perturbations $\bar{B}_1$ and $\bar{B}_2$ fail to generate C_o groups, and neither $\mathfrak{L}_0$ nor the $\mathfrak{L}(q,\tau)$ exponentiate to group representations on $\ell^p(\mathbb{Z})$ when $p \neq 2$ (Section 12D, Theorem 12.11). In particular, the $\bar{A}_k$ do not generate automorphism groups of $\ell^1(\mathbb{Z})$ for $k = 1,2$. On the other hand, as indicated in Section 1E above, we are able to show that $\mathfrak{L}_0$ and the $\mathfrak{L}(q,\tau)$ admit smeared exponentials on $\ell^p(\mathbb{Z})$ for $1 \leq p < \infty$ and on c_0 (Section 12D, Theorem 12.6) and that the family of smeared exponentials $V_{q,\tau}: \mathcal{D}(\tilde{G}) \to L(\ell^p(\mathbb{Z}))$ exhibits essentially the same analyticity, irreducibility and/or reduction properties encountered in $\ell^2(\mathbb{Z})$.

By $L^p(T)$, we mean here the L^p space on the unit circle, where D is naturally identified with the trigonometric polynomials. There, $\mathfrak{L}_0$ is easily interpreted as a Lie algebra of first-order differential operators on T as in Theorem 10.7, but the $\mathfrak{L}(q,\tau)$ contain less-tractable pseudodifferential operators and do not give rise to the kind of multiplier representations encountered in Chapter 10. However, here the $\mathfrak{L}(q,\tau)$ exhibit all of the main properties that are encountered in $\ell^2(\mathbb{Z})$ (or its unitary equivalent, $L^2(T)$). They yield analytic families of exponentials $V_{q\tau}$: $G \to \mathrm{Aut}(L^p)$ on all $L^p(T)$, $1 \leq p \neq 2 < \infty$, with reduction properties essentially as described for the ℓ^2 case.

In summary, the results in Chapter 12 go considerably beyond illustrative applications of theoretical techniques developed in the monograph. They introduce new phenomena in the area of application, upon which research is continuing. (Possible generalizations to other semisimple Lie groups G are discussed in Section 12A.)

Chapter 2

THE FINITE-DIMENSIONAL COMMUTATION CONDITION

In Chapter 1, we indicated that in order to obtain a useful commutation theory connecting operators A and B with their resolvents, semigroup-exponentials, and operational images, it is desirable to impose a finite-dimensionality condition upon their iterated commutators $[A,B] = AB - BA$, $[A, [A,B]]$, etc. (Here A and B are endomorphisms of a dense domain D in a Banach or locally convex space E.) Below in Section 2A we distinguish several technically different ways in which this condition enters into the development. Section 2B presents examples of differential operators which satisfy the condition.

Then in Section 2C we contrast two sources of finite-dimensional commutation relations for operators A and B in the enveloping algebra $E(\mathfrak{L})$ of a (finite dimensional) operator Lie algebra $\mathfrak{L}$: the universal case where only the abstract Lie structure of $\mathfrak{L}$ is involved, and the spatial case where the action of A and B on the domain D plays a fundamental rôle. The final section describes examples where a useful commutation theory can be obtained in the absence of this finite-dimensionality condition.

Unlike the exposition in the main developmental chapters 3-12, the treatment here presupposes familiarity with one of the detailed survey sections in Chapter 1: Section 1C. In particular, the reader is expected to be familiar with the "Sobolev C^1-norm" $\|\cdot\|_1$ and our notational conventions concerning operators in $\mathbf{A}(D)$ (endomorphisms of D), and L(E), as mappings from D_1 to D_1 or E. Of course, familiarity with Chapters 3-6 is also sufficient. The present chapter is not directly prerequisite to any of the other chapters.

A. Implications of Finite-dimensionality in Commutation Theory

In order to render tractable the discussion of the iterated commutators $\mathrm{adA}(B) = [A,B] = AB - BA$, $(\mathrm{adA})^2(B) = [A,[A,B]] = A^2B - 2ABA + BA^2$, ... $(\mathrm{adA})^{n+1}(B) = [A,(\mathrm{adA})^n(B)]$, we impose throughout most of the monograph the condition that:

$$\mathcal{O}_A(B) = \text{real span } \{(\text{ad}A)^k(B) : 0 \le k < \infty\} \tag{1}$$

be finite-dimensional. In fact, it is useful to note that the dimension of $\mathcal{O}_A(B)$ is exactly the smallest integer d such that $(\text{ad}A)^d(B)$ can be expressed as a linear combination of the $\{(\text{ad}A)^k(B) : 0 \le k < d\}$, or equivalently the smallest degree d such that a nontrivial polynomial p in adA annihilates $B : 0 = p(B) = \Sigma\{a_k(\text{ad}A)^k(B) : 0 \le k \le d\}$. (Verification of these equivalences is an elementary exercise in linear algebra closely related to computations in Appendix D)

Two rather different uses of this finite-dimensionality assumption are made in our development of commutation theory in Chapters 3-6. These can best be illustrated in the context of Chapter 3 and the beginning of Section 1C above. We assume conditions (E) and (D) so that $\overline{A}$ generates a strongly continuous semigroup which leaves D invariant for all $t \in [0,\infty)$:$V(t,A)D \subset D$. Then the semigroup commutation identity (1-4) reads

$$B\,V(t,A)u = V(t,A)\exp(-t\ \text{ad}\ A)(B)u \tag{2}$$

for all $u \in D$. The first use of finite-dimensionality is made in interpreting $\exp(-t\text{ad}A)(B)u = \Sigma\{(-t)^k/k!\,(\text{ad}A)^k(B)u : 0 \le k \le \infty\}$ as an analytic vector-valued function of t for each $u \in D$. That is, the finite-dimensional operator adA has a well-defined analytic power series exponential acting upon $\mathcal{O}_A(B)$. Also, the strong-operator topology induced on $\mathbf{A}(D)$ by D relativizes on $\mathcal{O}_A(B)$ to the unique finite-dimensional vector-topology there, which also agrees with the $L(D_1,E)$-norm topology defined by $\|\cdot\|$ (Notice also that finite-dimensionality is used in defining a single <u>norm</u> $\|\cdot\|_1$ on D, in terms of the finite basis $\mathbf{B} = \{B_1,\dots,B_d\}$, which makes each $C \in \mathcal{O}_A(B)$ bounded from $(D, \|\cdot\|_1)$ to $(E, \|\cdot\|)$ when E is Banach.) Indeed, the action map $\mathcal{O}_A(B) \times (D, \|\cdot\|_1) \to (E, \|\cdot\|)$ is jointly continuous when $\mathcal{O}_A(B)$ has the $L(D_1,E)$ operator norm, so that the map $t \to \exp(-t\text{ad}A)(B)u$ becomes analytic by composition in quite a strong sense. As we point out in Section D below, there are special cases where exp(-tadA)(B) admits a useful interpretation even where $\mathcal{O}_A(B)$ is infinite-dimensional, but the finite-dimensionality condition is the simplest and most natural way to ensure this interpretation while unifying a variety of applications.

In order to make sense of the corresponding infinite series expansions in the resolvent comutation relations and the operational calculus commutation relations (1-6)

$$B^{(1)}R_1(\lambda,A)u = \Sigma\{(-1)^k R(\lambda,A)^{k+1} [(adA)^k(B)]^{(1)}u : 0\leq k<\infty\} \quad (3)$$

and (1-8)

$$B^{(1)}\varphi_1(A)u = \Sigma\{(-1)^k/k!\ \varphi^{(k)}(A)[(adA)^k(B)]^{(1)}u : 0\leq k<\infty\} \quad (4)$$

we make explicit use of the operator norm of adA on $\mathcal{O}_A(B)$ to estimate the powers $(adA)^k(B)$. That is, if $\mathcal{O}_A(B)$ is equipped with the $L(D_1,E)$ norm, then the power-bound $\|[(adA)^k(B)^{(1)}\| \leq \|adA\|^k \|B^{(1)}\|$ can be used in combination with estimates on $\|R(\lambda,A)^{k+1}\|$ and $\|\varphi^{(k)}(A)\|$ to check convergence on the right in the identities (3) and (4). This use of finite-dimensionality is only a slight extension of its use in defining exp(-tadA)(B) in making sense of (2) above.

The second use of finite-dimensionality involves the spectral analysis of adA on the complexification $\tilde{\mathcal{O}}_A(B) = span_{\mathbb{C}}\{(adA)^k(B) : 0 \leq k < \infty\}$. Here, $\tilde{\mathcal{O}}_A(B)$ has a basis $\{B_1,\ldots,B_d\}$ of generalized eigenvectors for adA with eigenvalue α_k and ascent s_k : $(adA-\alpha_k)^{s_k}(B_k) = 0$. As we indicated in Section 1C, this kind of spectral resolution permits qualitatively different kinds of commutation relations, which can be viewed as translating either A or its spectrum by the eigenvalues α_k. The very special case where B itself satisfies $adA(B) = \alpha B$ (i.e., $(adA-\alpha)(B) = 0$), so that α is an eigenvalue of adA in the classical sense, the commutation identities become, in one form,

$$B^{(1)}R_1(\lambda,A) = R(\lambda,A-\alpha)B^{(1)} \quad (5)$$

$$B^{(1)}V_1(t,A) = V(t,A-\alpha)B^{(1)} \quad (6)$$

$$B^{(1)}\varphi_1(A) = \varphi(A-\alpha)B^{(1)} \quad (7)$$

These identities suggest that more general spectral properties or Wedderburn-decompositions of adA = S + N into semisimple and nilpotent parts could be used when $\tilde{\mathcal{O}}_A(B)$ is infinite-dimensional. But technical problems intervene in the details of formulation and proof, while at this writing all known applications of such decompositions and commutation relations are in the finite-dimensional setting.

B. Examples Involving Differential Operators

Some elementary examples involving ordinary and partial differential operators can serve to indicate the practical content of the finite-dimensionality condition. Since the process of forming commutators of variable-coefficient differential operators A, B in $\mathbb{R}^d$ produces new differential operators whose coefficients are (complicated) expressions in the derivatives of the coefficients of A and B, it is clear in a qualitative sense that the condition that $\mathcal{O}_A(B)$ be finite-dimensional is also equivalent to an elaborate system of nonlinear coupled differential equations in these coefficient functions. There seems to be very little of a general nature that can be said beyond this. However, a few special cases will illustrate the flavor. First, if $A = d/dx$ and $B = f(x)$ (pointwise multiplication operator) then $\mathrm{ad}A(B) = f'(x) = -\mathrm{ad}B(A)$, $(\mathrm{ad}A)^2(B) = f''(x)$ and $(\mathrm{ad}B)^2(A) = 0$, while $(\mathrm{ad}A)^k(B) = f^{(k)}(x)$ in general. Thus $\mathcal{O}_B(A)$ is two-dimensional without restriction on f, while $\mathcal{O}_A(B)$ is finite-dimensional if and only if for some polynomial p, $p(d/dx)f = 0$: f solves a homogeneous constant coefficient equation, i.e., is a linear combination of products of polynomials, exponentials and trigonometric functions. If $A = d/dx$ and $B = f(x)d/dx$, we similarly obtain $(\mathrm{ad}A)^k(B) = f^{(k)}(x)d/dx$ and $\mathcal{O}_A(B)$ is finite-dimensional if and only if f is as above. But here $\mathrm{ad}B(A) = -f'(x)d/dx$, $(\mathrm{ad}B)^2(A) = [f'(x)^2-ff''(x)]d/dx$ and $(\mathrm{ad}B)^k(A)$ is a linear combination of monomials in $f^{(0)}\ldots f^{(k)}$ of degree k multiplying d/dx. Thus the condition for $\mathcal{O}_B(A)$ to be d-dimensional for $d > 2$ becomes a nonlinear differential equation, as our first example illustrates. (See Example 1 of Chapter 9 for a similar second-order situation.)

Example 1. If $A = d/dx$ and $B = x^3\, d/dx$ then $\mathcal{O}_A(B)$ is four-dimensional, while $\mathcal{O}_B(A)$ is infinite-dimensional.

The first claim is immediate from our remarks above. For the second claim, it is useful first to observe that
$[x^m\, d/dx, x^n\, d/dx]u = x^m(nx^{n-1})du/dx - x^n(mx^{m-1})du/dx$
$= -(m-n)x^{m+n-1}du/dx$. Hence ad $B(A) = 3x^2\, d/dx$,
$(\text{ad } B)^2(A) = 3[x^3\, d/dx, x^2\, d/dx] = 3x^4\, d/dx$, and by an easy induction $(\text{ad } B)^k(A) = C(k)x^{2k}\, d/dx$ for $C(k) \neq 0$, whence no linear dependence occurs.

One interesting feature of Example 1 is the fact that $\mathcal{O}_A(B)$ is finite-dimensional while the operator Lie algebra $\mathcal{L}(A,B)$ generated by A and B is not: our commutation theory is not restricted to the setting of finite-dimensional Lie algebras. With the exception of examples like those above, most of the interesting examples of finite-dimensional $\mathcal{O}_A(B)$ for ordinary or partial differential operators do have a Lie-theoretic connection: A and B are generally elements of the enveloping associative algebra $E(\mathcal{L})$ associated with a Lie algebra $\mathcal{L}$ of vector fields or first-order differential operators of the type discussed in Chapter 1). As such, they are covered by the discussion in Section C below. Several specific examples are worked out elsewhere in the monograph, or follow trivially from generalities described in Section C. We summarize some of these here for completeness.

First, we recall the familiar differential operator representation for the canonical or Heisenberg commutation relations, with $P = d/dx$ and $Q = ix$, $[P,Q] = iI$. Then $\mathcal{O}_P(Q) = \mathbb{R}Q + i\mathbb{R}$ in the nilpotent operator Lie algebra $\mathcal{L} = \mathbb{R}\, P + \mathbb{R}\, Q + i\,\mathbb{R}$. It is slightly less well-known that the "second-order" elements of the enveloping algebra $E(\mathcal{L})$ of this operator Lie algebra form a finite-dimensional operator Lie algebra, so that operators such as $P^2 = d^2/dx^2$, $Q^2 = -x^2$ and $QP = ixd/dx$ all have finite-dimensional ad-orbits with respect to each other. These remarks are justified in Example 3 in Section C below, which focusses primarily upon a copy of the Lie algebra $s\ell(2,\mathbb{R})$ spanned by $i(P^2 + Q^2)$, $i(P^2 - Q^2)$ and $PQ + QP$.

We note also that in Chapter 12, another representation of semisimple Lie algebra $s\ell(2,\mathbb{R})$ (more precisely, su(1,1)) as a Lie algebra of vector fields on the unit circle is discussed, with $A_1 = d/d\theta$, $A_2 = \sin\theta\, d/d\theta$ and $A_3 = \cos\theta\, d/d\theta$. Each of these operators has 2-dimensional ad-orbits with respect to each of the others. See Section 12A for details. The solvable operator Lie algebras generated by xd/dx, ix and d/dx (contained in the second order Lie algebra mentioned above) provide different example types. An interesting family of nilpotent Lie algebras of partial differential operators in $\mathbb{R}^{d+1}$ is described in Section 10D, the simplest case takes $d = 1$, $A = \partial/\partial x_0$, $B = x_0\partial/\partial x_1$, so that $[A,B] = \partial/\partial x_1$ is central and $\mathcal{O}_A(B) = \mathbb{R}\, x_0\partial/\partial x_1 + \mathbb{R}\partial/\partial x_1$.

Of course, every Lie group G can supply examples of differential operator Lie algebras which contain finite-dimensional ad-orbits: one simply chooses some convenient system of local coordinates in G and writes the Lie algebra $\mathfrak{G}$ of G (viewed as right invariant vector fields) as first-order partial differential operators in those local coordinates on a suitable neighborhood of 0 in the Euclidian space coordinatizing G. These easy examples are less interesting than the few examples of finite-dimensional ad-orbits for higher-order partial differential operators in the enveloping algebras which can be constructed in special cases via the machinery of Section C below.

C. Examples from Universal and Operator Enveloping Algebras

The examples in this section concern operators A,B in the associative subalgebra $E(\mathfrak{L}) \subset \mathfrak{A}(D)$ generated by a finite-dimensional operator Lie algebra $\mathfrak{L}$ (i.e., $E(\mathfrak{L})$ is the enveloping algebra of $\mathfrak{L}$). In our earliest work [Mr 1-4] and [Jo 1], we were concerned only with pairs $A, B \in \mathfrak{L}$, where both $\mathcal{O}_A(B)$ and $\mathcal{O}_B(A)$ are trivially finite-dimensional since they are contained in $\mathfrak{L}$. An easy extension of that class of examples shows that if $A \in \mathfrak{L}$, then for any $B \in E(\mathfrak{L})$ we have $\mathcal{O}_A(B)$ finite-dimensional. (This is an easy special case of Proposition D1, in Appendix D, which shows that in any associative algebra $\mathfrak{U}$, the set $\mathfrak{h}_A = \{B \in \mathfrak{U} : \mathcal{O}_A(B)$ is finite-dimensional$\}$ determined by a given $A \in \mathfrak{U}$ is an associative and Lie subalgebra of $\mathfrak{U}$. We refer the reader to the appendix for the details.) This essentially familiar remark has some interest when applied to the contemporary quantum theory of elementary particles, in the invariance or symmetry formulation. There, one has a representation U of a large symmetry group G (or at least an infinitesimal representation dU of its Lie algebra $\mathfrak{g}$) such that $dU(\mathfrak{g}) = \mathfrak{L}$ is a Lie algebra of imaginary multiples of important physical observables (linear and angular momenta, isospin, etc.) including an infinitesimal time translation operator H which controls the time-evolution of other observables. Our remark shows that for any observable $B \in E(\mathfrak{L})$, $\mathcal{O}_H(B)$ is finite-dimensional. (Then $B(t) = \exp(t \text{ ad } H)(B(0))$ yields one formula for the time evolution of a large class of observables, $B(t) = V(t,H)B(0)V(-t,H)$ is an alternative version of this formula, which by various versions of identity (2) can be seen to agree with the first formula.)

This same matter of commutation theory for physical observables motivates the consideration of the reversed problem, with $A \in E(\mathfrak{L})$ of higher order and either $B \in \mathfrak{L}$ or $B \in E(\mathfrak{L})$ as well (e.g., A the mass operator and its generalizations as discussed in Section 1B and Chapter 4). A priori, even if $B \in \mathfrak{L}$, one sees that if A has "degree exactly k" ($A \in E_k(\mathfrak{L})$ but $A \notin E_{k-1}(\mathfrak{L})$) then $(\text{ad}A)(B)$ has degree k, $(\text{ad } A)^2(B)$ could have degree as large as $2k - 1$, and in general $(\text{ad } A)^{\ell}(B)$ might have degree as large as $\ell(k-1)+1$, so that finite-dimensionality of $\mathcal{O}_A(B)$ requires strong algebraic relations which prevent this growth in degree. One trivial sort of constraining relation occurs if D is finite-dimensional, so that $\mathbf{A}(D)$ and $E(\mathfrak{L})$ as a whole are finite-dimensional. Another occurs if $\mathfrak{L}$ is Abelian, or in general if $A \in Z(\mathfrak{L})$, the center of $E(\mathfrak{L})$. (Recall that for noncompact semisimple Lie algebras, which are in a sense maximally noncommutative, the center $Z(\mathfrak{L})$ is quite nontrivial and in fact can be used to classify "irreducible" representations.)

The following result of Segal and Goodman [Gd 2] is the most general nontrivial source of higher-dimensional ad-orbits in enveloping algebras known to the authors. It is universal in the sense that it relies only on the structure of $\mathfrak{L}$ as an abstract Lie algebra and not upon the nature of the spatial action of $\mathfrak{L}$ on D. Specifically, it describes finite-dimensional ad-orbits in the complex universal enveloping algebra $\mathfrak{U}(\mathfrak{g})$ of the abstract Lie algebra isomorphic to $\mathfrak{L}$, whence since $E(\mathfrak{L})$ is a homomorphic copy of $\mathfrak{U}(\mathfrak{g})$ (frequently with large kernel, e.g., D finite-dimensional) the Segal-Goodman Lemma yields finite-dimensional ad-orbits in every $E(\mathfrak{L})$.

Segal-Goodman Lemma. Let $\mathfrak{g}$ be a finite-dimensional Lie algebra, and let $\mathfrak{h}$ be an Abelian subalgebra that is nilpotently embedded in $\mathfrak{g}$ (i.e., for all $A \in \mathfrak{h}$, $B \in \mathfrak{g}$, $(\text{ad } A)^k(B) = 0$ for all large k). Then for every $A \in \mathfrak{U}(\mathfrak{h})$ (viewed as a subalgebra of $\mathfrak{U}(\mathfrak{g})$), ad A acts nilpotently upon $\mathfrak{g}$: $(\text{ad } A)^k(\mathfrak{g}) = \{0\}$ for large k. Hence $\mathcal{O}_A(B)$ is finite-dimensional for all $B \in \mathfrak{g}$.

This result can be read off as a corollary of calculations performed in Proposition 2.2, which in fact deliver the following extension (see also Section 4D). (A shorter alternative proof of 2.1 can be obtained from the Segal-Goodman Lemma using Proposition D1 from Appendix D as quoted above.)

2.1. Proposition

Let A be as above. Then ad A acts nilpotently upon every subspace $\mathfrak{U}_k(\mathfrak{g}) = \{B \in \mathfrak{U}(\mathfrak{g}) : \text{degree } (B) \leq k\}$, $(\text{ad } A)^n(\mathfrak{U}_k(\mathfrak{g})) = \{0\}$ for all large n, whence $\mathcal{O}_A(B)$ is finite-dimensional for all $B \in \mathfrak{U}(\mathfrak{g})$.

The simplest applications of these two results take $\mathfrak{h}$ to be one-dimensional nilpotently embedded, so that $A = p(C)$ is a polynomial in the basis element C for $\mathfrak{h}$, and the special case $A = C^2$ of a square will receive quite a bit of attention below (as well as in 1F and elsewhere). An analysis of squares is given in Appendix D, where necessary and sufficient conditions are described such that $\mathcal{O}_A(B)$ be finite-dimensional when $A = C^2$ for $C \in \mathfrak{h}$.

It is natural to ask whether the restriction to Abelian $\mathfrak{h}$ is necessary. Nilpotent embeddedness of course automatically requires (by finite-dimensionality) that $\mathfrak{h}$ be nilpotent, but Goodman has pointed out that simple examples refute the conjecture that nilpotence is sufficient [Gd 3]. In fact, we show below that "$\mathfrak{h}$ Abelian" is necessary, by showing that every non-Abelian nilpotent $\mathfrak{h}$ contains a copy of the simplest possible non-Abelian nilpotent "Heisenberg" Lie algebra, and that the failure occurs in that algebra for A a sum of two noncommuting squares.

In fact, it is useful to construct a general example-machine associated with the three-dimensional non-Abelian nilpotent Heisenberg Lie algebra $\mathfrak{h}_3$, which has a basis X, Y, Z with $[X,Y] = Z$ and Z central. From the (essentially known) calculations that we do here, two contrasting examples are derived in the present section in order to distinguish "universal" phenomena determined by the Lie structure of $\mathfrak{L} \cong \mathfrak{g}$ from those deriving from the spatial action of $\mathfrak{L}$ on D.

This machine also produces useful examples of successful commutation theory with infinite-dimensional ad-orbits in 1F, which are still later combined with other calculations here in order to shed light upon the Bony example in Section 10D (as mentioned in Section 1D above).

We shall at first be concerned with several quadratic operators in the complex universal enveloping algebra $\mathfrak{U} = \mathfrak{U}_{\mathbb{C}}(\mathfrak{h}_3)$: $S_1 = -i/4X^2$, $S_2 = -i/4Y^2$, $A = -i/4(X^2+Y^2)$, $B = -i/4(X^2-Y^2)$, $C = 1/4\, Z\{X,Y\} =_{\mathrm{Def}} 1/4\, Z(XY+YX) = 1/16\, Z[(X+Y)^2-(X-Y)^2]$, where the numerical factors are chosen to simplify subsequent calculations. It is useful for various purposes to view $\mathfrak{U}$ as the algebra of right-invariant differential operators on the Heisenberg group (cf. [Bk, Prop. 25 pg. 252]), acting upon smooth functions, where for our purposes that group will be $\mathbb{R}^3$ with the product

$$(x,y,z)\circ(x',y',z') = (x+x',\ y+y',\ z+z'+x'y). \tag{8}$$

(This formula is appropriate to canonical coordinates of the second kind, for the $\{X,Y,Z\}$ basis.)
An easy calculation yields for the basic right-invariant "derivatives from the left" that then

$$X = \partial/\partial x\ ,\ Y = \partial/\partial y + x\ \partial/\partial z\ ,\ Z = \partial/\partial z. \tag{9}$$

(For example, $Yu(x',y',z') = \partial/\partial y\ u((x,y,z)\circ(x',y',z'))|_{x=y=z=0} = u_2(x',y',z') + x'u_3(x',y',z')$, where the subscripts denote derivatives with respect to the second and third variables respectively.)

We also wish to consider infinitesimal representations of $\mathfrak{h}_3$ that send the central element Z into scalars ("quasi-simple" representations), but in order to simplify calculations we assume that $dV : \mathfrak{h}_3 \to \mathbb{C} \subset \mathfrak{A}(D)$ has $dV(Z) = i$. (Here D may be equipped with any locally convex topology τ.) We adopt the usual mathematician's variant upon physicists' notations by putting $P = dV(X)$, $Q = dV(Y)$, $H = dV(A)$, $L = dV(B)$ and $R = dV(C)$, P is intended to suggest the momentum observable, and Q the position observable, with $H = -i/4(P^2+Q^2)$ the Hamiltonian and $L = -i/4(P^2-Q^2)$ the Lagrangian for a harmonic oscillator (R eludes interpretation). In the usual representation on $D = C_c^\infty(\mathbb{R}) \subset L^2(\mathbb{R})$, $P = d/dx$ and $Q = ix$ are skew-symmetric, as are H, L, R, etc., so that the operators themselves (not their imaginary multiples) have exponentials. We recall that $\mathfrak{h}_3$ is usually called the Heisenberg Lie algebra because $[X,Y] = Z$ translated under dV to the Heisenberg (canonical) commutation relations

$$PQ - QP = [P,Q] = i. \tag{10}$$

Our "universal" example then indicates the kind of pathological infinite-dimensional commutation behavior which can occur among certain second-order elements of $\mathfrak{U}(\mathfrak{h}_3)$.

Example 2. Although S_1 and S_2 have finite-dimensional ad-orbits in $\mathfrak{U}$, both the linear combinations $A = S_1 + S_2$ and $B = S_1 - S_2$ and the commutator $[S_1,S_2] = 1/2\,C$ have infinite-dimensional adjoint-action upon each other and upon X and Y. Thus:

(1) nilpotent and finite-dimensional adjoint actions are not preserved by noncommutative linear combinations or formation of commutators, and

(2) although $\mathfrak{g} = \mathfrak{h}_3$ is nilpotently embedded in itself, the conclusion of the Segal-Goodman Lemma fails for $A \in \mathfrak{U}(\mathfrak{h}_3)$.

The second example essentially shows that these pathologies disappear under "quasi-simple" representations of $\mathfrak{h}_3$. (This improvement persists in weakened form whenever dV(Z) satisfies a suitable polynomial identity $p(dV(Z)) = 0$.)

Example 3. The operators H, L, R, P, Q and i span a six-dimensional real Lie algebra of operators with radical isomorphic to $\mathfrak{h}_3$, and semisimple Levi factor isomorphic to $s\ell(2,\mathbb{R})$ acting upon the copy of $\mathfrak{h}_3$ via the usual (local) automorphism action of $SL(2,\mathbb{R})$ on the canonical commutation relations (e.g., ad $H(P) = -1/2Q = -$ ad $L(P)$, ad $R(P) = -1/2P$, ad $R(Q) = 1/2Q$).

Verifications. Because of subsequent applications, we display a few more computational details than usual in an introductory section of this type. By the derivation property of [,] on associative products, $[S_1,Y] = -i/4(X[X,Y]+[X,Y]X)$ $= -i/4(XZ+ZX) = -i/2\,ZX$ and $[S_2,X] = i/2\,ZY$ in a similar way. Thus

$$[S_1,S_2] = -i/4[S_1,Y^2] = -i/4\{[S_1,Y],Y\} = (-i)^2/8\,Z\{X,Y\}$$
$$= -1/8\,Z\{X,Y\} = 1/2\,C \qquad (11)$$

while similar calculations using derivations on anti-commutators (left to the reader) yield

$$[C,S_1] = -1/4\,Z[\{X,Y\},S_1] = Z^2S_1$$
$$[C,S_2] = -Z^2S_2\,. \qquad (12)$$

Everything follows from these basic facts.

Considering adjoint actions on $\mathfrak{h}_3$ itself, we have

$$\text{ad } A(X) = [S_2,X] = i/2\,ZY,\ \text{ad } A(Y) = -i/2\,ZX \qquad (13)$$

whence by induction and centrality of Z

$$(\text{ad } A)^{2k}(X) = (Z/2)^{2k}X,\ (\text{ad } A)^{2k+1}(X) = i(Z/2)^{2k+1}Y,$$
$$(\text{ad } A)^{2k}(Y) = (Z/2)^{2k}Y,\ (\text{ad } A)^{2k+1}(Y) = -i(Z/2)^{2k+1}X. \quad (14)$$

The universality of $\mathfrak{U}$ shows that there is no linear dependence in the sequences (14) so that $\mathcal{O}_A(X)$ and $\mathcal{O}_A(Y)$ are infinite-dimensional. (Concretely, we recall that Z corresponds to $\partial/\partial z$ in $\mathbb{R}^3$. In fact, for each member of either sequence, one can manufacture a polynomial which yields 1 at 0 after application of the corresponding differential operator but is annihilated at 0 by all others.) For later purposes, it is instructive to notice that formally

$$\exp(t\ \text{ad } A)\ (X) = \cosh(tZ/2)X + i\ \sinh(tZ/2)Y$$
$$= \tfrac{1}{2}(\exp(t/2\ Z)+ \exp(-t/2\ Z))X$$
$$+\ i/2(\exp(t/2\ Z)-\exp(-t/2\ Z))Y. \qquad (15)$$

Acting upon functions, these exponential factors in Z can be interpreted as translation by the elements $\exp(\pm t/2\ Z)$ in the Heisenberg group. (See Section D below.)

Carrying these calculations over into $\mathbf{A}(D)$ via our quasi-simple dV (i.e., replacing X by P, Y by Q, Z by i, and A by H), we get $\operatorname{ad} H(P) = -1/2\, Q$, $\operatorname{ad} H(Q) = 1/2\, P$ and $\mathcal{O}_H(P) = \mathcal{O}_H(Q) = \mathbb{R}P + \mathbb{R}Q$. Moreover (14) and (15) lead to

$$\exp(t \operatorname{ad} H)(P) = \cos t/2\, P - \sin t/2\, Q \tag{14'}$$

and by a similar calculation

$$\exp(t \operatorname{ad} H)(Q) = \sin t/2\, P + \cos t/2\, Q\,. \tag{15'}$$

These are just the classical solutions to Heisenberg's version of Schrödinger's equation for the harmonic oscillator under discussion (mass 2, spring constant 1/2) which says that for any observable T, we expect

$$dT/dt = [H,T] = \operatorname{ad} H(T) \tag{16}$$

reflecting the fact that P(t) and Q(t) execute circles in the "phase plane" spanned by their initial values P and Q. Notice that if the group-commutation identities such as (2) hold, with H exponentiable of course, then these identities also yield the classical rigorous identity

$$V(t,H)PV(-t,H) = \cos(t/2)P - \sin(t/2)Q \tag{17}$$

which gives an alternative expression for the time-evolution of P.

Turning to orbit relations between second-order elements, first in $\mathfrak{U}$, we quickly obtain

$$[A,B] = [S_1+S_2,S_1-S_2] = [S_2,S_1]-[S_1,S_2] = -\,C$$
$$[A,C] = [S_1,C]+[S_2,C] = -\,Z^2(S_1-S_2) = -\,Z^2B$$
$$[B,C] = [S_1,C]-[S_2,C] = -\,Z^2A, \tag{18}$$

using (12). But then by induction

$$(\operatorname{ad} A)^{2k}(B) = Z^{2k}B,\quad (\operatorname{ad} A)^{2k+1}(B) = -\,Z^{2k}\,C$$
$$(\operatorname{ad} C)^{2k}(A) = Z^{4k}A,\quad (\operatorname{ad} C)^{2k+1}(A) = Z^{4k+2}\,B \tag{19}$$

etc. As before, the increasing powers of $Z = \partial/\partial z$ in these commutators prevents linear dependencies, so that $\mathcal{O}_A(B)$ and $\mathcal{O}_C(A)$ are infinite-dimensional. (The calculation for $\mathcal{O}_B(A)$

and $\mathcal{O}_C(B)$ are similar, and those for $\mathcal{O}_A(C)$ and $\mathcal{O}_B(C)$ are equally routine. All yield increasing powers of Z^2.)

When dV is applied to (19), Z^2 goes into $i^2 = -1$ to obtain

$$[H,L] = -R, \ [H,R] = L, \ [L,R] = H \tag{20}$$

With the identification $A_0 = H$, $A_1 = L$, $A_2 = R$ we recognize these as the commutation relations for a basis in $s\ell(2,\mathbb{R})$ (cf. Chapter 12, also Shale [Sh]).

These calculations confirm most of the algebraic claims in the two examples, and the remainder follow in a similar way.

We conclude with the promised converse to the Segal-Goodman Lemma, which also emphasizes the importance of squares of Lie algebra elements, a topic which is analyzed further for $E(\mathfrak{L})$ rather than $\mathfrak{U}(\mathfrak{g})$ in Appendix D.

2.2. Proposition

(a) Let $C \in \mathfrak{g}$, and let $A = C^2$. Then $\mathcal{O}_A(B)$ is finite-dimensional for all $B \in \mathfrak{g}$ if and only if ad C acts nilpotently on $\mathfrak{g}$, and then ad A acts nilpotently as well.

(b) Let $\mathfrak{h} \subset \mathfrak{g}$ be a Lie subalgebra such that for every $A \in \mathfrak{U}(\mathfrak{h})$ and $B \in \mathfrak{g}$, $\mathcal{O}_A(B)$ is finite-dimensional. Then $\mathfrak{h}$ is Abelian and nilpotently embedded.

Proof: (a) Suppose that ad C does not act nilpotently on $\mathfrak{g}$. Then in the complexification $\mathfrak{g}_{\mathbb{C}}$ of $\mathfrak{g}$, the natural extension of ad C must have a nonzero eigenvalue α and eigenvector B. But if ad A has finite-dimensional ad-orbits in $\mathfrak{g}$ it also does in $\mathfrak{g}_{\mathbb{C}}$ by linearity, and in particular $\mathcal{O}_A(B)$ is finite-dimensional. But we have ad $A(B) = [C^2,B] = C[C,B]+[C,B]C = \alpha(CB+BC) = 2\alpha CB - \alpha[B,C] = 2\alpha CB - \alpha^2 B = (2\alpha C-\alpha^2)B$. Since $2\alpha C - \alpha^2$ commutes with A, we get by the obvious induction that $(\text{ad } A)^k(B) = (2\alpha C-\alpha^2)^k(B) = 0$ in $\mathfrak{U}(\mathfrak{g})$, which contradicts universality. Thus ad C acts nilpotently. But in general (as above) we have ad $A(B) = C$ ad $C(B)$ + ad $C(B)C = 2C$ ad $C(B)$ $-(\text{ad } C)^2(B) = (2C-\text{ad } C)\text{ad } C(B)$. The obvious binomial calculation (given explicitly in Appendix D) then shows that $(\text{ad } A)^k$ contains a factor $(\text{ad } C)^k$ and annihilates $B \in \mathfrak{L}$ for large enough k.

(b) By (a), each $C \in \mathfrak{h}$ acts nilpotently on $\mathfrak{g}$, which means by our definition that $\mathfrak{h}$ is nilpotently embedded. But by an old result in Lie theory, since each $C \in \mathfrak{h}$ acts nilpotently on finite-dimensional $\mathfrak{h}$, $\mathfrak{h}$ is nilpotent. Then, as remarked above, if $\mathfrak{h}$ is not Abelian it contains a copy of the Heisenberg Lie algebra, whence Example 3 exhibits an element of $\mathfrak{U}(\mathfrak{h})$ with infinite-dimensional ad-orbit, proving by contradiction that $\mathfrak{h}$ must be Abelian. (To obtain the desired copy of $\mathfrak{h}_3$ in $\mathfrak{h}$, we recall that the lower central series $\mathfrak{h}^0 = \mathfrak{h}$, $\mathfrak{h}^{k+1} = [\mathfrak{h},\mathfrak{h}^k]$ must be properly decreasing to a nontrivial central ideal $\mathfrak{h}^s \neq \{0\}$ with $\mathfrak{h}^{s+1} = \{0\}$. Then some nonzero $Z \in \mathfrak{h}^s$ must be of the form $Z = [X,Y]$ for $X \in \mathfrak{h}$, $Y \in \mathfrak{h}^{s-1}$, both nonzero, and central Z commutes with both, so span $\{X,Y,Z\} \cong \mathfrak{h}_3$).

D. Relaxing the Finite-dimensionality Condition

At this writing, the finite-dimensional $\mathcal{O}_A(B)$ condition is essential to a satisfactory commutation theory for unbounded operators that includes the semigroup, resolvent, and operational calculus commutation identities (2), (3), and (4) above. However, interesting examples have been obtained for the semigroup commutation relations (2) which yield both an interpretation and a proof when $\mathcal{O}_A(B)$ is infinite-dimensional. Since the semigroup relation enters into most of the applications discussed in Chapter 10-12, we indicate here the three different settings in which generalizations have been identified.

In the first approach, one needs the operators A and B to be contained within a subalgebra $\mathbf{A} \subset \mathbf{A}(D)$ which admits a locally-multiplicatively convex or multiplicative norm topology stronger than the strong-operator topology. For example, if the topology τ on D is normable, $\mathbf{A}$ should be bounded operators, while in general $\mathbf{A}$ could be the Γ - finite operators [Mr 6] or the ultracontinuous operators [Mr 5].

The crucial point is that the powers of ad A can be estimated in terms of powers of the norm (or m-seminorms) of A itself, since $\|\mathrm{ad}\,A\| \leq 2\|A\|$, whence $\exp(t\ \mathrm{ad}\ A)(B) = \Sigma\ t^k/k!(\mathrm{ad}\ A)^k(B)$ is convergent and depends strong-operator analytically upon t. The main features of the theory can be pushed through in this setting, but we have not yet identified any interesting applications.

The second direction of generalization is indicated by a number of suggestive examples in which the (semi-) group generators are elements of the enveloping algebra $E(\mathfrak{L})$ of an operator Lie algebra $\mathfrak{L}$. The first example, worked out in detail in Section 3C (with full proofs), treats the square A^2 of an operator A with group-exponential. (Then A^2 is the pregenerator on D of a holomorphic semigroup, as discussed in Appendix C, and for many D, $V(t,A^2)$ leaves D invariant and serves as a semigroup exponential in $\mathbf{A}(D)$. This occurs if $\mathfrak{L} = dV(\mathfrak{g})$ for a group representation V and D consists of the "hyper-Schwartz" vectors [Jo 2].) For such squares, if B is an ad A - eigenelement ad $A(B) = \alpha B$ with real eigenvalue α, then the appropriate commutation relation reads

$$B\,V(t,A^2) = V(t,(A-\alpha)^2)B = V(t,A^2)V(-t,2\alpha A-\alpha^2)B. \qquad (21)$$

Here, since $(\text{ad } A^2)(B) = \{A,[A,B]\} = \alpha\{A,B\} = (2\alpha A-\alpha^2)B$ and $(\text{ad } A^2)^k(B) = (2\alpha A-\alpha^2)^k B$ by induction, $\exp(-t \text{ ad } A^2)(B)$ $= \Sigma\ (-t)^k(2\alpha A-\alpha^2)^k/k!\ B = \exp(-t(2\alpha A-\alpha^2))B$ formally, and the group $V(-t,2\alpha A-\alpha^2) = e^{\alpha^2 t}\ V(-2\alpha t,A)$ supplies a rigorous interpretation of the exponential expression that should appear on the right if the semigroup commutation relation (2) held in a literal sense.

Further examples in the same spirit can be derived from non-quasi-simple representations of the Heisenberg Lie algebra $\mathfrak{h}_3$, where (semi-) group exponentials in $\mathbf{A}(D)$ replace potentially non-convergent commutator series in $E(\mathfrak{L})$. For easy reference, let us denote the basis in $\mathfrak{L} \cong \mathfrak{h}_3$ by X, Y, Z and continue the notation in $\mathfrak{U}_{\mathbb{C}}(\mathfrak{h}_3)$ when carried (homomorphically) into $E(\mathfrak{L})$. If $\mathfrak{L} = dV(\mathfrak{g})$ comes from a group representation V then $X^2 + Y^2$ generates a holomorphic semigroup [Jo 2]. (In the special case where V is unitary on some Hilbert space completion of D, this holomorphic semigroup assumes continuous boundary values, which take the form of a unitary one-parameter group.) One can then prove (under analogous assumptions) that

$$XV(t,A) = V(t,A)(\cosh(t/2,Z)X-i \sinh(t/s,Z)Y) \qquad (22)$$

with the natural conventions that

$$\cosh(t/2,Z) = \tfrac{1}{2}(V(t/2,Z)+ V(-t/2,Z))$$
$$\sinh(t/2,Z) = \tfrac{1}{2}(V(t/2,Z)- V(-t/2,Z)) \qquad (23)$$

in terms of a group-exponential for the central operator Z. (This offers a rigorous interpretation of (15) that specializes to (15') and (17) when $Z = i$.) It is important that (17) remains true if A is replaced by the operator $-i/4(X^2+Y^2+Z^2) = -i/4\ \Delta$, since centrality of Z means ad A $=$ ad$(-i/4\ \Delta)$. (See Section 11D for related discussion.) Applied to the Bony example of Section 10D, where $Z = \partial/\partial x_1$ on $\mathbb{R}^2$, the expressions in (23) become sums and differences of translations in the x-direction.

It is also interesting that in this case, the commutation identities (23) for second-order operators allow semigroup commutation relations when the imaginary factors are removed, but these fail to continue analytically to group commutation relations as above. For example, consider the action of $\frac{1}{4}(X^2+Y^2) = iA$ on B. Then formally

$$\exp(-t \text{ ad } A)B$$
$$=[\Sigma(-1)^k(-tZ)^{2k}/(2k)!]B-i[\Sigma(-1)^k Z^{2k}(-t)^{2k+1}/(2k+1)!]C \qquad (24)$$

Here the first sum can be interpreted (for essentially skew-adjoint Z) as $\cos(-t,Z^2) = \cos(t,Z^2) = \frac{1}{2}[V(t,iZ^2)+ V(-t,iZ^2)]$ and the second is best viewed as $-\int_0^t \cos(rZ^2)dr$, yielding an intelligible (and provable) identity

$$B\,V(t,iA) = V(t,iA)[\cos(tZ^2)B + i\int_0^t \cos(rZ^2)dr\ C]. \qquad (25)$$

In the event that Z is a bounded operator, one can remove the i factor from A, replace $\cos(t,Z^2)$ by $\cosh(t,Z^2)$, and obtain a group variant, but this cannot generally be done if Z is unbounded.

There is considerable reason to believe that these examples, where power series which fail to "converge" in $\mathcal{E}(\mathfrak{L})$ controlled in part by splitting out bounded (semi-) group factors in $\mathfrak{A}(D)$ to absorb divergences, may not be mere isolated anomalies. By Cartan - Jordan - Wedderburn decomposition tricks as introduced in Chapters 5-7, one can analyze the adjoint action of $\mathfrak{L}$ as a mixture of adA - eigenvector and nilpotent behavior, suggesting that many exponentiable operators in $\mathcal{E}(\mathfrak{L})$ can be reached by this sort of machinery.

Finally, there is the possibility of finding useful functional-analytic abstractions of the sort of commutation relations which connect one C^∞ vector field B on a manifold

M with a one-parameter group V(t,A) induced on functions by the flow generated by a second vector field A (as in Chapter 10 and Section 1G). Even without the condition that $\mathcal{O}_A(B)$ be finite-dimensional, one is able to obtain the "commutation relation"

$$B\,V(t,A) = V(t,A)[dV_*(t,A)](B) \tag{26}$$

where $dV_*(t,A)$ denotes the differential of the diffeomorphism $V_*(t,A)$ on M which induces V(t,A) by composition as in 1G $(V(t,A)u)(x) = [u \circ V_*(t,A)](x) = u(V_*(t,A)x)$. The family $\{dV_*(t,A) : t \in \mathbb{R}\}$ is a one-parameter group of automorphisms of the infinite-dimensional Lie algebra $\mathcal{L}(M)$ of all vector-fields on M, and it is true that for each vector-field B, $d/dt\; dV_*(t,A)(B)|_{t=0} = -[A,B] = -\operatorname{ad} A(B)$ whence this group formally satisfies

$$dV_*(t,A)(B) = \exp(-t \operatorname{ad} A)(B). \tag{27}$$

The problem here is to find a rigorous functional-analytic interpretation of (26) which has applications in addition to the example which motivates it.

This last problem forms a small part of the more ambitious program: find a systematic formulation of commutation theory which unifies these three examples with the finite-dimensional theory. In view of the variety of potential applications, this problem seems worthy of attack.

PART II

COMMUTATION RELATIONS AND REGULARITY PROPERTIES FOR OPERATORS IN THE ENVELOPING ALGEBRA OF REPRESENTATIONS OF LIE GROUPS

So engrossed was the Butcher,
he heeded them not,
As he wrote with a pen in each
hand,
And explained all the while
in a popular style
Which the Beaver could well
understand.

The Hunting of the Snark
Fit the Fifth
LEWIS CARROLL

An *Axiom*, you know, is a thing that you accept without contradiction. For instance, if I were to say 'Here we are!', that would be accepted without any contradiction, and it's a nice sort of remark to *begin* a conversation with. So it would be an *Axiom*.

Sylvie and Bruno Concluded
The Professor's Lecture
LEWIS CARROLL

'Those who insisted on mathematical rigour, on clear definitions of the operators and well-defined equations obeyed by them could not take Heaviside's solution seriously. Against these objections Heaviside held that 'mathematics is an experimental science, and definitions do not come first, but later on' (Heaviside, 1893, p. 121).

INTRODUCTION TO PART II

Consider a pair of operators P,Q in a Hilbert space $\mathcal{H}$, and assume that $PQ - QP = -iI$ where I is the identity operator on $\mathcal{H}$. This is the well-known Heisenberg commutation relation from quantum mechanics. In this book we are concerned with a variety of systems, considerably more general, of infinitesimal commutation relations. There are a great number of solutions P,Q satisfying the Heisenberg relation, but a special class of solutions is of central interest in mathematics, and theoretical physics. That is, the operator pairs P,Q with the additional property that the unitary one-parameter groups e^{itP} and e^{isQ} exist. It is then of interest to find sufficient conditions, which can be checked in concrete examples, such that an exponentiation of Heisenberg's relation is possible, subject to these conditions. Two of the exponentiated versions are

$$Q\, e^{itP} = e^{itP}\, (Q + tI) \ ,$$

or the important Weyl commutation relation

$$e^{isQ}\, e^{itP} = e^{-ist}\, e^{itP}\, e^{isQ}.$$

Being able to exponentiate the infinitesimal Heisenberg relation is particularly essential to applications since the resulting exponentials (viz., the Weyl relations) can be solved explicitly in terms of known operators: The one-parameter groups take the form of translation and multiplication on vector valued L^2-functions. That is, we get translation, resp., spectral, representations for the operators P,Q given at the outset. Of course, a unitary equivalence is involved. (This is the Stone-von Neumann theorem.) This line of attack has proved effective in diverse applications (scattering theory, ergodic theory, etc.) involving the canonical commutation relations.

Some of the more successful tools for exponentiating the infinitesimal relation $PQ - QP = -iI$ are based on analytic vector methods [Dx 1, Fu 1, Nℓ 1, Ps 2]. In this book we shall be concerned with the exponentiation problem for a given infinitesimal commutation relation of a general nature.

Motivated by applications to representations of Lie groups, and to a certain spectral problem for the mass-operator in elementary particle physics, we have formulated our infinitesimal commutation relations in considerable generality, as dictated by the applications. In fact, the classical Heisenberg relations turn up only as a very special case in our development. Instead of operators in Hilbert space we shall consider operator theory in Banach space, and more general locally convex spaces.

In Chapter 10 of the monograph we use the locally convex generality in an essential manner to give a new functional analytic proof of Palais' differential geometric theorem on integrability of Lie algebras of smooth vector fields. In view of this application it is of interest that our theorems (and proofs) are formulated entirely in terms of operator theory. The different applications (including the Palais theorem) are then translated from geometry, or elementary particle physics, into a form where the operator theory is applicable.

For the given operator system under consideration we introduce spaces of C^∞-vectors, and we show that the C^∞-vector methods provide a unifying approach to commutation problems in operator theory. In particular, our method gives stronger results in the application to the mass-operator, relative to the known analytic vector methods that have been used earlier by Flato et al. [F St 1], Segal [Sg 2], and Goodman [Gd 3].

Now, the three operators, iP, iQ, and iI, span a three-dimensional real Lie algebra of unbounded operators $\mathfrak{L}$, and $\mathfrak{L}$ is the range of a well defined Lie algebra representation $\rho: \mathfrak{h}_3 \to \mathfrak{L}$ where $\mathfrak{h}_3$ is the three-dimensional Heisenberg matrix Lie algebra. More generally, we consider a normed linear space D with Banach space completion E, and a finite-dimensional real Lie algebra $\mathfrak{g}$. A representation, $\rho: \mathfrak{g} \to \mathfrak{A}(D) =$ (the algebra of all linear endomorphisms in D), is always associated with a Lie algebra, $\mathfrak{L} = \rho(\mathfrak{g})$, of possibly unbounded operators in E.

Let L(E) be the algebra of all bounded operators in E, and let $V:G \to L(E)$ be a strongly continuous representation of a Lie group G, with Lie algebra $\mathfrak{g}$. Let dV be the corresponding infinitesimal representation on the Gårding space [Gå]. We say that $\mathfrak{L}$ is <u>exponentiable</u> if a representation V exists such that $dV(\mathfrak{g}) = \mathfrak{L}$. The representation ρ is said to be exact if $\rho = dV$.

It follows that the exponentiation problem for the Heisenberg relations, as well as the integrability problem of Palais for Lie algebras of vector fields, may be naturally imbedded into the exactness/exponentiability problem, introduced above, and considered in Chapter 9 of the monograph.

A major step in the solution to the latter problem concerns the algebraic relations which are implicit on a given pair of operators A,B which belong to a Lie algebra $\mathfrak{L}$ of (generally unbounded) operators.

In Part II we focus on a pair of operators A,B which are both defined on a common, invariant, and dense domain D in a Banach space E. The applications in Chapter 4 dictate a mathematical setup which is, in fact, more general than the one which is imposed by a given operator Lie algebra $\mathfrak{L}$.

The operator A is assumed throughout to generate a strongly continuous semigroup $\{V(t,A) : 0 \leqq t < \infty\} \subset L(E)$, and we are concerned with the commutation relation

$$B\,V(t,A) = V(t,A)\exp(-t \text{ ad } A)(B)$$

which, of course, is the natural generalization of Weyl's commutation relation.

The right-hand side of this relation makes it clear that the operators

$$\text{ad } A(B) = AB - BA,\ldots,(\text{ad } A)^{k+1}(B) = (\text{ad } A)(\text{ad } A)^{k}(B),$$

are of special interest. In particular, the operator $\exp(-t \text{ ad } A)(B)$ may be defined through a convergent power series if the ad-orbit

$$\mathcal{O}_A(B) = \text{lin. span}\{(\text{ad } A)^{k}(B) : k \in \mathbb{N}\}$$

is finite-dimensional. On the other hand, this condition may well be satisfied even if the Lie algebra generated by the operator pair A,B is ∞-dimensional. The algebraic and analytic features of ad-orbits enter essentially in our applications to the elementary particle mass-operator, and are analyzed in detail in the appendices.

Chapter 3

DOMAIN REGULARITY AND SEMIGROUP COMMUTATION RELATIONS

In the present Chapter, we obtain several generalizations of a classical commutation relation from the theory of matrix Lie algebras and Lie groups, often called the adjoint representation identity. Specifically, if A and B are operators on a finite-dimensional space E, this relation takes the form

$$\exp(tA)\,B\exp(-tA) = \exp(t\,\mathrm{ad}A)(B) = \Sigma\{t^k/k!(\mathrm{ad}\ A)^k(B) : 0 \le k < \infty\} \quad (1)$$

where ad $A(C) = AC - CA$. Here, we shall primarily be concerned with extensions of (1) to cases where A and B are linear endomorphisms of infinite-dimensional spaces E_∞ or D for which the exponentials in (1) can still be interpreted reasonably in terms of other endomorphisms of these spaces. As is well-known (and essentially recapitulated in Chapter 2), the standard matrix arguments using rearrangements of power series apply equally well to bounded Banach space operators A, B. In the context of interest here, where A and B are continuous operators on locally convex spaces or closable unbounded Banach space operators, we must use equally classical but less well-known C^∞ differential and integral calculus arguments. The various settings chosen for the results below are essentially the most general for which these C^∞ methods are natural,and nontrivial applications are to be found.

Three levels of generality are discussed below in Sections 3A-D (results are numbered sequentially throughout from beginning to end). In Section 3A, we consider the case in which A, B are continuous operators (everywhere-defined) on a locally convex space E_∞, contained within a finite-dimensional real Lie algebra $\mathcal{L}$ of such operators. There, the group $\{\exp(t\ \mathrm{ad}\ A): t \in \mathbb{R}\}$ of Lie automorphisms on the right in (1) can be interpreted as usual. We assume there that A generates a differentiable one-parameter group $\{V(t,A): t \in \mathbb{R}\}$ of automorphisms of E_∞ which replaces the exponentials on the left in (1). The simple result obtained in 3A (3.1) applies primarily in the exponentiation problem for infinitesimal operator Lie algebras (Chapter 9), both in the proof of the Main Integration Lemma in Chapter 8 and in applications to transformation groups (Chapter 10).

In Section 3B, two generalizations over 3A are incorporated. First, A and B are assumed to lie in the endomorphism algebra $\mathbf{A}(D)$ of a dense linear subspace D in a locally convex space E (often Banach in applications), and a weaker finite-dimensionality condition is imposed: the ad-orbit of B under A, $\mathcal{O}_A(B)$ = real span $\{(\mathrm{ad}\,A)^k(B)\colon 0 \le k \le \infty\}$, is assumed to be a finite-dimensional subspace of $\mathbf{A}(D)$. (Then ad A acts as an endomorphism of $\mathcal{O}_A(B)$ and exp(t ad A) is well-defined in the usual way.) Secondly, A is assumed to be closable, and $\bar{A}$ is assumed to generate a strongly continuous one-parameter semigroup $\{V(t,A)\colon t \in [0,\infty)\}$ on all of E that leaves D invariant (hence restricts to a strongly C^∞ semigroup of endomorphisms of D.) A variety of technical functional-analytic conditions can then be imposed upon A, B and/or D in order to permit the following semigroup-variant of (1) to be established in $\mathbf{A}(D)$ for $t \in [0,\infty)$

$$B\,V(t,A) = V(t,A)\exp(-t\,\mathrm{ad}\,A)(B). \tag{2}$$

This extra generality serves several purposes. In Chapter 9 it leads to a link between the Main Integration Lemma for operator Lie algebras and the hypotheses of our first (domain-regularity) exponentiation theorem 9.1 (or 9.1' in 1D) for infinitesimal Lie group representations. The full semigroup-$\mathcal{O}_A(B)$ generality is needed in our subsequent analysis of the O'Raifeartaigh mass-splitting result [Gd 3] in Chapter 4, where A is frequently a holomorphic semigroup-generator in the enveloping algebra $\mathcal{E}(\mathcal{L})$ of an operator Lie algebra $\mathcal{L}$, $B \in \mathcal{L}$, and the pair (A,B) need not generate a finite-dimensional Lie algebra. (This is illustrated in Chapter 1 where the appropriate background is developed.)

Section 3C is technical in nature. It serves to sort out the various functional-analytic regularity assumptions on A, B and D that are required in order to push through the conceptually simple proof of (2) that is given in 3B. While the material in 3C is of considerable practical interest in applications, it can be skipped in a first reading without major loss of continuity.

Finally, we sample rather briefly in Section 3C a result in which even the finite-dimensionality restriction on $\mathcal{O}_A(B)$ is relaxed but a recognizable facsimile of (2) can still be obtained by appropriately stretching its interpretation. This result serves to illustrate the direction of probable future development for the results in this section, and to establish connections with the comparable (more powerful) method of operator graph-density that appears in Chapters 6, 7 and 11.

Before turning to Section 3A, we pause to fix more detailed language on one-parameter operator semigroups that is used in the subsequent discussion. (See Appendix B for a folklore review.)

In general, if E is a real or complex Hausdorff locally convex space (lcs) with operator algebra $L_s(E)$ (continuous endomorphisms of E with the 'strong operator' or pointwise convergence topology), a C_0 one-parameter locally equicontinuous (cle) semigroup $\{V(t): t \in [0,\infty)\} \subset L(E)$ is a continuous function V from $[0,\infty)$ to $L_s(E)$ which satisfies the functional equation $V(s+t) = V(s)V(t)$, the initial condition $V(0) = 1$, and the 'local boundedness' condition that for all compact $K \subset [0,\infty)$, $\{V(t): t \in K\}$ is an equicontinuous set of endomorphisms of E. (This last condition is redundant for such familiar barreled spaces as Banach, Fréchet or LF spaces.) The (infinitesimal) generator C of $\{V(t): t \in [0,\infty)\}$ is defined for those initial condition vectors $u \in E$ such that the orbit function $\tilde{u}(t) = V(t)u$ is right-differentiable at $t=0$, with $Cu = \tilde{u}'(0)$. For such $u \in D(C)$, $\tilde{u}$ is differentiable at all $t \in [0,\infty)$, $\tilde{u}(t) \in D(C)$ for all t, and $\tilde{u}'(t) = C\tilde{u}(t)$. The semigroup is differentiable if and only if $D(C) = E$ and $C \in L(E)$. (This second condition is again redundant for the barreled examples mentioned above.) If each V(t) is an automorphism of E, and if the extended function $\{V(t): t \in \mathbb{R}\}$ obtained by setting $V(-t) = V(t)^{-1}$ satisfies the same conditions on $\mathbb{R}$ that were assumed above on $[0,\infty)$, we refer to $\{V(t): t \in \mathbb{R}\}$ as a cle (respectively, differentiable) group.

3A. Lie algebras of continuous operators

3.1. Proposition

Let $\mathcal{L}$ be a finite-dimensional real Lie algebra of continuous endomorphisms of a locally convex space E such that $A(=\bar{A})$ generates a differentiable cle group $\{V(t,A): t \in \mathbb{R}\}$ on E.

Then the series $\Sigma\{(n!)^{-1}t^n(\mathrm{ad}A)^n(B): 0 \leq n < \infty\}$ converges in $\mathcal{L}$ for all $t \in \mathbb{R}$, defining a function exp(t ad A)(B) which is analytic with respect to all possible (Hausdorff) operator topologies (alias the unique finite-dimensional topology) and such that for all $t \in \mathbb{R}$,

$$V(t,A)BV(-t,A) = \exp(t\ \mathrm{ad}\ A)(B). \qquad (3)$$

Proof: The idea is to check that if $F(t) = V(t,A)BV(-t,A)$ then $F'(t) = \mathrm{ad}\, A(F(t))$ and $F(0) = B$, so that F must be the unique solution to this exponential differential equation: $F(t) = \exp(t\, \mathrm{ad}\, A)(B)$. This would be immediate if we knew in advance that $F(t)$ took its range in $\mathcal{L}$ (or perhaps some larger finite-dimensional subspace of the operator algebra.)

Fortunately, one of the simplest uniqueness proofs for the solutions of such a differential equation does not actually require any foreknowledge of the range of F, and it can be combined with the check that F satisfies the differential equation mentioned above. If we fix $t \in \mathbb{R}$ and put

$$H(s) = V(t-s,A)\exp(s\, \mathrm{ad}\, A)(B)V(s-t,A) \tag{4}$$

then it suffices to check that H is constant, since then we will have

$$V(t,A)BV(-t,A) = H(0) = H(t) = \exp(t\, \mathrm{ad}\, A)(B). \tag{3}$$

To establish the constancy claim in (3), it suffices in turn to check that H is differentiable as an $L_s(E)$-valued function, with $H'(s) \equiv 0$. This we do by a double application of the product rule (Appendix A), which justifies the following formal calculation for all $u \in E$ and fixed $s_0 \in \mathbb{R}$

$$\begin{aligned} H'(s_0) &= d/ds\, V(t-s,A)[\exp(s_0 \mathrm{ad}\, A)(B)V(s_0-t,A)u]\big|_{s=s_0} \\ &+ V(t-s_0,A)d/ds\, \exp(s\, \mathrm{ad}\, A)(B)[V(s_0-t,A)u]\big|_{s=s_0} \\ &+ V(t-s_0,A)\exp(s_0 \mathrm{ad}\, A)(B)d/ds\, V(s-t,A)u\big|_{s=s_0} \\ &= V(t-s_0,A)\{(-A)\exp(s_0 \mathrm{ad}\, A)(B) + \mathrm{ad}\, A(\exp(s_0 \mathrm{ad}\, A)(B)) \\ &+ \exp(s_0 \mathrm{ad}\, A)(B)A\}V(s_0-t,A)u \\ &= V(t-s_0,A)\{0\}\cdot V(s_0-t,A)u = 0. \end{aligned} \tag{5}$$

Since $V(\pm(t-s),A)$ is assumed strongly differentiable and locally equicontinuous, the only details to check involve $\exp(s\, \mathrm{ad}\, A)(B)$. But since $\mathcal{L}$ is a finite-dimensional subspace of $L_s(E)$, the relative topology is the unique finite-dimensional topology, and $\exp(s\, \mathrm{ad}\, A)(B)$ is analytic, hence certainly differentiable.
Moreover, $\{\exp(s\, \mathrm{ad}\, A)(B)\colon s \in K\}$ is compact when K is compact, and such compact subsets of finite-dimensional operator subspaces are well-known to be equicontinuous since the

coordinate functions with respect to a basis in $\mathfrak{L}$ are uniformly bounded. Thus exp(s ad A)(B)V(s-t,A)u and in turn H(s) are differentiable by the product rule, justifying (5) and completing the proof.

3B. Semigroups and ad-orbits

In the usual applications of Proposition 3.1 (e.g., Theorem 8.1) slight asymmetry between the conditions imposed upon A and B does not occur: B generates a differentiable cle group $\{V(t,B): t \in \mathbb{R}\}$ as well. The requirements of the applications envisioned in this section are much less symmetrical. In fact, for some variations the only common conditions satisfied by A and B are that both are in the algebra $\mathbf{A}(D)$ of linear (generally discontinuous) endomorphisms of a dense subspace $D \subset E$. We can distinguish four qualitatively different sorts of conditions to be imposed upon A, B or both

(E) global semi-exponentiability of A, in the sense that the closure $\bar{A}$ is an operator which generates a cle semigroup $\{V(t,A): t \in [0,\infty)\}$;

(D) global domain invariance, in the sense that $V(t,A)D \subset D$ for all $t \in [0,\infty)$;

(F) finite-dimensional 'infinitesimal compatibility', in the sense that $\mathcal{O}_A(B)$ = real span $\{(\mathrm{ad}\ A)^k(B): 0 \leq k < \infty\}$ is finite-dimensional; and

(R) infinitesimal/global regularity, which always (at least implicitly) involves the assumption that for all $u \in D$ the function $G(t) = BV(t,A)u$ is differentiable, with $G'(t) = BAV(t,A)u$.

The motivating examples for these conditions have already been discussed in Chapter 1, as have their potential generalizations. Conditions (F) and (R) will receive further attention below, in 3.7 and in 3.3 through 3.6, respectively.

3.2. Theorem

Let D be a dense subspace of a lcs E, and let $A, B \in \mathbf{A}(D)$. Suppose that conditions (E), (D), (F) and (R) hold. Then the formal power series

$$e(t) = \exp(-t\ \mathrm{ad}\ A)(B) = \Sigma\{(n!)^{-1}(-t\ \mathrm{ad}\ A)^n(B): 0 \leq n < \infty\} \quad (6)$$

converges for all $t \in \mathbb{R}$, in $\mathbf{A}(D)$ with the strong-operator topology induced by D, and we have in $\mathbf{A}(D)$

$$BV(t,A) = V(t,A)e(t) = V(t,A)\exp(-t\ \mathrm{ad}\ A)(B). \quad (2)$$

Proof: The claims concerning e(t) are obvious, since the strong operator topology on $\mathbf{A}(D)$ relativizes on the finite-dimensional $\mathcal{O}_A(B)$ to the unique finite-dimensional topology there.

The idea of our proof for (2) is an obvious variant of that used in 3.1. First fixing $t > 0$, we put

$$H(s) = V(t-s,A)e(t-s)V(s,A), \quad s \in [0,t] \tag{7}$$

noting that it suffices to prove that H(s) is constant on [0,t] since then

$$BV(t,A) = H(t) = H(0) = V(t,A)e(t) \tag{2'}$$

will follow. Defining $K(s)\colon \mathcal{O}_A(B) \to \mathbf{A}(D)$ by $K(s)C = V(t-s,A)CV(s,A)$ for all $C \in \mathcal{O}_A(B)$ we see that H(s) is a 'product'

$$H(s) = K(s)e(t-s). \tag{8}$$

Formally, $d/ds\, e(t-s) = \mathrm{ad}\, A\, e(t-s)$, as the reader can easily check, so if we can show that for any $C \in \mathcal{O}_A(B)$, K(s)C is differentiable with $K'(s)C = K(s)(-\mathrm{ad}\, A(C))$, and if the product rule for differentiation can be applied, we will have

$$\begin{aligned} H'(s) &= K'(s)e(t-s) + K(s)d/ds\, e(t-s) \\ &= K(s)(-\mathrm{ad}\, A)(e(t-s)) + K(s)\mathrm{ad}\, A(e(t-s)) = 0, \end{aligned} \tag{9}$$

and constancy will be established. The problem with this formal argument lies in finding a suitable topology on $\mathbf{A}(D)$ that will make it work. The one that is technically most convenient is a weak-operator-type topology induced on $\mathbf{A}(D)$ by the separated pairing of D with a certain 'Gårding domain' $G^*(A) \subset E^*$ discussed in Appendix B. Specifically, for any $\varphi \in C_c^\infty((0,\infty))$, the integral in $V_A(\varphi) = \int_0^\infty \varphi(t)V(t,A)dt$ converges in $L_s(E)$ to a continuous operator, and $G^*(A) = \mathrm{span}\,\{V_A(\varphi)^*f : \varphi \in C_c^\infty(0,\infty), f \in E^*\}$ is shown there to be weak-* $(\sigma(E^*,E))$ dense, and invariant both under all $V(t,A)^*$ and under the adjoint $A^* = \bar{A}^*$ of A (in particular, $D(A^*)\supset G^*(A)$). Hence $G^*(A)$ separates points on D, and each pair $f \in G^*(A)$, $u \in D$ defines a seminorm $p_{f,u}(C) = |f(Cu)|$ on all $C \in \mathbf{A}(D)$. These seminorms give $\mathbf{A}(D)$ a Hausdorff locally convex topology τ which induces the unique finite-dimensional topology on $\mathcal{O}_A(B)$. We check that then $\{K(s)\colon s \in [0,t]\}$ is a continuous, locally equicontinuous and differentiable family of maps from $(\mathcal{O}_A(B),\tau)$ to $(\mathbf{A}(D),\tau)$, with $K'(s)(C) = K(s)(-\mathrm{ad}\, A(C))$ as claimed above, whence the product rule (Appendix A) will justify the calculation (9). Differentiability is basic.

To check differentiability, then, we remark first that since $K(s)$ is obviously linear as a map from $\mathbf{A}(D)$ to $\mathbf{A}(D)$, it suffices to check differentiability of $K(s)(\mathrm{ad}\ A)^k(B)$ for all $k = 0,1,\ldots$. But in turn since every $(\mathrm{ad}\ A)^k(B) = \sum_{i+j=k} (-1)^i\binom{k}{i}A^iBA^j$ by the binomial theorem (cf. Appendix D), it suffices to check differentiability for each of the

$$K(s)A^iBA^j = V(t-s,A)A^iBA^jV(s,A) = A^iV(t-s,A)BV(s,A)A^j.$$

Picking $u \in D$, $f \in G^*(A)$, this amounts to the verification that $f((K(s)A^iBA^j)u) = ((A^*)^if)(V(t-s,A)BV(s,A)A^ju)$ is differentiable. But for each $v = A^ju \in D$, $BV(s,A)v$ is differentiable by (R), with values in D, while for each $w \in D$, $V(t-s,A)w$ is differentiable, and $V(t-s,A)$ is equicontinuous on $[0,t]$, so the product rule (Appendix A) implies that $V(t-s,A)BV(s,A)v$ is a differentiable E-valued function with

$$\begin{aligned} d/ds\ V(t-s,A)BV(s,A)v &= V(t-s,A)(-AB)V(s,A)v \\ &\quad + V(t-s,A)BAV(s,A)v \\ &= V(t-s,A)(-\mathrm{ad}\ A(B))V(s,A)v. \end{aligned}$$

Thus since $(A^*)^if \in E^*$, it composes with the above function to yield differentiability and

$$\begin{aligned} d/ds\ f((K(s)A^iBA^j)u) &= (A^*)^if(V(t-s,A)(-\mathrm{ad}\ A)(B)V(s,A)A^ju) \\ &= f(K(s)(-\mathrm{ad}\ A(A^iBA^j))u). \end{aligned}$$

Reassembling the various linear combinations, we obtain $K'(s)C = K(s)(-\mathrm{ad}\ A(C))$ as claimed. But then $K(s)$ is clearly strongly continuous from $\mathcal{O}_A(B)$ to $(\mathbf{A}(D),\tau)$, and $\{K(s)\colon s \in [0,t]\}$ is a strong-operator-compact family of operators from the (automatically barrelled) finite-dimensional $(\mathcal{O}_A(B),\tau)$ to $(\mathbf{A}(D),\tau)$, so it is equicontinuous. Thus the product rule indeed applies to $K(s)e(t-s)$ as required, and the calculation (9) is correct. This completes the proof.

<u>Remark on the proof</u>: The argument given above is radically simplified if we assume at the outset that for all $C \in \mathcal{O}_A(B)$ and $u \in D$ the function $t \to CV(t,A)u$ is differentiable, with the appropriate derivative. Both the use of the weak topology defined by $G^*(A) \subset E^*$ and the reduction of the $(\mathrm{ad}\ A)^k(B)$ to polynomials in A and B can then be avoided, and the resulting proof becomes a close imitation of that for 3.1. But the

extra generality achieved by this more complicated argument has proved useful sufficiently frequently to justify the added effort here.

3C. Variations upon the regularity condition

Next, we return to a detailed examination of the regularity condition (R): We assumed that the function $G: [0,\infty) \to \mathbf{A}_s(D)$ defined by $G(t) = BV(t,A)$ is differentiable (i.e. for all $u \in D$, $G(t)u$ is a differentiable E-valued function) and $G'(t) = BAV(t,A) = BV(t,A)A = G(t)A$. For the practical purpose of simplifying applications of Theorem 3.2. and its progeny (notably Theorem 9.1), it is expedient to weaken the regularity properties which must be checked for $G(t)$, while in exchange imposing conditions upon the operator B that are normally available in the context of these applications, or upon the domain D. There are some reasons to believe that under the strongest restrictions on B, and when $\bar{A}$ generates a group, the regularity condition (R) may be an automatic consequence of the invariance condition (D) (perhaps combined with the finite-dimensionality condition (F)). We discuss the evidence pro and con at the end of Section 3C.

3.3. Lemma

Suppose that the pair $A, B \in \mathbf{A}(D)$ satisfy conditions (E) and (D) (i.e. $\bar{A}$ generates a cle semigroup $\{V(t,A): t \in [0,\infty)\}$ which leaves D invariant). Then either of the following conditions is sufficient to ensure that condition (R) is also satisfied.

(CWC) The operator B is closable, and G is weakly continuous on $[0,\infty)$, in the sense that for all $u \in D$ and $f \in E^*$ the function $G_u^f(t) = f(G(t)u)$ is continuous. If $\bar{A}$ generates a group, it is sufficient that each G_u^f be continuous at $t = 0$ in $\mathbb{R}$.

(ELB) The closure $\bar{B}$ of B generates a cle semigroup $\{V(r,B): r \in [0,\infty)\}$ and G is locally bounded on $[0,\infty)$ in the sense that for all $u \in D$ and compact $K \subset [0,\infty)$, $\{G(t)u ;t \in K\}$ is bounded in E. Again, if $\bar{A}$ generates a group it suffices that for each $u \in D$ there exist a neighborhood $N_u(0) \subset \mathbb{R}$ of 0 such that $\{G(t)u: t \in N_u(0)\}$ is bounded.

<u>Remark</u>: It is not known at present whether (*a priori* one-sided) conditions at 0 suffice even in the case where $\bar{A}$ generates only a semigroup. The difficulties involved are much the same as those entailed by the more ambitious project

of reducing (R) to (E), (D), exponentiability of B (and perhaps (F)). No applications are presently known where settlement of this question in the affirmative would have an impact.

Proof: For both claims, we exploit the integral identity

$$V(t,A)u = u + \int_0^t V(s,A)Au\,ds, \tag{10}$$

which is valid for all $u \in D \subset D(\bar{A})$, convergent both with respect to the weak and the original topologies on $D \subset E$.

Treating first (CWC), we recall that by a standard application of Mazur's theorem to Graph $(\bar{B})$, $\bar{B}$ is weakly closed. Thus since $Au \in D$, $BV(t,A)Au = G(t)Au$ is weakly continuous, and the standard theorem on interchange of Riemann integrals with closed operators implies that when B is applied to both sides of (10) we obtain

$$G(t)u = BV(t,A)u = Bu + \int_0^t BV(s,A)Au\,ds. \tag{11}$$

The fundamental theorem of (vector-valued) calculus then immediately implies that $G(t)u$ is weakly differentiable, with $G'(t) = BV(t,A)Au = G(t)Au$. The obvious induction, using $A^k u \in D$ for all k, then shows that $G(t)u$ is weakly C^∞ with $G^{(k)}(t)u = G(t)A^k u$. But weakly C^∞ functions are C^∞ with respect to the original topology (by 'weak boundedness implies original boundedness'; the standard proof is not in fact a uniform-boundedness or barreledness fact).

Concluding the treatment of (CWC), we observe that if $\bar{A}$ generates a group and each G_u^f is continuous at 0, then by (D) $G_u^f(t) = f(BV(t-t_0,A)V(t_0,A)u) = G^f_{V(t_0,A)u}(t-t_0)$ is continuous at $t-t_0 = 0$, yielding continuity for all t. (The corresponding semigroup argument would yield only one-sided continuity.)

We reduce the sufficiency proof for (ELB) to that for (CWC). If we put $B_n = n(V(1/n,B)-1) \in L(E)$ for a sequence of continuous approximants to B, then each B_n may be applied to both sides of (10), followed by any $f \in E^*$, and both may be taken under the integral to obtain for any $u \in D$

$$f(B_n V(t,A)u) = f(B_n u) + \int_0^t f(B_n V(s,A)Au)ds. \tag{12}$$

We use the local boundedness hypothesis and the dominated convergence theorem in combination with (D) to obtain pointwise convergence of (12) as a whole to

$$G_u^f(t) = f(Bu) + \int_0^t G^f_{Au}(s)ds. \tag{13}$$

Continuity of $G^f_u(t)$ then follows easily from boundedness of the integrand in (13) (which actually yields a Lipschitz condition). In order to obtain (13), we expand the integrand in (12) using a formula like (10) for V(s,B):

$$f(B_n V(s,A)Au) = f(n\int_0^{1/n} V(r,B)BV(s,A)Au\, dr)$$
$$= n\int_0^{1/n} f(V(r,B)BV(s,A)Au)\, dr. \tag{14}$$

Our local boundedness condition implies that for any $T < \infty$, $\{BV(s,A)Au\colon s \in [0,T]\}$ is bounded in E, whence by local equicontinuity of V(r,B) and continuity of $f \in E^*$, $\{f(V(r,B)BV(s,A)Au\colon r \in [0,1],\ s \in [0,T]\}$ is bounded in $\mathbb{C}$. Thus since the integral in (14) lies in the convex hull of this set, it follows that the integrands in (12) are uniformly bounded in n. Since each $V(s,A)Au \in D$ and $B_n \to B$ on D, it follows that the integrand in (12) converges pointwise to that in (13), while the two integrated terms similarly converge, yielding (13) as desired.

If $\bar{A}$ generates a group and for each u there is an $N_u(0)$ on which $G_u(t)$ is bounded, then for each t_0 there is an $N_{V(t_0,A)u}(0)$ on which $G(t)V(t_0,A)u = BV(t,A)V(t_0,A)u =$

$BV(t+t_0,A)u$ is bounded so that G(t)u is bounded in a neighborhood $N_u(t_0) = N_{V(t_0,A)u}+t_0$ as well. Consequently, since any compact $K \subset \mathbb{R}$ can be covered by finitely many such intervals, it follows that G(t)u is bounded on each compact K. (Notice that in the semigroup case one can obtain only half-open intervals, for which the compactness argument does not apply.) E.O.P.

This lemma then leads to a variety of formal improvements upon Theorem 3.2. The following versions have proven most useful for the applications made in Chapters 4 and 9 of the monograph.

3.4. Theorem

Let A, B be closable operators in the algebra $\mathbf{A}(D)$ of endomorphisms of a dense subspace D in a lcs E, and suppose that $\mathcal{O}_A(B) = \text{real span } \{(\text{ad } A)^k(B)\colon 0 \leq k < \infty\}$ is finite-dimensional. Suppose further that A and B are the restrictions to D of the generators of cle semigroups $\{V(s,A)\colon s \in [0,\infty)\}$ and $\{V(s,B)\colon s \in [0,\infty)\}$, respectively. Finally, suppose that every $V(t,A)$ leaves D invariant and that for every $u \in D$ the function $G_u(t) = BV(t,A)u$ is locally bounded on $[0,\infty)$. Then

(1) the closure $\bar{A}$ of A is the generator of $\{V(t,A): t \in [0,\infty)\}$ and

(2) for all $t \in [0,\infty)$ the identity

$$B\,V(t,A) = V(t,A)\ \exp(-t \text{ ad } A)(B)$$

holds in $\mathbf{A}(D)$.

Moreover, if $\{V(t,A): t \in [0,\infty)\}$ extends to a cle group satisfying the invariance condition, then it suffices that for some fixed $t_0 \in \mathbb{R}$ we have that for all $u \in D$ there is a neighborhood of t_0 where $G_u(t)$ is bounded.

Proof: The first claim (1) is an immediate consequence of a general core result [Ps1] that in this special case is sometimes known as the 'pregenerator theorem'. We review this result in Appendix B in order to indicate that one of the known Banach space proofs applies in the locally convex setting as well. The second claim is immediate from 3.3 combined with 3.2. E.O.P

The following related corollary is also useful later on.

3.5. Corollary

Suppose that A and B are as above, where $\bar{A}$ generates a group $\{V(t,A): t \in \mathbb{R}\}$, and suppose in addition that $\{V(s,B): s \in [0,\infty)\}$ leaves D invariant. Then for all $t \in \mathbb{R}$ the element $B(t) = \exp(t \text{ ad } A)(B)$ is closable, and $\bar{B}(t)$ generates the cle semigroup $\{V(t,A)V(s,B)V(-t,A): s \in [0,\infty)\}$. That is, the second exponential commutation relation holds

$$V(s, \exp(t \text{ ad } A)(B)) = V(t,A)V(s,B)V(-t,A). \tag{15}$$

Proof: We first observe that with t fixed, $\{V(t,A)V(s,B)V(-t,A): s \in [0,\infty)\}$ is obviously a cle semigroup which leaves D invariant. Moreover, on D we have that $d/ds\ V(t,A)V(s,B)V(-t,A)u_{|s=0} = V(t,A)BV(-t,A)u$ $= \exp(t \text{ ad } A)(B)u$ by Theorem 3.2 (since $V(-t,A)u \in D$ as well). Thus $B(t)$ is the restriction to D of the generator of the semigroup under discussion, and invariance of D implies by the pregenerator theorem (Appendix B) that D is a core for this generator, so $\bar{B}(t)$ is exactly the generator. This concludes the proof.

Returning to the basic issue of removing condition (R) as an independent assumption, or replacing it by natural, easily checked conditions, we observe that one aspect of our treatment of condition (ELB) in the proof of 3.3 supplies a fairly strong kind of automatic regularity without the local boundedness assumption. Specifically, we showed there that the function $G_u(t) = BV(t,A)u$ is a pointwise limit of a sequence of continuous functions $B_nV(t,A)u$ and that its composites $G_u^f(t) = f(BV(t,A)u)$ with continuous functionals $f \in E^*$ have a similar character. It therefore follows automatically that for any locally convex space E these functions are at least weakly ($\sigma(E,E^*)$) measurable and in the first Baire class, while if E is Fréchet (in particular, Banach) they are in fact strongly measurable and Baire-1. Since the theory of group representations abounds with examples where measurability combined with group structure implies continuity, while the Bochner-Montgomery Theorem (Montgomery-Zippin [MZ] p. 208 Theorem 3) on automatic smoothness of 'continuous' diffeomorphism group actions successfully exploits the points of continuity supplied by the Baire-1 condition, one might reasonably expect to obtain automatic continuity of the $G_u(t)$ here as well. The following result is the best that is presently available along these lines, and it has interesting applications later on in the development. Its proof serves to suggest in detail how more definitive results might possibly be obtained.

3.6. Proposition

Suppose that E is a Banach space with topology τ_0, and that $D \subset E$ is a dense (proper) subspace that is Fréchet with respect to a topology τ_∞ that if finer than the relative τ_0 topology. Let A, B $\in \mathbf{A}(D)$ be closable operators such that

(1) $\bar{A}$ generates a cle group $\{V(t,A): t \in \mathbb{R}\}$ that satisfies (D): $V(t,A)D \subset D$ for all $t \in \mathbb{R}$, and

(2) $\bar{B}$ generates a cle semigroup $\{V(r,B): r \in [0,\infty)\}$.

Then for all $u \in D$, the function $G_u(t) = BV(t,A)u$ is differentiable on $\mathbb{R}$, with derivative $G'(t) = BV(t,A)Au$.

Remark: The proof actually requires only that $\mathbb{R} \times (D,\tau_\infty)$ be Baire, but we know of no interesting non-Fréchet examples. with the possible exception of the non-locally-convex F spaces that arise in noncommutative integration theory (cf.

Nelson [Nℓ 2]). The notation τ_∞ anticipates the primary examples, where Fréchet 'C^∞ vector' topologies are involved.

Proof: Here, as in 3.3, we replace B by the continuous operator $B_n = n(V(1/n,B)-1)$, but use it to construct a function $G_n: \mathbb{R} \times D \to E$ by

$$G_n(t,u) = B_n V(t,A)u. \tag{16}$$

Since $V(t,A)u$ is a τ_0-continuous function and B_n is $\tau_0 - \tau_0$ continuous it follows that G_n is separately continuous from $\mathbb{R} \times (D,\tau_\infty)$ into (E,τ_0). But on D, local $\tau_0 - \tau_0$ equicontinuity of $\{V(t,A): t \in K\}$ for any compact K first implies by composition that $\{G_n(t,\cdot) = B_n V(t,A): t \in K\}$ is $\tau_0 - \tau_0$ equicontinuous, hence certainly equicontinuous from (D,τ_∞) to (E,τ_0) since τ_∞ is assumed finer (stronger). Thus by a standard result from general topology, each G_n is (jointly) continuous from the product space $\mathbb{R} \times (D,\tau_\infty)$ into (E,τ_0).

Letting $n \to \infty$, we obtain just as in 3.3 that $G(t,u) = BV(t,A)u$ is the pointwise limit of continuous functions. But $\mathbb{R} \times (D,\tau_\infty)$ is itself complete metric, hence Baire, so there must exist at least one point (t_0,u_0) in the product at which G is (jointly) continuous. We argue by linearity, invariance of D, and the group property of $\{V(t,A): t \in \mathbb{R}\}$ that since E is Banach, G is actually uniformly bounded in t, on a uniform δ-neighborhood of 0 in $\mathbb{R}$, uniformly on a neighborhood N of 0 in (D,τ_∞). Then, since N absorbs D, we will have more than satisfied the group version of 3.3, and the conclusion will follow. Specifically: let Ω_ε be a convex neighborhood of the origin in E. Then there is a neighborhood N of the origin in (D,τ_∞) and an open interval J containing t_0 such that

$$G(J,u_0+N) \subseteqq G(t_0,u_0) + \Omega_\varepsilon.$$

For every $t \in J$ and $u \in N$ we have

$$G(t,u_0) + G(t,u) \in G(t_0,u_0) + \Omega_\varepsilon$$

by linearity in the second variable. In particular for $u = 0$

$$G(t,u_0) \in G(t_0,u_0) + \Omega_\varepsilon.$$

Combining these two inclusions we get

$$G(t,u) \in 2\Omega_\varepsilon$$

for all $(t,u) \in j \times N$. Hence $G(\ ,\ )$ is continuous at $(t_0,0)$ and norm bounded on $J \times N$.

Using finally that N is absorbing we get that for every $u \in D$ the set of vectors $\{BV(t,A)u;\ t \in J\}$ is norm bounded. Consequently, 3.3 applies as predicted and the proof is complete.

Remark on the Proof: A sufficiently careful reanalysis of the exact way in which local boundedness was employed in 3.3 reveals that a seminorm-by-seminorm argument combining the proofs of 3.3 and 3.6 can be pushed through without requiring that E be Banach. The proof above cannot be separately improved to arbitrary locally convex spaces, however: if carried out seminorm-by-seminorm, one finds instead that for each continuous seminorm p, there exists a $\delta_p > 0$ such that for $|t| < \delta_p$

$$p(B\,V(t,A)u) \leq 2\lambda$$

but there is no immediate guarantee that a p-independent $\delta > 0$ can be obtained. All of our applications below concern the case where (D,τ_∞) is a Fréchet space of C^∞ vectors for some family $S = \{B_1,\ldots,B_q\} \subset \mathbf{A}(D)$ of unbounded closable operators in a Banach space E, and 3.6 as stated is quite adequate for these cases. (See Chapters 7 and 8.)

It is still an open question as to whether this result can be pushed through without involving Baire properties of D as well as E and $\mathbb{R}$. Arguments of this type can be used to show that for each vector $u \in D$, there exists a dense open 'good set' $\Gamma(u) \subset \mathbb{R}$ such that for each $t_0 \in \Gamma(u)$ there is a neighborhood $N(t_0)$ upon which a local commutation relation

$$B\,V(t,A)u = V(t-t_0,A)\ \exp[(t_0-t)\ \mathrm{ad}\ A](B)V(t_0,A)u \qquad (17)$$

holds. If $\Gamma = \cap\{\Gamma(u):\ u \in D\}$ could be shown to be nonvoid, then it would be easy to verify that the machinery of 3.3 and 3.2 applied to obtain global commutation relations. But Nelson's two-dimensional Abelian counter-example ([Nℓ 1], [Pw 1]) gives a case where D is 'locally invariant')as would be required for (17) to make literal sense in $\mathbf{A}(D)$: for an open dense $\Gamma(u)$ and $t \in N(t_0)$, $V(t,A)u \in D$ and (17) holds, but global commutation relations fail. This example shows that if (R) is ever to be removed entirely, the global invariance $V(\mathbb{R},A)D \subset D$ must be brought into play in a fundamentally new way.

3D. Infinite-dimensional $\mathcal{O}_A(B)$

We conclude the section with a brief domain-invariance analysis of the 'infinite-dimensional' commutation relation mentioned in Chapter 2. The algebraic setting, we recall, involves $A, B \in \mathbf{A}(D)$ such that $\mathrm{ad}\, A(B) = \alpha B$ for some ad A-eigenvalue $\alpha \in \mathbb{C}$ (so that $\mathcal{O}_A(B) \in \mathbb{R}\, B + i\,\mathbb{R}\, B$) and $\bar{A}$ generates a cle group $\{V(t,A): t \in \mathbb{R}\}$. At least when E is Banach, it automatically follows that $\Delta = A^2$ is closable and that $\bar{\Delta}$ generates a C_0 semigroup $\{V(t,\Delta): t \in [0,\infty)\}$ which analytically continues into the open half-plane $\{V(\zeta,\Delta): \mathrm{Re}(\zeta) > 0\}$ to form a holomorphic semigroup which can be described on axis by

$$V(t,\Delta)u = \int_{-\infty}^{\infty} p_t(s)V(s,A)u\, ds \tag{18}$$

where p_t is the one-dimensional Gauss kernel. (cf. Chapter 4 and Appendix C for details.) Similarly, since $\bar{A}-\alpha$ generates the group $V(t,A-\alpha) = e^{-\alpha t}V(t,A)$, one obtains that $\Delta_\alpha = (A-\alpha)^2$ has a closure $\bar{\Delta}_\alpha$ which generates a holomorphic semigroup $\{V(\zeta,\Delta_\alpha): \mathrm{Re}(\zeta) > 0 \text{ or } \zeta = 0\}$ that is C_0 at $\zeta = 0$.

3.7. Proposition

Suppose that A, B, Δ, and Δ_α are as above, with B closable. Suppose further that $V(t,\Delta)D \subset D$ for all $t \in [0,\infty)$, and for all $u \in D$, the map $G_u(t) = BV(t,\Delta)u$ is continuous in t. Then for all $\mathrm{Re}(\zeta) > 0$ (and trivially for $\zeta = 0$).

$$BV(\zeta,\Delta) = V(\zeta,\Delta_\alpha)B. \tag{19}$$

Proof: By 3.3, G_u is differentiable, with $G'_u(t) = BV(t,\Delta)\Delta u = B\Delta V(t,\Delta)u$. We first check (19) for $\zeta = t \in (0,\infty)$, later extending to other ζ by analytic continuation. Here, as in the proof of 3.2 (but much simplified) we pick $t > 0$ and for $s \in [0,t]$ put

$$H(s) = V(t-s,\Delta_\alpha)\, BV(s,\Delta)u.$$

As before, since $V(s,\Delta)u \in D$ and $BV(s,\Delta)u \in D$

$$\begin{aligned} H'(s) &= -V(t-s,\Delta_\alpha)\Delta_\alpha BV(s,\Delta)u + V(t-s,\Delta_\alpha)B\Delta V(s,\Delta)u \\ &= V(t-s,\Delta_\alpha)(B\Delta-\Delta_\alpha B)V(s,\Delta)u \end{aligned} \tag{20}$$

using the product rule and local equicontinuity. We check directly here that $B\Delta - \Delta_\alpha B = 0$, by passing the identities used in Chapter 1 and Chapter 11 for similar purposes. That is

$$B\Delta = BA^2 = ABA - [A,B]A = (A^2B - A[A,B]) - \alpha BA$$
$$= (A^2B - \alpha AB) - (\alpha AB - \alpha[A,B])$$
$$= (A^2 - 2\alpha A + \alpha^2)B = \Delta_\alpha B\,. \tag{21}$$

This forces $H'(s) \equiv 0$, so that $H(0) = H(t)$ and (19) follows for $\zeta = t > 0$. To continue off-axis, note that the right-hand side of (19) is holomorphic in $\mathrm{Re}(\zeta) > 0$ as applied to any $u \in D$, while $V(\zeta,\Delta)u$ is similarly holomorphic (i.e. expandable in a power series about any $t > 0$, with disc of convergence tangent to $t = 0$). Moreover, one checks routinely by repeated differentiation of both sides with respect to $t \in [0,\infty)$ that the Taylor coefficients for $V(\zeta,\Delta_\alpha)Bu$ agree with those for $V(\zeta,\Delta)u$ with B multiplied on the left (use (19) on axis and (21) inductively). That is, if B is applied on the left to a partial sum for the expansion of $V(\zeta,\Delta)u$, the result agrees term-by-term with the corresponding truncated expansion for $V(\zeta,\Delta_\alpha)Bu$. Thus since the expansion for the right-hand side in (19) converges, B is closable, and the expansion for $V(\zeta,\Delta)u$ converges, we see that the expansion for the left-hand side of (19) must also converge, to the same limit as the right. Since every ζ with $\mathrm{Re}(\zeta) > 0$ lies interior to some disc with center on $(0,\infty)$ and boundary passing through 0, (19) is established in full generality.

Remark 1: We note here that the proposition is related to Proposition 4.3, but the idea in the proof of 3.7 seems to be entirely different from the one of 4.3. However, there is a variant of 3.7 whose proof is very similar to the proof of 4.3. It turns out that the assumption: $V(t,\Delta)D \subset D$ for all $t \in [0,\infty)$, can be replaced by the assumption that D is invariant under $V(x,A)$ for all $x \in \mathbb{R}$. One also needs that $V(x,A)$ be locally bounded with respect to B. Using the analytically continued Gauss kernel $p_\zeta(x)$ ($\mathrm{Re}\ \zeta > 0$, $x \in \mathbb{R}$) of Appendix C one gets formulas for $V(\zeta,\Delta)$ and $V(\zeta,\Delta_\alpha)$ similar to (18), only with p_t replaced by p_ζ. Moreover by 3.2 the commutation relation

$$B\,V(x,A) = \exp(-\alpha x)\ V(x,A)\,B$$

holds on D for all $x \in \mathbb{R}$. Now the closable operator B commutes with the p_ζ version of the integral (18) for $V(\zeta,\Delta)$, so (19) follows quite easily when the identities are combined in the right way (cf. proof of 4.3).

Remark 2: As pointed out in Chapter 2, the identity (19) can be interpreted formally as

$$B\,V(\zeta,\Delta) = V(\zeta,\Delta)\ \exp(-\zeta\ \mathrm{ad}\ \Delta)(B)u \qquad (22)$$

whenever $u \in D$ is an analytic vector for A, even though in general $\mathcal{O}_\Delta(B) = \mathrm{span}\ \{(\mathrm{ad}\ \Delta)^k B : 0 \le k < \infty\}$ is infinite-dimensional (c.f., Appendix D). It can be shown that the formula (22) makes sense as

$$\begin{aligned} B\,V(\zeta,\Delta)u &= V(\zeta,\Delta)\ \bar{B}\ \exp(-\zeta\alpha(\alpha+2A))u \\ &= \sum_{n=0}^{\infty} ((-\zeta)^n/n!)\ V(\zeta,\Delta)B(\alpha(\alpha+2A))^n u. \end{aligned} \qquad (23$$

Clearly if u is an analytic vector for A and $|\zeta|$ is sufficiently small, the exponential on the right-hand side of (23) defines an absolutely convergent power series in ζ. Since B is closable, $V(\zeta,\Delta)\bar{B}$ can be applied termwise to this series. Hence the series following (23) is convergent.

The tools of Chapters 5-7 enable us to see that the domain $D = D_\infty(\bar{B})$ behaves exactly as discussed above, and does in fact contain a very large collection of analytic vectors for A.

Chapter 4

INVARIANT-DOMAIN COMMUTATION THEORY APPLIED TO THE MASS-SPLITTING PRINCIPLE

This chapter serves primarily to illustrate the sorts of applications to commutation problems for physical observables which lie within the scope of the globally-invariant-domain (DR) commutation methods developed in Chapter 3.

The applications considered here concern a mass-splitting principle from the quantum-theoretic treatment of elementary particles. The formal ideas behind this principle were introduced by O'Raifeartaigh in [O'R], and the first rigorous treatment known to us was later given by Jost in [Jt]. Subsequent generalizations of Jost's work by Segal [Sg 2], Sternheimer [St 2] and Goodman [Gd 3] use a combination of C^{∞}-vector and analytic vector methods to obtain their results, with analytic vectors as an indispensable tool. By contrast, our treatment below makes exclusive use of C^{∞}-vector methods, more in the spirit of Jost's original treatment [Jt] and other ideas of Jost and Sternheimer as informally described in [St].

The mass-splitting phenomenon concerns the description of elementary particles by irreducible unitary representations of the Poincaré group (the semidirect product of four-dimensional space-time and the homogeneous Lorentz group). See [O'R 1,2], [FSt], and for background material [Sm].

We give a brief (incomplete and heuristic) summary of the theory.

Let π_k for $k = 0,1,2,3$ be the generators of four-dimensional space-time (Minkowski space) H. Let G be a global symmetry group which contains H as a subgroup. Suppose H is nilpotently embedded in G [Gd 3; assumption (iii)], and let U be a continuous representation of G in a Hilbert space $\mathcal{H}$ such that the restriction of U to Minkowski space H is unitary. In standard relativistic theory the (self adjoint) energy- (or mass-) operator M is given by

$$-dU(\pi_0)^2 + dU(\pi_1)^2 + dU(\pi_2)^2 + dU(\pi_3)^2. \qquad (1)$$

The mass of a particle (corresponding to U) is equal to the energy of the particle at rest, which in the model corresponds to a minimal eigenvalue m of the mass operator M. Let H_m be the corresponding eigenspace. Then H_m is closed, and, according to the O'Raifeartaigh Principle, invariant under the full symmetry group G, i.e. $U(g)H_m \subset H_m$ for all $g \in G$. So, in particular, if U is irreducible, then the only possible states of the corresponding particle are the ones represented by the vectors in H_m. This conclusion, however, contradicts observed mass differences in pure states (irreducible representations).

On the basis of the O'Raifeartaigh Principle and positivity of the energy, Segal suggested in [Sg 2, and 3] an alternative mathematical model in which Minkowski space is replaced by the compact group U(2) (covered by $\mathbb{R} \times S^3$), and U(2) is viewed as a subgroup of the conformal group O(2,4) $\approx$SU(2,2). In this new picture the four-dimensional space-time is not nilpotently embedded as a subgroup of the global symmetry group as one can easily check.

In view of these new ideas of Segal (currently described as 'chrono-geometry'), and considering the extent of the earlier controversy [FSt 1,2] concerning the exact scope and physical significance of negative meta-theorems of mass-splitting type, it is worthwhile to attack these spectral splitting problems from the alternative mathematical viewpoints developed below.

Our analysis essentially follows the general plan of Goodman's treatment [Gd 3], in that it makes fundamental use of the invariance of the C^∞-vector Fréchet domain $H^\infty(U)$ determined by the representation U of the big symmetry group G. The discussion is structured as follows. In the preparatory section 4A, we combine methods from [Jo 2] with the general 'finite-dimensional ad-orbit' machinery of Chapter 3 in order to establish invariance of $H^\infty(U)$ under certain heat-type semigroups generated by quadratic sub-Laplacians which appear as components of the usual mass-energy operator (1). Then in Section 4B, we formulate a version of the general mass-splitting theorem for reference in subsequent discussion. Section 4C then moves toward the proof of a special case of this theorem (for mass operators M of the form (1)), constructing the unitary group generated by iM = i(P-N) as a product of groups generated by the sub-Laplacian components iP and iN, thus deriving both invariance of $H^\infty(U)$ and commutation relations for exp(itM) from those previously obtained for exp(itP) and exp(-itN). In Section 4D we discuss some of the aspects of the treatment of generalized (not-necessarily-quadratic) mass-energy operators previously treated by Goodman [Gd 3]. (Our methods sharpen certain estimates of Goodman [op.cit.].) The method for a generalized mass operator M differs in one respect from the method used in Section 4C. Instead of expressing M as a difference of positive parts, we consider the operator $M_1 = -(-\Delta)^n + cM$ where n is an integer, chosen sufficiently large, and c is a complex number. We then construct heat-type kernels for M_1 and proceed as in Section 4C. In Section 4E the Fourier transform operational calculus is employed, in the spirit of [Gd 3], to give a pure C^∞-vector proof that the group commutation relations for e^{itM} imply spectral- (i.e., mass-) splitting.

4A. Global invariance/regularity for heat-type semigroups

The result of this section is more general than is needed for the rest of the chapter. It is included in order to demonstrate that the smoothing property always holds for the holomorphic semigroups generated by the second-order quadratic expressions in the enveloping algebra of a continuous Lie group representation. It is not assumed that the representation has a unitary restriction, and it is not assumed that the ad-orbits are finite-dimensional.

First we fix some notation concerning the C^n-vectors for a representation. Let V be a continuous representation of a Lie group G in a Banach space E. Let $\mathfrak{g}$, $E(\mathfrak{g})$, and $E_n(\mathfrak{g})$ denote, respectively, the Lie algebra of right invariant vector fields on G, the associative enveloping algebra of $\mathfrak{g}$, and the linear span of the elements in $E(\mathfrak{g})$ of order less than or equal to n, for each n = 0,1,2,....

Now, V may be viewed as a representation of the convolution algebra $\mathcal{D}(G)$ of test functions on G: a continuous homomorphism of $\mathcal{D}(G)$ into the algebra L(E) of bounded operators on E. It is known ([Sz] or [JM 1]) that V extends to a representation of the convolution algebra of compactly supported distributions on G into an algebra of (unbounded) closed operators with the Gårding vectors as a common invariant domain. In particular, each operator dV(Y) for $Y \in E(\mathfrak{g})$ has a closed extension. An element Y in $E(\mathfrak{g})$ is identified with the distribution $Y\delta$, that is the differential operator Y applied to the Dirac measure δ.

For each $n = 0,1,2,\ldots,\infty$ the C^n vectors for V,$E^n(V)$ is the space of vectors $u \in E$ such that the mapping $g \to V(g)u$ is of class C^n of G into E. The infinitesimal generator of the one-parameter group $\{V(\exp tY); -\infty<t<\infty\}$ in L(E) for $Y \in \mathfrak{g}$ is known to be equal to the closure of dV(Y) [Ps 1]. The symbol $\tilde{Y}$

will represent the closed operator $\overline{dV(Y)}$.

The spaces $E^n(V)$ have an operator theoretic characterization [Gd 3, Prop. 1.1 and 2]. If $Y_1,\ldots,Y_d$ is a basis for $\mathfrak{g}$ then $E^n(V)$ is equal to the intersection of the domains of all the operator monomials $\overline{dV(Y_{i_1})}\ \overline{dV(Y_{i_2})}\cdots\overline{dV(Y_{i_n})}$ with $1 \leq i_j \leq d$.

For $n<\infty$, $E^n(V)$ is a Banach space when equipped with the graph norm

$$\|u\|_n = \sum_{k=0}^{n} \max\{\|\tilde{Y}_{i_1}\tilde{Y}_{i_2}\cdots\tilde{Y}_{i_k}u\|;\ 1 \leq i_j \leq d\}.$$

The space $E^\infty(V)$ is equipped with the Fréchet space topology defined by all the norms $\|\cdot\|_n$ (and $\|\cdot\|_0 = \|\cdot\|$).

It is well known that the restriction of V to $E^n(V)$ is a continuous representation of G in $E^n(V)$ for each $n = 1,2,\cdots,\infty$. Let Ad denote the adjoint representation of G in $\mathfrak{g}$, and fix some operator norm $|\cdot|$ in the finite-dimensional space End($\mathfrak{g}$). We then have the precise estimate for each n

$$\|V(g)u\|_n \leq (1 + |Ad(g^{-1})|^n)\,\|V(g)\|\ \|u\|_n \tag{2}$$

for all $g\in G$ and $u\in E^n(V)$ [Gd 2(1.5)].

4.1. Lemma

Let V be a continuous representation of a Lie group G in a Banach space. Let $X_1,\cdots,X_r$ be elements in the Lie algebra of G, and denote by M the closure of the operator $dV(\Sigma X_j{}^2)$.

Then M is the infinitesimal generator of a holomorphic semigroup $e^{\zeta M}$ (for ζ in the open right half-plane) given by

$$e^{\zeta M} = \int_H p_\zeta(h)\ V(h)\ dh, \tag{3}$$

where H is the subgroup of G generated by $X_1,\ldots,X_r$, and $p_\zeta(.)$ is the generalized 'Gauss kernel' belonging to $C^\infty(H) \cap L^2(H)$ with respect to Haar measure dh on H.

Furthermore, the semigroup $\{e^{\zeta M};\ Re\zeta>0\}$ leaves the C^n vectors, $E^n(V)$, invariant for each $n = 1,2,\cdots,\infty$.

Proof: The first part of the lemma is contained in [Jo 2, Prop. 3.2 and 3.3]. It is shown in [Jo 2] that for each $\zeta \in \mathbb{C}$ with $Re\zeta>0$ the fundamental solution $p_\zeta(.)$ to the 'heat operator' $\partial/\partial\zeta - \Sigma X_j{}^2$ is supported on the subgroup H generated by the Lie algebra elements $X_1,\cdots,X_r$. Furthermore it is shown that $p_\zeta(.)$ belongs to $C^\infty(H) \cap L^2(H)$ (with respect to a left invariant Haar measure on H), and that the mapping $\zeta \to p_\zeta$ is holomorphic from the open half-plane into $L^2(H)$. The integral (3) is viewed as a vector-valued Bochner integral taking values in the Banach space L(E) of bounded operators on E. The representation V can grow exponentially, so we need to know that the kernel p_ζ decays fast at infinity in order for the integral (3) to be guaranteed convergent. More specifically, let $|h|$ be a fixed left invariant Riemannian distance from h to the origin e in H. By Proposition 3.1 of [Jo 2] the integrals

$$\int_H \exp(c|h|)\ |p_\zeta(h)|dh \tag{4}$$

are finite for all $c>0$ and all $\zeta>0$. In the proof of the second part of the present lemma, we shall need finiteness of the integrals for all non-real ζ in the open right half-plane.

The arguments which yield integrability for non-real ζ will not be given here, since that would lead to an unnecessary distraction from the main line of arguments. The idea is analogous to the reasoning given in [Jo 2], and in [Gℓ 2, Lemma 7.1]. But an extra difficulty is added in the case when the kernel p_ζ is non-real. The reader does not have to be concerned about this technicality, because for the cases we are dealing with, H is the additive group $\mathbb{R}^\ell$ for some ℓ, and in that case finiteness of (4) can (and will) be checked by direct computation.

We show that, for each $n<\infty$, the holomorphic semigroup given by (3) is bounded in $E^n(V)$. Let $n<\infty$ be given. Consider, for each $u \in E^n(V)$, the integral

$$\int_H p_\zeta(h)\ V(h)u\ dh. \tag{5}$$

The integrand clearly belongs to $E^n(V)$ since the restriction of V to $E^n(V)$ is continuous. According to (2) we have the following estimate for the n-norm of the integrand,

$$\|p_\zeta(h)\ V(h)u\|_n \leq |p_\zeta(h)|(1 + |Ad(h^{-1})|^n)\|V(h)\|\ \|u\|_n. \tag{6}$$

It is well known [Gℓ 2] that there are finite constants c_1 and c_2 such that

$$|Ad(h^{-1})| \leq \exp(c_1|h|),\ \text{and}\ \|V(h)\| \leq \exp(c_2(1+|h|))$$

for all $h \in H$. Hence, it follows from the finiteness of (4), with c equal to $nc_1 + c_2$, that the integral (5) is convergent in $E^n(V)$, and we have

$$\|e^{\zeta M}u\|_n \leq \int_H |p_\zeta(h)| \exp((nc_1+c_2)|h|)dh \, \|u\|_n$$

for all $u \in E^n(V)$.

Finally, we have to show that the restriction of $e^{\zeta M}$ to $E^n(V)$ is strongly continuous. This is a standard argument which uses strong continuity of the restriction of V to $E^n(V)$, together with the fact that

$$\lim_{\zeta\to 0} \int_{|h|>\delta} e^{c|h|}|p_\zeta(h)|dh = 0, \text{ for all positive } c \text{ and } \delta.$$

The limit, $\zeta \to 0$, is nontangential in the sense that, for some $a > 0$: $\mathrm{Re}\,\zeta > a|\zeta|$.

This last fact is not proved here. For the special case of the classical Euclidean kernels, to be treated below, the property is certainly well known and very easy to check by direct computation. The reader who is interested in the general problem is referred to [Jo 2, Lemma 3.4-5, Proposition 3.2] where the kernels p_ζ are constructed as limits of resolvent kernels.

Alternatively, it is shown in [Jo 2] that for all $u \in E^\infty$, $e^{tM}u$ converges to u in the E^∞ topology for real $t \to 0+$. Now, if $u \in E^\infty$ is fixed, the function $u(\zeta) = e^{\zeta M}u$ is holomorphic in the open right half-plane with values in E^∞ and uniformly bounded in every sector $-\pi+\delta < \mathrm{Arg}\,\zeta < \pi-\delta$, [Jo 2]. Then it follows from the construction of the kernel $p_\zeta(h)$ in [Jo 2] that the restricted limit for $\zeta\to 0$ with $\inf[\mathrm{Re}\,\zeta|\zeta|^{-1}]>0$ of $u(\zeta)$ is equal to u.

Modulo simple facts on the generalized Gauss kernel which we have taken for granted, this completes the proof of Lemma 4.1.

4B. Formulation of the generalized mass-splitting theorem

The two most recent and systematic treatments of the mass-splitting phenomenon are due to Sternheimer [St 1,2] and Goodman [Gd 3]. For the convenience of the reader, we reproduce below a variant of the main result obtained by these authors.

Let G be a connected Lie group with an Abelian nilpotently embedded subgroup H. Let $\mathfrak{g}$ and $\mathfrak{h}$ be the corresponding Lie algebras, and let ad be the adjoint representation of $\mathfrak{g}$. The nilpotency condition requires that for each $X \in \mathfrak{h}$ the operator adX is nilpotent in $\mathfrak{g}$. The Lie algebra $\mathfrak{h}$ is just $\mathbb{R}^{\ell}$ for some ℓ, and the complex enveloping algebra $E(\mathfrak{h})$ is the algebra of all constant coefficient differential operators on H.

The enveloping algebra $E(\mathfrak{g})$ is equipped with its usual involution $T \to T^*$ extending $X^* = -X$ for $X \in \mathfrak{g}$. Let U be a continuous representation of G in a Hilbert space H and suppose that the restriction of U to H is unitary. Then the operator dU(T) is essentially self adjoint in H for all Hermitian elements T in $E(\mathfrak{h})$, (i.e. $T^* = T$) [NS]. Let M be the closure of dU(T) for a fixed (but arbitrary Hermitian element T in $E(\mathfrak{h})$.

4.2. Theorem (Goodman)

Let U and M be as above, and let P be the spectral projection of M corresponding to a compact connected component of the spectrum of M.

Then

$$U(g)P = P\,U(g) \quad \text{for all } g \in G. \tag{7}$$

Remark: Originally Goodman required either that U be unitary, or else that U be topologically irreducible. The C^∞ commutation methods developed in this book enable us to replace the analytic vector arguments in [Gd 3] by C^∞ arguments and to remove the unitarity or irreducibility assumption in the original version of [Gd 3, Corollary 5.1].

We develop the proof of 4.2 by a series of steps roughly parallel to those of Goodman [Gd 3], but implementing each step by means of C^∞ commutation theory rather than analytic vectors. The proof is given in full for mass operators of the form (1). Generalizations of various algebraic lemmas are formulated, with indications of proof, that supply the basis for a fully general C^∞-vectors proof of 4.2 as stated.

4C. The mass-operator as a commuting difference of sub-Laplacians

The main idea of this section is to obtain commutation relations and domain invariance for the group e^{itM} determined by a mass operator $M = P - N$ with positive sub-Laplacian components

$$P = dU(T_1)^- = (\Sigma_1^q dU(X_i)^2)^-, \quad N = dU(T_2)^- = (\sum_{q+1}^{r} dU(X_j)^2)^-,$$

reducing the problems for iM to those for iP and iN, and

analyzing the commutation theory for e^{itP}, e^{itN} in terms of that for the unitary groups generated by the X_i, X_j whose squares are involved. These reductions make use of Hermite polynomials in Lie algebra elements as a basic tool related to the Gaussian kernels whose integrated forms (operational images) represent the groups e^{itP}, e^{itN}. Since P and N are easily shown to be strongly commutative, the identity $e^{itM} = e^{itP}\, e^{-itN}$ converts data for P and N into that for M.

More precisely, then, we assume that M is the closure of $dV(-T_1+T_2)$ for quadratic expressions T_1 and T_2 in the enveloping algebra $E(\mathfrak{h})$. There are elements $X_1,\ldots,X_q,\ldots,X_r \in \mathfrak{h}$ such that $T_1 = \Sigma_1^q\, X_j^{\,2}$ and $T_2 = \Sigma_{q+1}^r\, X_j^{\,2}$.

Let us at first treat the quadratic expressions T_1 and T_2 separately. For each $m = 0,1,2,\ldots$ and each $\zeta \in \mathbb{C}$ we define certain Hermite polynomials $H_m(\zeta,i\xi)$ by the equation

$$(i\ d/d\xi)^m\ e^{-\zeta\cdot\xi^2} = H_m(\zeta,i\xi)\ e^{-\zeta\cdot\xi^2} \quad \text{for } \xi \in \mathbb{R}. \tag{8}$$

That is, each $H_m(\zeta,i\xi)$ is a Hermite polynomial of degree m in the variable $i\xi$. An elementary and familiar computation using (8) shows that the generating function is given by

$$\sum_0^\infty H_m(\zeta,i\xi)\ \lambda^m/m! = \exp(-\zeta[2\ i\xi\ \lambda - \lambda^2]). \tag{9}$$

Now, if $X \in \mathfrak{h}$ and $Y \in E(\mathfrak{g})$, nilpotence of adX and standard functional calculus for polynomials yields

$$\sum_0^\infty H_m(\zeta,i\xi)\ (\mathrm{ad}X)^m\ (Y)/m! = \exp(-\zeta[2\ i\xi\ \mathrm{ad}X-(\mathrm{ad}X)^2])(Y)$$

upon substitution of adX for λ in (9) as then applied to Y. (Notice that both sides reduce to polynomials in ζ, $i\xi$, and $(\mathrm{ad}X)^m Y$). Then using the easily-derived identity

$$\mathrm{ad}(X^2) = 2\ L_X\ \mathrm{ad}X - (\mathrm{ad}X)^2\ , \tag{10}$$

where L_X denotes left multiplication by X, and putting $T = X^2$ we may replace $i\xi$ by L_X in these two polynomials to obtain the following (formally infinite) polynomial identity

$$\sum_0^\infty H_m(\zeta,X)\ (\mathrm{ad}X)^m(Y)/m! = \exp(-\zeta\ \mathrm{ad}T)(Y). \tag{11}$$

Since the right-hand side of (11) reads formally

$$\Sigma_0^\infty\ (-\zeta)^j/j!\ (\mathrm{ad}T)^j\ (Y)$$

and must reduce to a finite-degree polynomial in ζ by (11), we see that $(\mathrm{ad}T)^m(Y) = 0$ for large m.

The problem is roughly whether one can replace the variables $i\xi$ and λ in (9) by the operators L_X and adX, respectively, and apply the resulting operator to an element $Y\in\mathcal{E}(\mathfrak{g})$. The argument above justifies a formal substitution, and the right-hand side of (9) becomes $\exp(-\zeta[2X\,\mathrm{ad}X-(\mathrm{ad}X)^2](Y)$ which in view of (10) is just $\exp(-\zeta\,\mathrm{ad}(X^2))(Y)$.

Suppose T is an arbitrary quadratic expression $T = \Sigma_1^q X_j^2$. The following multi-index notation is convenient in order to generalize the basic identity (11):

$$\begin{aligned} m &= (m_1,m_2,\ldots,m_q), \quad |m| = m_1 + m_2 + \ldots + m_q, \\ m! &= m_1!\, m_2! \cdots m_q!, \quad X = (X_1,\ldots,X_q), \\ H_m(\zeta,X) &= H_{m_1}(\zeta,X_1)\cdots H_{m_q}(\zeta,X_q), \\ (\mathrm{ad}X)^m &= (\mathrm{ad}X_1)^{m_1}(\mathrm{ad}X_2)^{m_2}\cdots(\mathrm{ad}X_q)^{m_q}. \end{aligned} \tag{12}$$

Then formula (11) holds for $T = \Sigma_1^q X_j^2$ when $H_m(\zeta,X)$ and $(\mathrm{ad}X)^m$, and m! are given by (12). Hence the multivariable formula is formally identical to the one variable formula with the notation (12) above.

We say that the degree of nilpotency is k if k is the smallest integer such that $(\mathrm{ad}\mathfrak{h})^{k+1}(\mathfrak{g}) = 0$ that is $(\mathrm{ad}X_1)\cdots(\mathrm{ad}X_{k+1})(Y)=0$ for all $X_1,\ldots,X_{k+1}\in\mathfrak{h}$ and $Y\in\mathfrak{g}$. (In two important applications, Minkowski space is nilpotently embedded in the Poincaré group, or in the conformal group, the degree of nilpotency being 1 or 2, respectively. This follows immediately from inspection of Lie algebra commutators (see e.g. [B-B, page 9]). We have the tower of subgroups $H\subset G'\subset G$ where H, G' and G denote Minkowski space, the Poincaré group, and the conformal group, respectively. Furthermore, if $\mathfrak{g}'$ and $\mathfrak{g}$ denote the respective Lie algebras then $[\mathfrak{h},[\mathfrak{h},\mathfrak{g}]]\subset\mathfrak{h}$.

If the degree of nilpotency is k then the summands on the left-hand side of (11) vanish when $|m|>kn$ and $Y\in\mathcal{E}_n(\mathfrak{g})$, (as one easily checks by the Leibniz rule for the 'derivative' of a product). Since the polynomial $H_m(\zeta,X)$ is of degree $|m|$ for each multi index m, it follows that the left-hand side of (11) belongs to

$$\mathcal{E}_{kn}(\mathfrak{h})\cdot\mathcal{E}_n(\mathfrak{g})\subset\mathcal{E}_{(k+1)n}(\mathfrak{g})$$

for every $Y\in\mathcal{E}_n(\mathfrak{g})$. Recall that adX leaves $\mathcal{E}_n(\mathfrak{g})$ invariant for all X and all n.

These observations suffice for identities in the algebra $E(\mathfrak{g})$.

The identity (11) may also be viewed as an operator commutation relation for Hilbert space operators. For each $Y \in E(\mathfrak{g})$ the operator $dU(Y)$ (with the C^∞ vectors as domain) has a closure which we (for clarity) denote by $\tilde{Y}$. If we put a $\sim$ on the variables X, Y and T in (11), then (11) becomes an operator identity for operators with the C^∞ vectors $\mathcal{H}^\infty(U)$ as a common invariant domain.

Unless otherwise stated we assume in the following that G is a connected Lie group with an Abelian nilpotently embedded subgroup H, and that U is a continuous representation of G in a Hilbert space $\mathcal{H}$ such that the restriction of U to H is unitary.

4.3. Proposition

Let $\tilde{T}$ be the closure the operator $dU(\Sigma_1^q X_j^2)$ for elements $X_j \in \mathfrak{h}$. Let $e^{\zeta\tilde{T}}$ be the holomorphic semigroup generated by $\tilde{T}$.

(i) Then the unitary group $\{e^{it\tilde{T}}: -\infty < t < \infty\}$ leaves invariant the C^∞ vectors for U, i.e. $e^{it\tilde{T}}\mathcal{H}^\infty(U) \subset \mathcal{H}^\infty(U)$ for all $t \in \mathbb{R}$.

(ii) We have the commutation relations

$$\tilde{Y}e^{\zeta\tilde{T}} = e^{\zeta\tilde{T}}\exp(-\zeta\,\mathrm{ad}\tilde{T})(\tilde{Y}) \quad \text{on } \mathcal{H}^\infty(U) \tag{13}$$

for every $\zeta \in \mathbb{C}$ with $\mathrm{Re}\zeta > 0$.

(iii) The unitary group $e^{it\tilde{T}}$ maps $\mathcal{H}^{(k+1)n}$ continuously into $\mathcal{H}^n$ for all $n > 0$, where k is the degree of nilpotency.

Remark: Comparison with Lemma 4.1 shows that the axial limits of the holomorphic semigroup $e^{\zeta\tilde{T}}$ are in 'some sense' of degree kn (i.e., $e^{it\tilde{T}}$ loses kn derivatives), whereas $e^{\zeta\tilde{T}}$ for $\mathrm{Re}\zeta > 0$ is of degree zero. Examples (the heat semigroup on $L^2(\mathbb{R})$ say) show that part (iii) of the proposition is essentially best possible.

Proof: Consider first a value of the parameter ζ in the open right half-plane, $\mathrm{Re}\zeta > 0$. By Lemma 4.1, $e^{\zeta\tilde{T}}$ maps $\mathcal{H}^n$ into itself for all $n > 0$. The Gauss kernel $p_\zeta(.)$ can easily be computed explicitly because the subgroup H' generated by

$X_1,\ldots,X_q$ is isomorphic to $\mathbb{R}^q$ (if the X_j's are linearly independent).

For $x = (x_1,\ldots,x_q)\in\mathbb{R}^q$, and $|x|^2 = x_1^2 + \ldots + x_q^2$ we have

$$p_\zeta(x) = (4\pi\zeta)^{-q/2} \exp(-|x|^2/4\zeta)$$

where $\zeta^{-q/2}$ has the determination which is positive when ζ is positive. Let us introduce the notation $\pi(x) = U(\exp_G(x_1X_1 + x_2X_2 + \ldots + x_qX_q))$ for $x \in \mathbb{R}^q$ where $\exp_G$ is the exponential mapping $\mathfrak{g} \to G$ which maps $\mathfrak{h}$ onto H'. Haar measure on H' coincides with Lebesgue measure $dx = dx_1dx_2 \ldots dx_q$.

Then formula (3) takes the form

$$e^{\zeta\widetilde{T}} = \int_{\mathbb{R}^q} p_\zeta(x)\ \pi(x)\ dx. \tag{14}$$

For every $Y\in E_n(\mathfrak{g})$ and $u\in H^n$ the vector $e^{\zeta\widetilde{T}}u$ belongs to the domain of $\widetilde{Y}$, and

$$\widetilde{Y}\, e^{\zeta\widetilde{T}}\, u = \int_{\mathbb{R}^q} p_\zeta(x)\ \widetilde{Y}\ \pi(x)u\ dx. \tag{15}$$

Indeed, the operator $\widetilde{Y}$ is bounded from H^n to H, and hence commutes with the integral (14).

Let u be a C^∞ vector ($u\in H^\infty$) and let Y belong to $E(\mathfrak{g})$. Then for fixed ζ with $\mathrm{Re}\zeta > 0$

$$e^{\zeta\widetilde{T}}\exp(-\zeta\ \mathrm{ad}\widetilde{T})(\widetilde{Y})u = \sum_{|m|\geq 0} \int_{\mathbb{R}^q} p_\zeta(x)\pi(x)dx\ H_m(\zeta,\widetilde{X})(\mathrm{ad}\widetilde{X})^m(\widetilde{Y})u/m!$$

$$= \sum_{|m|\geq 0} \int_{\mathbb{R}^q} p_\zeta(x)(-x)^m\pi(x)dx\ (\mathrm{ad}\widetilde{X})^m(\widetilde{Y})\ u/m!$$

$$= \int_{\mathbb{R}^q} p_\zeta(x)\ \pi(x) \sum_{|m|\geq 0} (-x)^m/m!\ (\mathrm{ad}\widetilde{X})^m(\widetilde{Y})u\ dx$$

$$= \int_{\mathbb{R}^q} p_\zeta(x)\pi(x)\exp(-x\,\mathrm{ad}\widetilde{X})(\widetilde{Y})u\ dx = \int_{\mathbb{R}^q} p_\zeta(x)\ \widetilde{Y}\,\pi(x)u\,dx$$

$$= \widetilde{Y} \int_{\mathbb{R}^q} p_\zeta(x)\ \pi(x)u\ dx = \widetilde{Y}\ e^{\zeta\widetilde{T}}\ u. \tag{16}$$

where we have used the notation $(-x)^m = (-1)^{|m|}x_1^{m_1}\ x_2^{m_2}\ldots x_q^{m_q}$, and, $x\ \mathrm{ad}\widetilde{X} = x_1\ \mathrm{ad}\widetilde{X}_1 + \ldots + x_q\ \mathrm{ad}\widetilde{X}_q$. Furthermore, the following identities (14), (11), and (15), are used in this order. Only finite sums are involved in the computations above, so we can interchange integrals and summations freely. Finally, the Spectral Theorem has been employed in conjunction with the inverse Fourier transform $\hat{p}_\zeta(\xi)$ of $p_\zeta(x)$: one can easily check that $\hat{p}_\zeta(\xi) = \exp(-\zeta|\xi|^2)$ for all $\xi \in \mathbb{R}_q$, and (by (8))

$$\hat{p}_\zeta(\xi)\, H_m(\zeta, i\xi) = (i\; d/d\xi)^m\, \hat{p}_\zeta(\xi) = \widehat{(-x)^m p_\zeta(\cdot)}(\xi) \qquad (17)$$

for all multi indices m. Using the Functional Calculus for the commutative operator q-tuple $(\tilde{X}_1,\ldots,\tilde{X}_q)$ we obtain the corresponding operator identity

$$\int p_\zeta(x)\pi(x)dx\; H_m(\zeta,\tilde{X}) = \int p_\zeta(x)(-x)^m\, \pi(x)\, dx$$

(formal substitution of $\tilde{X}$ for $i\xi$ in (17)) which is also used above.

These remarks together then justify all the steps in the above derivation of the commutation relation (13) for $\mathrm{Re}\zeta > 0$ and $u \in H^\infty$.

Let $t \in \mathbb{R}$, so that it is a point on the imaginary axis, and let ζ_ν be a sequence of complex numbers in the open right half-plane such that $it = \lim_{\nu\to\infty} \zeta_\nu$ non-tangentially. We have

$$\tilde{Y}\, e^{\zeta_\nu \tilde{T}}\, u = e^{\zeta_\nu \tilde{T}} \exp(-\zeta_\nu\, \mathrm{ad}\tilde{T})(\tilde{Y})u \qquad (18)$$

for fixed $u \in H^\infty$. The right-hand side of (18) converges to $e^{it\tilde{T}} \exp(-it\, \mathrm{ad}\tilde{T})(\tilde{Y})u$ for $\nu\to\infty$, so the left-hand side is convergent too. Since $\tilde{Y}$ is closed it follows that $e^{it\tilde{T}}u$ belongs to the domain of $\tilde{Y}$ and that

$$\tilde{Y}\, e^{it\tilde{T}}u = e^{it\tilde{T}} \exp(-it\, \mathrm{ad}\tilde{T})(\tilde{Y})u \qquad (19)$$

This finishes the proof of parts (i) and (ii) of the proposition.

Part (iii) follows from the remarks after identity (11). For fixed $Y \in E_n(\mathfrak{g})$ one checks (by counting degrees) that the terms on the left-hand side of (11) are all of degree less than or equal to $(k+1)n$. Since the commutation relation (13) is established on H^∞ for $\zeta = it$ (eqn. 19) the conclusion (iii) is an immediate consequence. E.O.P.

<u>Remarks</u>: (a) For the proof of Proposition 4.3 our Lemma 4.1 is more general than is needed. In the application above, the restriction of V (or U) to H is unitary and H is nilpotently embedded.

If the degree of nilpotency is k then the adjoint representation grows at most like $|x|^k$. Let the notation be

$$\tilde{Y} e^{\zeta\tilde{T}} u = \int_{\mathbb{R}^q} p_\zeta(x)\pi(x)\ (\mathrm{Ad}(-x)\tilde{Y})u\ dx \text{ for } Y\in E_n(\mathfrak{g}) \text{ and}$$

$u\in H^n$. Then (compare with (6))

$$\|\tilde{Y}\ e^{\zeta\tilde{T}}u\| \leq \int_{\mathbb{R}^q} |p_\zeta(x)|(1+|\mathrm{Ad}(-x)|^n)dx\|u\|_n.$$

We seek an upper bound on the integral for the case q = 1. The integral $\int_{-\infty}^{\infty} p_\zeta(x)|\ |x|^{kn}dx$ is of the order of magnitude

$$C_{n,\zeta} = \Gamma(\tfrac{kn}{2} + 1)^{1/2}|\zeta|^{\frac{kn+1}{2}}\ (\mathrm{Re}\zeta)^{-\frac{kn+2}{4}},$$

so we have $\|e^{\zeta T}u\|_n \leq C_{n,\zeta}\|u\|_n$ for $\mathrm{Re}\zeta > 0$. Since $\mathrm{Re}\zeta$ appears in the denominator in $C_{n,\zeta}$ we get an intuitive reason why some derivatives 'are lost' in the limit $\mathrm{Re}\zeta\to 0$ of the operator $e^{\zeta\tilde{T}}$

(b) If the restriction of U to H is not unitary then the limits of $e^{\zeta T}$ for $\mathrm{Re}\zeta\to 0$ fail to exist in general. Consider $T = d^2/dx^2$ on $L^1(\mathbb{R})$. Then T is the infinitesimal generator of a holomorphic semigroup $e^{\zeta T}$ in the <u>open</u> right half-plane. Let $\|\cdot\|$ be the $L^1(\mathbb{R})$ norm. Then

$$\|e^{\zeta T}u\| \leq |\zeta|^{1/2}\ (\mathrm{Re}\zeta)^{-1/2}\ \|u\| \text{ for } u\in L^1(\mathbb{R}).$$

Thus, the difficulties of passing to the limit $\mathrm{Re}\zeta\to 0$ are of the same nature as in Remark (a). It is known that $i\,d^2/dx^2$ does not generate a continuous one-parameter group in $L^1(\mathbb{R})$. Although we do not know of a reference for the latter fact, it would distract our main line of arguments to give a proof in the present work. We are grateful to D. Ragozin for a proof of the fact that $i\ d^2/dx^2$ does <u>not</u> generate a continuous one-parameter group on L^1 of the circle.

Suffice it to say that the limit problems make it hard to formulate a plausible O'Raifeartaigh Principle for the case where the restriction of U to H is not unitary (Cf. [Sg 2, Gd 3]).

Inspection of this proof (and of comparable limiting arguments below) reveals that unitarity of V (or U) on H is used only to ensure that the boundary group $e^{it\tilde{T}}$ exists and is the strong limit of the holomorphic semigroup $e^{\zeta\widetilde{T}}$ as $Re(\zeta) \to 0$. Such limit phenomena have been discussed by Hille-Phillips [HP, Section 17.9] and could be supplied in 4.3 as an hypothesis replacing the unitarity assumption. In fact, one can supply non-unitary examples where such limiting behavior does occur (e.g. the natural representation of the Poincaré group on $L^p(\mathbb{R}^4)$, where $1 \neq p \neq 2$ and $\mathbb{R}^4$ is Minkowski space). But in general, the limit of $c^{\zeta\widetilde{T}}$ does not exist as $Re(\zeta) \to 0$, nor is there a boundary group $e^{it\widetilde{T}}$, e.g., the Poincaré groups acting upon $L^1(\mathbb{R}^4)$). We are not aware of any physically relevant non-unitary examples along the lines suggested by Segal [Sg 2] where the limit is known to exist, so we have confined our attention to the unitary-restriction case considered by earlier authors. E.O.R.

4.4. Corollary

Let M be the closure of the operator $dU(-\Sigma_1^q X_j^2 + \Sigma_{q+1}^r X_j^2)$ for elements $X_1,\dots,X_q,\dots,X_r \in \mathfrak{h}$.

(i) Then iM is the infinitesimal generator of a unitary one parameter group $\{e^{itM}\ ;\ -\infty<t<\infty\}$ in $L(H)$.

(ii) The unitary group $W(t) = e^{itM}$ satisfies the commutation relations

$$\widetilde{Y}\, W(t) = W(t)\, \exp(-it\ \mathrm{ad}M)(\widetilde{Y}) \quad \text{on } H^\infty(U) \tag{20}$$

and $W(t)$ maps $H^{(k+1)n}$ continuously into H^n for all $n > 0$ and $t \in \mathbb{R}$, where k is the degree of nilpotency.

Proof: Let $W_1(t)$ and $W_2(t)$ be the unitary groups generated by the closures of $idU(T_1)$ and $idU(T_2)$, respectively, where $T_1 = \Sigma_1^q X_j^2$ and $T_2 = \Sigma_{q+1}^r X_j^2$. Then according to the proposition $W_1(t)$ and $W_2(t)$ both satisfy the conclusion of the corollary. If $W(t) = W_1(-t)\, W_2(t)$ for $t \in \mathbb{R}$ then $W(t)$ maps $H^{(2k+1)n}$ into H^n for all $n > 0$. Part (ii) of the corollary yields the stronger conclusion that $W(t)$ maps $H^{(k+1)n}$ into H^n.

For technical reasons it is assumed in the proof that the elements $X_1,\dots,X_q,\dots,X_r$ are linearly independent. Let us define $T = -T_1 + T_2$, and apply formula (11) to T_1 and T_2 separately. Due to the commutativity of $E(\mathfrak{h})$ the identities

(11) for T_1 and T_2 ($\zeta = \pm it$) can easily be combined. The resulting commutation relation is

$$\sum_m H_{m_1}(-it,X_1)\cdots H_{m_q}(-it,X_q)H_{m_{q+1}}(it,X_{q+1})\cdots H_{m_r}(it,X_r)(\mathrm{ad}X)^m(Y)/m!$$

$$= \exp(-it\ \mathrm{ad}T)(Y) \quad \text{for all } Y\in E(\mathfrak{g}) \qquad (21)$$

where the notation $m = (m_1,\ldots,m_q,m_{q+1},\ldots,m_r)$ and $(\mathrm{ad}X)^m(Y)$

$= (\mathrm{ad}X_1)^{m_1}\ldots(\mathrm{ad}X_q)^{m_q}\ldots(\mathrm{ad}X_r)^{m_r}(Y)$ is employed. The

left-hand side of (21) is easily seen to be of degree $(k+1)n$ for every $Y\in E_n(\mathfrak{g})$.

Suppose for the moment that (20) is proved. Let Y belong to some $E_n(\mathfrak{g})$. Then the operator $\exp(-it\ \mathrm{ad}M)(\tilde{Y})$ is of degree $(k+1)n$, and $\|\tilde{Y}\ W(t)u\| \le C_{n,t}\|u\|_{(k+1)n}$ for all $u\in H^\infty$ and $t\in\mathbb{R}$. Here $C_{n,t}$ is a finite constant which is independent of u. Hence $W(t)$ maps $H^{(k+1)n}$ into H^n which is the desired conclusion.

It is however an easy consequence of Theorem 3.5 that the unitary groups $W_1(t)$ and $W_2(t)$ commute. Then we do in fact have $W(t) = W_1(-t)W_2(t)$ for all $t\in\mathbb{R}$, such that (20) follows from a double application of Proposition 4.3.

In order to invoke Theorem 3.5 it must be verified that the local boundedness condition is satisfied for the operator pair $\tilde{T}_1$ and $\tilde{T}_2$. Now, $W_1(t)$ maps $H^{(k+1)2}$ continuously into H^2. Since $\tilde{T}_2$ is of degree 2 it is clear that the mapping $t \to \tilde{T}_2\ W_1(t)u$ is continuous from $\mathbb{R}$ into H for every $u\in H^\infty$. It follows that Theorem 3.5 does indeed apply, and the proof is completed.

4D. Remarks on general Minkowskian observables

We have restricted our attention to the particular mass operator of the form given in the corollary, because the commutation relations (11) for quadratic expressions involve the well-known Hermite polynomials.

We describe below (with details of proof) a generalized line of argument, analogous to the above, which applies to arbitrary Hermitian elements in the algebra $E(\mathfrak{h})$.
Let $X_1,\ldots,X_\ell$ be a basis for $\mathfrak{h}$. An element $T\in E(\mathfrak{h})$ is a constant coefficient polynomial in the commuting operators $(X_1,\ldots,X_\ell)$. For $T\in E(\mathfrak{h})$ let $T(i\xi)$ be the polynomial obtained by substituting the vector $i\xi = (i\xi_1,\ldots,i\xi_\ell)$ for $X=(X_1,\ldots,X_\ell)$

in T. If T is Hermitian one easily checks that $T(i\xi)$ is real for all $\xi \in \mathbb{R}^\ell$

Let T be a fixed Hermitian element in $E(\mathfrak{h})$. As in the quadratic case there is associated with T a sequence of generalized Hermite polynomials $Q_m(\zeta,i\xi)$ in the variables $\zeta\in\mathbb{C}$ and $i\xi$ for every multi index $m = (m_1,m_2,\ldots,m_\ell)$. The polynomials are given by

$$(i\ d/d\xi)^m \exp(\zeta T(i\xi)) = Q_m(\zeta,i\xi)\exp(\zeta T(i\xi)) \tag{22}$$

for all $\xi\in\mathbb{R}$. The analogue to (11) is the commutation relation

$$\sum_{|m|\geq 0} Q_m(\zeta,X)\ (\mathrm{ad}X)^m(Y)/m! = \exp(-\zeta\ \mathrm{ad}T)(Y) \tag{23}$$

for $\zeta\in\mathbb{C}$ and $Y\in E(\mathfrak{g})$, where the multi index notation is used. (Note that Lemma 5.1 of [Gd 3] is a consequence of formula (23).)

Suppose that T is of degree $\mu > 2$. Then (22) implies that the polynomial Q_m is of degree $|m|(\mu-1)$ in the variable $i\xi$ for each m.

Let k be the degree of nilpotency, and let Y be a fixed element in some $E_n(\mathfrak{g})$. Then it is clear that the summands on the left-hand side of (23) vanish for $|m|>kn$. Hence the expression in (23) is of degree at most $kn(\mu-1) + n$.

For fixed $T \in E(\mathfrak{h})$ we denote by M the self-adjoint operator $\overline{dU(T)}$; (cf. [NS]). We show that the unitary one-parameter group $\{\exp(itM) :- \infty< t < \infty\}$ generated by iM leaves $H^\infty(U)$ invariant and satisfies commutation relations similar to (19).

Let $X_1,\ldots,X_\ell$ be a basis for $\mathfrak{h}$, and put $L = \Sigma_1^\ell X_j{}^2$. Then of course the closure Δ of $dU(L)$ is a negative self-adjoint operator (again by [NS]). Elements $S \in E(\mathfrak{h})$ are simply polynomials in the commuting basis elements $X = (X_1,\ldots,X_\ell)$, or equivalently constant coefficient differential operators on $H \approx \mathbb{R}^\ell$. We denote by $S(i\xi)$ the polynomial obtained from S by substituting $i\xi = (i\xi_1,\ldots,i\xi_\ell)$ for X. If S is Hermitian then $S(i\xi)$ is real for all $\xi \in \mathbb{R}^\ell$.

Suppose the degree of T is equal to μ. Let N be an integer $N > \mu$ and define

$$f(\zeta,\xi) = \exp(\zeta(-|\xi|^{2N} + T(i\xi)))$$

on $\mathbb{C} \times \mathbb{R}^\ell$. For every ζ with $\mathrm{Re}\ \zeta > 0$, $f(\zeta,\cdot)$ belongs to the Schwartz space $S(\mathbb{R}^\ell)$ of rapidly decreasing functions on $\mathbb{R}^\ell$.

Consequently the (inverse) Fourier transform

$$P_\zeta(x) = (2\pi)^{-\ell} \int_{\mathbb{R}^\ell} e^{ix\xi} f(\zeta,\xi)\, d\xi$$

belongs to $\mathcal{S}(\mathbb{R}^\ell)$ as well. If $S = -(-L)^N + T$ then, clearly, $S(i\xi) = -|\xi|^{2N} + T(i\xi)$. As in the proof of 4.3 we put $\pi(x) = U(\exp_G(x_1X_1 +\ldots+x_\ell X_\ell))$ for $x \in \mathbb{R}^\ell$. Then the integral $W(\zeta) = \int_{\mathbb{R}^\ell} P_\zeta(x)\pi(x)dx$ is convergent when Re $\zeta > 0$ and defines a holomorphic semigroup in the open right half-plane with infinitesimal generator equal to $\tilde{S} = -(-\Delta)^N + M$. By Remark (a), following the proof of 4.3, we have for every $u \in H^\infty(U)$ and $n = 0,1,2,\ldots$

$$\|\pi(x)u\|_n \leq \text{Const.}(1+|x|^{kn}).$$

Since $P_\zeta(.)$ belongs to $\mathcal{S}(\mathbb{R}^\ell)$ it follows that the operator $W(\zeta)$ defined above leaves $H^\infty(U)$ invariant when Re $\zeta > 0$. As in the proof of 4.3 we get on $H^\infty(U)$, $\tilde{Y}\, W(\zeta) = (W(\zeta)\exp(-\zeta \text{ ad } \tilde{S})(\tilde{Y})$ for all $Y \in E(\mathfrak{g})$ and all ζ with Re $\zeta > 0$.

Let $W_0(\zeta)$ be the semigroup generated by $\tilde{S}_0 = -(-\Delta)^N$ (i.e., $T = 0$). Then the above conclusions hold for $W_0(\zeta)$ as well.

Now let $t \in \mathbb{R}$ such that it is a point on the imaginary axis and pick a sequence of numbers ζ_ν with positive real parts such that $\zeta_\nu \to it$. We have in the strong operator topology

$$\lim_\nu W_0(\zeta_\nu) = \exp(-it(-\Delta)^N)$$

and

$$\lim_\nu W(\zeta_\nu) = \exp(-it(-\Delta)^N)\exp(itM).$$

It follows as in the proof of 4.3 (eqn. (18)) that each of the commuting unitary one-parameter groups generated by $i\tilde{S}$ and $i\tilde{S}_0$ leave $H^\infty(U)$ invariant and satisfy commutation relations similar to (19). Since $\exp(itM) = \exp(it\tilde{S})\exp(-it\tilde{S}_0)$ the same is true for $\exp(itM)$.

Suppose the spectrum of M is negative. Then by the spectral theorem $\exp(\zeta M)$ defines a holomorphic semigroup in the right half-plane with boundary values on the imaginary axis. We show that $\exp(\zeta M)$ leaves $H^\infty(U)$ invariant and satisfies commutation relations similar to (13). Since this is already known for $\exp(ikM)$ for $k \in \mathbb{R}$ it is enough to consider $\exp(kM)$ for $k > 0$.

The element $S_1 = S_0 + iT$ is not Hermitian, but the

closure $\widetilde{S}_1$ of $dU(S_1)$ is the infinitesimal generator of a holomorphic semigroup $W_1(\zeta)$ in the open right half-plane. Indeed, the kernel

$$Q_\zeta(x) = (2\pi)^{-\ell} \int_{\mathbb{R}^\ell} e^{ix\xi} \exp(\zeta S_1(i\xi))\, d\xi$$

belongs to $S(\mathbb{R}^\ell)$ when Re $\zeta > 0$, and

$$W_1(\zeta) = \int_{\mathbb{R}^\ell} Q_\zeta(x)\pi(x)dx.$$

One checks as in the proof of 4.3 that $W_1(\zeta)$ leaves $H^\infty(U)$ invariant and satisfies commutation relations similar to (13). The same property is satisfied by exp(kM) as can be seen from the following identity. Let $k > 0$ be given, and let ζ_ν be a sequence with Re $\zeta_\nu > 0$ and $\zeta_\nu \to -ik$. Then

$$\exp(kM) = \exp(ik\, \widetilde{S}_0)\lim_\nu W_1(\zeta_\nu)$$

where the limit exists in the strong operator topology.

These remarks indicate the main ideas in a proof of the following result.

4.5. Proposition

Let T be a Hermitian element of degree μ in $E(\mathbf{h})$.

Let M denote the closure of the operator dU(T).

Then iM is the infinitesimal generator of a unitary strongly continuous one-parameter group $\{e^{itM}: -\infty<t<\infty\}$ in $L(H)$.

(i) The C^∞ vectors $H^\infty(U)$ are invariant under e^{itM} for all $t\in\mathbb{R}$, and the commutation relations

$$\widetilde{Y}\, e^{itM} = e^{itM} \exp(-it\, \mathrm{ad}M)(\widetilde{Y}) \tag{24}$$

hold on H^∞ for all $Y \in E(\mathbf{g})$.

(ii) The operators e^{itM} map $H^{kn(\mu-1)+n}$ continuously into H^n for all $n > 0$, where k is the degree of nilpotency.

(iii) If the spectrum of M is negative then M generates a holomorphic semigroup $e^{\zeta M}$ in the right half-plane whose boundary values coincide with e^{itM}. Properties (i) and (ii) extend to $e^{\zeta M}$ for Re$\zeta > 0$.

Remark: Part (ii) improves (3.10) in [Gd 3]. Goodman shows that e^{itM} is continuous from $H^{\alpha+n}$ into H^n when $\alpha = 2\mu[(k-1)n+\ell]$. We emphasize that the proposition is only needed in case one is interested in testing some of the suggested, slightly 'non-standard', mass operators. (Recall that the standard mass operator leads to the paradox.)

4E. Fourier transform calculus and centrality of isolated projections

Let us give a C^∞ proof of Theorem 4.2. Suppose that $M = dU(T)$ for some Hermitian element $T \in E(\mathfrak{h})$, and let $\int \lambda E(d\lambda)$ be the spectral decomposition of M. Let P be the spectral projection of M corresponding to a compact connected component $[\alpha,\beta]$ of the spectrum of M. Let $\varphi \in C_c^\infty(\mathbb{R})$ be a test function which is identically equal to one on $[\alpha,\beta]$ and vanishes at all other points of the spectrum of M.

If $\varphi^{(j)}$ for $j = 0,1,2,\ldots$ denote the derivatives of φ, then we have by the Fourier inversion formula and the Spectral Theorem

$$\varphi^{(j)}(M) = \int_{-\infty}^{\infty}(it)^j \hat{\varphi}(t)\, e^{itM}\, dt. \tag{25}$$

It is a consequence of the commutation relations (20), or (24), that the integral (25) defines a continuous endomorphism in H^∞ for all j. By the particular choice of φ we have $\varphi(M) = P$, and $\varphi^{(j)}(M) = 0$ for $j=1,2,\ldots$ Furthermore, we show essentially as in [Gd 3] that

$$P\,\tilde{Y} \subset \tilde{Y}\,P \tag{26}$$

for all $Y \in \mathfrak{g}$. Indeed, let $u \in H^\infty$. Then $Pu \in H^\infty$ and

$$\begin{aligned}\tilde{Y}\,Pu &= \int \varphi(t)\, \tilde{Y}\, e^{itM} u\, dt\\ &= \int(\hat{\varphi}(t)\Sigma(it)^j/j!\, e^{itM}\, dt)(\mathrm{ad}M)^j(\tilde{Y})u\\ &= \Sigma(1/j!)\, \varphi^{(j)}(M)\, (\mathrm{ad}M)^j(\tilde{Y})u\\ &= \varphi(M)\, \tilde{Y}u, \text{ where (25) is used.}\end{aligned}$$

Now P is bounded in H and leaves H^∞ invariant. Then Theorem 3.4 applies to the pair $(\tilde{Y},P)$ for each $Y \in \mathfrak{g}$ for fixed $Y \in \mathfrak{g}$ and $u \in H^\infty$ the mapping $t \to P\,U(\exp tY)u$ is locally bounded (with values in H), and $\tilde{Y}$ is the infinitesimal generator of a c.l.e. group $\{U(\exp tY)\colon -\infty<t<\infty\}$ in $L(H)$. Hence by Theorem 3.4 the commutation relation (26) exponentiates to $P\,U(\exp Y) = U(\exp Y)\,P$. Since G is connected the desired conclusion (7) of Theorem 4.2 follows. E.O.P.

There is a quick way of getting the commutation relations (24) which applies to the case where it is known a priori that the C^∞ vectors are invariant under e^{itM}. In many examples it is clear from the nature of the representation U that the C^∞ vectors are invariant under the unitary groups generated by elements in the enveloping algebra of Minkowski space. In those cases the identity (23) is not needed because Chapter 3 yields commutation relations directly.

The details of this application of our C^∞ methods are given below. The result is rather a main lemma for Theorem 4.2, but the ideas are of separate interest so they are isolated in the following proposition.

As above, let U be a representation of G whose restriction to H is unitary in a Hilbert space $\mathcal{H}$, and let $\mathcal{H}^\infty$ be the C^∞ vectors of U.

4.6. Proposition

Let T be a Hermitian element in $E(\mathfrak{h})$ and let M be the self adjoint closure of the operator dU(T). Suppose $\mathcal{H}^\infty$is invariant under the unitary group $\{W(t): -\infty<t<\infty\}$ generated by iM.

Then the commutation relations (24) hold on $\mathcal{H}^\infty$ and the conclusion of Theorem 4.2 follows.

Proof: We have already derived 4.2 from the commutation relation relations (24). Therefore, we shall only indicate below how the commutation relations may be obtained from the invariance assumption of the proposition.

The present proof is a complete analogue to the proof of Theorem 8.6. It is verified below that the ad-orbit of Y under adT is finite-dimensional and that the local boundedness condition of Theorem 3.4 is satisfied for the pair iM and $\tilde{Y}$.

The space of C^∞ vectors $\mathcal{H}^\infty(U)$ is certainly a Fréchet space. For $Y\in\mathfrak{g}$ the function $f(t,u) = \tilde{Y}\, W(t)u$ from the complete metric space $\mathbb{R} \times \mathcal{H}^\infty$ into $\mathcal{H}$ can be approximated by a sequence of continuous functions $F_n(t,u) = \tilde{Y}_n W(t)u$ as in the proof of Lemma 3.3. It is shown in Proposition 3.6 that there is a non-trivial interval J such that the set of vectors $\{\tilde{Y}\, W(t)u:\ t\in J\}$ is bounded in $\mathcal{H}$ for all $u\in \mathcal{H}^\infty$.

As for the ad-orbit, it is clear from (23) that $\mathcal{O}_{iM}(\tilde{Y})$ is finite-dimensional. In fact, we get the precise information that if $Y\in E_n(\mathfrak{g})$ and k is the degree of nilpotency, then

$$(\mathrm{ad}T)^{(k+1)n}Y = 0 \text{ for all } T\in E(\mathfrak{h}).$$

(The dimension of $\mathcal{O}_T(Y)$ does not depend on the degree of the polynomial T.) Indeed, (22) easily implies that $Q_m(\zeta,\cdot)$ is of degree $|m|$ in the variable ζ. Hence, the coefficient to $\zeta^{(k+1)n}$ (on the left-hand side of (23)) vanishes, and so does the corresponding coefficient on the right-hand side.

Theorem 3.4 now yields the commutation relations (24),

(at least for Y of degree one). Using that conjugation by W(t) respects the multiplicative structure of $E(\mathfrak{L})$, (24) immediately extends to higher order Y as well, by a now standard argument. E.O.P.

PART III

CONDITIONS FOR A SYSTEM OF UNBOUNDED OPERATORS TO SATISFY A GIVEN COMMUTATION RELATION

"I'll give you the Axioms of Science. After that I shall exhibit some Specimens. Then I shall explain a Process or two. And I shall conclude with a few Experiments. ..."

Silvie and Bruno Concluded
The Professor's Lecture
LEWIS CARROLL

"I say, look here, you know," said the Emperor, who was getting a little restless. "How many Axioms are you going to give us? At *this* rate, we shan't get to the *Experiments* till to-morrow week!"

"Oh, sooner than *that*, I assure you!" the Professor replied, looking up in alarm. "There are only", (he referred to his notes again)" only *two* more, that are really *necessary*."

"Read'em out, and get on to the *Specimens*!" grumbled the Emperor.

Silvie and Bruno Concluded
The Professor's Lecture
LEWIS CARROLL

In the final general discussion at the Solvay Conference, Poincaré drew attention to the necessity of investigating the question of the equations used in any application of quantum theory. "In this context one must keep in mind", he said, "that one can probably prove every theorem without too much effort if one bases the proof on two mutually contradictory premises" (Poincaré in Eucken, 1914, p. 364).

INTRODUCTION TO PART III

Let V be a strongly continuous representation of a Lie group G in a Banach space E, and let dV be the derived representation of the corresponding Lie algebra $\mathfrak{g}$. The Gårding vectors [Gå 1] serve as a common invariant core domain, $D \subset E$, for the individual operators in $dV(\mathfrak{g})$. Hence the operator Lie algebra $\mathfrak{L} = dV(\mathfrak{g})$ is a finite-dimensional real Lie subalgebra of the endomorphism algebra End(D). Recall that an operator Lie algebra $\mathfrak{L}$ on a dense domain D in a Banach space E is said to be <u>exact</u>, or <u>exponentiable</u>, if there is a C_0 representation V of the simply connected Lie group G, with Lie algebra $\mathfrak{g}$ isomorphic to $\mathfrak{L}$ (Ado's theorem), such that $dV(\mathfrak{g}) = \mathfrak{L}$. The exactness assumption is, as it turns out, too restrictive for our applications. But for convenience we proceed with a discussion of commutation relations for the case where $\mathfrak{L}$ is exact. The point of view is then quickly reversed.

For $X \in \mathfrak{g}$, and $A = dV(X)$, we write $V(t,A)$ for the one-parameter group $V(\exp(tX))$ arising from a given C_0 representation V of G, where $\exp : \mathfrak{g} \to G$ is the exponential mapping of Lie theory. If $\lambda \in \mathbb{C}$ is a point in the resolvent set $\rho(A)$ of A, we adopt the notation $R(\lambda,A) = (\lambda - \bar{A})^{-1}$ for the bounded resolvent operator. If finally φ is a scalar function holomorphic in a neighborhood of $\sigma^*(A) = (\mathbb{C} \setminus \rho(A)) \cup \{\infty\}$, then the holomorphic functional calculus $\varphi(A)$ is defined, through a well known integral formula, in terms of $R(\lambda,A)$.

The following known commutation formulas have been derived and discussed in Chapters 1 through 3:

(1) $$B\,V(t,A) = V(t,A)\exp(-t \operatorname{ad} A)(B),$$

(2) $$B\,R(\lambda,A) = \Sigma_{k=0}^{\infty} R(\lambda,A)^{k+1}(-\operatorname{ad} A)^k(B),$$

(3) $$B\,\varphi(A) = \Sigma_{k=0}^{\infty}(k!)^{-1}\varphi^{(k)}(A)(-\operatorname{ad} A)^k(B).$$

These identities are valid on the Gårding space D, and for operators A,B in $\mathfrak{L}$. (We shall later consider A in the enveloping algebra of $\mathfrak{L}$.) Note that

$(\mathrm{ad}\ A)(B) = [A,B] \in \mathcal{L}$, $(\mathrm{ad}\ A)^2(B) = [A,[A,B]] \in \mathcal{L}$ etc., since $\mathcal{L}$ is a Lie algebra. Since $\mathcal{L}$ is also finite-dimensional, the power series $\Sigma_{k=0}(k!)^{-1}(\mathrm{ad}\ A)^k(B)$ is convergent, and the sum, $C = \exp(\mathrm{ad}\ A)(B)$, is a well defined element in $\mathcal{L}$. As such, C is also regarded as an operator in E with the same dense domain, viz., D.

Our main interest here will be a set of much more general commutation relations, although similar to identities (1) through (3) above. But our point of view will be reversed from the outset, relative to the one above which starts from a given C_0 representation of a Lie group G. Typically we shall assume, as given from the beginning, a Lie algebra $\mathcal{L}$ of unbounded operators in a Banach space E. (In particular $\mathcal{L}$ is not assumed exact!) The domain D of $\mathcal{L}$ is given implicitly as a dense linear subspace, $D \subset E$, and $\mathcal{L} \subset \mathrm{End}(D)$ = the linear endomorphisms of D. As demonstrated in Chapter 4 through 12, the special case where $\mathcal{L}$ is exact, i.e., $\mathcal{L} = dV(\mathfrak{g})$, is much too restrictive for the applications. Our starting point is instead a pair of operators $A,B \in \mathrm{End}(D)$, and a finite-dimensional complex linear subspace $M \subset \mathrm{End}(D)$. It will be assumed only that M is invariant under ad A, and that $B \in M$. (Such complex ad-modules (A,M) encompass the important structure of shift-operators in mathematical physics). In this set-up, the Jordan-Wedderburn decomposition is available for ad A, when regarded as a linear endomorphism of M. Specifically let $\sigma_M(\mathrm{ad}\ A) = \{\alpha_1,\ldots,\alpha_p\}$ be the eigenvalue spectrum, and let P_j be the projection in M onto the generalized eigenspace

$$M_j = \{C \in M: (\mathrm{ad}\ A-\alpha_j)^{s_j+1}(C) = 0 \text{ for some integer } s_j\},$$

$$j = 1,\ldots,p.$$

The ascent s_j of M_j is given by $(\mathrm{ad}\ A-\alpha_j)^{s_j} \neq 0 = (\mathrm{ad}\ A-\alpha_j)^{s_j+1}$.

If $\lambda \notin \sigma(A) \cup \{\mu-\alpha_j : \mu \in \sigma(A)\ , 1 \le j \le p\}$, (the M-augmented spectrum), then we give infinitesimal sufficient conditions for the validity of the following type of commutation relations:

$$(1')\quad B\,V(t,A) = \Sigma_{j=1}^{p} e^{-t\alpha_j} \Sigma_{k=0}^{s_j} \frac{(-t)^k}{k!}\, V(t,A)(\mathrm{ad}\ A-\alpha_j)^k(P_jB)\ ,$$

$$(2') \quad B\,R(\lambda,A) = \Sigma_{j=1}^{p}\Sigma_{k=0}^{s_j} R(\lambda+\alpha_j,A)^{k+1}(\alpha_j-\text{ad }A)^k(P_jB),$$

$$(3') \quad B\,\varphi(A) = \Sigma_{j=1}^{p}\Sigma_{k=0}^{s_j}\varphi^{(k)}(A-\alpha_j)(\alpha_j-\text{ad }A)^k(P_jB)(k!)^{-1}.$$

If, for example, A is an element in the associative universal enveloping algebra $\mathfrak{U}(\mathfrak{L})$ of $\mathfrak{L}$, anyone of the relations (2') and (3') may hold in cases where the others do not hold, or fail to make sense. For example, the bounded resolvent $R(\lambda,A)$ may exist for operators A such that $\bar{A}$ fails to generate a one-parameter semigroup, let alone a group. A typical area of applications for (3') is the one where the operator $\varphi(A)$ is a spectral projection P_σ (i.e., a bounded idempotent) associated with an isolated subset σ in the spectrum of A, (σ closed and relatively open in $\sigma^*(A)$.) In fact, the O'Raifeartaigh theorem (treated in detail in Chapter 4) is a special instance of this application of (3'). Here, the mass operator $A \in \mathfrak{U}(\mathfrak{L})$, and $B \in \mathfrak{L} \subset M_0$ (= the generalized ad A - eigenspace corresponding to the point $\alpha = 0$ in $\sigma_M(\text{ad }A)$). Hence relation (3') reduces to $B\,\varphi(A) = \Sigma_{k=0}^{s}(k!)^{-1}\varphi^{(k)}(A)(-\text{ad }A)^k(B)$, where $\varphi(A)$ is any holomorphic functional calculus for the operator A. In the O'Raifeartaigh application, A is the mass operator, and we consider a fixed compact connected component σ of the spectrum of A. We may then pick φ analytic such that $\varphi \equiv 1$ on σ, and $\varphi \equiv 0$ on the other components. Then $\varphi(A) = P_\sigma$, and $\varphi^{(k)}(A) \equiv 0$ for $k \geq 1$, and as a consequence, identity (3') reduces further to the identity

$$B\,P_\sigma = P_\sigma\,B.$$

Finally, as it turns out in Chapter 4, all of the known versions of the O'Raifeartaigh theorem follow from this: If $\mathfrak{L}$ is the derived operator Lie algebra $dV(\mathfrak{g})$ of some irreducible representation V, then the mass spectrum can have only one isolated component; which is to say that a known and observed "mass-splitting" in the spectrum of A cannot be explained in any one of the elementary particle models which is based on the Poincaré group with $dV(\mathfrak{g}) = \mathfrak{L} \subset M_0(\text{ad }A)$. (Since our commutation theory, in Chapters 5 through 7, is valid also for non-exact operator Lie algebras $\mathfrak{L}$ it follows that exponentiability is not essential in explaining "mass-splitting" [O'R 1, St 2].)

Segal's model, which is based instead on the conformal group [Sg 3], represents here the most convincing resolution of the mass-splitting "paradox", and our general formula (3') can now be used in a positive manner in analyzing the structure of the mass spectrum in this, as well as in other, models for elementary particles.

For a given operator $A \in \mathrm{End}(D)$, it may be possible to find a complex linear subspace $M \subset \mathrm{End}(D)$, with a semisimple action of ad A on it. If $\sigma_M(\mathrm{ad}\, A) = \{\alpha_1,\dots,\alpha_p\}$, the semisimplicity corresponds to the ascent zero conditions $s_1 = \dots = s_p = 0$, which are characteristic for a variety of ladder operators. In this case, the three commutation formulas simplify to:

$$(1'') \quad B\,V(t,A) = \Sigma_{j=1}^{p} V(t,A-\alpha_j)P_j(B),$$

$$(2'') \quad B\,R(\lambda,A) = \Sigma_{j=1}^{p} R(\lambda+\alpha_j,A)P_j(B),$$

$$(3'') \quad B\,\varphi(A) = \Sigma_{j=1}^{p} \varphi(A-\alpha_j)P_j(B).$$

Here the commutation identities are considered for elements $B \in M$, and $P_j(B)$ is the component of B in the α_j-eigenspace for the action of ad A on M. The identities (1") - (3") apply, for example, to the analysis of the energy spectrum, as well as the mass spectrum, in the above mentioned elementary particle model, based on the 15-dimensional conformal group, and due to Segal.

If A is <u>the energy operator</u> in the Segal model ([Sg 3], and [Ørs]) with $\mathfrak{L}$ isomorphic to the 15-dimensional conformal Lie algebra $\approx so(2,4)$, then the spectral theory of A follows from one of the relations (1") - (3"), (depending on the available generality: exactness, global integrability of the Cauchy problem, e.t.c.). This was observed in [Ørs] in a particular case where A can be chosen as the O(1) generator in a certain Cartan subalgebra $\mathfrak{h} \subset \mathfrak{L}$.

There are different possibilities for <u>the mass operator</u>: To apply (1") - (3") instead to the mass operator A in the model, one may choose a 10-dimensional semisimple subalgebra $\mathfrak{g} \subset \mathfrak{L}$, and let A be a certain Casimir element (second order) for $\mathfrak{g}$. Introducing a third algebra $\approx s\ell(2,\mathbb{R})$, with a shift type action on $\mathfrak{g}$, it is possible to find enough eigen-elements

B_j to determine the discrete spectrum for A, via formulas (1") - (3"). (We are grateful to B. Ørsted for enlightening discussions on the mass spectrum in the Segal model, and we refer the reader to [Ørs] for detailed calculations).

Our present brief discussion of elementary particle spectral theory has been included here only for the purpose of indicating the wide applicability of our algebraic formalism, and to make the connection to other chapters in the monograph. In fact, the main focus, in Chapters 5-7, is on the introduction of a certain analytic condition, on a given pair (A,M), which implies the commutation relations in integrated form. Since the elements in M are unbounded operators in a Banach space E, it is possible to define a first order graph topology on the domain space D : viz., a first-order generalized Sobolev space $D_1(M)$. The condition which we call graph density, and denote by (GD), is the requirement that the range space $(\lambda - A)D$ be dense in $D_1(M)$ for λ in the resolvent set of A. (So, this is a zero-deficiency condition!).

If A is a pregenerator for a C_0 semigroup $\{V(t,A) : 0 \leq t < \infty\}$ of bounded operators in E, then we show that the semigroup commutation identities (1), (1'), or (1") (depending on the algebraic setting), follow from the GD-condition on A. For more general operators A, we derive instead the commutation identities of the second, or the third type.

In Chapter 7, we then apply the theory to a new construction of generalized C^∞ Sobolev spaces D_∞ which are invariant under the semigroup $V(t,A)$, pregenerated by the operator A in a given operator system (A,M,D). The invariance question for the constructed D_∞ turns out to depend heavily on the commutation theory developed in Chapters 5 and 6. Specifically: If E is a Banach space completion of D, and if the closure $\bar{A}$ of the operator A exists, and is the infinitesimal generator of a C_0 semigroup $V(t,A)$ in E, then we examine the possibility of constructing C^∞ domains D_∞, $D \subset D_\infty \subset E$, such that $V(t,A)$ leaves invariant D_∞ for all $t \geq 0$. In more general cases where the semigroup is not available, the issue is instead invariance of D_∞ under $R(\lambda,A)$, or $\varphi(A)$.

In Chapters 8 and 9, the commutation theory is applied to the special case of Lie algebras $\mathfrak{L}$ of unbounded operators. For a given finite-dimensional real Lie subalgebra $\mathfrak{L} \subset \mathrm{End}(D)$, one is interested in sufficient conditions on a Lie generating subset $S \subset \mathcal{L}$ which imply exactness of $\mathcal{L}$, i.e., exponentiability to a C_0 Lie group representation in a particular Banach space completion E of D.

Assuming from the outset, in Theorem 9.2, that each element A in S (some given Lie generating subset) satisfies a GD-condition vis a vis $\mathcal{L}$, we proceed to derive the commutation relations (2), and then (1), for $B \in \mathcal{L}$. Here the domain considerations (which are quite delicate) have been dealt with in Sections 5B, 6A and 6C.

A second major technical tool in the proof of exactness is then the construction of C^∞ domains (i.e., generalized Sobolev spaces of infinite order) which are invariant under the one-parameter exponentials $\{V(t,A): t \in \mathbb{R}\}$ for $A \in S$.

In each of the applications, mentioned above, the mathematical development begins with a normed linear space D, $\|\cdot\|$. The linear endomorphisms in D form an associative algebra End(D) which gets the structure of a Lie algebra when equipped with the commutator bracket $[A,B] = AB - BA$ for $A, B \in \text{End}(D)$. But the elements in End(D) may also be regarded as densely defined, unbounded operators in the Banach space completion E of D. To specify when a given element A in End(D) is regarded as a partially defined operator in E we shall write $A \in \mathbf{A}(D)$. That is, $\mathbf{A}(D)$ coincides with End(D) as a set. The $\mathbf{A}$-terminology is used only to remind the reader that unbounded operators in a larger Banach space E are involved.

This distinction is important if, in a given setup, the starting point is the derived Lie algebra $\mathcal{L}$, or the associative enveloping algebra $\mathfrak{U}(\mathcal{L})$, of a strongly continuous (C_0) representation of a Lie group in a Banach space E. In this set-up we may take D to be the Gårding space, and the norm to be the initially given norm on E. Then $\mathcal{L}$, as well as $\mathfrak{U}(\mathcal{L})$, become Lie subalgebras of $\mathbf{A}(D)$, regarded either as a Lie algebra of unbounded operators, or as an associative operator algebra.

Let D, $\|\cdot\|$ be a normed space with Banach completion E, and let A,B be a pair of operators in $\mathbf{A}(D)$. For $k = 1,2,\ldots$ the operators $(\text{ad } A)^k(B) = [A,[\ldots[A,B]\ldots]$ also fall in $\mathbf{A}(D)$. Let $M \subset \mathbf{A}(D)$ be the smallest linear subspace containing B, and invariant under the derivation ad $A : \mathbf{A}(D) \to \mathbf{A}(D)$. Then, of course, M is spanned by the elements $\{(\text{ad } A)^k(B) : k = 0,1,\ldots\}$, and ad A restricts to a linear endomorphism in $\mathcal{M}$. This subspace is denoted $\mathcal{O}_A(B)$, and is called the ad-orbit. But we shall consider more general pairs $(A,\mathcal{M})$ satisfying the two requirements:

(i) $\mathcal{M}$ is a complex linear subspace of $\mathbf{A}(D)$

and

(ii) $\mathcal{M}$ is invariant for the derivation, ad A.

If λ is in the resolvent set of A, the operator $R(\lambda,A) = (\lambda - \bar{A})^{-1}$ is a bounded and everywhere defined inverse to $\lambda - \bar{A}$ where $\bar{A}$ is the operator closure. (Here it is important that we think of A as an operator in E). Now, the right-hand side of the commutation relation (2) involves the ad-orbit $\mathcal{O}_A(B)$ directly. If, for example, $\mathcal{O}_A(B)$ is finite-dimensional there is no convergence problem in the (ad A)-infinite series on the right-hand side of each of the commutation relations (1) - (3). We focus on this case. But the applications frequently involve the two operators A, and A^2 together, and conditions for finite-dimensionality of $\mathcal{O}_{A^2}(B)$ are important. Such a condition is analyzed in Appendix D.

In the present introduction, we have considered three separate aspects (listed below) of operator commutation relations.

(1) Commutation theory for the operators and semigroups on dense invariant C^∞-domains. (The invariance here refers to the bounded operator semigroups!)

(2) Infinitesimal commutation theory for resolvents, semigroups, and operational calculi. (Graph-density)

(3) Strongly continuous exponentials for operator Lie algebras.

Chapter 5

GRAPH-DENSITY APPLIED TO RESOLVENT COMMUTATION, AND OPERATIONAL CALCULUS

The primary objective of this chapter is to obtain purely infinitesimal conditions which are equivalent to the validity of resolvent commutation relations, which connect the resolvents $R(\lambda,A) = (\lambda - \bar{A})^{-1}$ of one closable endomorphism $A \in \mathbf{A}(D)$ with another such operator $B \in \mathbf{A}(D)$.[1] Here, D is taken to be a dense subspace of a Banach space E, and the most easily stated example of such a commutation relation is the geometrically convergent identity for $u \in D$

$$\bar{B}\,R(\lambda,A)u = \Sigma\{(-1)^k R(\lambda,A)^{k+1}(\text{ad } A)^k(B)u\colon\ 0 \le k < \infty\}. \tag{1}$$

(We pointed out in Chapter 2 that this identity seems to have no classical precursors, but that it can readily be recognized as the formal Laplace transform of a well-known semigroup commutation identity associated with adjoint representations of Lie groups on Lie algebras:

$$\begin{aligned}\bar{B}\,V(t,A)u &= V(t,A)\exp(-t\text{ ad } A)(B)u \\ &= \Sigma\{(-1)^k t^k/k!\ V(t,A)(\text{ad } A)^k(B)u\colon\ 0 \le k < \infty\}. \end{aligned}\tag{2}$$

This relationship is explored in detail in Chapter 6.)

In Section 5A we establish sufficient conditions for (1), obtaining in the process an alternative version of this identity which replaces the infinite series in $R(\lambda,A)$ on the right by a finite polynomial in translates $R(\lambda+\alpha_j,A)$ of the resolvent (by eigenvalues α_j of the operator ad A). Two background conditions are imposed in order to ensure that the expressions in these commutation relations are meaningful: the operator B is assumed to lie in a finite-dimensional ad A-invariant

---

1) $\mathbf{A}(D)$ = the algebra of all linear endomorphisms of the vector space D.

complex subspace $M \subset \mathbf{A}(D)$, and λ is constrained to lie far enough from the spectrum $\sigma(A)$ of A that the resolvents $R(\lambda+\alpha_j, A)$ exist and that the series in (1) converges. The principal analytic condition, called the graph-density condition, can be viewed as a zero-deficiency condition for the number $\lambda \in \rho(A)$ with respect to a stronger graph or C^1 topology: the space $D_\lambda = (\lambda-A)D$ is assumed to be dense in D with respect to the weakest (norm-) topology τ_1 on D, stronger than the relative topology from E, such that every $B \in M$ is continuous from (D,τ_1) into the Banach space E. This topology can be normed by selecting a basis $B_1,\ldots,B_d$ for M, setting $B_0 = I$ the identity operator, and putting

$$\|u\|_1 = \max\{\|B_i u\| : 0 \le i \le d\}. \tag{3}$$

Section 5B serves two purposes. First, it establishes that the graph density condition is (essentially)[2] necessary as well as sufficient for commutation relations such as (1). Second, it relates these commutation relations to the spectral theory of A on a natural Banach space $(D_1,\|\cdot\|_1)$ of C^1-vectors for M (the completion of $(D,\|\cdot\|_1)$, in fact) and reformulates the commutation relations as identities in bounded operators from $(D_1,\|\cdot\|_1)$ to $(E,\|\cdot\|)$.

Section 5C takes up the λ-dependence of the graph-density condition and of these (equivalent) commutation relations, pointing out that both can be analytically continued from a given λ into an entire component of a slightly shrunken version of the resolvent set $\rho(A)$. Such continuation results are essential to the operational calculi treated later. They also provide technical simplifications in the applications of these results to semigroup commutation relations (Chapter 6) and Lie algebra exponentiation (Chapter 9).

In Section 5D the commutation identities (e.g., (1)) are applied to derive commutation relations connecting $B \in M$ to images $\varphi(A)$ of A under the holomorphic operational calculus. For example we show that if φ is holomorphic in a sufficiently large neighborhood of $\sigma(A) \cup \{\infty\}$ then (1) leads to

---

2) If the commutation relation is known then graph density follows for some domain D' containing D (cf., Proposition 5.5).

$$\bar{B}\,\varphi(A)u = \Sigma\{(-1)^k/k!\ \varphi^{(k)}(A)(\mathrm{ad}\ A)^k(B)u\colon 0 \leq k < \infty\}. \tag{4}$$

(It also follows that $\varphi(A)$ restricts to a bounded operator on $(D_1, \|\cdot\|_1)$.) When applied to the study of the spectral projections P_σ attached to "sufficiently isolated" spectral sets $\sigma \subset \sigma^*(A) = \sigma(A) \cup \{\infty\}$, these identities show that P_σ reduces every $B \in M$. This remark supplies spectral splitting results similar to those involved in the O'Raifeartaigh mass-splitting effect, by an operational calculus approach similar to that used by Goodman [Gd 3] (recalled in detail in Chapter 4).

Before proceeding to the details, we recall briefly the relationship between the results in the present chapter and the three main components of this monograph. Obviously, this Chapter serves as the keystone for the portion of our work which derives commutation relations from assumptions most directly connected with resolvents and spectra, leading to results concerning operational calculi, spectral structure, and the like. In Chapter 6, the present ideas are generalized, first by extending the operational calculus to the exponential functions $\varphi(A) = \exp(tA) = V(t,A)$ when A generates a semigroup. (Notice that (2) is the formal consequence of (4) as applied to the function $\varphi(\lambda) = e^{\lambda t}$, but the fact that ∞ is an essential singularity of φ precludes the direct application of (4) when A is unbounded.) Then the various operational calculi for semigroup and group generators can be used to supply further extensions of the commutation relations for $\varphi(A)$. These lead in Chapter 11 to a number of suggestive examples indicating the way in which commutation relations, of classical interest in physics, necessarily restrict the spectral behavior of the operators involved.

Chapter 7 then provides the bridge connecting the resolvent spectral component of the monograph to the other main components: invariant domain commutation theory and construction of Lie group representations as integrals or exponentials of operator Lie algebras. In it, hypotheses of graph-density type are used to construct dense semigroup-invariant domains of C^∞ vectors of the type systematically exploited in Chapters 3 and 4. These C^∞ domains are also used to show that graph-density hypotheses yield exponentiability criteria for operator Lie algebras (Chapters 8, 9, 12). As indicated in Chapters 9 and 12, the graph-density exponentiation theorem thus obtained is more general than the invariant-domain exponentiation result which follows from the development in Chapter 3. In particular, it permits problems in analytic continuation of group representations to be reduced to previously-studied questions in the perturbation theory of semigroups.

5A. Augmented spectra and resolvent commutation relations

This section concerns the interplay between the spectral behavior of A on E and that of ad A on suitably-chosen subspaces $\mathcal{M}$ of $\mathbf{A}(D)$. The underlying requirement is that ad A have a reasonable spectral theory on $\mathcal{M}$. For the applications in group representations and theoretical physics envisioned here, the restriction to finite-dimensional $\mathcal{M}$ is more natural and leads to more satisfactory results. We suppose here that $\mathcal{M} \subset \mathbf{A}(D)$ is complex in order to guarantee that the endomorphism ad A has a complete spectral reduction theory on $\mathcal{M}$. In practice, this means that the real ad-orbits

$\mathcal{O}_A(B)$ = real span $\{(\text{ad } A)^k(B) : 0 \le k < \infty\}$ in Chapter 4 or the real Lie algebras $\mathcal{L}$ in various chapters (e.g. 9) must be complexified: $\mathcal{M} = \mathcal{O}_A^{\mathbb{C}}(B)$ = complex span $\{(\text{ad } A)^k(B)\}$ $= \mathcal{O}_A(B) + i\mathcal{O}_A(B)$ or $\mathcal{M} = \mathcal{L}_{\mathbb{C}} = \mathcal{L} + i\mathcal{L}$.

Working in a space $\mathcal{M}$ of operators over an algebraically-closed ground-field turns out to supply sharper and technically more convenient variants of the commutation relation (1), based upon the Jordan-Wedderburn decomposition for ad A. Specifically, let $\sigma_{\mathcal{M}}(\text{ad } A) = \{\alpha_1,\dots,\alpha_p\}$ denote the eigenvalues of ad A on $\mathcal{M}$, and let P_j denote the projection of $\mathcal{M}$ onto the jth generalized eigenspace $\mathcal{M}_j = \{B \in \mathcal{M}: (\text{ad } A-\alpha_j)^s(B)=0$ for some integer s$\}$, while s_j will denote the non-negative integer such that $(\text{ad } A-\alpha_j)^{s_j} \neq 0 = (\text{ad } A-\alpha_j)^{s_j+1}$ on $\mathcal{M}_j$ (the ascent or step of α_j.) Then the infinite sum (and attendant convergence difficulties) in (1) can be replaced by an analytically simpler (if notationally more cumbersome) algebraic relation

$$\bar{B}\, R(\lambda,A)u = \Sigma\{(-1)^k R(\lambda+\alpha_j,A)^{k+1}(\text{ad } A-\alpha_j)^k(P_jB)u:\ 1 \le j \le p,\ 0 \le k \le s_j\} \tag{5}$$

This relation has meaning only for λ in the complement of the <u>M-augmented spectrum</u> of A, viz.,

$\sigma(A;M) = \sigma(A) \cup \{\lambda-\alpha_j: \lambda \in \sigma(A), \alpha_j \in \sigma_M(\text{ad } A)\}$. That is, λ must lie in the <u>M-diminished resolvent set</u>

$\rho(A;M) = \mathbb{C} \sim \sigma(A;M)$. Two examples, of major current interest in applications to Lie theory, will serve to illustrate the point of (5). First, suppose that ad A is nilpotent of step s on M: $(\text{ad } A)^s \neq 0 = (\text{ad } A)^{s+1}$ on M. Then $\sigma_M(\text{ad } A) = \{0\}$ and $\sigma(A;M) = \sigma(A)$, while (5) reduces to the obvious truncated version of (1): for $u \in D$

$$\bar{B}R(\lambda,A)u = \Sigma\{(-1)^k R(\lambda,A)^{k+1}(\text{ad } A)^k(B)u: 0 \le k \le s\} \qquad (5')$$

At the opposite extreme, suppose that $M = \mathfrak{L}_{\mathbb{C}}$ for $\mathfrak{L}$ (algebraically isomorphic to) a semisimple Lie algebra, and suppose $A \in \mathfrak{L}$ is regular. There the Cartan decomposition theory for $\mathfrak{L}_{\mathbb{C}}$ ensures that all eigenspaces have ascent 0 (they are direct sums of root subspaces for the Cartan subalgebra $\mathfrak{h}$ containing A, associated with sets of roots that agree on A), and (5) then becomes a formula concerning translated resolvents acting upon $u \in D$:

$$\bar{B}\,R(\lambda,A)u = \Sigma\ \{\ R(\lambda+\alpha_j,A)\ P_j(B)u: 1 \le j \le p\}. \qquad (5'')$$

The general form of (5) is obtained as follows.

<u>5.1. Theorem</u>

Let D be a dense subspace of a complex Banach space E, let $A \in \mathbf{A}(D)$ be a closable endomorphism of D, and let $M \subset \mathbf{A}(D)$ be a finite-dimensional complex subspace of $\mathbf{A}(D)$ that is invariant under ad A. Suppose that for some $\lambda \notin \sigma(A;M)$ (notations as above) the space $D_\lambda = (\lambda-A)D$ is dense in D with respect to the topology τ_1 normed by

$\|u\|_1 = \max\{\|B_i u\|: 0 \le i \le d\}$, $B_0 = I, B_1,\ldots,B_d$ a basis for M.

Then for every closable $B \in M$

(1) the resolvent $R(\lambda,A) = (\lambda-\bar{A})^{-1}$ maps D into $D(\bar{B})$

and

(2) $$\bar{B}\,R(\lambda,A)u = \Sigma\ \{(-1)^k R(\lambda+\alpha_j,A)^{k+1}(\text{ad } A-\alpha_j)^k(P_jB)u: 1 \le j \le p,\ 0 \le k \le s_j\}\ . \qquad (5)$$

Remark: This result is not in fact confined to Banach-space operators. The proof turns out to require only that all of the operators $\lambda + \alpha_j - \bar{A}$ have continuous two-sided inverses on E. However, most of the rest of the theory requires that the resolvents $(\lambda-\bar{A})^{-1}$ take their values in an operator Banach algebra, which restricts the setting either to the classical case where E is Banach or to the slightly more general situation treated independently by Chilana [Ch] and the second author [Mr 6]. In the interests of clarity and uniformity of setting, we have omitted the obvious and routine extensions just suggested.

Proof: Beginning algebraically, we first check that whenever $B \in M$, $\alpha \in \mathbb{C}$ with $\lambda + \alpha \in \rho(A)$, and $u \in D_\lambda = (\lambda-A)D$, then

$$B\,R(\lambda,A)u = R(\lambda+\alpha,A)Bu - R(\lambda+\alpha,A)(\mathrm{ad}\,A - \alpha)(B)R(\lambda,A)u. \qquad (6)$$

In fact, putting $u = (\lambda-A)v$ for $v \in D$, so that $v = R(\lambda,A)u$,

$$\begin{aligned} R(\lambda+\alpha,A)\,Bu &= R(\lambda+\alpha,A)B(\lambda-A)v = R(\lambda+\alpha,A)[(\lambda+\alpha-A)B+[A,B]-\alpha B]v \\ &= Bv + R(\lambda+\alpha,A)(\mathrm{ad}\,A - \alpha)(B)v \\ &= B\,R(\lambda,A)u + R(\lambda+\alpha,A)(\mathrm{ad}\,A - \alpha)(B)\,R(\lambda,A)u \end{aligned}$$

and (6) follows.

Next, we extend (6) inductively to obtain

$$\begin{aligned} B\,R(\lambda,A)u &= \Sigma\{(-1)^k\, R(\lambda+\alpha,A)^{k+1}(\mathrm{ad}\,A - \alpha)^k(B)u\colon 0 \le k \le s\} \\ &\quad + (-1)^{s+1}\, R(\lambda+\alpha,A)^{s+1}(\mathrm{ad}\,A - \alpha)^{s+1}(B)\,R(\lambda,A)u \qquad (7) \end{aligned}$$

when $u \in D$ and s is any integer. (The induction step is accomplished by applying (6), with $(\mathrm{ad}\,A-\alpha)^{s+1}(B)$ replacing B, to the "remainder term" in (7), splitting that term into a commuted part that extends the sum and a remainder term of one order higher.)

In particular, (7) can be applied for $\alpha = \alpha_j$ in the spectrum $\sigma_M(\mathrm{ad}\,A)$, with $s = s_j$ and B replaced by P_jB. Then for $u \in D$

$$P_jB\,R(\lambda,A)u = \Sigma\{(-1)^k\, R(\lambda+\alpha_j,A)^{k+1}(\mathrm{ad}\,A - \alpha_j)^k(P_jB)u\colon 0 \le k \le s_j\} \qquad (8)$$

since the remainder vanishes. But for arbitrary B, we have $B = \Sigma\{P_jB\colon 1 \le j \le p\}$, so that summation of (8) over all $1 \le j \le p$ yields (5) with u restricted to D. This completes the algebraic part of the proof.

The analytic portion is routine. For any $u \in D$, the graph-density condition supplies a sequence $u_n \in D$ with $\|u_n-u\|_1 \to 0$. But by the (clearly equivalent) description of τ_1 as a topology such that all $B \in M$ map (D,τ_1) continuously into E, each $(\mathrm{ad}\ A-\alpha_j)^k(P_jB)u_n \to (\mathrm{ad}\ A-\alpha_j)^k(P_jB)u$ in E. Consequently, since the powers $R(\lambda+\alpha_j,A)^{k+1}$ of the resolvents are bounded, every summand $(-1)^k\ R(\lambda+\alpha_j,A)^{k+1}(\mathrm{ad}\ A-\alpha_j)^k(P_jB)u_n$ converges to the corresponding limit, and the right-hand side of (5) for u is seen to be the limit of the corresponding expressions for the u_n. But $\|u_n-u\|_1 \to 0$ implies that $\|u_n-u\| \to 0$ in E, so that $R(\lambda,A)u_n \to R(\lambda,A)u$ in E too. Hence, if B is closable, the validity of (5) for the u_n and the convergence of the right-hand side puts $R(\lambda,A)u \in D(\bar{B})$, establishing both claims in the theorem.

Returning to identity (1), we check by similar methods that slightly stronger restrictions upon λ yield that identity as well.

5.2. Theorem

Let $A \in \mathbf{A}(D)$ and $M \subset \mathbf{A}(D)$ be as in Theorem 5.1. Suppose that for some $\lambda \in \mathbb{C}$ such that $\mathrm{dist}(\lambda,\sigma(A)) > \max\{|\alpha_j|\colon \alpha_j \in \sigma_M(\mathrm{ad}\ A)\}$ the space $D_\lambda = (\lambda-A)D$ is τ_1-dense in D. Then for all $u \in D$ and closable $B \in M$ we have $R(\lambda,A)u \in D(\bar{B})$ and

$$\bar{B}\,R(\lambda,A)u = \Sigma\{(-1)^k R(\lambda,A)^{k+1}(\mathrm{ad}\ A)^k(B)u\colon 0 \le k < \infty\}. \tag{1}$$

Proof: As in Theorem 5.1, the first step is algebraic. Applying (7) with $\alpha = 0$ and $u \in D_\lambda$, we have for every integer n

$$B\,R(\lambda,A)u = \Sigma\{(-1)^k\ R(\lambda,A)^{k+1}(\mathrm{ad}\ A)^k(B)u\colon 0 \le k \le n\}$$
$$+ (-1)^{n+1}\ R(\lambda,A)^{n+1}(\mathrm{ad}\ A)^{n+1}(B)\ R(\lambda,A)u. \tag{9}$$

Next, we check that as $n \to \infty$ the remainder converges too, yielding (1) for the special case $u \in D_\lambda$. In fact, we show that for suitable $M < \infty$ and $r < 1$, it is true for all $v \in D$ and $k = 1,2,\ldots$ that

$$\|R(\lambda,A)^k (\text{ad } A)^k (B)v\| \leq Mr^k \|v\|_1. \tag{10}$$

For $v = R(\lambda,A)u$ and $k = n+1$, this immediately supplies the desired geometric vanishing of the remainder as $n \to \infty$. It is convenient to norm $\mathcal{M}$ by selecting a basis $B_1,\ldots,B_d$ and putting $|\Sigma \beta_i B_i| = \Sigma|\beta_i|$, defining $|(\text{ad } A)^k|$ as the corresponding operator norm. Then for $v \in D$ we get from $(\text{ad } A)^k(B) = \Sigma[(\text{ad } A)^k(B)]_i B_i$ that

$$\|R(\lambda,A)^k(\text{ad } A)^k(B)v\| \leq \|R(\lambda,A)^k\| \, \|\Sigma[(\text{ad } A)^k(B)]_i B_i v\|$$
$$\leq \|R(\lambda,A)^k\| \Sigma |[(\text{ad } A)^k(B)]_i| \, \|B_i v\| \leq \|R(\lambda,A)^k\| \, |(\text{ad } A)^k(B)| \, \|v\|_1$$
$$\leq \|R(\lambda,A)^k\| \, |(\text{ad } A)^k| \, |B| \, \|v\|_1. \tag{11}$$

Thus, taking $M = |B|$, it suffices to check that eventually $\|R(\lambda,A)^k\| \, |(\text{ad } A)^k| < r^k$, or $\|R(\lambda,A)^k\|^{1/k} |(\text{ad } A)^k|^{1/k} < r$, for some $r < 1$. By Gelfand's spectral radius formula, this is clearly equivalent to the check that the product of the spectral radius $\nu(R(\lambda,A)) = \lim\{\|R(\lambda,A)^k\|^{1/k} : k \to \infty\}$ and $\nu(\text{ad } A) = \lim\{|(\text{ad } A)^k|^{1/k} : k \to \infty\}$ is strictly dominated by 1. Now, $\nu(\text{ad } A) = \max\{|\alpha_j| : \alpha_j \in \sigma_{\mathcal{M}}(\text{ad } A)\}$, and we have by hypothesis that $1 > \text{dist}(\lambda,\sigma(A))^{-1} \max\{|\alpha_j|\}$ so the problem reduces to the verification that

$$\nu(R(\lambda,A)) = \text{dist}(\lambda,\sigma(A))^{-1} \tag{12}$$

This folk-identity can be read off from the fact that both sides are in fact the reciprocal of the radius of convergence of the Neumann expansion for $R(\mu,A)$ about the point λ. (On the left, use Cauchy's root test, and on the right note that the spectral point on the circle of radius $\text{dist}(\lambda,\sigma(A))$ is the nearest singularity of the function $\mu \to R(\mu,A)$.)

Having established (1) for $u \in D_\lambda$, it remains to extend this identity to all of D by τ_1-limits using the graph-density condition. By the argument used in Theorem 5.1, we see that if $u \in D$ and $\|u_n - u\|_1 \to 0$ for $u_n \in D$, then all terms $(-1)^k R(\lambda,A)^{k+1}(\text{ad } A)^k(B)u_n$ in the series on the right in (1) converge to the appropriate limits. But the estimate (10), applied to $u_n = v$, shows that the sum converges geometrically like Σr^k, uniformly in n (since $\{u_n\}$ is $\|\cdot\|_1$-Cauchy), so a routine application of the vector-valued Lebesque Dominated Convergence Theorem (for the counting measure on $\{0,1,\ldots\}$) shows that the sums for the u_n must converge to the sum for u. Closedness of $\bar{B}$ and boundedness of $R(\lambda,A)$ then combine as in Theorem 5.1 to ensure the convergence of $B\,R(\lambda,A)u_n$ to $\bar{B}\,R(\lambda,A)u$ on the left, establishing the full generality of (1).

Several of the algebraic identities obtained in the proofs of Theorems 5.1 and 5.2 are of sufficient interest that we extract them and record their extensions formally for later reference.

5.3. Corollary

Suppose that A, B and λ satisfy the conditions of Theorem 5.1 (or 5.2).
(a) For every $u \in D$, the third resolvent identity holds:

$$[\bar{B}, R(\lambda,A)]u = \bar{B}\,R(\lambda,A)u - R(\lambda,A)Bu = -R(\lambda,A)[\bar{A},\bar{B}]R(\lambda,A)u. \quad (13)$$

(b) If $\mu \in \rho(B)$ as well, then

$$[R(\lambda,A),R(\mu,B)] \supseteq R(\mu,B)R(\lambda,A)[\bar{A},\bar{B}]R(\lambda,A)R(\mu,B) \quad (14)$$

and the operator on the right is densely defined.

Proof: (a) Given the fact that $R(\lambda,A)$ maps D into $D(\bar{B})$, and that it sends all of E onto $D(\bar{A})$, (13) can be obtained essentially as in (6). Alternatively, we recall that

$$\bar{A}\,R(\lambda,A) = -(\lambda-\bar{A})R(\lambda,A)+\lambda R(\lambda,A)$$

$$= -I+\lambda R(\lambda,A) \supseteq -R(\lambda,A)(\lambda-A)+\lambda R(\lambda,A) = R(\lambda,A)A. \quad (15)$$

Consequently, for $u \in D$

$$R(\lambda,A)[\bar{A},\bar{B}]R(\lambda,A)u = ((-I+\lambda R(\lambda,A))\bar{B}R(\lambda,A)-R(\lambda,A)\bar{B}(-I+\lambda R(\lambda,A)))u. \tag{16}$$

Cancellation on the right and division by (-1) yields (13).

In (b), (14) is actually true without any hypotheses of graph-density type so the interest is contained in the claim that the right-hand side is densely-defined (in fact, its domain contains $(\mu-B)D$, which is dense since $(\mu-\bar{B})D(\bar{B})=E$ and D is a core for $\bar{B}$, cf. Lemma 5.6). The computation is an elaboration of the one given above. That is, v is in the domain of the right-hand side of (14) if and only if $R(\lambda,A)R(\mu,B)v \in D(\bar{A}\bar{B}) \cap D(\bar{B}\bar{A})$. Then

$$R(\mu,B)R(\lambda,A)(\bar{A}\bar{B}-\bar{B}\bar{A})R(\lambda,A)R(\mu,B)v$$

$$= R(\mu,B)\{(-I+\lambda R(\lambda,A))\bar{B}R(\lambda,A)-R(\lambda,A)\bar{B}(-I+\lambda R(\lambda,A))\}R(\mu,B)v$$

$$\overset{(*)}{=} R(\mu,B)\{R(\lambda,A)\bar{B}-\bar{B}R(\lambda,A)\}R(\mu,B)v$$

$$= R(\mu,B)R(\lambda,A)(-I+\mu R(\mu,B))v-(-I+\mu R(\mu,B))R(\lambda,A)R(\mu,B)v$$

$$\overset{(+)}{=} -[R(\mu,B),R(\lambda,A)]v. \tag{17}$$

The crucial steps in these calculations (*) and (+), where operator sums have been distributed, are all justified by observing that the vectors produced by the summand are separately in the domains of the unbounded operators to their left. For example, both $(-1)R(\mu,B)v$ and $\lambda R(\lambda,A)R(\mu,B)v$ are in $D(\bar{B})$, since $R(\mu,B)v$ automatically is and $R(\lambda,A)R(\mu,B)v$ must be in order for that vector to be in $D(\bar{A}\bar{B})$. We leave remaining details to the reader. E.O.P.

<u>Remark</u>: The "resolvent commutation relation with remainder", obtained from (7) when $\alpha = 0$, has an extension to all of D essentially as in (13). It can be obtained from (13) or from (7). We omit details.

5B. Commutation relations on D_1

Theorems 5.1 and 5.2 contain all of the algebraic information required for commutation theory. But they omit certain topological and geometric aspects of the subject that are essential for a clean transition to semigroups, operational calculus, and Lie algebra exponentials. The additional information is couched in terms of a generalized Sobolev space $(D_1, \|\cdot\|_1)$ of "C^1-vectors" with a topology τ_1 of "convergence of first derivatives". If M has a basis of closable operators $B_1, \ldots, B_d$ (as it does in many applications) this space is easily described, as we do below. A more technical substitute is available even if M is not known to have such a basis, but we defer its discussion to Chapter 6 for expository reasons.

If M has a basis $\mathbf{B} = \{B_1, \ldots, B_d\}$ of closable operators, we put $E_1(\mathbf{B}) = \cap\{D(\bar{B}_i): 1 \le i \le d\}$ and norm this space as usual by $\|u\|_1 = \max\{\|\bar{B}_i u\| : 0 \le i \le d\}$ with $B_0 u = u$. Then D_1 is the closure of D in $E_1(\mathbf{B})$ with respect to this norm. It is well-known that since the $\bar{B}_i$ are closed, both $(E_1(\mathbf{B}), \|\cdot\|_1)$ and its subspace D_1 are Banach spaces, and that D_1 is independent of the particular closable basis chosen for M. (Recall that a sequence $\{u_n\}$ is $\|\cdot\|_1$-Cauchy if and only if u_n and all $\bar{B}_i(u_n)$ are $\|\cdot\|_1$-Cauchy, and a comparable remark holds for $\|\cdot\|_1$-convergence to $u \in E_1(\mathbf{B})$. Hence, the $\|\cdot\|$-limit of a $\|\cdot\|_1$-Cauchy sequence is easily seen by closedness to be the $\|\cdot\|_1$-limit. The fact that D_1 is independent of the choice of $\mathbf{B}$ is an easy consequence of the observation that every $B \in M$ is a bounded map from $(D, \|\cdot\|_1)$ to $E, \|\cdot\|)$ with respect to any one of the basis norms $\|\cdot\|_1$, whence any norm defined by a second basis can be estimated in terms of the given one. Consequently all are equivalent on D and yield the same completion.)

We shall also need the fact that any $\|\cdot\|$-closable operator $A \in \mathbf{A}(D)$ also has a $\|\cdot\|_1$-closure $\bar{A}_1$ in D_1 when D is viewed as a subspace of D_1. (If we form the relative $\|\cdot\| \times \|\cdot\|$ closure of Graph (A) in $D_1 \times D_1$, the result is the graph of a restriction of $\bar{A}$. But this contains the necessarily smaller $\|\cdot\|_1 \times \|\cdot\|_1$-closure of Graph (A) in $D_1 \times D_1$, so the latter is the graph of an operator, namely $\bar{A}_1$.)

The main result of this section then follows in a relatively straightforward way from Theorems 5.1 and 5.2.

5.4. Theorem

Let $A \in \mathbf{A}(D)$ be a closable operator, and let $M \subset \mathbf{A}(D)$ be a finite-dimensional complex ad A-invariant subspace of $\mathbf{A}(D)$. Suppose that M has a closable basis and that $D_\lambda = (\lambda-A)D$ is τ_1-dense in D.

(1) If $\lambda \in \rho(A;M)$, then the resolvent $R(\lambda,A) = (\lambda-\bar{A})^{-1}$ leaves D_1 invariant and restricts there to a $\|\cdot\|_1$-bounded resolvent $R_1(\lambda,A) = (\lambda-\bar{A}_1)^{-1}$ for $\bar{A}_1$ that satisfies the following commutation relation in $L(D_1,E)$ with respect to the bounded extension $B^{(1)}$ of any $B \in M$ to D_1:

$$B^{(1)}R_1(\lambda,A) = \Sigma\{(-1)^k\ R(\lambda+\alpha_j,A)^{k+1}[(\text{ad }A-\alpha_j)^k(P_jB)]^{(1)}:\ 1 \le j \le p\ ,\ 0 \le k \le s_j\}. \tag{18}$$

(2) If instead dist $(\lambda,\sigma(A)) > \max\{|\alpha_j|: 1 \le j \le p\}$ then

$$B^{(1)}R_1(\lambda,A) = \Sigma\{(-1)^k\ R(\lambda,A)^{k+1}[(\text{ad }A)^k(B)]^{(1)}:\ 0 \le k < \infty\}. \tag{19}$$

(3) Moreover, if $\|R(\lambda,A)\|\ |\text{ad }A| < 1$, then the operator norm of $R_1(\lambda,A)$ in $L(D_1)$ admits the following estimate

$$\|R_1(\lambda,A)\|_1 \le \|R(\lambda,A)\|(1 - \|R(\lambda,A)\|\ |\text{ad }A|)^{-1}. \tag{20}$$

Remark: A variety of weaker estimates for $\|R_1(\lambda,A)\|_1$ can be obtained in the greater generality of (1) and (2) above, but (20) is essentially best-possible. It is needed in Chapter 6 in connection with the lifting of sharp Hille-Yosida generation conditions on E to comparable conditions on D_1.

Proof: Applying (5) of Theorem 5.1 to a basis element B_i and a vector $u \in D$, we get

$$\begin{aligned}\|B_iR(\lambda,A)u\| &\le \Sigma\{\|R(\lambda+\alpha_j,A)^{k+1}\|\ |(\text{ad }A-\alpha_j)^k|\ |P_jB_i|\ \|u\|_1:\ 1 \le j \le p\ ,\ 0 \le k \le s_j\} \\ &\le C(i)\|u\|_1\end{aligned} \tag{21}$$

essentially as in the verification of (11). Since $\|u\| \leq \|u\|_1$, we get by taking maxima that

$$\|R(\lambda,A)u\|_1 \leq \max\{\|R(\lambda,A)\|, C(i): 1 \leq i \leq d\}\|u\|_1. \tag{22}$$

Thus $R(\lambda,A)$ acts as a $\|\cdot\|_1$-bounded operator from D_λ onto D. By the graph-density condition, it is clear that D_λ is $\|\cdot\|_1$-dense in D_1, whence since D_1 is complete, $R(\lambda,A)$ admits a $\|\cdot\|_1$-bounded extension-by-limits to an endomorphism $R_1(\lambda,A)$ of $(D_1,\|\cdot\|_1)$. But this stronger $\|\cdot\|_1$-limit $R_1(\lambda,A)u$ must agree with the $\|\cdot\|$-limit $R(\lambda,A)u$ on any $u \in D_1$, whence it follows that $R(\lambda,A)$ leaves D_1 invariant.

Similar limit arguments confirm that $R_1(\lambda,A)$ is a two-sided inverse for the $\|\cdot\|_1$-closed operator $\lambda - \bar{A}_1$. For example, if $u \in D(\bar{A}_1)$, then there exists a sequence $\{u_n\} \subset D$ with $\|u_n-u\|_1 \to 0$ and $\|Au_n-\bar{A}_1u\|_1 \to 0$. But since each $u_n \in D$ we have

$$R_1(\lambda,A)(\lambda-\bar{A}_1)u_n = R(\lambda,A)(\lambda-A)u_n = u_n \to u \tag{23}$$

(in the $\|\cdot\|_1$-sense) while since $(\lambda-\bar{A}_1)u_n \to (\lambda-\bar{A}_1)u$ ($\|\cdot\|_1$-sense), and $\|\cdot\|_1$-boundedness of $R_1(\lambda,A)$ thus ensures that $R_1(\lambda,A)(\lambda-\bar{A}_1)u_n \to R_1(\lambda,A)(\lambda-\bar{A}_1)u$ ($\|\cdot\|_1$-sense). Thus $R_1(\lambda,A)$ is a left-inverse for $(\lambda-\bar{A}_1)$. The right inverse verification proceeds instead by approaching an arbitrary $u \in D_1$ by a sequence $\{u_n\}$ in $\|\cdot\|_1$-dense D so that $(\lambda-\bar{A}_1)R_1(\lambda,A)u_n = (\lambda-A)R(\lambda,A)u_n = u_n \to u(\|\cdot\|_1)$. Then $\|\cdot\|_1$-boundedness of $R_1(\lambda,A)$ and $\|\cdot\|_1$-closedness of $\lambda - \bar{A}_1$ shows as before that $(\lambda-\bar{A}_1)R_1(\lambda,A)u_n \to (\lambda-\bar{A}_1)R_1(\lambda,A)u$ as needed.

Identity (18) can be obtained from (5) in Theorem 5.1 either by extension from D_λ or by representing each $B^{(1)} = \Sigma\{\beta_i\bar{B}_i: 1 \leq i \leq d\}$ in terms of the closable basis elements. Identity (19) follows in the same way from (1) in Theorem 5.2. We leave details to the reader.

To obtain the estimate (20), we bound the terms in the series (19) for a basis element B_i as follows:

$$\|(-1)^k R(\lambda,A)^{k+1}(\text{ad } A)^k(B_i)u\| \le \|R(\lambda,A)\|^{k+1}|\text{ad } A|^k|B_i|\,\|u\|_1. \quad (24)$$

Then since $B_i^{(1)}v = \bar{B}_i v$ for $v \in D_1$ and $|B_i| = 1$ we get

$$\|\bar{B}_i R_1(\lambda,A)u\| \le \|R(\lambda,A)\|\Sigma\{(\|R(\lambda,A)\|\,|\text{ad } A|)^k : 0 \le k < \infty\}\,\|u\|_1$$
$$= \|R(\lambda,A)\|(1 - \|R(\lambda,A)\|\,|\text{ad } A|)^{-1}\|u\|_1. \quad (25)$$

But

$$\|\bar{B}_0 R_1(\lambda,A)u\| = \|R(\lambda,A)u\| \le \|R(\lambda,A)\|\,\|u\|$$
$$\le \|R(\lambda,A)\|(1 - \|R(\lambda,A)\|\,|\text{ad } A|)^{-1}\|u\|_1 \quad (26)$$

trivially so that upon taking maxima over $0 \le i \le d$ we get

$$\|R_1(\lambda,A)u\|_1 \le \|R(\lambda,A)\|(1 - \|R(\lambda,A)\|\,|\text{ad } A|)^{-1}\|u\|_1 \quad (27)$$

and (20) follows. E.O.P.

Using this result, we are able to obtain the following necessity-sufficiency result. It serves to indicate that various subsequent theorems are best-possible, and it facilitates the analytic continuation argument in the next section.

5.5. Proposition

Let A and M be as in Theorem 5.4. Suppose that $\lambda \in \rho(A;M)$. Then the following are equivalent.

(a) $D_\lambda = (\lambda - A)D$ is $\|\cdot\|_1$-dense in D.

(b) $R(\lambda,A)$ leaves D_1 invariant and restricts to a two-sided inverse for $\lambda - \bar{A}_1$ there.

Remark: Note that this does not quite imply that the graph-density condition is equivalent to (1), since (1) by itself ensures only that $R(\lambda,A)$ sends D_1 into $E_1(\mathbf{B})$, which may be properly larger.

Proof: That (a) implies (b) is precisely the first claim in Theorem 5.4 (1). For the converse, we first observe that since $R(\lambda,A)$ is bounded, it is relatively $\|\cdot\| \times \|\cdot\|$-closed ond D_1, hence its restriction $R_1(\lambda,A)$ is necessarily $\|\cdot\|_1 \times \|\cdot\|_1$-closed as well. It follows from the closed graph theorem that $R_1(\lambda,A)$ is bounded on D_1 and is the resolvent of $\bar{A}_1$ at λ. The rest of the argument uses the following folk-lemma, which we also need in Chapter 11.

5.6. Lemma

Let T be a closed, densely-defined operator in a Banach space F, and suppose that T has a bounded inverse. Let D be a dense subspace of D(T). Then the following are equivalent.
(1) D is a core for T (i.e. T is the closure of its restriction to D).
(2) D is dense in D(T) with respect to the graph norm $\|u\|_T = \max\{\|u\|, \|Tu\|\}$.
(3) TD is dense in F.
(4) $D = T^{-1}F_0$ for some dense subspace F_0 of F.

Proof: (1) $\Leftrightarrow$ (2) is immediate: $\|u_n - u\|_T \to 0$ for $u \in D(T)$, $u_n \in D$ if and only if $u_n \to u$ and $Tu_n \to Tu$. If (2) holds, then the facts that $T(D(T)) = F$ and T is bounded from $(D(T), \|\cdot\|_T)$ to F combine to show that T sends D into a $\|\cdot\|$-dense subspace. Claim (4) is just a rephrasing of (3). Closing the circle, if (4) holds, let $u \in D(T)$, $v = Tu$ and $v_n \in F_0$ with $v_n \to v$. Then $u_n = T^{-1}v_n \to u = T^{-1}v$ by boundedness of T^{-1}, so $Tu_n = v_n \to v = Tu$ and T is the closure of its restriction to D.

Completing the proof of Proposition 5.5, we take $T = \lambda - \bar{A}_1$ in $F = D_1$. By definition, D is a core for $\bar{A}_1$, hence for T and (1) holds in Lemma 5.6, whence by (3) there, $TD = (\lambda - \bar{A}_1)D = D_\lambda$ is dense.

5C. Analytic continuation of commutation relations

This section concerns the dependence of the results in the last two upon the choice of $\lambda \in \rho(A;M)$, observing that these propogate to nearby μ, hence into components of $\rho(A;M)$.

5.7. Corollary

Let A and M be as in Theorem 5.4.
(a) The set of points $\lambda \in \rho(A;M)$ where the graph-density condition holds is open in $\mathbb{C}$ and relatively closed in $\rho(A;M)$.

(b) In any component U of $\rho(A;M)$, the conclusions of Theorem 5.4 hold either for all $\lambda \in U$ or for none.

Proof: (a) Openness follows from Proposition 5.5, since the graph-density condition holds for λ if and only if λ is in the (open) resolvent set $\rho(\bar{A}_1)$ of $\bar{A}_1$. For closedness, suppose that $\lambda \in \rho(A;M)$ and that $\lambda_n \to \lambda$ with each λ_n satisfying the graph-density condition. Then by Proposition 5.5 (a) ⇒ (b) the sequence $\{\lambda_n\}$ is contained in $\rho(\bar{A}_1)$, and by the first resolvent identity in $L(D_1)$,

$$R_1(\lambda_n,A) - R_1(\lambda_m,A) = (\lambda_m-\lambda_n)R_1(\lambda_n,A)R_1(\lambda_m,A).$$

Now, by (20) the $\|\cdot\|_1$-operator norm of $R_1(\lambda_n,A)$ is bounded uniformly in n, since $\lambda \in \rho(A) \supset \rho(A;M)$. Whence the sequence of operators $\{R_1(\lambda_n,A)\} \subseteq L(D_1)$ is $\|\cdot\|_1$-Cauchy, and hence convergent by completeness of the Banach algebra $L(D_1)$. Finally, one uses closedness of $\bar{A}_1$ to show that the limit is in fact a two-sided inverse for $\lambda - \bar{A}_1$, so by Proposition 5.5 (b) ⇒ (a), D_λ is $\|\cdot\|_1$-dense in D, and all other conclusions of Theorem 5.4 follow from this. The component claim in (b) is immediate from (a). E.O.P.

We leave to the reader to verify that the corollary is in fact a consequence of the following (more or less well known) one-operator Lemma, which is going to be used later.

5.7'. Lemma

Let T be closed operator in a Banach space F.
(i) Then the resolvent set

$$\rho(T) = \{\lambda \in \mathbb{C} : (\lambda-T)^{-1} \in L(F)\}$$

is open.
(ii) Let D be a core for T, and let h be a locally bounded function defined on a subset Ω of the complex plane (i.e., h is bounded on every compact $\subset\Omega$). Suppose that

$$h(\lambda)\|(\lambda-T)u\| \geq \|u\|$$

for all $u \in D$ and $\lambda \in \Omega$.

Then the set of points $\lambda \in \Omega$ where $(\lambda - T)D$ is dense in F is relatively closed in Ω. E.O.L.

In most applications of Lemma 5.7' F is D_1 and T is $\bar{A}_1$.

Remarks: (1) Notice that in the important special case where ad A is nilpotent on M, $\rho(A;M) = \rho(A)$ whence it emerges that the resolvent set $\rho(\bar{A}_1)$ of A on D_1 contains a union of components of $\rho(A)$. (Frequently, the two resolvent sets coincide.)
(2) It is still an open question as to whether the resolvent set of $\bar{A}_1$ might in some cases be larger than the present result suggests. Could there be points in $\rho(A) \cap \sigma(A;M)$ that are in $\rho(\bar{A}_1)$? Indeed, could $\rho(\bar{A}_1)$ contain points in $\sigma(A)$?

5D. Commutation relations for the holomorphic operational calculus

Our purpose in this section is to apply resolvent commutation results and the extended holomorphic operational calculus to obtain a rigorous formulation and proof of the commutation identity

$$B\varphi(A) = \Sigma\{(-1)^k/k!\ \varphi^{(k)}(A)(\text{ad } A)^k(B) : 0 \le k < \infty\}, \tag{28}$$

for the operational calculus in Banach algebras $\mathfrak{U}$. Again, the basic reference is Chapter VII of Dunford and Schwartz [DS 1]. We recall for reference that the extended spectrum $\sigma^*(A) = \sigma(A) \cup \{\infty\}$ if A is unbounded, and that for φ holomorphic in a neighborhood U of $\sigma^*(A)$ in the Riemann sphere $\mathbb{C}^*$ we have for $A \in \mathbf{A}(D)$ with $R(\lambda,A) = (\lambda-\bar{A})^{-1}$ that

$$\varphi(A) = \varphi(\infty)+(2\pi i)^{-1}\int_\Gamma \varphi(\lambda)R(\lambda,A)d\lambda \tag{29}$$

for any choice of Γ a finite union of Jordan arcs positively enclosing $\sigma^*(A)$ (index +1 about each $\lambda \in \sigma^*(A)$), lying within $\rho(A) \cap U$. The following commutation relations for $\varphi(A)$ drop quite quickly out of the $L(D_1,E)$ forms of the resolvent commutation relations as given in Theorem 5.4, illustrating some of the ways in which that set-up permits bounded operator arguments to be carried over from the Banach algebra case.

5.8. Theorem

Let $A \in \mathbf{A}(D)$ be closable and let $M \subset \mathbf{A}(D)$ be a finite-dimensional complex ad A-invariant subspace with closable basis. Let φ be holomorphic in a neighborhood U of $\sigma^*(A)$.
(a) If U contains $\sigma(A;M)$ and if there exists at least one Γ in U as above that is contained entirely within components of $\rho(A;M)$ where the graph-density condition holds, then $\varphi(A)$ leaves D_1 invariant and restricts there to a $\|\cdot\|_1$-bounded operator $\varphi_1(A) \in L(D_1)$ which satisfies

$$B^{(1)}\varphi_1(A) = \Sigma\{(-1)^k/k!\ \varphi^{(k)}(A-\alpha_j)[(\text{ad } A-\alpha_j)^k(P_jB)]^{(1)} : 1 \leq j \leq p\ ,\ 0 \leq k \leq s_j\} \tag{30}$$

for every $B \in M$.
(b) If in fact U contains a Γ such that the graph-density condition holds everywhere on Γ and $\text{dist}(\Gamma,\sigma(A)) > \max\{|\alpha_j| : 1 \leq j \leq p\}$, then in addition

$$B^{(1)}\varphi_1(A) = \Sigma\{(-1)^k/k!\ \varphi^{(k)}(A)[(\text{ad } A)^k(B)]^{(1)} : 0 \leq k < \infty\} \tag{31}$$

for all $B \in M$.

Proof: (a) By Proposition 5.5, the fact that the graph-density condition holds for all $\lambda \in \Gamma$ implies that Γ is in the resolvent set $\rho(\bar{A}_1)$ of $\bar{A}_1$ as an operator in D_1, whence formula (29) restricts on D_1 to the corresponding integral for $\varphi(\bar{A}_1) = \varphi_1(A) \in L(D_1)$:

$$\varphi_1(A) = \varphi(\infty)+(2\pi i)^{-1}\int_\Gamma \varphi(\lambda)R_1(\lambda,A)d\lambda. \tag{29_1}$$

This $\|\cdot\|_1$-operator-norm-convergent integral converges $\|\cdot\|_1$-strongly and hence $\|\cdot\|$-strongly as applied to any $u \in D_1$, hence must agree on D_1 with (29) defining $\varphi(A)$, so $\varphi(A)$ restricts to $\varphi_1(A)$ on D_1.

Applying the bounded operator $B^{(1)} \in L(D_1,E)$ to both sides in (29), we see immediately that it may be taken inside the integral. Substituting (18) from Theorem 5.4 and applying the linearity of integration, we get

$$B^{(1)}\varphi_1(A) = \varphi(\infty)B^{(1)}+\Sigma\{(-1)^k[(2\pi i)^{-1}\int_\Gamma \varphi(\lambda)R(\lambda+\alpha_j,A)^{k+1}d\lambda]$$
$$\times [(\text{ad } A-\alpha_j)(P_jB)]^{(1)}: 1 \le j \le p,\ 0 \le k < \infty\}. \quad (32)$$

In order to transform (32) into (30), we first observe that $B^{(1)} = \Sigma\{(P_jB^{(1)}: 1 \le j \le p\}$ since $\Sigma' P_j = 1$ on M, and $R(\lambda+\alpha_j,A) = (\lambda+\alpha_j-\bar{A})^{-1} = R(\lambda,A-\alpha_j)$, whence by combining the $\varphi(\infty)B^{(1)}$ term with the $k = 0$ summands in the second expression and applying (29) to $A - \alpha_j$, (32) becomes

$$B^{(1)}\varphi_1(A) = \Sigma\{\varphi(A-\alpha_j)(P_jB)^{(1)}: 1 \le j \le p\}$$
$$+ \Sigma\{(-1)^k[(2\pi i)^{-1}\int_\Gamma \varphi(\lambda)R(\lambda,A-\alpha_j)^{k+1}d\lambda]$$
$$\times [(\text{ad } A-\alpha_j)^k(P_jB)]^{(1)}: 1 \le j \le p,\ 1 \le k \le s_j\}. \quad (33)$$

The final step involves the substitution of the identity

$$(2\pi i)^{-1}\int_\Gamma \varphi(\lambda)R(\lambda,A-\alpha_j)^{k+1}d\lambda = 1/k!\ \varphi^{(k)}(A-\alpha_j) \quad (34)$$

for $1 \le k \le s_j$. This identity is the unbounded operator version of the identity given on p. 591 of [DS 1], but the treatment of the point $\infty \in \sigma^*(A-\alpha_j)$ requires special care in the unbounded case, so we indicate details here for an alternative argument. Put $C = A - \alpha_j$ and observe that since U includes $\sigma(A;M)$ and Γ avoids $\sigma(A;M)$ it follows that $\sigma(C) \subset U$, $\Gamma \cap \sigma(C) = \emptyset$. First, note that by (29)

$$\varphi^{(k)}(C) = (2\pi i)^{-1}\int_\Gamma \varphi^{(k)}(\lambda)R(\lambda,C)d\lambda \quad (35)$$

since analyticity of φ at ∞ forces $\varphi^{(k)}(\infty) = 0$ for $k \ge 1$. (If $\varphi(\lambda) \to \varphi(\infty)$ as $\lambda \to \infty$, the Laurent series for $\varphi(\lambda)$ can have no positive powers of λ, hence those for $\varphi^{(k)}(\lambda)$ vanish to (k-1)st order at ∞.) Taking Γ' within the domain of homomorphy of φ and bounding a neighborhood of $\Gamma \cup \sigma(C)$, we can apply the Cauchy integral formula and Fubini's theorem to obtain

$$\varphi^{(k)}(C) = (2\pi i)^{-2}\int_{\Gamma}\int_{\Gamma'} k!\ \varphi(\mu)(\mu-\lambda)^{-(k+1)}d\mu\ R(\lambda,C)d\lambda$$
$$= (2\pi i)^{-2}\int_{\Gamma'} k!\ \varphi(\mu)[\int_{\Gamma}(\mu-\lambda)^{-(k+1)}R(\lambda,C)d\lambda]d\mu$$
$$= (2\pi i)^{-1}\int_{\Gamma'} k!\ \varphi(\mu)R(\mu,A)^{k+1}d\mu. \tag{36}$$

(Here, the function $\psi_{\mu}^{k}(\lambda) = (\mu-\lambda)^{-(k+1)}$ is analytic inside Γ', hence on a neighborhood of Γ, and $\psi_{\mu}^{k}(A) = R(\mu,A)^{k+1}$ is easily read off from arguments in VII.9 of [DS 1].) Then (34) follows from (36) by substitution and division by k!, whence (30) follows in turn from (33) and (34).

The argument for (b) uses (19) in place of (18), and the term-by-term substititions proceed exactly as in (a). The new feature in (31) involves convergence of the series and a Fubini argument for interchange of sums and integrals. That is, one argues just as in (33) that

$$B^{(1)}\varphi_1(A) = \varphi(\infty)B^{(1)}+(2\pi i)^{-1}\int_{\Gamma}\varphi(\lambda)R(\lambda,A)d\lambda\ B^{(1)}$$
$$+\ (2\pi i)^{-1}\int_{\Gamma}\Sigma(-1)^k\ \varphi(\lambda)R(\lambda,A)^{k+1}[(\text{ad } A)^k(B)]^{(1)}:$$
$$1 \le k < \infty\}. \tag{37}$$

Interchange of the integral and sum, followed by substitution from (29) and (34), will yield (31) as claimed. The necessary convergence data are obtained exactly as in Lemma VII.6.11 of [DS 1], where it is established for normally <u>bounded</u> A that if $\text{dist}(\Gamma,\sigma(A)) > \varepsilon$ then there exists a uniform bound M such that $\|R(\lambda,A)^k\| \le M\varepsilon^{-k}$ for all $k = 0,1,\ldots$. Inspection of the proof there reveals that boundedness of A is not required. (One simply expresses $R(\lambda,A)^k = (2\pi i)^{-1}\int_{\Gamma'}R(\lambda-\mu)^k R(\mu,A)d\mu$ as above, and estimates $\|R(\lambda,A)^k\|$ in terms of $\sup\{\|R(\mu,A)\|: \mu \in \Gamma'\}$.) Here, we use $\text{dist}(\Gamma,\sigma(A)) > \max\{|\alpha_j|: 1 \le j \le p\}$ to choose ε as follows. Noting that the spectral radius $\max\{|\alpha_j|\} = \nu(\text{ad } A) = \lim\{|(\text{ad } A)^n|^{1/n}\}$, we can find t so that $\text{dist}(\Gamma,\sigma(A)) > t > |(\text{ad } A)^n|^{1/n}$ eventually $(n > n_0)$. If we then choose ε by $\operatorname{dist}(\Gamma,\sigma(A)) > \varepsilon > t$ and argue as in the calculation of (11) (proof of Theorem 5.3), we get the estimate for $u \in D_1$:

$$\|R(\lambda,A)^{k+1}[(\operatorname{ad} A)^k(B)]^{(1)}u\| \le M\varepsilon^{-(k+1)}|\operatorname{ad} A^k| \, |B| \, \|u\|_1$$
$$\le M\varepsilon^{-1}(t/\varepsilon)^k|B| \, \|u\|_1$$

for $k > n_0$, uniformly in $\lambda \in \Gamma$. Since $|\varphi(\lambda)|$ is bounded on Γ, this yields a geometric bound on the norms of the terms in the series (as applied to $u \in D_1$) that is uniform in $\lambda \in \Gamma$ and is valid beyond a λ-independent integer n_0. Since Γ is of finite measure, this means that Fubini's theorem applies, the interchange of sums and integrals is justified, and (31) follows. E.O.P.

This result then applies to "characteristic functions of spectral sets", yielding spectral reduction. We recall that if $\sigma \subset \sigma^*(A)$ is closed and relatively open in $\sigma^*(A)$, then there exists a finite union Γ of Jordan arcs in $\rho(A)$ which positively encircle σ(index + 1) but enclose no points of its complement $\sigma_C = \sigma(A) \sim \sigma$. Then the operator

$$P_\sigma = (2\pi i)^{-1}\int_\Gamma R(\lambda,A)d\lambda + \chi_\sigma(\infty) \tag{38}$$

(where χ_σ denotes the characteristic function of σ) is a bounded idempotent. It projects onto a subspace E_σ such that $D_\sigma(\bar{A}) = E_\sigma \cap D(\bar{A})$ is dense in E_σ, and the restriction A_σ of A to $D_\sigma(\bar{A})$has exactly σ as spectrum. One ordinarily interprets (38) as an application of (29) to a suitable analytic continuation φ_σ of χ_σ. (Given disjoint neighborhoods $U_\sigma \supset \sigma$ and $U_C \supset \sigma_C$, let $\varphi_\sigma(\lambda) \equiv 1$ on U_σ, $\varphi_\sigma(\lambda) \equiv 0$ on U_C. Then $P_\sigma = \varphi_\sigma(A)$ is easily checked.)

Our results apply to this situation only when σ is <u>sufficient isolated</u>, in the technical sense that no translate $\sigma - \alpha_j$ meets any other translate $\sigma_C - \alpha_k$ of its complement, when α_j, α_k run through the eigenvalues of ad A on M. (As usual, if ad A is nilpotent these technicalities do not arise.)

5.9. Proposition

Let A and M be as in Theorem 5.8. Suppose that σ is a sufficiently isolated spectral set of A. Further, suppose that there exists a contour Γ in $\rho(A)$ enclosing σ and all translates $\sigma - \alpha_j$ of σ positively, but enclosing or meeting no points of σ_C or any translate $\sigma_C - \alpha_k$ of its complement, upon which the graph density condition holds. Then P_σ maps D_1 into D_1 and restricts to a bounded projection $(P_\sigma)_1$ there, such that for all $B \in M$

$$B^{(1)} (P_\sigma)_1 = P_\sigma B^{(1)}. \tag{39}$$

That is, P_σ simultaneously reduces the $B \in M$.

Proof: As usual, we may replace $R(\lambda,A)$ by $R_1(\lambda,A)$ in (38) to obtain $(P_\sigma)_1$ as a bounded operator on D_1, using Theorem 5.4 and Proposition 5.5. The relation (39) is obtained by taking φ to be identically 1 on and inside Γ (hence on σ and all of its translates) and identically 0 on some neighborhood U_C of σ_C and all of its translates, where no point of U_C lies on or inside σ. Such a φ satisfies the condition of Theorem 5.8. Moreover $\varphi^{(k)} = 0$ for all $k > 0$, so that (30) yields

$$B^{(1)} (P_\sigma)_1 = B^{(1)} \varphi_1(A) = \Sigma\{\varphi^{(0)}(A-\alpha_j)(P_jB)^{(1)}: 1 \le j \le p\}. \tag{40}$$

But $\varphi(A-\alpha_j) = P_\sigma$, as we show below, whence since $\Sigma(P_jB)^{(1)} = B^{(1)}$ we get (39). To see that $\varphi(A-\alpha_j) = P_\sigma$, write $\varphi_j(\lambda) = \varphi(\lambda-\alpha_j)$ and note that $\varphi(A-\alpha_j) = \varphi_j(A)$. But φ_j is by construction constantly 1 on a neighborhood of σ and constantly 0 on a neighborhood of σ_C, whence $\varphi_j(A)$ is simply another realization of P_σ. E.O.P.

Remark on the proof: Identity (39) can also be read off from (31). A direct proof of (39) from Theorem 5.4 can be given using only the defining integral (38) for P_σ and a specialization of the proof of Theorem 5.8; such a proof is more economical and instructive than the one given above. However, our primary point in including Proposition 5.9 is to illustrate the potential uses of Theorem 5.8.

In closing, we note that most of the simplifications mentioned above actually occur in the context of the O'Raifeartaigh mass-splitting phenomena discussed in Chapter 4: ad A is nilpotent and if $\sigma(A) \subset \mathbb{R}$ has any nontrivial spectral sets, then its complement has only one component so the graph-density condition holds either everywhere or nowhere. At this writing, we have been able to recover the results of Chapter 4 independently by the methods described above, but no examples are yet known where the potential extra generality of the present methods actually materializes.

Chapter 6

GRAPH-DENSITY APPLIED TO SEMIGROUP COMMUTATION RELATIONS

In this chapter, Laplace inversion techniques of Hille-Yosida type are used to transform the C^1-vector invariance and commutation properties of resolvents (obtained in Chapter 5) into comparable results for semigroups and groups. Modulo the last section, we thus obtain equivalence of infinitesimal 'zero-deficiency' graph-density conditions with the property that the (semi-)groups $\{V(t,A)\}$ restrict to C_0 (semi-)groups $\{V_1(t,A)\}$ acting upon the C^1-vector Banach space $(D_1,\|\cdot\|_1)$. The results obtained here are applied both in constructing globally invariant C^∞ domains in Chapter 7 and in obtaining the graph-density exponentiation Theorem 9.2 in Chapter 9.

This chapter is organized as follows. Section 6A continues the setting employed throughout most of Chapter 5: $A \in \mathbf{A}(D)$ is closable and $M \subset \mathbf{A}(D)$ is a finite-dimensional complex ad A-invariant subspace of $\mathbf{A}(D)$ with a closable basis. The results 5.4 and 5.5 are directly Laplace-inverted to obtain semigroup and group versions, which (as we pointed out in Chapter 5) can be viewed as a further extension of the operational calculus of Section 5D to the exponential function.

In Section 6B, we show how to remodel the results of Chapter 5 and Section 6A in order to treat the case where M is not a-priori known to possess a closable basis. This necessitates the treatment of $(D_1,\|\cdot\|_1)$ as an abstract Banach space which is mapped into E by a continuous homomorphism J_1 with possibly-nontrivial kernel (rather than as an algebraic subspace of E with a stronger topology). Proofs in this section are for the most part sketched, since they follow the same plan as their more natural prototypes treated earlier.

In Section 6C, the machinery of the second section is used bootstrap-fashion to make itself obsolete in the case where A generates a group: it is shown that without loss of generality one may assume that a closable basis is available. One of the remaining open problems of the section concerns the possibility of somehow short-cutting this inefficient-seeming plan of argument: for example, can one show directly without the machinery of 6B, that there must exist a closable basis for the complex ad-orbit $\mathcal{O}_A^{\mathbb{C}}(B)$ of a closable operator B under a group-pregenerator A?

Finally, Section 6D provides an informal discussion of the commutation relations obtained for the extended operational calculi (of Laplace-and Fourier-transform type) which are available when $\bar{A}$ generates a semigroup or group. Formally, these imitate the results in Section 5D.

6A. Semigroup commutation relations with a closable basis

As mentioned above, the present section concerns a closable $A \in \mathbf{A}(D)$ and a finite-dimensional complex ad(A) invariant subspace of $\mathbf{A}(D)$ which is assumed to possess a closable basis $\mathbf{B} = \{B_1,\ldots,B_d\}$ (i.e., each $B_i \in \mathbf{B}$ is closable). We recall that for $u \in E_1(\mathbf{B}) = \cap\{D(B_i): 1 \leq i \leq d\}$, $\|u\|_1 = \max\{\|\bar{B}_i u\|: 0 \leq i \leq d\}$ under the convention $B_0 = I$, and that D_1 is the $\|\cdot\|_1$-closure of D in $(E_1(\mathbf{B}),\|\cdot\|_1)$. Further, $\bar{A}_1$ denotes the $\|\cdot\|_1$-closure of A in $(D_1,\|\cdot\|_1)$ and for $B \in M$, $\bar{B}^{(1)}$ denotes the extension of B by limits to a bounded operator from $(D_1,\|\cdot\|_1)$ to $(E,\|\cdot\|)$. (Recall also that $\|\bar{B}^{(1)}u\| \leq |B|\,\|u\|_1$ for $u \in D_1$, where $|B| = \Sigma|\beta_i|$ when $B = \Sigma\,\beta_i B_i$ gives the norm on M. See Section 5B for details.)

Unlike in Section 5B, we assume here that $\bar{A}$ generates at least a C_0 semigroup $\{V(t,A): t \in [0,\infty)\}$ on the Banach space E. Recall that this semigroup is of type ω if and only if there exists a constant $M < \infty$ such that $\|V(t,A)\| \leq Me^{\omega t}$ for all $t \in [0,\infty)$. Then $\sigma(A) \subset \{\lambda \in \mathbb{C} : \mathrm{Re}(\lambda) \leq \omega\}$. If we write $\nu(\mathrm{ad}\,A) = \max\{|\alpha(j)|: \alpha(j) \in \sigma_M(\mathrm{ad}\,A)\}$ $= \lim\{|\mathrm{ad}\,A^k|^{1/k}: k \to \infty\}$ for the spectral radius of ad A on M, this means that the augmented spectrum $\sigma(A,M)$ $(= \sigma(A) \cup \{\lambda-\alpha(j): \lambda \in \sigma(A), \alpha(j) \in \sigma_M(\mathrm{ad}\,A)$ an eigenvalue of ad A on $M\})$ is entirely contained in the half-plane $H_{\omega+\nu(\mathrm{ad}\,A)} = \{\lambda \in \mathbb{C} : \mathrm{Re}(\lambda) \leq \omega + \nu(\mathrm{ad}\,A)\}$, hence that the complementary half-plane is entirely contained in a single component of the diminished resolvent set $\rho(A;M)$ $= \mathbb{C} \setminus \sigma(A;M)$. Therefore, Corollary 5.7 ensures that if a single λ with $\mathrm{Re}(\lambda) > \omega + \nu(\mathrm{ad}\,A)$ satisfies the graph-density-condition then all λ satisfying this inequality must have the graph-density property. We shall apply this remark below (without explicit citation) at several points.

6.1. Theorem

Suppose that A and M are as described above. Suppose further that for some λ with $\mathrm{Re}(\lambda) > \omega + \nu(\mathrm{ad}\ A)$ the subspace $D_\lambda = (\lambda - A)D$ is $\|\cdot\|_1$-dense in D.

(i) Then each of the operators $V(t,A)$ $(t \in [0,\infty))$ leaves D_1 invariant and restricts to a $\|\cdot\|_1$-bounded operator $V_1(t,A) \in L(D_1)$. Moreover, $\{V_1(t,A): t \in [0,\infty)\}$ acts as a C_0 semigroup on $(D_1,\|\cdot\|_1)$ whose infinitesimal generator is $\bar{A}_1$.

(ii) The semigroups $\{V(t,A): t \in [0,\infty)\}$ and $\{V_1(t,A): t \in [0,\infty)\}$ satisfy the following commutation relations with respect to each $B \in M$, as identities in $L(D_1,E)$

$$\begin{aligned} B^{(1)}V_1(t,A) &= V(t,A)[\exp(-t\ \mathrm{ad}\ A)(B)]^{(1)} \\ &= \Sigma\{(-t)^k/k!\ \exp(-t\alpha(j))V(t,A)[(\mathrm{ad}\ A-\alpha(j))^k(P_jB)]^{(1)}: \\ &\quad 1 \le j \le p\ ;\ 0 \le k \le s_j\} \end{aligned}$$

for all $t \in [0,\infty)$. (1)

(iii) On $(D_1,\|\cdot\|_1)$, $V_1(t,A)$ is bounded by

$$\|V_1(t,A)\|_1 \le M \exp((\omega + |\mathrm{ad}\ A|)t). \tag{2}$$

Remark: For most purposes, the first identity in (1) is the more useful, although in some proofs the more complicated second form is helpful, too. Recall that s_j is the ascent of the eigenvalue $\alpha(j)$ as ad A acts on M.

Proof: The most economical proof, which we use here, proceeds by three steps. First, the $\|\cdot\|_1$-closure $\bar{A}_1$ of A in D_1 is shown to satisfy the Hille-Yosida conditions and to generate a C_0 semigroup $\{V_1(t,A): t \in [0,\infty)\}$ on the Banach space $(D_1,\|\cdot\|_1)$. Second, $V_1(t,A)$ is shown by a differentiation argument to agree with $V(t,A)$ on D_1. Finally, the three expressions in (1) are shown to have the corresponding expressions in Equations (5.18) and (5.19) (Theorem 5.4) as Laplace transforms, whence by uniqueness of Laplace transforms, they agree. We later sketch a direct proof that yields all three of these facts at once by recapitulating the Laplace inversion steps used in the proof of the Hille-Yosida theorem, rather than by invoking that theorem.

We actually require the following folk-variant of the Hille-Yosida theorem, due essentially to Feller [Fℓ], who first described the necessary re-norming techniques for semigroups and resolvents that are used in its proof.

Hille-Yosida-Feller (HYF) Theorem. Let $\bar{A}$ be a closed, densely-defined operator on a complete complex normable (Banachable?) space E. Then $\bar{A}$ generates a C_0 semigroup $\{V(t,A): t \in [0,\infty)\}$ on E if and only if there exists a norm $\|.\|$ for the topology of E and a constant $\omega \geq 0$ such that every $\lambda > \omega$ is in the resolvent set $\rho(\bar{A})$ and $\|(\lambda-\bar{A})^{-1}\| \leq (\lambda-\omega)^{-1}$. With respect to this norm, $\|V(t,A)\| \leq \exp(\omega t)$ for all $t \in [0,\infty)$. E.O.T.

For our purposes, the point of this result is that the single estimate $\|(\lambda-\bar{A})^{-1}\| \leq (\lambda-\omega)^{-1}$ implies that, for all k, $\|(\lambda-\bar{A})^{-k}\| \leq (\lambda-\omega)^{-k}$, while if the norm is not so carefully chosen, one must find an $M < \infty$ and perform separate checks to ensure that $\|(\lambda-\bar{A})^{-k}\| \leq M(\lambda-\omega)^{-k}$ for each k. Specifically suppose that $\|.\|$ and ω have been chosen by the (HYF) Theorem for the operator A in Theorem 6.1. Then our graph-density hypothesis yields by Theorem 5.4 that closed A_1 exists in $(D_1, \|.\|_1)$ with a resolvent $R_1(\lambda,A) = (\lambda-A_1)^{-1}$ satisfying $\|R_1(\lambda,A)\|_1 \leq (\|R(\lambda,A)\|^{-1}-|\text{ad } A|)^{-1} \leq (\lambda-\omega-|\text{ad } A|)^{-1}$ when $\lambda-\omega > |\text{ad } A|$ (so that $\|R(\lambda,A)\| |\text{ad } A| < 1$).

Applying the converse of the (HYF) Theorem, this time in $(D_1,\|.\|_1)$, we obtain the asserted generation properties of A_1 and the estimate in (2) on $\|V_1(t,A)\|_1$ for the semigroup $\{V_1(t,A):t \in [0,\infty)\}$ that it generates.

To check in the second step that $V_1(t,A)$ agrees on D_1 with $V(t,A)$, we adapt the standard uniqueness argument for semigroups generated by a given densely-defined A. That is, take $u \in D(\bar{A}_1)$ and form $F(s) = V(t-s,A)V_1(s,A)u$ for $s \in [0,t]$. When $t > 0$, $V_1(s,A)u \in D(\bar{A}_1)$ by Hille-Yosida theory, and $D(\bar{A}_1) \subset D(\bar{A})$ by the proof of 5.6, so $V_1(s,A)u \in D(\bar{A})$ and F can be differentiated by the product rule in the usual way. But then since $\bar{A}$ and $\bar{A}_1$ agree on vectors $V_1(s,A)u \in D(\bar{A}_1)$, $F'(s)$ vanishes identically on $[0,t]$ (as the reader may verify). Then $V(t,A)u = F(0) = F(t) = V_1(t,A)u$ for all $u \in D(\bar{A}_1)$, and this identity extends to all $u \in D_1$ by $\|.\|_1$-density of D (and $D(\bar{A}_1)$) in D_1. (Note that since $\bar{V}(t,A)$ agrees with a $\|.\|_1$-bounded operator on $D(\bar{A}_1)$, it must send $\|.\|_1$-limits in D_1 of sequences in $D(\bar{A}_1)$ back into D_1.)

For the third step, we may use the uniqueness theorem for vector-valued Laplace transforms or reduce the matter to the scalar theory via the Hahn-Banach theorem. That is, for the values of λ under discussion, standard Hille-Yosida theory shows that

$$R_1(\lambda,A) = \int_0^\infty e^{-\lambda t} V_1(t,A)dt$$

as a strongly $\|.\|_1$-convergent integral. Consequently, since each $B^{(1)}$ is continuous from $(D_1,\|.\|_1)$ to $(E,\|.\|)$, it follows that

$$B^{(1)}R_1(\lambda,A) = \int_0^\infty e^{-\lambda t} B^{(1)}V_1(t,A)dt \tag{3}$$

as a strongly $\|.\|$-convergent integral. Similarly, the last expression in (1) has a Laplace tranform in E of the form

$$\int_0^\infty e^{-\lambda t} \Sigma\{(-t)^k/k!\ \exp(-t\alpha(j))V(t,A)(\text{ad } A-\alpha(j))^k(P_jB): \\ 1 \le j \le p\ ,\ 0 \le k \le s_j\}dt$$

$$= \Sigma\{\int_0^\infty(-t)^k/k!\ \exp(-t(\lambda+\alpha(j)))V(t,A)dt(\text{ad } A-\alpha(j))^k(P_jB)\}$$

$$= \Sigma\{1/k!\ d^k/d\lambda^k\ R(\lambda+\alpha(j),A)(\text{ad } A-\alpha(j))^k(P_jB)\}$$

$$= \Sigma\{(-1)^kR(\lambda+\alpha(j),A)^{k+1}(\text{ad } A-\alpha(j))^k(P_jB):\ 1\le j\le p\ ,\ 0\le k\le s_j\}. \tag{4}$$

By the first equation in (5.18) (Theorem 5.4) (3) and (4) agree, so the first and the third expression in (1) must agree by the uniqueness theorem.

The remaining identity in (1) is obtained in a similar manner, but there a vector-valued Fubini theorem is needed in order to interchange the Laplace transform integral and the infinite series when transforming the second expression. (Again, the Hahn-Banach theorem reduces matters to the scalar case.) That is, for each term in the series $V(t,A)\exp(-t \text{ ad } A)(B) = \Sigma\{(-t)^k/k!\ V(t,A)(\text{ad } A)^k(B):\ 0\le k<\infty\}$ we get

$$\int_0^\infty(-t)^k/k!\ e^{-\lambda t}V(t,A)dt(\text{ad } A)^k(B)$$

$$= (k!)^{-1}\ d^k/d\lambda^k\ R(\lambda,A)(\text{ad } A)^k(B)$$

$$= (-1)^k\ R(\lambda,A)^{k+1}(\text{ad } A)^k(B). \tag{5}$$

But as we have already seen, this term-by-term integrated series converges geometrically in norm, while the series prior to integration converges exponentially, whence the integration and summation processes commute. Thus the second expression in (1) has the second expression in (5.19) as transform, so the uniqueness theorem implies that the first and second expression in (1) agree. This completes the proof.

Remarks on the proof: (1) A more constructive direct proof of this theorem proceeds as follows. Recall that in Hille's proof of the Hille-Yosida theorem (cf. [HP]), one approximates V(t,A) strongly (uniformly with respect to t in compact sets) by

$$V_n(t,A) = [n/t\ R(n/t,A)]^n = (1-t/n\ \bar{A})^{-n}. \qquad (6)$$

Since (6) represents the approximate semigroups as powers of resolvents, one can apply the resolvent commutation relations from 5.4 inductively to commute elements $B \in M$ around the $V_n(t,A)$. When computed in detail, this method yields expressions on the right in which $BV_n(t,A)$ is reduced to a polynomial or infinite series with terms containing products of approximating groups $V_n(t,A)$, approximate identities $n/t\ R(n/t,A)$, and elements of M, multiplied together in the order mentioned. When limits are taken, these commuted expressions are seen to converge to the corresponding middle and right-hand expressions in (1). The resulting convergence on the left of the $B_iV_n(t,A)$ where B_i runs through the basis, then yields both invariance of D_1 under V(t,A) and the identity (1). The estimate in (2) is obtained from (1) as applied to the basis.

In a variant upon this approach, one uses Yosida's inversion instead, putting $A_n = nAR(n,A)$ as a bounded approximant to A, and $V_n(t,A) = \exp(tA_n)$. While this formula approximates V(t,A) by a well-behaved sequence of analytic semigroups, each $V_n(t,A)$ is an infinite series in resolvents. Thus, although one can in fact calculate a double-series expression for each $BV_n(t,A)$ using 5.4, and then take limits as $n \to \infty$ just as sketched above, the computations are horrendous combinatorial exercises whose complexity disguises the mechanism of the proof.

Whatever interest these remarks may have lies in the light that they shed upon the differences among various standard semigroup techniques when they are applied in mildly noncommutative multiple-operator contexts. It is surprising that minor technical variations of this sort can exhibit such radical differences in difficulty when applied in this setting.

(2) The exponential growth estimate in 6.1 (2) can be improved slightly, in that it can be shown that for any $\sigma > \max\{|\alpha(j)|: 1 \le j \le p\} = \nu(\mathrm{ad}\, A)$ (the spectral radius of ad A on M) there exists an $M_\sigma < \infty$ such that $\|V_1(t,A)\|_1 \le M_\sigma \exp((\omega+\sigma)t)$ for all $t \in [0,\infty)$. But the following example shows that in general no such M_σ exists for $\sigma < \max\{|\alpha(j)|\}$.

Example: An essentially skew-adjoint operator $A \in \mathbf{A}(D)$ on a Hilbert space E, and a space $M = \mathrm{span}\{B_+,B_-\} \subset \mathbf{A}(D)$ invariant under ad A such that $\mathrm{ad}\,A(B_+) = B_+$, $\mathrm{ad}\,A(B_-) = -B_-$ (ad A has eigenvalues ± 1), where the estimate $\|V_1(t,A)\|_1 \le \exp(|t|) = \exp(|\mathrm{ad}\,A|\,|t|)$ is best-possible for all large $t \in \mathbb{R}$.

Here, we take $E = L^2(SL(2,\mathbb{R}))$ and $D = C_c^\infty(SL(2,\mathbb{R}))$, recalling that there exists a noncompact Cartan basis in the Lie algebra: H, X_+ and X_- with $X_\pm$ root vectors for the Cartan subalgebra $\mathbb{R}\,H$ such that $[H,X_\pm] = \pm X_\pm$. The group acts unitarily on L^2 by left translation (cf. L^2 is formed with respect to left Haar measure) and we put $A = dU(H)$, $B_\pm = dU(X_\pm)$. An easy calculation with the commutation relations (Theorem 3.2 or Theorem 6.1 will apply) shows that $B_\pm^{(1)}V(t,A) = \exp(\pm t)V(t,A)B_\pm^{(1)}$. From this it easily follows that for $u \in D_1$

$$\begin{aligned}\|V_1(t,A)u\|_1 &= \max(\|u\|,\|B_+^{(1)}V(t,A)u\|,\|B_-^{(1)}V(t,A)u\|)\\ &= \max\{\|u\|,e^t\|V(t,A)B_+u\|,e^{-t}\|V(t,A)B_-^{(1)}u\|\}\\ &= \max\{\|u\|,e^t\|B_+u\|,e^{-t}\|B_-u\|\}.\end{aligned}$$

This explicitly confirms that 6.1 (2) holds, but also shows that $\|V_1(t,A)u\|_1$ grows asymptotically like $e^{|t|}$ at $\pm\infty$, so that no estimate of the form $\|V_1(t,A)\|_1 \le M_\sigma \exp(\sigma|t|)$ can work for $\sigma < 1$ (the spectral radius). E.O.E.

Extending the analysis of necessary and sufficient conditions for commutation relations begun in 5.5, we combine that result with the statement and proof methods from 6.1 above to obtain the following three-part equivalence.

6.2. Corollary

Let $A \in \mathbf{A}(D)$ be closable and let M be a finite-dimensional ad A-invariant subspace with closable basis as in 6.1. Suppose that $\bar{A}$ generates a C_0 semigroup $\{V(t,A): t \in [0,\infty)\}$ and that $\|V(t,A)\| \leq Me^{\omega t}$ for given constants M, $\omega < \infty$ and all $t \in [0,\infty)$, so that $\lambda > \omega$ implies that $\lambda \in \rho(A)$. If $\nu(\text{ad } A)$ denotes the spectral radius of ad A with respect to any desired norm for the finite-dimensional topology of M, then the following are equivalent.

(a) For some $\lambda > \omega + \nu(\text{ad } A)$, $D_\lambda = (\lambda - A)D$ is $\|\cdot\|_1$-dense in D.

(b) For all $\lambda > \omega + \nu(\text{ad } A)$, the resolvent $R(\lambda,A)$ leaves D_1 invariant, and restricts to a ($\|\cdot\|_1$)-bounded) two-sided inverse $R_1(\lambda,A)$ for $\lambda - \bar{A}_1$.

(c) For all $t \in [0,\infty)$, the semigroup operators $V(t,A)$ leave D_1 invariant, and their restrictions $V_1(t,A)$ define a C_0 semigroup $\{V_1(t,A): t \in [0,\infty)\}$ on $(D_1, \|\cdot\|_1)$ whose infinitesimal generator is $\bar{A}_1$.

Proof: The equivalence of (a) and (b) is just 5.5. That (a) implies (c) is contained in 6.1. The proof that (c) implies (b) is also implicit in the proof of 6.1: $R_1(\lambda,A)$ is the strongly $\|\cdot\|_1$-convergent (hence relatively $\|\cdot\|$-convergent Laplace transform of $V_1(t,A)$ on D_1, while $R(\lambda,A)$ is the strongly $\|\cdot\|$-convergent Laplace transform of $V(t,A)$ on all of E. Hence if $V_1(t,A)u = V(t,A)u$ for all $u \in D_1$, it follows that $R(\lambda,A)$ must agree on D_1 with the (necessarily $\|\cdot\|_1$-bounded) resolvent $R_1(\lambda,A)$ of the generator $\bar{A}_1$ of $V_1(t,A)$, and (b) follows from (c).

Remarks: (1) Results 6.1 and 6.2 have obvious generalizations to pregenerators A of groups of type ω (where $\|V(t,A)\| \leq M \exp(\omega|t|)$ for all $t \in \mathbb{R}$, replacing $\text{Re}(\lambda) > \omega + \nu(\text{ad } A)$ by $|\text{Re}(\lambda)| > \omega + \nu(\text{ad } A)$ throughout). We omit details.

(2) Just as we remarked after 5.5, the closed graph theorem can be used here to show that if each $V(t,A)$ leaves D_1 invariant, then its restriction $V_1(t,A)$ to D_1 is automatically $\|\cdot\|_1$-bounded. But even more is true here: if $\bar{A}$ generates a group $\{V(t,A): t \in \mathbb{R}\}$ that leaves D_1 invariant and if M has a basis $\mathbf{B} = \{B_i\}$ of restrictions to D of semigroup-generators, then $\{V_1(t,A): t \in \mathbb{R}\}$ can be shown to be a C_0 group on D_1. (That is, $t \to V_1(t,A)u$ is automatically $\|\cdot\|_1$-continuous when $u \in D_1$.) The proof is a category argument on the complete metric space $\mathbb{R} \times D_1$, rather in the spirit of our remarks at the end of Chapter 3. A more useful conclusion is obtained by

this line of argument at the end of Chapter 7, where D_1 is replaced by the Fréchet space of C^∞ vectors $D_\infty(M)$. We refer the reader to 7.5 for details.

6B. Variants of Sections 5B and 6A for general M

For many purposes, particularly in the theory of unitary groups and representations in Hilbert spaces E, the theory as developed in 5.4 and 6.1 is quite adequate. There, at least A and a basis $\mathbf{B} = \{B_i : 1 \leq i \leq d\}$ for M will usually consist of skew-symmetric (or symmetric) operators, which are automatically closable. (For example, if A and B are skew-symmetric, then $\mathcal{O}_A(B)$ = real span $\{(\mathrm{ad}\ A)^k(B) : 0 \leq k < \infty\}$ consists of skew-symmetric operators which span M = complexification $(\mathcal{O}_A(B))$. Similarly, if $\mathfrak{L}$ is a Lie algebra of skew-symmetrics on D, $A \in \mathfrak{L}$ and M = complexification $(\mathfrak{L})$ this is true.)

By contrast, in typical non-unitary problems one begins at the outset with a pair A, B or a collection S of naturally-given closable operators in terms of which all problem-data are given. In the more accessible cases, one discovers upon calculation either that $\mathcal{O}_A(B)$ is finite-dimensional or that S generates a real finite-dimensional Lie algebra of operators, whence the problem is potentially within the scope of the methods considered here. But in general, closability is preserved neither by products nor by linear combinations (folklore), so that propagation of closability to a basis for $\mathcal{O}_A(B)$ or L becomes a highly nontrivial question. (The same remark is true for the complexifications.) This difficulty arises with sufficient frequency (generally in applications deeper than those illustrated in Chapters 11 and 12) that it is worth considering the possible pathologies involved in abandoning the closable basis hypothesis.

In practice, a bootstrap situation often arises, as we prove below: one does not know at the outset that M has a closable basis, but after a theory for general M is developed and applied, one is able to use the resulting generalized commutation relations to show that the M in question actually must have had such a basis, and the possibility of pathology is then banished after-the-fact. In other cases, the question is still open (after commutation relations are obtained) as to whether the ad A-invariant subspace M has a closable basis. Consequently in this section we carry the full generality necessary for the worst cases.

The next lemma, and the example which follows it,

indicate the potential pathologies involved in the possible absence of a closable basis. They describe the substitutes for the space $(D_1, \|\cdot\|_1)$ and for the operators $B^{(1)} \in L(D_1, E)$ and $\bar{A}_1$ in D_1 that were so easily introduced prior to the statement of Theorem 5.4.

6.3. Lemma

Let $M \subset \mathbf{A}(D)$ be a finite-dimensional complex subspace. Let τ_1 be the weakest (normable) topology on D, stronger than the initial topology τ_0, which renders all $B \in M$ continuous from (D, τ_1) into E, and let D_1 be the abstract completion of this space.

(a) There exists a bounded mapping $J_1 \colon D_1 \to E$ which extends the natural injection of (D, τ_1) into $(\bar{E}, \tau_0)$. If M has a closable basis, then J_1 is injective.

(b) For each $B \in M$, viewed as a map from $D \subset D_1$ into E, there exists a unique bounded extension-by-limits to an operator $B^{(1)} \colon (D_1, \|\cdot\|_1) \to (E, \|\cdot\|)$. If B is closable, then kernel $(B^{(1)}) \supset$ kernel (J_1), and $B^{(1)} \supseteq \bar{B}J_1 \supseteq J_1 B$ (where B is viewed as acting as an endomorphism of $D \subset D_1$ in the last product).

(c) Suppose that $A \in \mathbf{A}(D)$ is closable in E, and that M is ad A-invariant. Then when A is viewed as an endomorphism of $D \subset D_1$, it is closable in D_1 with closure $\bar{A}_1$, and the intertwining relation

$$\bar{A}J_1 \supseteq J_1\bar{A}_1 \tag{7}$$

holds, so that in particular $\bar{A}_1(\ker(J_1) \cap D(\bar{A}_1)) \subset \ker(J_1)$.

Example: A space $M = \mathbb{C}B \subset \mathbf{A}(E)$ on a Banach space E (i.e. $D = E$) such that $J_1 \colon D_1 \to E$ is not injective and in addition the natural quotient Banach space $D_1/\text{kernel}(J_1)$ is topologically isomorphic to $(E, \|\cdot\|)$, so that the quotient norm is strictly weaker on $D = E$ than the C^1 norm $\|u\|_1 = \|u\| + \|Bu\|$. (This example shows that the otherwise-plausible move of using $J_1(D_1)$ with the quotient norm as the 'Banach subspace of C^1 vectors in E' will fail topologically, since the $B \in M$ cannot be expected then to yield bounded maps from this quotient into E.)

The example is obtained by taking B to be any non-closed, everywhere-defined operator, and recognizing that D_1 is naturally identifiable with the closure of the graph of B while J_1 then becomes the first-coordinate projection.

(Specifically, let F be a dense proper subspace, f a nonzero linear functional with F as kernel, $u_0 \neq 0 \in E$, and $Bu = f(u)u_0$. Then f is not closed, so neither is B.) The map $u \to (u,Bu)$ then takes $(D,\|\cdot\|_1)$ isometrically into Graph (B), so that $(D_1,\|\cdot\|_1)$ can be identified with Graph $(B)^-$ in $E \times E$, while since J_1 corresponds with the map $(u,Bu) \to u$ on Graph (B), it corresponds with $(u,v) \to u$ on Graph $(B)^-$. But Graph $(B)^-$ is the graph of an operator if and only if this map is injective, so the map cannot be injective. Since J_1 is bounded so is the natural lift-down map $\tilde{J}_1$: $D_1/\text{kernel}(J_1) \to E$. But $\tilde{J}_1$ is then a bounded algebraic isomorphism carrying the quotient Banach space into $(E,\|\cdot\|)$, so it must be a topological isomorphism. It is clear that $\|\cdot\|_1$ must be strictly stronger than the quotient norm, since B is bounded from $(E,\|\cdot\|_1)$ into $(E,\|\cdot\|)$ but not from $(E,\|\cdot\|)$ into itself. E.O.E.

Proof: If $\mathbf{B} = \{B_1,\ldots,B_d\}$ is any basis for M, τ_1 is normed as usual by $\|u\|_1 = \max\{\|B_iu\|: 0 \leq i \leq d\}(B_0 = I)$. The natural injection J_1^0: $(D,\|\cdot\|_1) \to (E,\|\cdot\|)$ is then a contraction mapping, and extends to a contraction J_1: $(D_1,\|\cdot\|_1) \to (E,\|\cdot\|)$. The basis elements $B_i \in \mathbf{B}$ similarly induce contraction mappings $B_i^{(1)}$: $(D_1,\|\cdot\|_1) \to (E,\|\cdot\|)$, whence by linear combinations every $B_1 \in M$ induces a bounded mapping. (In fact, if we give M the ℓ^1-norm $|B| = \Sigma|b_i|$ for $B = \Sigma\, b_iB_i$, then $|B| \geq \|B^{(1)}\|$ for the operator norm in $L(D_1,E.)$

In order to complete (a) and (b), let $B \in M$ be closable in E and suppose that $\{u_n\}$ is a sequence in D with $\|u_n-u\|_1 \to 0$ for some $u \in D_1$. Then $Bu_n = B^{(1)}u_n \to B^{(1)}u$ in E, while $u_n = J_1u_n \to J_1u \in E$. It follows that $J_1u \in D(\bar{B})$ and that $B^{(1)}u = \bar{B}J_1u$, establishing $B^{(1)} \supset \bar{B}J_1$ and kernel $(B^{(1)}) \supset$ kernel (J_1). Consequently, (b) is proved. But if M has a basis of such closable operators and $u \in \ker(J_1)$, then for some $\{u_n\} \subset D$ we have $\|u_n-u\|_1 \to 0$ and $\|u_n\| = \|J_1u_n-J_1u\| \to 0$, whence $\bar{B}_iu_n \to \bar{B}_iJ_1u = 0$ and $\|u_n\|_1 \to 0$ as well. Hence, $u = 0$ and J_1 is injective.

For the proof of (c) we prepare a sublemma that supplies both closability and the intertwining formula. Let $\{u_n\} \subset D$ with $\|u_n-u\|_1 \to 0$ and $\|Au_n-v\|_1 \to 0$ for $u, v \in D_1$. Then since J_1 is the identity mapping from $D \subset D_1$ to $D \subset E$ we have

$u_n - J_1u = J_1(u_n-u) \to 0$ and $Au_n - J_1v = J_1(Au_n-v) \to 0$ in E, so that $J_1u \in D(\bar{A})$ and $\bar{A}J_1u = J_1v$. But we also have

$$AB_iu_n = B_iAu_n + (\text{ad } A)(B_i)u_n \to B_i^{(1)}v + [\text{ad } A(B_i)]^{(1)}u$$

and $B_iu_n \to B_i^{(1)}u$, so it follows that $B_i^{(1)}u \in D(\bar{A})$ and (solving for $B_i^{(1)}v$)

$$\lim B_iAu_n = B_i^{(1)}v = \bar{A}B_i^{(1)}u - [\text{ad } A(B_i)]^{(1)}u.$$

In particular, if $\{u_n\}$ is a $\|\cdot\|_1$-null sequence, so that $u = 0$ and $\|u_n\| \to 0$ in E as well, then we see that $B_iAu_n \to \bar{A}B_i^{(1)}0 - [\text{ad } A(B_i)]^{(1)}0 = 0$ and $Au_n \to \bar{A}J_10 = 0$, whence $\|Au_n\|_1 \to 0$ and $v = 0$ as well. That is, if $\|u_n\|_1 \to 0$ and $\|Au_n-v\|_1 \to 0$ then $v = 0$, so A satisfies a standard closability criterion in D_1 and $\bar{A}_1$ exists. But then for the general sequence $\{u_n\}$ discussed at the outset, we have by definition that $v = \bar{A}_1u$, so the intertwining relation also follows. E.O.P.

With this construction of a generalized $(D_1,\|\cdot\|_1)$ and an intertwining map J_1: $D_1 \to E$ in hand, we proceed to generalize 5.4 and 6.1 together. In the interests of brevity, we proceed by describing the changes-in-proof needed to establish this generalization, avoiding a detailed recapitulation of the arguments.

6.4. Theorem

Let $A \in \mathbf{A}(D)$ and $M \subset \mathbf{A}(D)$ be as described in the preamble to 6.1, excepting the provision that M has a closable basis. Then the following intertwining identities hold.

(a) For all large $\lambda > \omega_1$, there exists on $(D_1,\|\cdot\|_1)$ a bounded resolvent $R_1(\lambda,A)$ for $\bar{A}_1$ satisfying the intertwining relation

$$R(\lambda,A)J_1 = J_1R_1(\lambda,A) \tag{8}$$

and the resolvent commutation relations

$$B^{(1)}R_1(\lambda,A) = \Sigma\{(-1)^k R(\lambda,A)^{k+1}[(\text{ad } A)^k(B)]^{(1)}: 0 \le k < \infty\}$$
$$= \Sigma\{(-1)^k R(\lambda+\alpha(j),A)^{k+1}[(\text{ad } A-\alpha(j))^k(P_jB)]^{(1)};\ 1 \le j \le p,\ 0 \le k \le s_j\} \tag{9}$$

hold between bounded operators from $(D_1,\|\cdot\|_1)$ to $(E,\|\cdot\|)$.

(b) The $\|\cdot\|_1$-closure $\bar{A}_1$ of A in $(D_1,\|\cdot\|_1)$ is the infinitesimal generator of a C_0 semigroup $\{V_1(t,A): t \in [0,\infty)\}$ on $(D_1,\|\cdot\|_1)$ which satisfies the intertwining relation

$$V(t,A)J_1 = J_1V_1(t,A), \ t \in [0,\infty) \tag{10}$$

and the semigroup commutation relations

$$B^{(1)}V_1(t,A) = V(t,A)[\exp(-t \text{ ad } A)(B)]^{(1)}$$

$$= \Sigma\{(-t)^k/k! \exp(-t\alpha(j)) \ V(t,A)[(\text{ad } A-\alpha(j))^k(P_j(B))]^{(1)}: \quad 1 \leq j \leq p, \ \ 0 \leq k \leq s_j\} \tag{11}$$

as bounded operators from $(D_1,\|\cdot\|_1)$ to $(E,\|\cdot\|)$.

Proof sketch: (a) On D_λ, viewed as a dense subspace of D_1, we proceed precisely as in the proof of 5.4 to check that $R(\lambda,A)$ acts as a $\|\cdot\|_1$-bounded operator, while checking as in 6.1 that $\|R(\lambda,A)u\|_1 \leq (\lambda-\omega-|\text{ad } A|)^{-1}\|u\|_1$ for a suitably chosen Feller-norm on E. Here, however, this simply implies that $R(\lambda,A)$ extends by limits to a bounded operator $R_1(\lambda,A)$ on $(D_1,\|\cdot\|_1)$. Despite the slight risk of confusion, it is best not to distinguish between D as a subspace of E and D as a subspace of D_1, so that the identity $J_1u = u$ holds for all $u \in D$. Then, for $u \in D_\lambda$ we get $R(\lambda,A)J_1u = R(\lambda,A)u = J_1R(\lambda,A)u = J_1R_1(\lambda,A)u$, and this identity extends by $\|\cdot\|_1$-limits to all $u \in D_1$, yielding (8). (Note that $R(\lambda,A)u \in D$, too.) A similar argument begins with the correctness of (9) as applied to $u \in D_\lambda$ by judicious sprinkling of J_1 throughout Equations (5.5) in 5.1 and (5.9) in 5.2 extending the resulting identities to all $u \in D_1$ by limits as in 5.4. The only slight novelty and technicality here lies in the check that $R_1(\lambda,A)$ is a two-sided inverse for $\lambda - \bar{A}_1$ (already a lengthy-but-routine check in 5.4). But here again, for $u_n \in D_\lambda$ we have by notational agreement that $u_n = (\lambda-A)R(\lambda,A)u_n = (\lambda-\bar{A}_1)R_1(\lambda,A)u_n$ and as $u_n \to u \in D_1$ in the $\|\cdot\|_1$-sense we obtain that $R_1(\lambda,A)u_n \to R_1(\lambda,A)u$, so by closedness $R_1(\lambda,A)u \in D(\bar{A}_1)$ and $u = (\lambda-\bar{A}_1)R_1(\lambda,A)u$. Similarly, for $u_n \in D$, $u_n = R(\lambda,A)(\lambda-A)u_n = R_1(\lambda,A)(\lambda-\bar{A}_1)u_n$ and if the sequence is chosen so that $u_n \to u$ and $\bar{A}_1u_n \to \bar{A}_1u$ for some $u \in D(\bar{A}_1)$ then in the limit $u = R_1(\lambda,A)(\lambda-\bar{A}_1)u$.

Given (a) and the Hille-Yosida-Feller estimate on $R_1(\lambda,A)$ described above, the existence of $\{V_1(t,A): t \in [0,\infty)\}$ as a C_0 semigroup generated by $\bar{A}_1$ follows as in 6.1 but on the new D_1. Here, (10) follows from (8) by boundedness of J_1 and uniqueness of Laplace transforms. (That is,

$$R(\lambda,A)J_1u = \int_0^\infty e^{-\lambda t} V(t,A)J_1u\, dt$$

and

$$J_1R_1(\lambda,A)u = J_1 \int_0^\infty e^{-\lambda t} V_1(t,A)u\, dt = \int_0^\infty e^{-\lambda t} J_1V_1(t,A)u\, dt$$

as integrals convergent in E.) Then (11) follows by a similar argument from (9), precisely as in the proof of 6.1. E.O.P.

Remark: In the proof-sketch given above, we have implicity used 5.5 to extend the graph-density condition from the λ given in the hypotheses to all λ with $\text{Re}(\lambda) > \omega + \nu(\text{ad } A)$, essentially as indicated at the beginning of Section 6A. Notice that the argument in 5.5 makes no essential use of the assumption that $D_1 \subset E$; it relies upon the fact that $\lambda - \bar{A}_1$ must map the core set $D \subset D_1$ into a dense subspace D_λ of D_1, but since $D_\lambda \subset D$, the map J_1 identifies $\|\cdot\|_1$-density of D_λ in D (as subsets of D_1) with the same condition in E.

6C. Automatic availability of a closable basis

The following results describe situations in which a closable basis for M can be obtained without a-priori assumptions to that effect.

6.5. Lemma

Let $A \in \mathbf{A}(D)$ be a closable operator, and let $\mathcal{L} \subset \mathbf{A}(D)$ be a finite-dimensional real adA-invariant subspace. Suppose that $\bar{A}$ generates a C_0 one-parameter group $\{V(t,A): t \in \mathbb{R}\}$ of type ω, and that there exist $\lambda_\pm$ with $\text{Re}(\lambda_+) > \omega + \nu(\text{ad } A)$, $\text{Re}(\lambda_-) < -(\omega+\nu(\text{ad } A))$ such that $D_\pm = (\lambda_\pm - A)D$ is $\|\cdot\|_1$-dense in D. Then

(1) for all closable $B \in \mathcal{L}$, $O_A(B)$ has a closable basis, and

(2) the real span $\mathcal{L}_0$ of all closable elements in $\mathcal{L}$ is ad A-invariant.

Proof: That (1) implies (2) is an obvious consequence of the linearity of ad A. We prove (1) by checking that the group $\{V(t,A): t \in \mathbb{R}\}$ leaves $J_1(D_1)$ invariant and that for all $u \in D$,

$$V(t,A)\bar{B}V(-t,A)u = \exp(t \text{ ad } A)(B)u. \tag{12}$$

Once (12) is obtained, every $\exp(t \text{ ad } A)(B) \in \mathbf{A}(D)$ is seen to be the restriction to D of the conjugate of closable B with respect to a boundedly invertible operator $V = V(t,A) \in L(E)$. It is routine to verify that $V\bar{B}V^{-1}$ is closed on the domain $D(V\bar{B}V^{-1}) = VD(\bar{B})$, whence (12) shows that all $\exp(t \text{ ad } A)(B)$ are closable, and we need only check that these span $\mathcal{O}_A(B)$. But $\mathcal{O}' = $ real span$\{\exp(t \text{ ad } A)(B): t \in \mathbb{R}\}$ contains B, and it is finite-dimensional, hence closed with respect to the relative strong operator topology from $\mathbf{A}_s(D)$. Consequently, the limits ad $A(\exp(t \text{ ad } A)(B) = d/dt \exp(t \text{ ad } A)(B)$ are in $\mathcal{O}'$, and $\mathcal{O}'$ is an ad A-invariant subspace containing B. It follows that $\mathcal{O}' \supset \mathcal{O}_A(B)$, but the reverse inclusion is obvious, so in fact the closable $\exp(t \text{ ad } A)(B)$ span $\mathcal{O}_A(B)$.

It remains to check (12). But the hypotheses in 6.5 imply that both A and (-A) satisfy the conditions in 6.4. (Here, $R(-\lambda,-A) = -R(\lambda,A)$ and $V(-t,-A) = V(t,A)$.) Consequently, 6.4(b) extends to all $t \in \mathbb{R}$. In particular, we get $V(t,A)J_1u = J_1V_1(t,A)u$ for all $u \in D_1$ and all $t \in \mathbb{R}$, so that $V(t,A)$ leaves $J_1(D_1)$ invariant. Hence for $u \in D$, $V(-t,A)u \in J_1(D_1)$, and (by 6.3) is in the domain $D(\bar{B})$ of any closable B. Application of the intertwining relations then yields by 6.3 and 6.4 that

$$\bar{B}V(-t,A)J_1u = \bar{B}J_1V_1(-t,A)u = B^{(1)}V_1(-t,A)u$$

$$= V(-t,A)[\exp(t \text{ ad } A)(B)]^{(1)}u = V(-t,A)\exp(-t \text{ ad } A)(B)u$$

since $u \in D$. Applying $V(t,A)$ to both sides, we get (12).

One of the cases of interest concerns $\mathbf{\mathcal{L}} = \mathcal{O}_A(B)$, in Chapter 11: under the circumstances described there, $\mathcal{O}_A(B)$ has a closable basis (as does its complexification M), whence $J_1 : D_1 \to E$ is an injection and the simplified theory of 6.1-6.2 applies. The other application lies in Lie algebra exponentiation theory, where the following corollary is needed.

6.6. Corollary

Let $\mathfrak{L} \subset \mathfrak{A}(D)$ be a finite-dimensional real Lie algebra, and let $S \subset \mathfrak{L}$ be a Lie generating set consisting of closable operators A such that A and $\mathfrak{L}$ satisfy the conditions in 6.5. Then $\mathfrak{L}$ has a closable basis.

Proof: We check that $\mathfrak{L}_C$ in (2) of 6.5 is the Lie subalgebra $\mathfrak{L}(S)$ generated by S, namely $\mathfrak{L}$ itself. Clearly $\mathfrak{L}_C$ contains S, and since it is ad A-invariant for all $A \in S$, it can be seen by the obvious induction to contain all iterated commutators $[A_1,[\ldots,[A_{k-1},A_k]\ldots]]$ for $\{A_i\} \subset S$. But the span of these iterated commutators is exactly $\mathfrak{L}(S)$.

6D. Remarks on operational calculi

The (semi-) group commutation relations obtained in 6A can be used in two basically different ways to extend the commutation theory of Section 5D to more powerful operational calculi. The more classical approach combines integration theory with Fourier or Laplace transforms to treat broader classes of functions φ where $\varphi(A)$ is still a bounded operator on E. By contrast, the other approach (due essentially to Schwartz and implicit in [Sz 1]) uses distribution theory and permits extension to functions φ where $\varphi(A)$ may only be closable.

For the classical examples, φ is taken to be the Laplace or Fourier transform of a complex measure Φ which vanishes sufficiently rapidly at ∞. A reasonable result can be obtained for semigroup-generators $\bar{A}$ using the operational calculus of Hille and Phillips ([HP] Ch. 15), but their discussion is unnecessarily general and technical for our purposes. In the interests of expository clarity, we confine our attention here to Bade's operational calculus for the case where $\bar{A}$ generates a group $\{V(t,A)\colon t \in \mathbb{R}\}$ of type ω: $\|V(t,A)\| \leq Me^{\omega|t|}$ for $M < \infty$, $t \in \mathbb{R}$. (This is concisely summarized in VII.2 of [DS 1]; we follow their notation as closely as possible.) There $S(A)$ denotes the space of complex measures Φ such that for some $\varepsilon > 0$ (possibly Φ-dependent) $\int_{-\infty}^{\infty}\exp[(\omega+\varepsilon)|t|]d|\Phi|(t)<\infty$, where $|\Phi|$ denotes the total variation of Φ. Then the class of functions φ to which the calculus applies, $\upsilon(A)$, consists of the bilateral Laplace transforms of these Φ:

$$\varphi(\lambda) = \int_{-\infty}^{\infty} e^{-\lambda t}\, d\Phi(t), \quad |\mathrm{Re}(\lambda)| < \omega + \varepsilon \,, \tag{13}$$

which are analytic in a strip containing $\sigma(A)$ but not necessarily analytic at ∞. With φ and Φ related by (13), one then defines $\varphi(A)$ by the strongly convergent integral

$$\varphi(A)u = \int_{-\infty}^{\infty} V(-t,A)u \, d\Phi(t) \quad u \in E. \tag{14}$$

Now, assuming that M is a finite-dimensional ad A-invariant subspace and that the hypotheses of the group versions of 6.1 and 6.2 hold, we obtain that $V(t,A)$ restricts to a C_0 group $\{V_1(t,A): t \in \mathbb{R}\}$ on $(D_1, \|\cdot\|_1)$. If Φ has an ε as above with $\varepsilon > |\mathrm{ad}\, A|$, this means by estimate (2) in 6.1 that the calculus applies on D_1 as well, to produce via (14) an operator $\varphi_1(A)$ which acts boundedly on D_1 and agrees with $\varphi(A)$ there. Applying the bounded operator $B^{(1)} \in L(D_1,E)$ determined by any $B \in M$, we may compute for any $u \in D_1$

$$B^{(1)}\varphi_1(A)u = B^{(1)} \int_{-\infty}^{\infty} V_1(-t,A)u \, d\Phi(t) = \int_{-\infty}^{\infty} B^{(1)} V_1(-t,A)u \, d\Phi(t). \tag{15}$$

But then the finite-sum commutation identity (1) can be applied to the integrand, yielding (upon distributing the integral over the finite sum)

$$\begin{aligned} B^{(1)}\varphi_1(A)u &= \Sigma\{(-1)^k/k! \int_{-\infty}^{\infty} V(-t,A-\alpha(j))[(\mathrm{ad}\, A-\alpha(j))^k(P_jB)]^{(1)}u \\ &\qquad \times (-t)^k \, d\Phi(t): 1 \le j \le p \,, \, 0 \le k \le s_j\} \\ &= \Sigma\{(-1)^k/k! \int_{-\infty}^{\infty} V(-t,A)u_{jk} \exp(\alpha(j)t)(-t)^k \, d\Phi(t)\} \end{aligned} \tag{16}$$

where we have put $u_{jk} = [(\mathrm{ad}\, A-\alpha(j))^k(P_jB)]^{(1)}u$ for brevity. But an easy calculation shows then that the measure $\Phi_{jk}(s) = \int_s \exp(\alpha(j)t)(-t)^k \, d\Phi(t)$ is also in $S(A)$ and that its bilateral Laplace transform $\varphi_{jk}(\lambda)$ satisfies

$$\varphi_{jk}(\lambda) = \int_{-\infty}^{\infty} e^{-\lambda t} e^{\alpha(j)t}(-t)^k \, d\Phi(t) = \varphi^{(k)}(\lambda-\alpha(j)). \tag{17}$$

Consequently, each of the integrated terms in (16) can be replaced by

$$\varphi_{jk}(A)u_{jk} = \varphi^{(k)}(A-\alpha(j))[(\mathrm{ad}\, A-\alpha(j))^k(P_jB)]^{(1)}u. \tag{18}$$

This substitution yields

$$\begin{aligned} B^{(1)}\varphi_1(A)u = \Sigma\{(-1)^k/k! \; \varphi^{(k)}(A-\alpha(j))[(\mathrm{ad}\, A-\alpha(j))^k(P_jB)]^{(1)}u: \\ 1 \le j \le p \,;\, 0 \le k \le s_j\} \end{aligned} \tag{19}$$

which is formally identical to Equation (5.30).

A slightly more delicate argument, using the first (infinite-sum) identity in (1) rather than the second, yields

$$B^{(1)}\varphi_1(A)u = \Sigma\{(-1)^k/k!\ \int_{-\infty}^{\infty} V(-t,A)[(\mathrm{ad}\ A)^k(B)]^{(1)}u(-t)^k\ d\Phi(t):\ 0 \le k < \infty\}$$

$$= \Sigma\{(-1)^k/k!\ \varphi^{(k)}(A)[\ (\mathrm{ad}\ A)^k(B)\ ]^{(1)}u]:\ 0 \le k < \infty\}. \qquad (20)$$

Here, a Fubini argument using estimates of the moments $\int_{-\infty}^{\infty}|t|^k\ d|\Phi|(t)$ is required to interchange summation and integration.

When the group $\{V(t,A):\ t \in \mathbb{R}\}$ is uniformly norm-bounded (in particular, unitary on a Hilbert space) the bilateral Laplace transform is most often replaced by the <u>Fourier-transform</u>: φ is taken to be the Fourier transform of a measure Φ with finite moments. In the unitary case, standard arguments from spectral theory and the Plancherel formula show that the resulting $\varphi(A)$ agrees with that usually given by the spectral theorem:

$$\varphi(A) = \int_{\sigma(A)}\varphi(\lambda)dE(\lambda). \qquad (21)$$

(This is the conceptual framework that underlies Goodman's main preparatory Theorem 5.2 in his treatment of the Generalized O'Raifeartaigh theorem [Gd]. Our discussion above lays the groundwork for possible future non-unitary generalizations, to not-necessarily-integrable Lie algebras $\mathcal{L}$, of the work of Goodman.)

The alternative distribution-theoretic approach to operational calculi is closely related to matters discussed in [JM], and to the Lions-Chazarain theory of distribution (semi-) groups. It also shows promise of clarifying spectral reduction phenomena of O'Raifeartaigh type. We defer detailed discussion until [JM] is in print, merely sketching the formalities here. The ideas go as follows. Given that $\bar{A}$ generates a semigroup $\{V(t,A):\ t \in [0,\infty)\}$ and that 6.1 applies, one first integrates ('smears') the semigroup to a convolution algebra representation via

$$V(f,A) = \int_0^{\infty} f(t)V(t,A)dt \qquad (22)$$

to obtain an operator valued distribution $f \to V(f,A)$ sending compactly-supported test functions $f \in \mathcal{D}([0,\infty))$ into bounded operators. As usual, one takes $D_0 \subset E$ to be

$D_0 = \mathrm{span}\{V(f,A)u: u \in E,\ f \in \mathcal{D}([0,\infty))\}$, the Gårding vectors, and represents all compactly-supported distributions Φ on $[0,\infty)$ as closable operators $V(\Phi,A)$ on D_0 via convolution action of Φ on $\mathcal{D}$:

$$V(\Phi,A)\Sigma\, V(f_i,A)u_i = \Sigma\, V(\Phi * f_i,A)u_i. \tag{23}$$

If φ is the conjugate Laplace-transform of Φ, in the sense of distributions, then we put

$$\varphi(A) = V(\Phi,A) \tag{24}$$

(That is, if $\hat{f}(\lambda) = \int_0^\infty e^{\lambda t} f(t)dt$, one defines φ by $\langle\varphi,\hat{f}\rangle = \langle\Phi,f\rangle$ for all test functions f. Cf. [Li] or [Ko] for details.) Commutation relations are then derived in this setting by applying 6.1, (1) first to (22) to obtain commutation relations for the $B^{(1)}$ and the $V_1(f,A)$ obtained by carrying out the integration process of (22) in $(D_1,\|\cdot\|_1)$. Formally, one obtains

$$\begin{aligned} B^{(1)}V_1(f,A) &= \int_0^\infty f(t)B^{(1)}V_1(t,A)dt \\ &= \int_0^\infty \Sigma\{(-1)^k/k!\ t^k f(t)V(t,A)[(\mathrm{ad}\ A)^k(B)]\} \\ &= \Sigma\{(-1)^k/k!\ V(M^k f,A)[(\mathrm{ad}\ A)^k(B)]^{(1)}: 0 \le k < \infty\} \end{aligned} \tag{25}$$

where $(M^k f)(t) = t^k f(t)$. When this identity is chased through the transform identifications and (23), (24), one sees first that in some sense

$$B^{(1)}V_1(\Phi,A) = \Sigma\{(-1)^k/k!\ V(M^k\Phi,A)[(\mathrm{ad}\ A)^k(B)]^{(1)}$$

and that $M^k\Phi$ has $\varphi^{(k)}$ as conjugate transform, whence

$$B^{(1)}\varphi_1(A) = \Sigma\{(-1)^k/k!\ \varphi^{(k)}(A)[(\mathrm{ad}\ A)^k(B)]^{(1)}\}$$

emerges here too, even when $\varphi(A)$ is merely closed in $(D_1,\|\cdot\|_1)$.

More useful results are obtained, particularly in the uniformly bounded group case, when the compactly-supported test functions $\mathcal{D}(\mathbb{R})$ are replaced by the rapidly-decreasing functions $\mathcal{S}(\mathbb{R})$ or the faster-than-exponentially-decreasing 'hyper-schwartz' functions used in [Jo 2]. Then the convolving distribution Φ need not have compact support, and it becomes possible in well-behaved cases to treat $\varphi(A)$ where φ is (essentially) the characteristic function of a compact set in $\sigma(A)$. We defer publication of details pending further investigation of these matters.

Chapter 7

CONSTRUCTION OF GLOBALLY SEMIGROUP-INVARIANT C^{∞}-DOMAINS

In this chapter, the 'infinitesimal' spectral-type commutation theory on Banach spaces (Chapters 5 and 6) is connected with the global invariant-domain commutation theory on locally convex spaces (Chapter 3). In particular, we obtain from the hypotheses of Chapter 3 (domain regularity) or of Chapter 5 (graph-density) that whenever the underlying space E is Banach (or Fréchet) there exists a domain D_{∞} of C^{∞}-vectors which is a Fréchet space with respect to a stronger C^{∞}-topology. This space is invariant under the action at least of the (semi) group operators V(t,A) involved in the problem and under the operational images $\varphi(A)$ of A constructed via a suitably restricted semigroup/transform operational calculus. In suitable instances, it is also invariant under certain of the resolvents $R(\lambda,A)$ for A as well. The net effect of these developments is to unify the different types of commutation relations, while preparing for a unified, economical treatment of smooth integrals and continuous exponentials of operator Lie algebras in Chapters 8 and 9, and laying the groundwork for our subsequent analysis of smeared (distribution) exponentials. (In addition, the present developments constitute a first step in the direction of a more ambitious resolvent-based program for the production of invariant C^{∞}-core-domains [Mr 9].)

More specifically, we recall that Chapter 3 made fundamental use of the technically-convenient assumption that the given domain D is invariant not only under the infinitesimal operators $A,B \in \mathbf{A}(D)$ (so that D is a 'C^{∞}-domain') but also under the global semigroup 'exponentials' $\{V(t,A): t \in [0,\infty)\}$. As we pointed out there, this hypothesis places strong restrictions upon the given domain D. Here, we show that the infinitesimal conditions from Chapter 5 and 6 (which have been shown to be necessary as well as sufficient for commutation theory of the type discussed here) always permit the given domain D to be enlarged in a standard way to supply such a globally-invariant C^{∞}-domain. Essentially, the existence of suitable globally-invariant C^{1}-domains supplied by 6.1, and (implicitly) 6.3, makes it possible to check that the minimal globally-invariant subspace D_I containing D actually possesses all of the other properties exploited in Chapter 3.

For expository reasons, it is preferable to separate the development into two technically different cases which are tailored for radically different sorts of applications. (One can supply a unifying generalization, but its formulation is clumsy and artificial while its proofs are needlessly technical and complicated. A few remarks in this direction are given below in Section 7A.)

The first case, which involves a pair A, B $\in \mathbf{A}(D)$ of closable endomorphisms of a dense subspace $D \subset E$ in a Banach space E, is most closely related to operator theory and is treated in Section 7B. There, we assume that $\mathcal{O}_A(B)$ = real span $\{(\mathrm{ad}\ A)^k(B): 0 \leq k < \infty\}$ is finite-dimensional, that $\bar{A}$ generates only a <u>semigroup</u> $\{V(t,A): t \in [0,\infty)\}$ and that $D_\lambda = (\lambda-A)D$ is dense in D with respect to the $\mathcal{O}_A(B)$ graph-norm $\|\cdot\|_1$ for some (hence all) λ with $\mathrm{Re}(\lambda)$ large and <u>positive</u>.

The second case involving a finite-dimensional real Lie algebra $\mathcal{L} \subset \mathbf{A}(D)$ (for D and E as above), serves as an intermediate step in the theory of continuous group representations, and is developed in Section 7C. There, A runs through a Lie-generating set $S \subset \mathcal{L}$, and each A (for $A \in S$) is assumed to generate a <u>group</u> $\{V(t,A): t \in \mathbb{R}\}$. Here $D_\lambda = (\lambda-A)D$ is assumed to be dense in D with respect to the $\mathcal{L}$-graph norm $\|\cdot\|_1$, for some (hence all) λ with $\mathrm{Re}(\lambda)$ large and <u>positive</u> or <u>negative</u>.

Finally, Section 7D discusses the invariance properties of the Fréchet C^∞-domains with respect to other bounded operators involved in the theory: resolvents $R(\lambda,A)$, operational images $\varphi(A)$ and spectral projections P_σ. This informal discussion, like that of Section 6D, sheds some light upon the exact role of nilpotency in the Goodman treatment of mass splitting ([Gd] and Chapter 4). It suggests possible directions of generalization, but the absence at present of any definitive results in these directions precludes further detailed development in this monograph.

7A. Fréchet C^∞-domains in Banach spaces

In this section we review some of the standard notions involved in the construction of Fréchet C^∞-domains, and then contrast the technical ways in which these are involved in Cases 1 and 2 as treated in Sections 7B and 7C, respectively.

The purely algebraic aspects of the matters under

discussion here are most easily formulated within the following framework.

7.1. Definition

Let $\bar{S}$ be a collection of closed operators $\bar{A}$, $\bar{B}$, etc., in a locally convex space E.
(a) A subspace $F \subset E$ is a C^∞-domain for $\bar{S}$ if and only if $F \subset D(\bar{A})$ for all $\bar{A} \in \bar{S}$ and $\bar{A}F \subset F$.
(b) The space $E_\infty(\bar{S}) = \cup\{F \subset E : F \text{ is a } C^\infty\text{-domain for } \bar{S}\}$ is the space of C^∞-vectors for $\bar{S}$.

It is obvious that $E_\infty(\bar{S})$ is itself a C^∞-domain for $\bar{S}$; indeed, it is the maximal C^∞-domain. In pathological cases, $E_\infty(\bar{S})$ may consist only of the zero vector, and the main problem in the general theory of C^∞-vectors is to give conditions sufficient to ensure that this space is 'big enough' The appropriate criteria for largeness are: (1) $E_\infty(\bar{S})$ should be dense in E, and (2) $E_\infty(\bar{S})$ should be a core for each $\bar{A} \in \bar{S}$, in the sense that $\bar{A}$ is the closure of its restriction to $E_\infty(\bar{S})$.

As applied in Sections 7B and 7C below, this largeness problem does not arise. In Section 7B, we assume that $\mathcal{O}_A(B)$ has a basis $S = \{B_1,\ldots,B_d\}$ consisting of closable operators, and $\bar{S} = \{\bar{B}_j : 1 \le j \le d\}$. Then D itself is, by hypothesis, a C^∞-domain for $\bar{S}$, so $E_\infty(\bar{S}) \supset D$. But D is a dense core domain for $\bar{S}$, so $E_\infty(\bar{S})$ certainly retains these properties. Similarly, in Section 7C we take S to be the Lie generating set for $\mathfrak{L}$, and $\bar{S} = \{\bar{A} : A \in S\}$, so the same reasoning shows that $E_\infty(\bar{S})$ is a dense core domain.

Turning to the topology appropriate for $E_\infty(\bar{S})$, we confine our attention to the case where E is Banach and S is finite. (All remarks made below are true without change if E is Fréchet and S is countable, while suitably modified versions persist in general, but we shall have use only for the finite-Banach case here.) The ideas involved are entirely standard, but we deviate somewhat from the usual discussion for technical and expository reasons. Since it is awkward to separate definitions from enabling lemmas, we proceed informally.

Specifically, putting $\bar{S} = \{\bar{B}_1,\ldots,\bar{B}_d\}$, we proceed by induction to construct a projective sequence $(E_n(\bar{S}),\|\cdot\|_n)$ of Banach spaces, where for all n the space $(E_{n+1},\|\cdot\|_{n+1})$ is obtained from $(E_n, \|\cdot\|_n)$ by precisely the same process that was used to build $(E_1, \|\cdot\|_1)$ from $(E_0, \|\cdot\|_0) = (E, \|\cdot\|)$ in Chapter 5. Then $(E^\infty(\bar{S}), \tau_\infty)$ is obtained as a projective limit. (This tactic avoids juggling multi-indices, and it sets matters up for easy inductive proofs of the results in Sections 7B and 7C). As usual, put $B_0 = 1$, $E_0 = E$, $\|\cdot\|_0 = \|\cdot\|$ (the norm for E). Suppose inductively for a given $n \geq 1$ subspaces $E_k = E_k(\bar{S})$ and norms $\|\cdot\|_k$ have been constructed when $1 \leq k \leq n$, in such a way that

(1) $E_{k+1} \subset E_k \subset D(\bar{B}_j)$ for $1 \leq k < n$, $1 \leq j \leq d$;

(2) each $\bar{B}_j$ maps E_{k+1} into E_k, $0 \leq k < n$, $1 \leq j \leq d$;

(3) $\|u\|_{k+1} = \max\{\|\bar{B}_j u\|_k : 0 \leq j \leq d\}$; and

(4) $(E_k, \|\cdot\|_k)$ is a Banach space.

For the induction step, we first choose $E_{n+1} = \{u \in E_n : \bar{B}_j u \in E_n$ $1 \leq j \leq d\}$, extending (1) and (2), and then define $\|u\|_{n+1} = \max\{\|\bar{B}_i u\|_n\}$ so as to extend (3). Since it is evident that the topology on E_n is stronger than the initial topology τ_0 relativized from $E_0 = E$, it is easy to see that the operators $B_j^{(n)}$ in E_n with domain E_{n+1} are all $\|\cdot\|_n$-closable and that E_{n+1} is exactly the intersection of the domains of their $\|\cdot\|_n$-closures $\bar{B}_j^{(n)}$. The closability claim is established just as we established closability in $(D_1, \|\cdot\|_1)$ of a $\|\cdot\|$-closable A prior to Theorem 5.4 in Section 5B. (The $\|\cdot\|_n$-closure of the graph of $B_j^{(n)}$ in $E_n \times E_n$ is naturally contained in the graph of $\bar{B}_j$ in $E \times E$.) In fact, it is clear that $D(\bar{B}_j^{(n)}) = \{u \in E_n : \bar{B}_j u \in E_n\}$ by this argument, whence the characterization of E_{n+1} as the intersection of the $D(\bar{B}_j^{(n)})$ is immediate. Then the completeness argument used for $E_1(\mathbf{B})$ in Section 5B applies without change to see that $(E_{n+1}, \|\cdot\|_{n+1})$ is complete.

With this accomplished, it is evident that $F = \cap\{E_n : 0 \leq n < \infty\}$ acquires a natural projective Fréchet structure from the Banach spaces $(E_n, \|\cdot\|_n)$. By construction, the operators $\bar{B}_j$ send each E_{n+1} boundedly into E_n, hence they leave F_∞ invariant and act continuously there. Thus $F_\infty \subset E_\infty(\bar{S})$ and it is necessary only to check that $F_\infty \supset E_\infty(\bar{S})$ to see that they are identical. But if F is any C^∞-domain for $\bar{S}$, it is clear by induction on the construction $E_{n+1} = \cap\{u \in E_n : \bar{B}_j u \in E_n\}$ that $F \subset E_n$ for all n, whence $F \subset F_\infty$. Applied to $F = E_\infty(\bar{S})$, this yields the desired containment.

Finally, we contrast the uses of these constructions in Section 7B and 7C. In Section 7B, we apply the construction to obtain $E_\infty(\bar{S})$ for $\bar{S}$ derived from a basis for $\mathcal{O}_A(B)$. But $\bar{S}$ will in general omit the generator $\bar{A}$ of the semigroup $\{V(t,A): t \in [0,\infty)\}$. The task there is to isolate a closed subspace, denoted $D_\infty(\mathcal{O}_A(B))$, which is invariant under this semigroup, and to check that the restriction $\{V_\infty(t,A): t \in [0,\infty)\}$ acts as a C_0-semigroup there. The 'extrinsic' nature of this result can be illustrated by the case where B is an ad A-eigenvector: ad $A(B) = \alpha B$ for $\alpha \in \mathbb{C}$, so that $\mathcal{O}_A(B)$ is spanned by B and iB. Then $E_\infty(\bar{S}) = E_\infty(\{\bar{B}\})$ is exactly the maximal C^∞-domain for $\bar{B}$, and our result will confirm that there exists $D_\infty(\{\bar{B}\})$ (actually $= E_\infty(\{\bar{B}\})$) that is invariant under $V(t,A)$, yielding a nontrivial compatibility relation between $\bar{B}$ and the semigroup generated by A.

In the Lie-algebraic context of Section 7C, on the other hand, the set $\bar{S} = \{\bar{A} : A \in S\}$ determined by the Lie generating set $S \subset \mathcal{L}$ is at once the set of operators defining the C^∞-vector space $E_\infty(\bar{S})$ and the collection of generators of the groups $\{V(t,A): t \in \mathbb{R}\}$ which are intended to leave the ultimate Fréchet subspace $D_\infty(\mathcal{L})$ invariant. Since a non-commutative collection of group-operators $\{V(t,A): t \in \mathbb{R}, A \in S\}$ is involved, the desired domain is more difficult to describe, and its properties are harder to check. Moreover, an extra bit of juggling is involved, because the commutation theory of Chapter 6 requires consideration of the entire Lie algebra $\mathcal{L}$ and its complexification $\mathcal{L}_{\mathbb{C}} = M$. We check, in fact, that $\mathcal{L}$ (hence M) has a closable basis $\mathbf{B}$, and that $E_\infty(\bar{S}) = E_\infty(\bar{\mathbf{B}})$, using the more difficult theory of Sections 6B and 6C. In one respect, the theory is simpler and more satisfactory here: the group-generators $\bar{A}$ restrict to continuous, everywhere-defined endomorphisms of $E_\infty(\bar{S})$ and of $D_\infty(\mathcal{L})$, so that the restricted groups $\{V_\infty(t,A): t \in \mathbb{R}\}$ become differentiable groups on the latter Fréchet space. Ultimately, the results of Chapters 8 and 9 show in this section that every $A \in \mathcal{L}$ pregenerates a group ($\bar{A}$ is the generator), that all of these groups act differentiably upon $D_\infty(\mathcal{L})$, and that $D_\infty(\mathcal{L}) = E_\infty(\bar{S})$.

The general framework for a unifying generalization involves a family F of semigroup pregenerators $A \in \mathbf{A}(D)$, and a finite-dimensional real subspace $\mathfrak{L} \subset \mathbf{A}(D)$ invariant under ad A for all $A \in F$, where $\mathfrak{L}$ must be assumed to have a closable basis S which determines $E_\infty(\bar{S})$. Several additional technical assumptions must be imposed in order to deal with the presence of a noncommutative family $\{V(t,A): t \in [0,\infty), A \in F\}$ of semigroups. We omit details.

7B. The extrinsic two-operator case

We proceed here to treat the situation where $A, B \in \mathbf{A}(D)$ are closable operators, $\mathcal{O}_A(B)$ is finite-dimensional with a closable basis $S = \{B_j : 1 \leq j \leq d\}$, and the graph-density condition holds for at least one λ with $\mathrm{Re}(\lambda)$ sufficiently large: $D_\lambda = (\lambda - A)D$ is dense in D with respect to the norm $\|u\|_1 = \max\{\|B_j u\| : 0 \leq j \leq d\}$ for some $\mathrm{Re}(\lambda) > \omega + |\mathrm{ad}\ A|$, where $\|V(t,A)\| \leq Me^{\omega t}$ for $M < \infty$ and all $t \in \mathbb{R}_+$. As in Section 6A, we obtain from Section 5C that this density condition continues analytically into the half-plane $\mathrm{Re}\,(\lambda) > \omega + \nu(\mathrm{ad}\ A)$ (and possibly beyond).

Our first result, which can be established without a closable basis in which the full strength of Section 6B is used, supplies a not-necessarily-complete globally-invariant domain of exactly the sort discussed in Chapters 3 and 4.

7.2. Theorem

Let $A, B \in \mathbf{A}(D)$ be as described above (where $\mathcal{O}_A(B)$ need not have a closable basis). Then the subspace $D_A = \mathrm{span}\{V(t,A)u: u \in D,\ t \in [0,\infty)\}$ has the following properties:

(a) it is invariant under the operator $V(t,A)$ for all $t \in [0,\infty)$;

(b) it is a C^∞ domain for $\bar{S} = \{\bar{B}\}$ (also for $\bar{S}' = \{\bar{A},\bar{B}\}$ and for $\bar{S}'' = \{\bar{C} : C \in \mathcal{O}_A(B) \text{ closable}\}$);

(c) if $\tilde{A}$ is the restriction of $\bar{A}$ to D_A, $\tilde{B}$ is the comparable restriction of $\bar{B}$, and $\tilde{\mathcal{O}}_A(B) = $ real span$\{(\mathrm{ad}\ \tilde{A})^k(\tilde{B}): 0 \leq k < \infty\}$. Then $\sim$ extends to a natural isomorphism of $\mathcal{O}_A(B)$ as an ad A-module onto $\tilde{\mathcal{O}}_A(B)$ as an ad $\tilde{A}$ module; and

(d) for every $\tilde{C} \in \tilde{\mathcal{O}}_A(B)$ and every $t \in [0,\infty)$

$$\tilde{C}\,V(t,A) = V(t,A)\exp(-t\ \mathrm{ad}\ \tilde{A})(\tilde{C}) \tag{1}$$

as endomorphisms of D_A.

Remark: Strictly speaking, $\mathcal{O}_A(B)$ (respectively, $\tilde{\mathcal{O}}_A(B)$) is a module with respect to the ring of real polynomials p(ad A) (respectively, p(ad $\tilde{A}$)) generated by ad A(ad $\tilde{A}$).

Proof: All claims are first established for actions of operators on the spanning vectors $v = V(s,A)u$ supplied by $s \in [0,\infty)$, $u \in D$. They then extend by linearity to all $v \in D_A$.

Claim (a) is trivial: $V(t,A)v = V(t+s,A)u \in D_A$ whence $V(t,A)D_A \subset D_A$ by linearity.

Similarly, the fact that D_A is a C^∞-domain for $\{A\}$ is trivial: $D \subset D(\bar{A})$ and $V(t,A)D(\bar{A}) \subset D(\bar{A})$ by semigroup theory, so $D_A \subset D(\bar{A})$. But $\bar{A}v = \bar{A}\,V(s,A)u = V(s,A)Au \in D_A$ since $Au \in D$, so by linearity $\bar{A}D_A \subset D_A$. That D_A is a C^∞-domain for B is only slightly deeper. First, note by 6.4 that $J_1V_1(t,A) = V(t,A)J_1$, whence $J_1(D_1)$ is invariant under $V(t,A)$, and since $D \subset J_1(D_1)$ by construction, $D_A \subset J_1(D_1)$. (Here M = complexification $\mathcal{O}_A(B)$). But by 6.3, we have $\bar{B}J_1 \supset B^{(1)}$ and, in particular, $J_1(D_1) \subset D(\bar{B})$ (the same holds for any other closable $C \in \mathcal{O}_A(B)$ as well). Hence for $v = V(s,A)u$, $u \in D$, we have by "$J_1u = u$" that

$$\bar{B}v = \bar{B}V(s,A)u = \bar{B}V(s,A)J_1u = \bar{B}J_1V_1(s,A)u = B^{(1)}V_1(s,A)u$$
$$= V(s,A)[\exp(-s \text{ ad } A)(B)]^{(1)}u = V(s,A)\exp(-s \text{ ad } A)(B)u. \quad (2)$$

Here, we have used 6.4(b), and we have identified D as a subspace of E with D as a subspace of D_1, so that $u = J_1u$ (J_1 is the identity map on D) and $[\exp(-s \text{ ad } A)(B)]^{(1)}u = [\exp(-s \text{ ad } A)(B)]u$, since $[\ ]^{(1)}$ denotes the extension of $[\ ]$ to all of D_1 as a bounded operator into $(E,\|\cdot\|)$. Since $\exp(-s \text{ ad } A)(B)u \in D$, this expresses $\bar{B}v$ as a member of D_A so $\bar{B}D_A \subset D_A$ (the same applies to any closable $C \in \mathcal{O}_A(B)$). By linearity, this completes the proof of (b).

Next, we prove by induction on k that

$$(\text{ad } \tilde{A})^k(\tilde{B})v = (\text{ad } \tilde{A})^k(\tilde{B})V(s,A)u$$
$$= V(s,A)\exp(-s \text{ ad } A)((\text{ad } A)^k(B))u. \quad (3)$$

The case $k = 0$ is exactly (2). Abbreviating $\tilde{C}_k = (\text{ad } \tilde{A})^k(\tilde{B})$

(similarly $C_k = (\mathrm{ad}\ A)^k(B)$), we note that on D_A, $\mathrm{ad}\ \tilde{A}(\tilde{C}_k)v = (\bar{A}\tilde{C}_k - \tilde{C}_k\bar{A})v$, so that here

$$\tilde{C}_{k+1}v = \mathrm{ad}\ \tilde{A}(\tilde{C}_k)v = \bar{A}\tilde{C}_k V(s,A)u - \tilde{C}_k\bar{A}V(s,A)u$$

$$= \bar{A}V(s,A)\exp(-s\ \mathrm{ad}\ A)(C_k)u - \tilde{C}_k V(s,A)Au$$

$$= V(s,A)[A\ \exp(-s\ \mathrm{ad}\ A)(C_k)u - \exp\ (-s\ \mathrm{ad}\ A)(C_k)A]u$$

$$= V(s,A)\exp(-s\ \mathrm{ad}\ A)(\mathrm{ad}\ A(C_k))u = V(s,A)\exp(-s\ \mathrm{ad}\ A)(C_{k+1})u$$

where we have employed both the induction hypothesis and the fact that $\bar{A}u = Au \in D$ when $u \in D$. It follows easily by linearity from this that for any polynomial p, we have

$$p(\mathrm{ad}\ \tilde{A})(\tilde{B})v = V(s,A)\exp(-s\ \mathrm{ad}\ A)(p(\mathrm{ad}\ A)(B))u \qquad (4)$$

and

$$\mathrm{ad}\ \tilde{A}(p(\mathrm{ad}\ \tilde{A})(B))v = V(s,A)\exp(-s\ \mathrm{ad}\ A)(\mathrm{ad}\ A(p(\mathrm{ad}\ A)(B))v. \quad (5)$$

Then since the action of ad A on $\mathcal{O}_A(B)$ is algebraically determined by the monic minimal polynomial p (cf. Appendix D) such that $p(\mathrm{ad}\ A)(B) = 0$ (i.e., leading coefficient 1), (4) quickly shows that $p(\mathrm{ad}\ \tilde{A})(\tilde{B}) = 0$ on D_A if and only if $p(\mathrm{ad}\ A)(B) = 0$, so that the action of ad A (and its polynomials) on $\mathcal{O}_A(B)$ is identical to that of ad $\tilde{A}$ on $\tilde{\mathcal{O}}_A(B)$. Hence (c) is proved.

Finally, for (d) we first observe that (4) may be interpreted now as

$$\tilde{C}V(s,A)u = V(s,A)\exp(-s\ \mathrm{ad}\ A)(C)u \qquad (6)$$

for all $s \in [0,\infty)$, $u \in D$, and corresponding operators $\tilde{C}$ and C. Hence for $v = V(s,A)u$ and any $t \in [0,\infty)$

$$\tilde{C}V(t,A)v = \tilde{C}V(s+t,A)u = V(s+t,A)\exp(-(s+t)\mathrm{ad}\ A)(C)u$$

$$= V(t,A)[V(s,A)\exp(-s\ \mathrm{ad}\ A)(\exp(-t\ \mathrm{ad}\ A))(C)]u$$

$$= V(t,A)\exp(-t\ \mathrm{ad}\ \tilde{A})(\tilde{C})V(s,A)u$$

$$= V(t,A)\exp(-t\ \mathrm{ad}\ \tilde{A})(\tilde{C})v \qquad (7)$$

by a double application of (6). (Notice that since $\exp(-t\ \mathrm{ad}\ \tilde{A})(\tilde{C})$ is a polynomial in $(\mathrm{ad}\ \tilde{A})$ applied to $\tilde{B}$, by finite-dimensionality of $\mathcal{O}_A(B)$, (4) applies.) But then (7) extends by linearity to all $v \in D_A$ and (1) is proved.

The following corollary then moves from D_A to its closure in $E_\infty(\bar{S}) = E_\infty(\bar{\mathbf{B}})$ as described in Section 7A.

7.3. Corollary

Let A, B $\in \mathbf{A}(D)$ be as in 7.2, and suppose in addition that $\mathcal{O}_A(B)$ has a closable basis $\mathbf{B} = \{B_i : 1 \le i \le d\}$ (e.g. by 6.5(1)). Let $D_\infty(\mathcal{O}_A(B)) \subset D_1$ be the closure of D_A in $E_\infty(\bar{\mathbf{B}})$ as described above, with projective Fréchet topology τ_∞.

(a) The semigroup $\{V(t,A): t \in [0,\infty)\}$ on E leaves $D_\infty(\mathcal{O}_A(B))$ invariant, and restricts to a continuous, locally equicontinuous (c.l.e.) semigroup $\{V_\infty(t,A): t \in [0,\infty)\}$ on $(D_\infty(\mathcal{O}_A(B)),\tau_\infty)$.

(b) Every operator $C \in \mathcal{O}_A(B)$ has an extension C_∞ to $D_\infty = D_\infty(\mathcal{O}_A(B))$ via $C_\infty u = \Sigma b_i \bar{B}_i u = C^{(1)}u$ for $u \in D_\infty$ when $C = \Sigma b_i B_i$ in $\mathcal{O}_A(B)$. For each such C_∞, we have in $\mathbf{A}(D_\infty)$

$$C_\infty V_\infty(t,A) = V_\infty(t,A)[\exp(-t \text{ ad } A)(C)]_\infty. \tag{8}$$

(c) As a result, if $\|V(t,A)\| \le M \exp(\omega t)$ on E, for $M, \omega < \infty$, then for each integer n and each $u \in D_\infty$

$$\|V_\infty(t,A)u\|_n \le M \exp((\omega+n|\text{ad } A|)t) \|u\|_n . \tag{9}$$

Remarks: (1) The continuity and local equicontinuity of the restriction semigroup $\{V_\infty(t,A): t \in [0,\infty)\}$ in (a) sometimes follows automatically, by category arguments, from invariance above. The method is related to that in 3.7 and in 7.5 below, but we omit details. Estimates of the sort in (9) need not hold without (b).

(2) Notice that in (b), we do not in general know that D_∞ is a C^∞-domain for $\bar{A}$, or even that it is contained in $D(\bar{A})$. Hence $[\exp(-t \text{ ad } A)(C)]_\infty$ has meaning only as an analytic function of t with values in the (isomorphic) space $\mathcal{O}_A^\infty(B)$ of extensions to $\mathbf{A}(D_\infty)$ of elements in $\mathcal{O}_A(B)$ or as a restriction to D_∞ of an analytic operator-valued function with values in the Banach space $L(D_1,E)$. That is, "ad A" makes sense on $\mathcal{O}_A^\infty(B)$ only as the isomorphic copy of ad A on $\mathcal{O}_A(B)$, but unlike the situation on D_A in 7.2, it does not necessarily arise as the operator-commutator of an operator $A_\infty \in \mathbf{A}(D_\infty)$ with operators in $\mathcal{O}_A^\infty(B)$.

(3) The estimates (9) show that the semigroup operators $V_\infty(t,A)$ are members of a locally multiplicatively convex algebra of operators on D_∞ described in [Mc] and termed

"ultracontinuous" by Moore [Mr 6]. Babalola [Bb 1] has given an abstract treatment of semigroups of such operators, including a Hille-Yosida Theorem and some perturbation results. The recent papers of Giles and co-authors ([GK], [GJKS]) are also pertinent to future analysis of such semigroups.

Proof: We first apply 7.2 (particularly eqn. (1)) to obtain (8) and (9) specifically, for $u \in D_A$, $t \in [0,\infty)$ and $B_i \in \mathbf{B}$, we have

$$\tilde{B}_i V(t,A)u = V(t,A)\exp(-t \text{ ad } \tilde{A})(\tilde{B}_i)u. \tag{10}$$

Inductively, beginning with n = 0, suppose that already

$$\|V(t,A)u\|_n \leq M \exp(\omega + n|\text{ad } A|)t \ \|u\|_n \tag{11}$$

is known. Then

$$\begin{aligned} \|\tilde{B}_i V(t,A)u\|_n &\leq M \exp((\omega+n|\text{ad } A|)t)\|\exp(-t \text{ ad } \tilde{A})(\tilde{B}_i)u\|_n \\ &\leq M \exp((\omega+n|\text{ad } A|)t)\exp(t|\text{ad } \tilde{A}|)|\tilde{B}_i| \ \|u\|_{n+1} \\ &\leq M \exp[(\omega+(n+1)|\text{ad } A|)t]\|u\|_{n+1} \end{aligned} \tag{12}$$

since by 7.2(b), ad $\tilde{A}$ acts on $\tilde{\mathcal{O}}_A(B)$ as ad A does on $\mathcal{O}_A(B)$, and since estimates of the form $\|\tilde{C}u\|_n \leq |\tilde{C}| \ \|u\|_{n+1}$ follow by the same reasoning used to establish $\|Cu\| \leq |C| \ \|u\|_1$ in Chapter 5. But then by definition since $(\bar{B}_i)_n \supset \tilde{B}_i$,

$$\begin{aligned} \|V(t,A)u\|_{n+1} &= \max\{\|\bar{B}_i V(t,A)u\|_n : 0 \leq i \leq d\} \\ &\leq M \exp[(\omega+(n+1)|\text{ad } A|)t]\|u\|_{n+1} \end{aligned} \tag{13}$$

since $\exp((\omega+n|\text{ad } A|)t) \leq \exp[(\omega+(n+1)|\text{ad } A|)t]$ (for the i=0 case). Hence the induction is complete.

Using (13), we see that each V(t,A) extends by limits to $V_n(t,A) \in L(D_n)$ where D_n is the $\|\cdot\|_n$-closure of D_A in $E_n(\mathbf{B})$. The algebraic semigroup properties also extend, along with the norm-estimate (13). In addition, (10) shows inductively that if $t \to V(t,A)u = V_n(t,A)$ is $\|\cdot\|_n$-continuous for a given n when $u \in D_A$, then $t \to \tilde{B}_i V_n(t,A)u$ is also $\|\cdot\|_n$-continuous, whence easily $t \to V(t,A)u = V_{n+1}(t,A)u$ is $\|\cdot\|_{n+1}$-continuous. But continuity on dense D_A when combined with the extended estimates (13) then ensures by the usual 3-ε argument that $V_{n+1}(t,A)$ is strongly continuous on D_{n+1}. But we also see that since $V_n(t,A)$ and $V(t,A)$ agree (by construction) on D_A and since $\|\cdot\|_n$-limits are also $\|\cdot\|_0 = \|\cdot\|$-limits, $V_n(t,A)$ and $V(t,A)$ must agree on all of D_n, so each D_n is invariant under $V(t,A)$. It is easy to check that $D_\infty = D_\infty(\mathcal{O}_A(B)) = \cap\{D_n : 0 \le n < \infty\}$, whence by the above information, $V(t,A)$ leaves D_∞ invariant. By the nature of projective topologies $t \to V(t,A)u = V_\infty(t,A)u$ is continuous with respect to the limit Fréchet topology τ_∞, and the extended estimates from (13) restrict on D_∞ to (9), so (a) and (c) follow. Moreover, (10) easily extends by τ_∞ limits to all $u \in D_\infty$, yielding

$$\bar{B}_i V_\infty(t,A)u = V_\infty(t,A)[\exp(-t \text{ ad } A)(B_i)]_\infty u \tag{14}$$

since τ_∞-limits on the left are certainly τ_0 (or $\|\cdot\|$) limits and closedness of the $\bar{B}_i$ with respect to $\|\cdot\|$ does the rest. Then (8) follows from (14) by linear combinations. (We observe that the operators C_∞ are also the continuous endomorphisms of (D_∞,τ_∞) obtained by τ_∞-limit-extention from the $\tilde{C} \in \mathcal{O}_A(\tilde{B}) \subset \mathbf{A}(D_A)$. We have used this interpretation on the right in (14). E.O.P.

<u>Remark</u>: Since we made no direct use of the construction of D_A (from 7.2) in the course of the proof of 7.3, precisely the same argument can be applied to pairs A, $B \in \mathbf{A}(D)$ satisfying the conditions of Chapter 3, with D replacing D_A: there $\mathcal{O}_A(B)$ is finite-dimensional, $V(t,A)D \subset D$ by hypothesis, and for each $u \in D$ the function $G_u(t) = BV(t,A)u$ is differentiable with $G_u'(t) = BV(t,A)Au$ (either by hypothesis or via Section 3C). Then 3.4 supplies a commutation relation replacing (1) above, and the argument of 7.3 supplies a $V(t,A)$-invariant Fréchet completion $D_\infty(\mathcal{O}_A(B))$ (when E is Banach or Fréchet) which satisfies (a) - (d) of 7.3. It is unnecessary to record the result formally.

7C. <u>The Lie algebra case</u>

In the interests of brevity, we combine the analogs of 7.2 and 7.3 together in a single result concerning Lie-generating subsets S of finite-dimensional real Lie algebras $\mathfrak{L} \subset \mathbf{A}(D)$.

7.4. Theorem

Let S be a (finite) Lie-generating set for a finite-dimensional Lie subalgebra $\mathcal{L} \subset \mathbf{A}(D)$ on a dense domain D in a Banach space E. Suppose that each $A \in S$ is a pregenerator of a C_0 group $\{V(t,A)\colon t \in \mathbb{R}\}$ on E and that each $D_\lambda(A) = (\lambda - A)D$ is $\|\cdot\|_1$-dense in D for all $A \in S$ and $|\lambda|$ sufficiently large. Let D_S be the smallest subspace of E that contains D and is invariant under all $V(t,A)$ for $A \in S$ and $t \in \mathbb{R}$.

(a) Then $D_S \subset E_\infty(\bar{S})$, the closure $D_\infty(S)$ of D_S in $E_\infty(\bar{S})$ is invariant under every $V(t,A)$ for $A \in S$ and $t \in \mathbb{R}$, and the restriction groups $\{V_\infty(t,A)\colon t \in \mathbb{R}\}$ act as differentiable locally equicontinuous groups on this Fréchet space.

(b) The spaces $D_\infty(S)$ and D_S are both C^∞-domains for S.

(c) If S_∞ denotes the set of restrictions A_∞ to $D_\infty(S)$ of the closures $\bar{A}$ of elements $A \in S$, then S_∞ generates a Lie algebra $\mathcal{L}_\infty$ of continuous endomorphisms of the Fréchet space $D_\infty(S)$, and $\mathcal{L}_\infty$ is isomorphic to $\mathcal{L}$ via the obvious map from $\mathcal{L}$ to $\mathcal{L}_\infty$ induced by this identification of generating sets.

(d) For all $C_\infty \in \mathcal{L}_\infty$ and $A_\infty \in S_\infty$ (corresponding to $A \in S$) we have

$$C_\infty V_\infty(t,A) = V_\infty(t,A)\exp(-t \text{ ad } A_\infty)(C_\infty) \tag{15}$$

as an identity in the algebra $L(D_\infty(S))$ of continuous operators on the Fréchet space $D_\infty(S)$. This identity restricts to a comparable identity in $\mathbf{A}(D_{\mathcal{L}})$.

(e) If $\|V(t,A)\| \le M \exp(\omega|t|)$ on E and $|\text{ad } A|$ denotes any norm for the action of ad $A \in \text{ad}(S)$ on $\mathcal{L}$, then we have for each n and $t \in \mathbb{R}$,

$$\|V_\infty(t,A)\|_n \le M \exp[(\omega + n|\text{ad } A|)|t|]. \tag{16}$$

Proof: As previously indicated, Corollary 6.6 shows that $\mathcal{L}$ (and its complexification M) has a closable basis $\mathbf{B}$, so that our proof can be carried out in the setting of 6.1. Since the operators $\pm A$ satisfy the conditions of 6.1 for each $A \in S$, it follows from 6.1(1) that D_1 is a subspace of E which is invariant with respect to all of the $V(t,A)$ for $t \in \mathbb{R}$ and $A \in S$ (recall that $V(t,A) = V(-t,-A)$ for $t \in (-\infty, 0]$). Hence $D_S \subset D_1$ and by eqn. (6.1)

$$C^{(1)}V(t,A)u = V(t,A)[\exp(-t \text{ ad } A)(C)]^{(1)}u \tag{17}$$

for all $C \in \mathfrak{L}$ and $u \in D_1$ (hence all u in D_S or $D_\infty(S)$). Once we check that D_S is a C^∞-domain, (17) quite quickly leads to (15) by τ_∞-limits.

In order to check quickly that D_S is a C^∞-domain for S, it is convenient to realize it constructively as the union of an inductively-defined sequence of C^∞-domains as follows: $D_S^0 = D$ and $D_S^{n+1} = \text{span}\{V(t,A)u : u \in D_S^n, t \in \mathbb{R}, A \in S\}$. Clearly, $V(t,A)D_S^n \subset D_S^{n+1}$ for all n, $t \in \mathbb{R}$ and $A \in S$, so that each $V(t,A)$ sends $D_S^\infty = \cup\{D_S^n : 0 \leq n < \infty\}$ into itself. Then since $D_S^\infty \supset D$, we have $D_S^\infty \supset D_S$. But since $D \subset D_S$ and D_S is $V(t,A)$-invariant, it follows by induction that $D_S^n \subset D_S$, which yields the reverse inclusion $D_S^\infty \subset D_S$.

Now, by hypothesis $D_S^0 = D$ is a C^∞ domain (i.e., $S \subset \mathbf{A}(D)$), and for $C \in S$ (or any closable $C \in \mathfrak{L}$) we know that $C^{(1)}u = \bar{C}u$ for all $u \in D_1$. Thus, if we suppose inductively that D_S^n is a C^∞ domain, for any $v = V(t,A)u$ in the spanning set for D_S^{n+1} (i.e., $u \in D_S^n$, $A \in S$, $t \in \mathbb{R}$) and $C \in S$ we have

$$\bar{C}v = \bar{C}V(t,A)u = C^{(1)}V(t,A)u = V(t,A)[\exp(-t\text{ ad }A)(C)]^{(1)}u$$

$$= V(t,A)[\exp(-t\text{ ad }A)(C)]^{-}u. \tag{18}$$

Hence, since $u \in D_S^n$ and $[\exp(-t\text{ ad }A)(C)]^{-}D_S^n \subset D_S^n$, it follows that the last expression in (19) lies in D_S^{n+1}, whence $\bar{C}v \in D_S^{n+1}$ and by linearity $\bar{C}D_S^{n+1} \subset D_S^{n+1}$. We conclude from this induction that D_S, as a union of C^∞-domains, is itself a C^∞-domain. Consequently, $D_S \subset E_\infty(S)$ and $D_\infty(S)$ can be formed as described.

The next stage in the argument uses (17) to establish by induction that for $u \in D_S$, $A \in S$ and $t \in \mathbb{R}$

$$\|V(t,A)u\|_n \leq M \exp[(\omega + n|\text{ad }A|)|t|] \tag{19}$$

where M and ω are as in (e). The calculation exactly repeats (10)-(13) in the proof of 7.3 and will not be recapitulated here. As in 7.3 these estimates extend by τ_∞-limits to all $u \in D_\infty(S)$, establishing (e) and the invariance and local equicontinuity claims in (a). Continuity of the map $t \to V_\infty(t,A)u \in (D_\infty(S),\tau_\infty)$ also follows by induction exactly as it did in 7.3. Hence (as in 3.3) we obtain the integral formula for $u \in D_\infty(S)$

$$V_\infty(t,A)u-u = \int_0^t V_\infty(s,A)A_\infty u\, ds. \qquad (20)$$

This identity holds as a $\|.\|$-convergent integral formula since $u \in D(\bar{A})$, but the integral is τ_∞-convergent by the continuity established above. Hence, since the integrand is τ_∞-continuous, $V_\infty(t,A)u$ is τ_∞-differentiable and $(d/dt)V_\infty(t,A)u = V_\infty(t,A)A_\infty u$, an operator in $L(D_\infty(S))$.

To obtain (c), we observe that since all elements of $E(\mathfrak{L})$ are polynomials in members of S, hence τ_∞-continuous as endomorphisms of D, all algebraic relations in $E(\mathfrak{L})$ extend by τ_∞-limits to the formally identical relations in $E(\mathfrak{L}_\infty) \subset L(D_\infty(S))$.

In particular, $\mathfrak{L}_\infty$ has the same structure as $\mathfrak{L}$. Moreover, since $\exp(-t \text{ ad } A)(C)$ is expressible as a polynomial in elements of S (by linear dependence of commutators) it follows that $[\exp(-t \text{ ad } A)(C)]_\infty = \exp(-t \text{ ad } A_\infty)(C_\infty)$. Since this τ_∞-extension $[\ ]_\infty$ must agree on $D_\infty(S)$ with the closure $[\ ]^-$ in E and the bounded extension $[\ ]^{(1)}$ from D_1 to E, we obtain (15) from (17) and the proof is complete.

The following results indicates that all of the algebraic and quantitative results in the preceding theorem are actually automatic consequences of the invariance property in one important special case.

7.5. Proposition

Let S be a Lie-generating set for $\mathfrak{L} \subset \mathbf{A}(D)$, and suppose that each $A \in S$ is a pregenerator. Suppose that the smallest subspace $D_S \subset E$ that is invariant under all $\{V(t,A): t \in \mathbb{R}, A \in S\}$ is contained in $E_\infty(\bar{S})$, and that its τ_∞-closure $D_\infty(S)$ in $E_\infty(\bar{S})$ is invariant under all $V(t,A)$. Then all other conclusions in 7.4 follow automatically.

Proof: First we observe, as in the proof of 7.4(c) above, that $\mathfrak{L}$ extends isomorphically to $\mathfrak{L}_\infty \subset L(D_\infty(S))$ by τ_∞-limits. (That argument depended only upon the construction of $E_\infty(\bar{S})$ and the fact that D_S is a C^∞-domain, not upon consequences of the $D_\lambda(A)$-density hypothesis). Hence, we may replace S by S_∞ and $\mathfrak{L}$ by $\mathfrak{L}_\infty$ in what follows. Moreover, since $A \subset A_\infty \subset \bar{A}$ for all $A \in S$, A_∞ is a pregenerator on $D_\infty(S)$.

Consequently, we may pick A_∞, B_∞ in S_∞ and apply Proposition 3.6 to conclude that for all $u \in D_\infty(S)$, $B_\infty V_\infty(t,A)u = \bar{B}V(t,A)u$ is differentiable in t with the appropriate derivative. Thus, since $\mathcal{L}_\infty$ is finite-dimensional, $\mathcal{O}_{A_\infty}(B_\infty)$ also must be, and Theorem 3.2 (or 3.5) applies to yield the commutation relation (15) directly when $C_\infty \in \mathcal{S}_\infty$. But then for $B_\infty, C_\infty \in \mathcal{S}_\infty$ we get

$$\begin{aligned}[B_\infty,C_\infty]V_\infty(t,A) &= B_\infty V_\infty(t,A)\exp(-t \text{ ad } A_\infty)(C_\infty) \\ &\qquad - C_\infty V_\infty(t,A)\exp(-t \text{ ad } A_\infty)(B_\infty) \\ &= V_\infty(t,A)[\exp(-t \text{ ad } A_\infty)(B_\infty), \exp(-t \text{ ad } A_\infty)(C_\infty)] \\ &= V_\infty(t,A)\exp(-t \text{ ad } A_\infty)([B_\infty,C_\infty]), \qquad (21)\end{aligned}$$

using the well-known automorphism action of $\exp(t \text{ ad } A_\infty)$ in a finite-dimensional Lie algebra $\mathcal{L}_\infty$ to justify the last identity. Applying this calculation inductively, we obtain (15) for any iterated commutator $C_\infty = [A^1_\infty,[A^2_\infty \dots,[A^{k-1}_\infty, A^k_\infty]\dots]]$ and by linearity, for any $C_\infty \in \mathcal{L}_\infty$. The extended (15) is then applied as in the proof of 7.4 to establish the estimates (16) and other claims in 7.4.

7D. C^∞-action of resolvents, projections and operational calculus

This section supplies an informal discussion of the behavior of resolvents $R(\lambda,A)$, projections P_σ associated with spectral sets $\sigma \subset \sigma(A)$, and operational images $\varphi(A)$, as these act upon the projective sequence $(D_n,\|\cdot\|_n)$ of C^n-vector Banach spaces and on the limit Fréchet space D_∞. For concreteness, we discuss the two-operator situation of Section 7B, but assuming that $\bar{A}$ generates a group (so that the discussion in Section 6D applies). Similar phenomena arise in the Lie algebra setting of Section 7C, and for semigroups.

The main thrust of our remarks can be summarized as follows. If ad A acts on $M = \mathcal{O}^{\mathbb{C}}_A(B) = \text{complex span}\{(\text{ad } A)^k(B)\}$ as a nilpotent transformation, then the bounded operators $R(\lambda,A)$, P_σ and $\varphi(A)$ exhibit the same properties on all of the spaces $D_n = D_n(\mathcal{O}_A(B))$ that they had on E, and these properties carry over in a reasonable way on $D_\infty(\mathcal{O}_A(B))$. With minor (but important) qualifications, the same remains true if all of the eigenvalues of ad A are imaginary. (The qualifications concern the spectral projections, which retain their properties only for those spectral sets σ which are invariant under translation by the additive semigroup generated by $\sigma(\text{ad } A)$.) But if $\sigma(\text{ad } A)$ contains points with nontrivial real part, the behavior of resolvents, projections and functional images tends to deteriorate on the higher order $D_n(\mathcal{O}_A(B))$, and neither resolvents nor spectral projections need leave $D_\infty(\mathcal{O}_A(B))$ invariant. Moreover, the group $\{V_\infty(t,A): t \in \mathbb{R}\}$ on $D_\infty(\mathcal{O}_A(B))$ typically exhibits faster-than-exponential growth in t, so that the (Fourier/Laplace) operational calculus on D_∞ generally applies only to transforms φ of measures Φ which vanish at ∞ faster than any negative exponential.

Beginning with the nilpotent case, we observe first by 7.2 that without loss of generality we may assume that D is $V(t,A)$-invariant. (Otherwise, replace it first by D_A and then in turn by $(D_A)_{-A}$ = complex span $\{V(t,A)u: u \in D, t \in \mathbb{R}\}$ to obtain invariance under $V(t,A)$ for $\pm t \in [0,\infty)$. Notice that D_1 is the $\|\cdot\|_1$-closure of this enlarged domain.) The idea is to follow the inductive process down the sequence $D_n = D_n(\mathcal{O}_A(B))$, first for the theory of Chapter 5 and then for that of Chapter 6.

In the setting of Chapter 5, we have already observed that if ad A is nilpotent, $\sigma(\text{ad } A) = \{0\}$ and the augmented spectrum $\sigma(A;M)$ coincides with $\sigma(A)$. Consequently, if some (hence all) λ with $|\text{Re}(\lambda)|$ large satisfy the graph-density condition, then $\rho(\bar{A}_1)$ (the resolvent set of the $\|\cdot\|_1$-closure of A in D_1) includes at least the unbounded right and left components of $\rho(A)$ (the resolvent set of A on E) and $R(\lambda,A)$ leaves D_1 invariant for these λ. Since D_2 is related to D_1 as D_1 is related to E, it will follow that the resolvent set $\rho(\bar{A}_2)$ of the $\|\cdot\|_2$-closure of A also contains these left and right components, once we check that $D_\lambda = (\lambda - A)D$ is $\|\cdot\|_2$-dense in D for suitable λ. But by the proof of Theorem 7.3, $\{V(t,A)\}$ restricts to a C_0 group $\{V_2(t,A): t \in \mathbb{R}\}$ on D_2, whence by Laplace transforms the resolvents $R_2(\lambda,A) \in L(D_2)$ exist for $|\text{Re}(\lambda)|$ large, and 5.7 then implies the desired density property when $\|\cdot\|_2$ is viewed as the new "$\|\cdot\|_1$ norm on the Banach-space $(D_1,\|\cdot\|_1)$". The induction is obvious: the closure $\bar{A}_n$ of A in $(D_n,\|\cdot\|_n)$ has at least the right and left components of $\rho(A)$ in its resolvent set, and for all λ in these components, the resolvent $R(\lambda,A)$ on E leaves D_n

invariant and acts boundedly there. Consequently, all such $R(\lambda,A)$ leave D_∞ invariant as well, and act continuously there. A slightly more detailed examination of resolvent estimates, using renorming techniques, shows that these right and left components of $\rho(A)$ are contained in the "finite resolvent set" for the closure $\bar{A}_\infty$ of A on the Fréchet space D_∞, as described in [Mr 6].

The behavior of spectral projections P_σ is less accessible unless $\sigma(A) \subset i\,\mathbb{R}$. If σ is guaranteed to be entirely surrounded by the union of the left and right components of $\rho(A)$ then P_σ can be shown (by the same inductive process on 5.9) to restrict to a bounded projection on every D_n and to a continuous projection on D_∞. When $\sigma(A) \subset i\,\mathbb{R}$, then the presence of any nontrivial spectral set σ ensures that there are gaps in $\sigma(A)$ and thus that $\rho(A)$ has a single component, whence <u>all</u> P_σ exhibit the behavior just described.

Similar remarks apply to the holomorphic operational calculus, of which the P_σ construction is a special case. The more powerful transform calculi require a more detailed examination of the results in Chapter 6, notably Theorem 6.1. When ad A is nilpotent, both forms of identity (6-1) collapse into a single formula for each basis element $B_i \in \mathcal{O}_A(B)$ and each $u \in D_1$:

$$\bar{B}_i V(t,A)u = B_i^{(1)} V_1(t,A)u$$
$$= V(t,A)\Sigma\{(-t)^k/k![(\mathrm{ad}\ A)^k(B_i)]^{(1)}u\colon 0 \le k \le s\}. \quad (22)$$

it follows easily from this that if $\|V(t,A)\| \le Me^{\omega|t|}$,

$$\|\bar{B}_i V(t,A)u\| \le Me^{\omega|t|}\Sigma\{|t|^k/k!\,|\mathrm{ad}\ A|^k|B_i|\,\|u\|_1\colon 0 \le k \le s\} \quad (23)$$

so that by the definition of $\|\cdot\|_1$ we get

$$\|\exp(-\omega|t|)V_1(t,A)\|_1 \le \Sigma\{C_k|t|^k\colon 0 \le k \le s\} \quad (24)$$

for appropriate coefficients C_k. That is, $V_1(t,A)$ on D_1 grows in $|t|$ at a rate only polynomially greater than the growth of $V(t,A)$ on E, where the polynomial has degree at most s. Inductive use of the estimation procedure in (23) reveals that, on D_n, $\|\exp(-\omega|t|)V_n(t,A)\|_n \le pol_n(|t|)$, where the positive polynomial pol_n has degree at most ns. Consequently, for any $\varepsilon > 0$, there is an $M_n^\varepsilon < \infty$ such that $\|V_n(t,A)\|_n \le M_n^\varepsilon \exp((\omega+\varepsilon)|t|)$ $t \in \mathbb{R}$. Thus for general $\omega \ge 0$, the bilateral Laplace transform calculus applies on all D_n for precisely the functions φ to which it applies on E, and $\varphi(A)$ maps each D_n $\|\cdot\|_n$-boundedly into itself. In the limit, φ maps D_∞ continuously into itself (actually acting there as a "finite" operator in the sense of [Mr 6].) For $\omega = 0$, particularly $\bar{A}$ skew-adjoint on a Hilbert space, the polynomial estimates show that whenever the measure Φ has finite moments of all orders, its Fourier transform φ acts as described upon D_n and D_∞. (One argues essentially as in [Gd 3].)

Turning to the imaginary eigenvalue situation ($\sigma(\mathrm{ad}\ A) \subset i\,\mathbb{R}$). Suppose that $\|V(t,A)\| \le Me^{\omega|t|}$ for all $t \in \mathbb{R}$. Then $\sigma(A) \subset S_\omega = \{\lambda \in \mathbb{C} : |\mathrm{Re}(\lambda)| \le \omega\}$, the strip of width 2ω centered on the imaginary axis. Although the augmented spectrum $\sigma(A;M) = \sigma(A) \cup \{\lambda - \alpha(j),\ \lambda \in \sigma(A),\ \alpha(j) \in \sigma(\mathrm{ad}\ A)\}$ may well be larger than $\sigma(A)$ in this case, it will still be contained in this strip, and if all λ with large $|\mathrm{Re}(\lambda)|$ satisfy the graph-density condition, then Theorem 5.4 ensures that $\sigma(\bar{A}_1)$ in D_1 must also lie in S_ω. The induction down the sequence $(D_n, \|\cdot\|_n)$ then proceeds essentially as in the nilpotent case, showing that $\sigma(\bar{A}_n) \subset S_\omega$ for all n and that if $\lambda \notin S_\omega$ then $R(\lambda,A)$ leaves D_n invariant with bounded restriction there. Consequently, all such $R(\lambda,A)$ leave the C^∞ vectors D_∞ invariant and act continuously there. In fact, if we let $\sigma_\infty(A;M)$ denote the smallest set continuing $\sigma(A)$ and invariant under translation by $-\alpha(j)$ for all $\alpha(j) \in \sigma(\mathrm{ad}\ A)$, this resolvent behavior extends to all $\lambda \notin \sigma_\infty(A;M)$.

As mentioned previously, projection-behavior is more complicated in this case. If σ_∞ is any closed, relatively open subset of $\sigma_\infty(A;M)$, one can show that $\sigma = \sigma_\infty \cap \sigma(A)$ has a spectral projection P_σ that acts boundedly upon all D_n and continuously on D_∞, provided that σ_∞ is surrounded by the union of the unbounded left and right components of $\mathbb{C} \sim \sigma_\infty(A;M)$. It is not clear as to the possible usefulness of a detailed result of this kind. The same remarks apply to the holomorphic operational calculus.

In the imaginary-eigenvalue case, the group-behavior and transform operational calculi turn out to behave just as in the nilpotent case, but for computationally messier reasons. In fact the estimate (23) is replaced here by

$$\begin{aligned}\|\bar{B}_i V(t,A)u\| &= \|B^{(1)}V_1(t,A)u\| \\ &\le \|V(t,A)\|\,\|\Sigma\{(-t)^k/k!\,e^{-t\alpha(j)}[(\mathrm{ad}\ A-\alpha(j))^k(B_i)]^{(1)}u: \\ &\qquad 1 \le j \le p,\ 0 \le k \le s_j\} \\ &\le Me^{\omega|t|}\Sigma\{|t|^k/k!\,|\mathrm{ad}\ A-\alpha(j)|^k|B_i|\,\|u\|_1: \\ &\qquad 1 \le j \le p,\ 0 \le k \le s_j\} \qquad (25)\end{aligned}$$

since $\alpha(j) \in i\,\mathbb{R}$ implies $|e^{-t\alpha(j)}| = 1$ for all t, j. Thus if $s = \max\{s_j\}$ is the maximum ascent of the eigenvalues, we obtain as before $\|\exp(-\omega|t|)V_1(t,A)\|_1 \le \mathrm{po}\ell_1(|t|)$ with $\mathrm{po}\ell_1(|t|)$ of degree s, and in general $\|\exp(-\omega|t|)V_n(t,A)\|_n \le \mathrm{po}\ell_n(|t|)$ with degree ns just as before. Consequently, all remarks concerning the Laplace and Fourier calculi in the nilpotent case apply here as well, for the same reasons. Notice that the remarks here apply for A, B $\in \mathfrak{L} \subset \mathbf{A}(D)$ a Lie algebra whenever $\mathfrak{L}$ is compact semisimple, or more generally $\mathfrak{L} = \mathfrak{L}_C \oplus N$ with nilpotent radical N and compact Levi factor $\mathfrak{L}_C$. (Since some solvable Lie algebras also give rise to the imaginary ad A-eigenvalue situation as well-notably solvable subalgebras of compact algebras - the cases above do not exhaust the Lie algebraic applications.)

Finally, we sketch the sort of pathological behavior that can occur if some eigenvalues $\alpha(j)$ for ad A have nonzero real part. Then, the augmented spectrum $\alpha(A;M)$ moves outside the ω-strip S_ω, and the repeated augmentation involved in the induction process can lead in some cases to $-\alpha(j)$-translation-invariant sets $\sigma_\infty(A;M)$ (notation as in the pure imaginary case above) which meet every half-plane. It can occur that the sequence of spectra $\sigma(\bar{A}_n)$ of the $\|\cdot\|_n$-closures $\bar{A}_n$ of A in D_n can form an expanding family of strips which in the limit cover all of $\mathbb{C}$. Thus <u>no</u> resolvent $R(\lambda,A)$ need leave every D_n invariant (no matter how large $|\mathrm{Re}(\lambda)|$ may be) and no resolvents of A need leave D_∞ invariant in the limit. This pathology will be reflected in the holomorphic operational calculus and its associated treatment of spectral projections.

From the group perspective, one sees by the analog of (25) when some $\mathrm{Re}(\alpha(j)) \ne 0$ (or as in the example of Section 6B) that $\|V_1(t,A)\|_1$ must grow at least as fast as

$\exp((\omega+r)|t|)$, where $r = \max\{|\mathrm{Re}(\alpha(j))| : \alpha(j) \in \sigma(\mathrm{ad}\ A)\}$. Inductively, $\|V_n(t,A)\|_n$ grows at least as fast as $\exp((\omega+nr)|t|)$ and as a result, $V_\infty(t,A)$ exhibits "faster than exponential" growth on D_∞ (i.e., for no number ν is it the case that $\{\exp(-\nu|t|)V_\infty(t,A): t \in \mathbb{R}\}$ is equicontinuous on D_∞ - see [Mr 9].) Consequently, even the transform calculi are restricted in this case to transforms φ of measures Φ which "vanish faster than any exponential at ∞" (e.g. Φ compactly supported). These restrictions lead to difficulties in any attempt to apply these calculi in obtaining spectral reduction on D_∞.

It is important to observe that the pathologies under discussion are typical of the case in which A, B $\in \mathfrak{L}$ for $\mathfrak{L}$ noncompact semisimple. Consequently, classical spectral-theoretic behavior is not to be expected on C^∞-vector spaces for noncompact semisimple Lie algebras and classical resolvent/Laplace transform methods will in general fail for the treatment of the groups $\{V_\infty(t,A): t \in \mathbb{R}\}$ on D_∞. Such examples supply one of several reasons for interest in the locally convex (semi) group theories of Babalola [Bb 1], Dembart [Db], Kōmura [Ko], Ouchi [Ou], and Waelbroeck [Wb 2].

Remark: Many of the spectral aspects of the preceding discussion have close contact with ideas developed by Sternheimer in [St 2], particularly as his ideas apply to parabolic (nilpotent), elliptic (compact) and hyperbolic (noncompact regular) elements of semisimple Lie algebras.

E.O.R.

PART IV

CONDITIONS FOR A LIE ALGEBRA OF UNBOUNDED OPERATORS TO GENERATE A STRONGLY CONTINUOUS REPRESENTATION OF THE LIE GROUP

We have sailed many months,
  we have sailed many weeks,
(Four weeks to the month
  you may mark),
But never as yet ('tis your
  Captain who speaks)
Have we caught the least glimpse
  of a Snark!

The Hunting of the Snark
Fit the Second
LEWIS CARROLL

'To be sure, the most important thing is still missing, namely the actual integration [of the equations of matrix mechanics] in the case of hydrogen' (Heisenberg to Pauli, 18 September 1925, quoted in Mehra-Rechenberg, 1982, vol. 3, p. 96).

To study the group we may work with the infinitesimal operators. There are ten independent ones, four translation operators P_μ ($\mu = 0,1,2,3$) and six rotation operators about a point $M_{\mu\nu} = -M_{\nu\mu}$. They satisfy definite commutation relations,

$$[P_\mu, P_\nu] = 0 \quad [P_\mu, M_{\nu\rho}] = g_{\mu\nu}P_\rho - g_{\mu\rho}P_\nu$$

$$[M_{\mu\nu}, M_{\rho\sigma}] = -g_{\mu\rho}M_{\nu\sigma} + g_{\nu\rho}M_{\mu\sigma} + g_{\mu\sigma}M_{\nu\rho} - g_{\nu\sigma}M_{\mu\rho}.$$

Any representation of the Poincaré group provides ten operators satisfying these commutation relations. Conversely, any set of ten operators satisfying these relations gives a representation of the Poincaré group (provided a certain global condition is also satisfied, that continual application of an infinitesimal rotation such as M_{12} so as to build up a complete revolution gives 1 or -1). The ten operators then provide the mathematical basis for a dynamical system in relativistic quantum mechanics.

The whole problem of relativistic quantum mechanics, which has been holding up the development of theoretical physics for decades, reduces to finding suitable sets of ten operators satisfying the commutation relations.

One can look at the problem purely mathematically. Mathematicians have worked out all the irreducible representations of the group. But the irreducible representations just correspond to isolated particles and do not take us very far.

For dealing with interesting physical systems involving particles in interaction we need representations which are far from irreducible. The problem of finding the physically important representations is the big unsolved problem. (Dirac 1977, quoted in Marlow, 1978.)

'If one finds a difficulty in a calculation which is otherwise quite convincing, one should not push the difficulty away: one should rather try to make it the centre of the whole thing' (Heisenberg, Conversations. pp. 163-164, quoted in Mehra-Rechenberg, 1982, vol. 3, p. 227).

INTRODUCTION TO PART IV

When vector fields, or more generally, higher-order partial differential operators are considered in a given space of functions, or sections, then the commutators $[A,B] = AB - BA$ for different operators A and B are often known. In the theory of infinite-dimensional group representations ([Dx 1-3], [Nℓ 1],[Wr]), in the analysis of partial differential equations ([Hö]), and in mathematical physics ([Lr], [O'R], [Sg 3], [Sm]), a careful analysis of the commutators, and the commutator Lie algebras, involved, has been a major ingredient in the solution of diverse analytic problems.

We isolate three types of operator relations involving such commutators. As it turns out, the interesting applications lead to commutators, and Lie algebras, of <u>unbounded</u> operators. But the individual operators will typically generate one-parameter groups, or semigroups, of <u>bounded</u> operators. If A is such an infinitesimal generator then the resolvent operator $R(\lambda,A) = (\lambda-A)^{-1}$ is defined and bounded for $\mathrm{Re}(\lambda)$ sufficiently large. Similarly, a well known functional calculus $\varphi(A)$ is defined for a certain class of scalar functions φ. (Of course, $R(\lambda,A)$, and $\varphi(A)$, may be defined under more general conditions, even when A is not necessarily the generator of an operator semigroup.) If A and B are elements in a given system of unbounded operators, and if the commutators
$[A,B] = (\mathrm{ad}\ A)(B) = AB - BA, \ldots, (\mathrm{ad}\ A)^{k+1}(B) = (\mathrm{ad}\ A)(\mathrm{ad}\ A)^k(B)$,
$k \in \mathbb{N}$ are well defined and satisfy a certain regularity condition, then we are concerned with a rigorous formulation of the following formal commutation relations:

$$B\ R(\lambda,A) = \Sigma_{n=0}^{\infty}(-1)^n\ R(\lambda,A)^{n+1}(\mathrm{ad}\ A)^n(B)$$

and

$$B\ \varphi(A) = \Sigma_{n=0}^{\infty}(-1)^n(n!)^{-1}\ \varphi^{(n)}(A)(\mathrm{ad}\ A)^n(B)\ .$$

If A is further known to be the generator of an operator semigroup $\{V(t,A)\colon 0 \leq t < \infty\}$, then we introduced the formal relation

$$B\ V(t,A) = V(t,A)\ \exp(-t\ \mathrm{ad}\ A)(B)\ .$$

We apply here the commutation relations to the exponentiation problem for Lie algebras of unbounded operators. This problem was first formulated in the 1950 Thesis of I.M.Singer [Sr 1], and since then in a number of papers: E. Nelson [Nℓ 1], R.T. Moore [Mr 1], M. Flato et al. [FSSS], P. Jorgensen [Jo 1], and in special cases: J. Dixmier [Dx 1 and 2], B. Fuglede [Fu 1], T. Kato [Kt 1], J. Tits and L. Waelbroeck [TW], T. Yao [Ya], Kisynski [Ki], Rusinek [Ru] - to mention only a sample.

While the early investigations (going back to Bargmann [Bg]) were mainly concerned with unitary representations of Lie groups in Hilbert space, and the derived infinitesimal representations of the Lie algebra, an important development, in 1957, within geometry (viz., Palais' global formulation of the Lie theory of transformation groups, [Pℓ]) suggested the need for a theory which applies to representations of Lie groups in Banach spaces and, in fact, in more general topological linear spaces. (This need was also dictated by other applications: see, for example, [BB], [F], [Gd 4], [H-Ch], [KS], [Lr], and [Wr].)

The Palais Theorem is of particular interest in this connection, since it was formulated and proved within geometry, the theory of foliations being the main tool. In contrast, our first exponentiation theorem (Theorem 9.1, below) utilizes only operator theory in formulation and proof. Nonetheless, in Chapter 10 we show that Palais' theorem may be derived as a corollary of Theorem 9.1. For this purpose, the present generality of locally convex linear spaces in our exponentiation theory is essential.

A well-known construction of L. Gårding [Gå 1] associates to any strongly continuous representation V (of a given Lie group G) a derived infinitesimal representation dV of the associated Lie algebra $\mathfrak{g}$. Moreover, dV extends canonically to the universal associative enveloping algebra $\mathfrak{U}(\mathfrak{g})$. Let E be the locally convex linear space of the representation V, and let L(E) be the algebra of all continuous linear operators in E. Then V maps G into the group of invertible elements in L(E), but, for elements X in the Lie algebra $\mathfrak{g}$, the operator dV(X) is generally discontinuous, unbounded, and only partially defined. Nonetheless, Gårding showed that the operator Lie algebra $\mathfrak{L} = dV(\mathfrak{g})$ is a useful object to study: The different operators dV(X) have a common dense and invariant domain of definition, the so called Gårding domain D_G. Since the specification of the domain is essential for a rigorous analysis of unbounded operators, the pair $(\mathfrak{L}, D_G)$ constitutes an operator Lie algebra. We say that it is exponentiable because there is a continuous representation V of the Lie group such that $dV(\mathfrak{g}) = \mathfrak{L}$. Analogously, we say that a representation ρ of a given Lie algebra, $\mathfrak{g}$ say, is exact if a Lie group representation V exists such that $\rho = dV$.

Essential to Gårding's construction is the observation that, for an exact representation $\rho = dV$ (of a Lie algebra $\mathfrak{g}$ with Lie group G, and exponential mapping $\exp : \mathfrak{g} \to G$), we know that the individual operators, $dV(X)$, generate strongly continuous one-parameter groups of operators on E. In fact, for vectors u in the dense domain D_G, we have

$$dV(X)u = \lim_{t \to 0} t^{-1}(V(\exp(tX))u - u)$$

and, $t \to V(\exp(tX))$, is the one-parameter group $(\subset L(E))$ in question. (However, this individual exponentiability is not a sufficient condition for our problem.)

In most of the applications (starting, for example, with Bargmann [Bg]), the Lie group representation V is not given explicitly. Instead the elements (alias operators) in the infinitesimal operator Lie algebras $\mathfrak{L}$ are given by simple formulas, for example various combinations of raising and lowering operators. But if the representation V is required for the analysis, we are faced with the exponentiation problem (EP):

(1) For a given Lie algebra of unbounded operators, how do we decide if it is the derived infinitesimal Lie algebra of some (global) Lie group representation V? (If it is, then it can be seen quite easily that the representation V is essentially unique.)

(2) If $\mathfrak{L}$ is exponentiable, are there methods and formulas for constructing the exponential (alias representation) V?

While earlier methods (starting with Nelson [Nℓ 1]) were mainly based on analytic vector techniques, the present treatment is based entirely on integration methods carried out on spaces of smooth (i.e., C^∞) vectors.

In Chapter 12, our two exponentiation theorems, 9.1 and 9.2, are applied to Lie algebras $\mathfrak{L}$ isomorphic to $s\ell(2,\mathbb{R})$. But, even then, it turns out that the assumptions in Theorem 9.1 and 9.2 are not always easy to check directly. Hence, we develop, in Theorems 9.3 through 9.7, a perturbation theory for Lie algebras of unbounded operators. If $\mathcal{L}$ is such a Lie algebra, A is an operator in $\mathcal{L}$, and P is a "regular" operator (specified in Theorem 9.3), then the one-parameter group exp(t(A+P)) = V(t,A+P) is given by the well-known Dyson expansion. Under general conditions, we show that the operators A+P generate a perturbed Lie algebra, $\mathcal{L}_p$, and moreover that exponentiability of $\mathcal{L}$ implies that of $\mathcal{L}_p$. This perturbation method is a new contribution (due to the co-authors) to part (2) of problem (EP). In applications, a regular class of such perturbations P can indeed be found, such that Theorems 9.1 and 9.2 can be applied. The point of view is frequently reversed such that 9.1 or 9.2, are instead applied to a certain base-point Lie algebra $\mathcal{L}_o$. For $G = SL(2,\mathbb{R})$, it turns out that, when P runs through the class of regular perturbations, then the Lie group representations $\exp((\mathcal{L}_o)_p)$, obtained from a suitable chosen base-point Lie algebra, include "essentially all" of the non-unitary dual object to G.

Our objective here is to solve the exponentiation problem. Given at the outset is a Lie algebra $\mathcal{L}$ of unbounded operators in a Banach space E. We introduce two types of sufficient analytic conditions which ensure that $\mathcal{L}$ exponentiates to strongly continuous Lie group representation on E. With our perturbation techniques the conditions in Theorems 9.1 and 9.2 (on the problem (EP)) can now be checked straightforwardly on the major classes of operator Lie algebras which have appeared in the literature. (References given above. The reader is also referred to Chapter 12).

The first part of the theory (corresponding to Chapter 8) is technical. The reader who wishes to begin with continuous representations may prefer to go directly to the second part (Chapter 9) and then return to the theorems from Chapter 8 at the places where they are applied. Other results which are used in the proofs are collected in an appendix to Part IV.

Chapter 8

INTEGRATION OF SMOOTH OPERATOR LIE ALGEBRAS

In this chapter, we formulate and prove an infinite-dimensional operator-theoretic version of the well-known classical theorem stating that every finite-dimensional Lie algebra is, in fact, associated to a Lie group as the infinitesimal (derived) Lie algebra. A number of applications to infinite-dimensional representations of Lie algebras and Lie groups are given in Chapters 9, 10 and 12.

We introduce the concept of a smooth operator Lie algebra and prove that every such Lie algebra integrates in a natural sense to a differentiable group representation. Our result concerns "smooth" Lie algebras of continuous everywhere-defined operators on locally convex spaces upon which "one-parameter integrability" restrictions are placed. It shows that every such smooth Lie algebra integrates to a differentiable group representation. (The classical prototype, often described as the global version of Sophus Lie's Third Local Theorem, is sometimes attributed to Élie Cartan in 1936. In many ways our result is equally closely related to two theorems of I.D. Ado on matrix representations of Lie groups in 1934, and on comparable representations of Lie algebras in 1947.)

The setting of our main theorem is abstract: a locally convex space D, and a smooth Lie algebra $\mathcal{L} \subset L(D)$ of continuous linear endomorphisms upon it, are given at the outset. But in the applications D is frequently a space of C^∞-vectors for a Lie algebra of unbounded operators in a Banach space E which contains D as a dense linear subspace.

Apart from being of independent interest, the Integration Theorem 8.1 serves to separate two important aspects of the exponentiation problem. On the one hand, there are the (for the most part) differential geometrical and algebraic questions related to the commutation relations, the adjoint representation, and different variants of the product rule for differentiation. On the other hand, there are the functional analytic questions concerning domains of the unbounded infinitesimal operators, and the problem of constructing a C^∞-domain invariant under the one-parameter groups of operators generated by the infinitesimal operators in the Lie algebra.

Finally, we show by an example that not all differentiable group representations (as defined below) are restrictions of continuous group representations to the space of C^∞-vectors.

8A. Smooth Lie algebras and differentiable representations

Let D be a locally convex space, and denote by L(D) the space of continuous linear endomorphisms of D. Equipped with the strong operator topology L(D) is again a locally convex space, and is denoted by $L_s(D)$.

Let G be a Lie group with the Lie algebra $\mathfrak{g}$. A differentiable representation of G in D is a differentiable locally equicontinuous homomorphism V of G into $L_s(D)$. That is, we require that V as a mapping from G to $L_s(D)$ has derivatives of all orders, and that the derivatives belong to L(D). (If D is barreled the last condition is known to be automatic [Mr 5].) Furthermore, when K is a compact subset of G the set of operators $V(K) \subset L(D)$ is then equicontinuous. If we denote the differential of V (evaluated at the identity in G) by dV, then dV is a homomorphism of the Lie algebra $\mathfrak{g}$ into the commutator Lie algebra L(D).

Let $\mathfrak{L}$ be a finite-dimensional real Lie algebra contained in L(D). Suppose there is a differentiable representation V of G in D such that dV is an isomorphism and $dV(\mathfrak{g}) = \mathfrak{L}$. Then $\mathfrak{L}$ is said to be integrable.

An operator $A \in L(D)$ (continuous and linear) is called a smooth infinitesimal generator (or just integrable) if there is a differentiable representation $t \to V(t,A)$ of the additive group of the real line $\mathbb{R}$ in D such that $A = d/dt\, V(t,A)\big|_{t=0}$. That is, the difference quotients $t^{-1}(V(t,A)-I)$ converge to A in the topology of $L_s(D)$ as $t \to 0_+$.

The following natural (and non-trivial) question will be answered to the affirmative in this section: If A and B are two smooth infinitesimal generators belonging to a finite-dimensional commutator Lie algebra contained in L(D), are then the sum A + B and the commutator $[A,B] = AB - BA$ again smooth infinitesimal generators? (Some vector-field examples of Palais [Pℓ, p. 90] illustrate the nontriviality of such questions.)

The affirmative answer is a consequence of the following:

8.1. Theorem

Let $\mathfrak{L}$ be a finite-dimensional real Lie algebra contained in L(D). Suppose $\mathfrak{L}$ is generated as a Lie algebra by a subset which consists of smooth infinitesimal generators.

Then $\mathfrak{L}$ integrates to a differentiable Lie group representation.

Remarks: A subset S of $\mathfrak{L}$ is said to generate $\mathfrak{L}$ as a Lie algebra if every element in $\mathfrak{L}$ is a real linear combination of commutators of elements from S (i.e. the smallest subalgebra of $\mathfrak{L}$ containing S is $\mathfrak{L}$ itself).

The proof of Theorem 8.1 is based on the product rule for differentiation of operator valued functions (Appendix A), so local equicontinuity is crucial.

Of course, if D is a barreled space then the local equicontinuity assumptions are redundant.

Operator Lie algebras which satisfy the assumptions of Theorem 8.1 are called smooth. The theorem then says that every smooth operator Lie algebra is integrable.

In what follows, $\mathfrak{L} \subset L(D)$ will be a fixed finite-dimensional real Lie algebra of continuous linear endomorphisms in a locally convex space D. An element $A \in \mathfrak{L}$ is integrable if the one-dimensional Lie subalgebra $\mathbb{R}A$ is integrable. In other words, A is a smooth infinitesimal generator (cf. the definition above). With this terminology, Theorem 8.1 has the equivalent formulation: The set of integrable elements in $\mathfrak{L}$ form a sub Lie algebra $\mathfrak{h}$, and $\mathfrak{L}$ is integrable if and only if $\mathfrak{h} = \mathfrak{L}$.

The theorem is thus essentially equivalent to The Theorem in [T-W]. But our proof is conceptually rather simple, being based solely on the operator commutation relations (1) below, combined with a certain identity (9) in the Lie algebra $\mathfrak{g}$ giving a coordinate expression for the adjoint representation Ad of G in $\mathfrak{g}$. (We do not use any other concepts from geometry. Integrable subgroups and foliations play no role in our proof.)

Let G be an abstract connected and simply connected Lie group whose Lie algebra $\mathfrak{g}$ of right invariant vector fields is isomorphic to the operator Lie algebra $\mathfrak{L}$.

The following lemma is an immediate consequence of Proposition 3.1 and will be used without comment.

8.2. Lemma

Let A be an integrable element in $\mathfrak{L}$, generating a differentiable group $\{V(t,A): -\infty<t<\infty\}$. Then the commutation relations

$$V(t,A)\,B\,V(-t,A) = \exp(t \text{ ad } A)(B) \tag{1}$$

hold for all $B \in \mathfrak{L}$ and $t \in \mathbb{R}$, where the right-hand side is defined by the usual power series expansion
$\sum_{0}^{\infty} t^n/n!\ (\text{ad } A)^n(B)$.

Proof of Theorem 8.1: Let ρ be a specified Lie algebra isomorphism of $\mathfrak{g}$ onto $\mathfrak{L}$. We first prove a simple version of the theorem. It is first assumed that there is a basis $B_1, B_2,\ldots,B_d$ for $\mathfrak{L}$ consisting of integrable elements. Let $\{V(t,B_j): -\infty<t<\infty\}$ be the differentiable one-parameter group generated by B_j for each $j = 1,2,\ldots,d$. Let $X_1, X_2,\ldots,X_d$ be a basis for $\mathfrak{g}$ such that $\rho(X_j) = B_j$ for $1 \leq j \leq d$, and let $\exp: \mathfrak{g} \to G$ be the exponential mapping of the Lie algebra $\mathfrak{g}$ onto a neighborhood of the identity e in G. It is well known that there is a compact spherical neighborhood Ω of the identity in G and coordinate functions of the second kind $g \to (t_1(g), t_2(g),\ldots,t_d(g))$ mapping Ω diffeomorphically onto a neighborhood of the origin in $\mathbb{R}^d$ such that

$$g = \exp t_1(g)X_1 \exp t_2(g)X_2 \cdots \exp t_d(g)X_d \tag{2}$$

for all $g \in \Omega$. Let us define a mapping V of Ω into L(D) in terms of these coordinates. That is

$$V(g) = V(t_1(g),B_1)V(t_2(g),B_2)\cdots V(t_d(g),B_d) \tag{3}$$

for $g \in \Omega$. The program is to show, using the commutation relations (1), that (3) defines a local homomorphism of Ω into L(D). Extension of the local homomorphism is always possible by ([Hc], Theorem 3.1, Chap. IV, p. 54) since G is assumed simply connected.

For given elements $X \in \mathfrak{g}$ and $g \in \Omega$ one can find a nontrivial open interval I on the line containing the origin such that:

$$\exp tX \in \Omega \text{ and } \exp(tX)g \in \Omega \quad \text{for all } t \in I. \tag{4}$$

We proceed to prove the identity

$$V(\exp(tX)g) = V(\exp(tX))V(g) \tag{5}$$

for all $t \in I$, or equivalently

$$V(\exp(tX))^{-1} V(\exp(tX)g = V(g) \tag{6}$$

where both sides of (5) (and (6)) are defined via the coordinate expression (3).

The proof of (6) is based on the following Lie algebra identity connecting the adjoint representation Ad of G in $\mathfrak{g}$ to the coordinates $(t_1, t_2, \ldots, t_d)$.

8.3. Lemma

Let G be a Lie group with Lie algebra $\mathfrak{g}$. Let Ω be a coordinate neighborhood with coordinates $(t_1,\ldots,t_d)$ given by (2).

Let $X \in \mathfrak{g}$ and $g \in \Omega$ be fixed elements, and let I be an interval which satisfies condition (4) with respect to X and g.

(a) Then there exists analytic functions $t \to \tau_k(t)$ on I such that

$$\exp(tX)g = \exp(\tau_1(t)X_1) \ldots \exp(\tau_d(t)X_d). \tag{7}$$

(b) Moreover, if we define group elements

$$g_1 = e, \text{ and } g_k(t) = \exp(\tau_1(t)X_1)\cdots\exp(\tau_{k-1}(t)X_{k-1}) \tag{8}$$

for $2 \leq k \leq d$ and $t \in I$, then we have the identity

$$X = \sum_{k=1}^{d} d\tau_k(t)/dt \; \mathrm{Ad}(g_k(t))X_k \tag{9}$$

or equivalently

$$X = \sum_{k=1}^{d} d\tau_k(t)/dt \; \exp(\tau_1 \mathrm{ad}\, X_1)\cdots\exp(\tau_{k-1}\mathrm{ad}\, X_{k-1})(X_k)$$

for $t \in I$ where the functions $\tau_k = \tau_k(t)$ are defined in (a), and ad X is the adjoint representation of $\mathfrak{g}$.

Proof: Part (a) is a triviality. The functions τ_k are defined by $\tau_k(t) = t_k(\exp(tX)g)$ for $t \in I$. For the proof of part (b) the following notation is convenient. Let left (and right) multiplication by an element $g \in G$ be given by $L(g)a = g\cdot a$ ($R(g)a = a\cdot g$) for all $a \in G$. The differential of the diffeomorphism L(g) (and R(g)), evaluated at the identity e, is a linear isomorphism of the tangent space $T_e(G)$ to G at e onto tangent space $T_g(G)$ to G at g. For an element X in the Lie algebra $\mathfrak{g}$ (which is isomorphic to $T_e(G)$) and $g \in G$ we use the notation

$$g\cdot X = dL(g)_e(X) \text{ and } X\cdot g = dR(g)_e(X). \qquad (10)$$

With this notation, the product rule for differentiation takes the following handy form.

Let $t \to f(t)$ and $t \to g(t)$ be differentiable functions defined on an open interval J and taking values in the group G. Using group multiplication we define the mapping $h(t) = f(t)\cdot g(t)$ for all $t \in J$. It is clear that h is differentiable. Now, the derivative $f'(t)$ belongs to the tangent space $T_{f(t)}(G)$ to G at the point $f(t)$, and similarly for $g'(t)$ and $h'(t)$. The following product rule is verified below

$$h'(t) = f'(t)\cdot g(t) + f(t)\cdot g'(t) \qquad (11)$$

for $t \in J$. (Note that multiplication by $g(t)$ on the right ($f(t)$ on the left) maps $T_{f(t)}(G)$ ($T_{g(t)}(G)$) onto $T_{f(t)g(t)}(G)$ so (11) is an identity in the latter tangent space when the vector-group element multiplication is defined via (10).) Indeed let $\pi: G \times G \to G$ $(a,b) \to a\cdot b$ denote group multiplication in G. Then for every point $(a,b) \in G \times G$ the differential $d\pi(a,b\,;-,-)$ is a bilinear mapping of $T_a(G) \times T_b(G)$ into $T_{ab}(G)$.

Using the well-known identity $\exp(tX)\cdot\exp(tY) = \exp[t(X + Y) + O(t^2)]$ one gets

$$d\pi(e,e;X,Y) = X + Y \qquad (12)$$

for $(X,Y) \in \mathfrak{g} \times \mathfrak{g}$.

As usual, the local behaviour of π at a general point $(a,b) \in G \times G$ can be lifted back to that near (e,e) by the following factorization of π as the composite of four mappings

$$\pi = L(a)\circ R(b) \circ \pi \circ [L(a^{-1}) \times R(b^{-1})]. \qquad (13)$$

As a consequence of (12), (13) and the convention (10) one gets for all $(\xi,\eta) \in T_a(G) \times T_b(G)$

$$d\pi(a,b;\xi,\eta) = \xi\cdot b + a\cdot\eta. \qquad (14)$$

Now, $h: J \to G$ is given as the composite of two mappings $h = \pi\circ[f \times g]$, so by the chain rule for differentiable mappings and substitution into (14) one gets

$$h'(t) = d\pi(f(t),g(t);f'(t),g'(t))$$
$$= f'(t)\cdot g(t) + f(t)\cdot g'(t).$$

This completes the proof of (11).

The following identities are now easily verified

$$d/dt\ \exp(tX)g = X\cdot(\exp(tX)g) \tag{15}$$

and

$$d/dt\ \exp\tau_1 X_1 \cdots \exp\tau_d X_d = \sum_{k=1}^{d} d\tau_k/dt\ (\mathrm{Ad}(g_k)X_k)\cdot\exp\tau_1 X_1 \cdots \exp\tau_d X_d, \tag{16}$$

where g_k is given by (8). Note that the t dependence is suppressed in the coordinate functions $\tau_k = \tau_k(t)$ given by (7).

The identity (15) is trivial, and (16) can be proved as follows. By the product rule (11) one verifies that the left-hand side of (16) is

$$\sum_{k=1}^{d} \exp\tau_1 X_1 \ldots \exp\tau_{k-1}X_{k-1}(d/dt\ \exp\tau_k(t)X_k)\cdots\exp\tau_d X_d$$

$$= \Sigma_1^d\ d\tau_k/dt\ \exp\tau_1 X_1 \cdots \exp\tau_{k-1}X_{k-1}(X_k\cdot\exp\tau_k X_k)\cdots\exp\tau_d X_d.$$

This last expression is equal to the right-hand side of (16) by the simple identities $g_k\cdot X_k = [\mathrm{Ad}(g_k)X_k]\cdot g_k$. This completes the proof of the lemma.

The identities of the lemma correspond to operator identities. Recall that $\mathfrak{g}$ is the Lie algebra of right invariant vector fields on G. For clarity, we specify a Lie algebra isomorphism ρ of $\mathfrak{g}$ onto the operator Lie algebra $\mathfrak{L} \in L(D)$. The adjoint representation Ad of G in $\mathfrak{g}$ induces a representation Ad of G in $\mathfrak{L}$ via the isomorphism ρ. Specifically

$$\widetilde{\mathrm{Ad}}(g)(B) = [\rho\circ\mathrm{Ad}(g)\circ\rho^{-1}](B)$$

for all $g \in G$ and $B \in \mathfrak{L}$. In this notation

$$\exp(\tau_1 \mathrm{ad}\ B_1)\cdots\exp(\tau_{k-1}\mathrm{ad}\ B_{k-1})(B_k)$$

$$= \rho(\exp(\tau_1 \mathrm{ad}X_1)\cdots\exp(\tau_{k-1}\mathrm{ad}X_{k-1})(X_k)) = \rho(\mathrm{Ad}(g_k)X_k)$$

$$= \rho(\widetilde{\mathrm{Ad}}(g_k)X_k) = \mathrm{Ad}(g_k)(B_k) \tag{17}$$

with g_k given by (8).

Going back to the identity (6) we observe that the left-hand side of (6) defines a differentiable mapping

$$F(t,X) = V(\exp(tX))^{-1}V(\exp(tX)g)$$

in the variable t from I into $L_s(D)$ for all fixed X. (The g dependence is suppressed.) Now, the function F is actually defined by the coordinate expression (3), so differentiability of F involves repeated application of the product rule for operator valued differentiation (Appendix A). Moreover the value of the derivative d/dt F(t,X) is given by the product rule.

It is clear from Lemma 8.2 that the commutation relations

$$V(t,B_j)B = \exp(t \text{ ad } B_j)(B)\ V(t,B_j) \tag{18}$$

hold for all $B \in \mathfrak{L}$ and $1 \leq j \leq d$ since the basis elements B_j are assumed integrable.

The program is to show that d/dt F(t,X) vanishes, and hence that F(t,X) is constant as a function of t for each X. Suppose the constancy is already proved for X belonging to the basis $X_j = \rho^{-1}(B_j)$ for $1 \leq j \leq d$. Then $V(s,B_j)V(g) = V(\exp(sX_j)g)$ for $s \in I$ and $1 \leq j \leq d$. Since g is an arbitrary element in Ω we conclude that for every

$a = \exp(s_1X_1) \cdots \exp(s_dX_d) \in \Omega$ such that $a\cdot g \in \Omega$:

$$\begin{aligned} V(a)V(g) &= V(s_1,B_1)\cdots V(s_d,B_d)V(g) \\ &= V(s_1,B_1)\cdots V(s_{d-1},B_{d-1})V(\exp(s_dX_d)g) \\ &= \cdots \\ &= V(\exp(s_1X_1)\cdots\exp(s_dX_d)g) \\ &= V(a\cdot g). \end{aligned}$$

This certainly shows that V is a local homomorphism. Hence, it is enough to show that d/dt F(t,X) vanishes when X is equal to one of the basis elements X_j. In that case, the differentiation simplifies slightly since

$$d/dt\ V(\exp(tX_j))^{-1} = -V(-t,B_j)\rho(X_j). \tag{19}$$

The resulting computation is

$$\begin{aligned} d/dt\,F(t,X) &= -V(-t,B)\rho(X)V(\tau_1,B_1)\cdots V(\tau_d,B_d) \\ &\quad + \Sigma_{k=1}^{d} d\tau_k/dt\ V(-t,B)V(\tau_1,B_1)\cdots V(\tau_{k-1},B_{k-1})B_kV(\tau_k,B_k)\cdots V(\tau_d,B_d) \\ &= V(-t,B)[-\rho(X)+\Sigma_{k=1}^{d} d\tau_k/dt\ \rho(Ad(g_k)X_k)]V(\tau_1,B_1)\cdots V(\tau_d,B_d) \\ &= 0 \end{aligned} \tag{20}$$

where the product rule (Appendix A), identity (19), the commutation relations (18), identity (17), and Lemma 8.3 have been used in the indicated order.

This finishes the proof of the local homomorphism property. The extension of V from Ω to the whole group is also denoted by V. Now, the extension V is a group homomorphism from G into the group of invertible elements in L(D). Its restriction to Ω is differentiable by (3) and the product rule. Hence V is differentiable on all of G, as one can see by using analyticity (or just smoothness) of group multiplication in G.

The verification that V is locally equicontinuous is also routine from the local defining formula for V as a finite product of locally equicontinuous one-parameter groups. Indeed, let K be a compact subset of G. Then there is a finite subset $\{a_\nu;\ 1 \le \nu \le N\}$ of K such that K is contained in the union $\cup_{\nu=1}^{N} a_\nu \Omega$. But then $V(K) \subset \cup_{\nu=1}^{N} V(a_\nu)V(\Omega)$. Since Ω is compact and each of the factors $V(t_j,B_j)$ in (3) is locally equicontinuous the set of operators $V(\Omega)$, and hence each of the sets $V(a_\nu)V(\Omega)$, is equicontinuous. Equicontinuity of V(K) is now clear.

It still has to be checked that the differentiable representation V is an integral of $\mathfrak{L}$, i.e., $dV(\mathfrak{g}) = \mathfrak{L}$, or equivalently $d/dt\, V(\exp tX)|_{t=0} = \sigma(X)$ for all $X \in \mathfrak{g}$. This last identity is a consequence of a computation completely analogous to (20) above. Only this time the coordinates are given by $\exp tX = \exp\tau_1 X_1 \cdots \exp\tau_d X_d$ and the product rule is applied to $V(\exp tX) = V(\tau_1,B_1)\cdots V(\tau_d,B_d)$.

These last details conclude the proof of Theorem 8.1 under the assumption that $\mathfrak{L}$ has a basis consisting of integrable elements.

If $\mathfrak{L}$ is assumed only to have a set S of Lie generators consisting of integrable elements, we indicate below how one obtains a basis of integrable elements.

Suppose that all the elements in a set S of Lie generators are integrable. Let M be the real linear span of the elements

$$\exp(t_1 \mathrm{ad}\, A_1) \cdots \exp(t_k \mathrm{ad}\, A_k)(A_{k+1}) \tag{21}$$

with $t_1, t_2, \ldots \in \mathbb{R}$, $A_1, A_2, \ldots \in S$, and $k = 0,1,2,\ldots$, For a fixed set of elements $A_1, A_2, \ldots, A_{k+1} \in S$ the expansion (21) may be viewed as an analytic function $F(t_1,\ldots,t_k)$ of k real variables with values in the finite-dimensional (and hence closed) subspace M of $\mathcal{L}$. The partial derivative $\partial/\partial t = \partial/\partial t_1 \ldots \partial/\partial t_k$ of F evaluated at t=0 is equal to $[A_1,[A_2,[\cdot,\ldots,[A_k,A_{k+1}]\ldots]]]$. Since M is closed it contains all these commutators, so M must be equal to $\mathcal{L}$. Hence a basis for $\mathcal{L}$ can be selected from among the elements of the form (21).

It is shown below by induction that each such element is integrable, i.e. the infinitesimal generator of a differentiable one-parameter group. The case k=0 reduces to the assumption that every element in S is integrable. Let k be a positive integer, and suppose that every element of the form (21) is integrable (where k is now held fixed.) Consider an arbitrary element

$$B_1 = \exp(t_1 \text{ad } A_1)\ldots\exp(t_{k+1}\text{ad } A_{k+1})(A_{k+2}).$$

The operator $\exp(t_1\text{ad } A_1)$ can be factored out so that $B_1 = \exp(t_1\text{ad } A_1)(B_o)$, where B_o is now an element to which the induction assumption applies. Let $s \to V(s,B_o)$ denote the corresponding differentiable one-parameter group, and define

$$V(s,B_1)=V(t_1,A_1)V(s,B_o)V(t_1,A_1)^{-1} \text{ for all } s \in \mathbb{R}. \quad (22)$$

Then clearly $\{V(s,B_1):-\infty<s<\infty\}$ is a differentiable one-parameter group, since the class of differentiable groups in $L_s(D)$ is invariant under conjugation by invertible elements in L(D). Differentiation of (22) at s=0 using $B_1V(t_1,A_1) = V(t_1,A_1)B_o$ yields immediately $d/ds\, V(s,B_1)|_{s=0} = B_1$. Hence B_1 is integrable and the induction is completed; and the situation where integrability is assumed only for a set of Lie generators is reduced to the case where there is a basis for $\mathcal{L}$ of integrable elements. Since the latter case is treated above, the proof of Theorem 8.1 is completed. E.O.P.

8B. Applications in C^∞-vector spaces

The rest of the chapter relates differentiable group representations and smooth Lie algebras, on the one hand, to spaces of C^∞-vectors for continuous group representations on the other. We first recall some notation. (See Chapter 7.)

Let E be a locally convex space (which is, in most applications, a Banach space), and let D be a dense linear subspace of E. We denote by $\mathbf{A}(D)$ the algebra of linear endomorphisms of D. That is, elements of $\mathbf{A}(D)$ are (in general unbounded) densely defined operators having D as a common invariant domain. The algebra $\mathbf{A}(D)$ is associative and has the structure of a Lie algebra equipped with the commutator bracket $[A,B] = AB-BA$.

Let $\mathfrak{L}$ be a finite-dimensional real Lie subalgebra of $\mathbf{A}(D)$, and let $E(\mathfrak{L}) \subset \mathbf{A}(D)$ be the associative enveloping algebra of $\mathfrak{L}$. The linear space D is always thought of as a locally convex space equipped with the projective topology τ_{∞} defined by the requirement that every operator in the enveloping algebra $E(\mathfrak{L})$ be continuous. If $\Gamma = \{p\}$ is a calibration of seminorms defining the topology of E, then the projective τ_{∞} topology on D is given by the seminorms $u \to p(Bu)$, where p varies in Γ and B in $E(\mathfrak{L})$. (When B varies, only in the space $E_n(\mathfrak{L})$ of elements of order less than or equal to n, then we get an abstract C^n-topology on D, which is denoted by τ_n.) Then $\mathfrak{L}$ becomes a Lie algebra of continuous linear endomorphisms in (D,τ_{∞}), in fact $E(\mathfrak{L}) \subset L(D,\tau_{\infty})$.

The locally convex spaces (D,τ_n) are, in general, not complete (as simple examples show: d/dx on $L^2(0,1)$ say). In case $\mathfrak{L}$ is the infinitesimal Lie algebra of a continuous group representation, or of a smeared group representation, then every operator in $E(\mathfrak{L})$ has a closed extension (see Chapter 4). The following easy observation is used without comment. If E is a Fréchet space, and if every operator in $E_n(\mathfrak{L})$ has a closed extension then the completion of (D,τ_n) is a Fréchet space.

Example: Let E be the Hilbert space $L^2(\mathbb{R})$ and let D be the compactly supported C^{∞} functions $C_0^{\infty}(R)$. Let $\mathfrak{L} \subset \mathbf{A}(D)$ be the Heisenberg Lie algebra spanned by $\{d/dx, ix, iI\}$. Then the completion of (D,τ_{∞}) is equal to the Fréchet space $S(\mathbb{R})$ of rapidly decreasing functions on the line (S stands for Schwartz) [Ps 1, Ex.5.1].

Given a Lie algebra $\mathfrak{L} \subset \mathbf{A}(D)$ where D is a dense linear subspace of a l.c.s. E, then it is frequently given that some operators $A \in \mathfrak{L}$ have closures $\bar{A}$, and that each $\bar{A}$ is the infinitesimal generator of a continuous locally equicontinuous (c.l.e.) one-parameter group $\{V(t,A): -\infty<t<\infty\}$ in $L(E)$. We say that A is a pregenerator of a c.l.e. group in L(E). Suppose the bounded operators V(t,A) leave the domain D invariant for all $t \in \mathbb{R}$. We show below that a necessary and sufficient condition for the restriction of $\{V(t,A): -\infty<t<\infty\}$ to D to be a differentiable representation of the additive group of the line $\mathbb{R}$ in (D,τ_{∞}) is that the commutation relations (23) hold.

The result serves as a lemma for the exponentiation theorems in Chapter 9.

8.4. Lemma

Let D be a dense linear subspace of a l.c.s. E. Let $\mathfrak{L} \subset \mathbf{A}(D)$ be a finite-dimensional real Lie algebra, and let D be equipped with its projective C^∞-topology.

(a) Suppose that an operator $A \in \mathfrak{L}$ is a pregenerator of a c.l.e. group $\{V(t,A): -\infty<t<\infty\}$ in L(E). Then A is an integrable element if and only if the commutation relations

$$V(t,A)\, B\, V(-t,A) = \exp(t \text{ ad } A)(B) \tag{23}$$

hold for all B in the associative enveloping algebra $E(\mathfrak{L})$.

(b) Let G be a Lie group with Lie algebra $\mathfrak{g}$, and let V be a continuous locally equicontinuous representation of G in E. Suppose that $dV(\mathfrak{g}) \subset \mathbf{A}(D)$ and that the group V(G) of bounded operators in E leaves D invariant.

Then the restriction of V to D defines a differentiable representation of G in D.

Proof of part (a): Let $A \in \mathfrak{L}$ be a pregenerator of a c.l.e. group V(t,A) in L(E), and suppose D is a group invariant domain on which the commutation relations (23) hold. Let $V_\infty(t,A)$ be the restriction of V(t,A) to D. Then for every compact $K \subset \mathbb{R}$, and for every seminorm $p \in \Gamma$ there is a $q \in \Gamma$ such that $p(V(t,A)u) \leq q(u)$ for all $t \in K$ and $u \in E$. Let B be an element in $E(\mathfrak{L})$, of order n, say. Then there is a basis B_j for $E_n(\mathfrak{L})$ and analytic coordinate functions $\alpha_j(t)$ such that $\exp(-t \text{ ad } A)(B) = \sum_j \alpha_j(t) B_j$. Let c be the finite supremum of the values $|\alpha_j(t)|$ for $j=1,2,\ldots,\dim E_n(\mathfrak{L})$ and $t \in K$. Then the following estimate holds for all $t \in K$ and $u \in D$

$$\begin{aligned} p(B\, V(t,A)u) &= p(V(t,A) \textstyle\sum_j \alpha_j(t) B_j u) \\ &\leq c \sum_j p(V(t,A)B_j u) \leq c \sum_j q(B_j u) \end{aligned} \tag{24}$$

showing that the set of operators $V_\infty(t,A); t \in K$ is equicontinuous with respect to the τ_∞ topology on D. Moreover the identity

$$B\, V_\infty(t,A)u = \sum_j \alpha_j(t)\; V(t,A) B_j u \tag{25}$$

for $u \in D$ and $t \in \mathbb{R}$, shows that the mapping $t \to V_\infty(t,A)$ is smooth from $\mathbb{R}$ into $L_s(D)$. Recall that the vectors $B_j u$ all belong to D which, in turn, is contained in the domain of $\bar{A}$ and invariant under $V(t,A)$. In other words, $d/dt\ V_\infty(t,A)|_{t=0}=A$ (the difference quotients being convergent in the topology of $L_s(D)$). Hence the element A is integrable.

Conversely, suppose that A is integrable, and let $\{V(t): -\infty<t<\infty\}$ be a differentiable one-parameter group in $L_s(D)$ such that $d/dt\ V_\infty(t)|_{t=0}=A$. Then Lemma 8.2 implies the commutation relations

$$V_\infty(t)\ B\ V_\infty(-t) = \exp(t\ \mathrm{ad}\ A)B \tag{26}$$

hold for $B \in \mathfrak{L}$. To extend (26) to higher-order elements, $B \in \mathcal{E}_n(\mathfrak{L})$ for $n > 1$, we remark that conjugation by $V_\infty(t)$, i.e. the left-hand side of (26), is a multiplicative isomorphism in the associative algebra $\mathcal{E}(\mathfrak{L})$, and it leaves $\mathfrak{L}$ invariant. It follows that the set of elements B, for which (26) holds, forms an associative subalgebra of $\mathcal{E}(\mathfrak{L})$, and it contains $\mathfrak{L}$. Hence, (26) holds for all $B \in \mathcal{E}(\mathfrak{L})$. To get from (26) to (23) we note that $V_\infty(t)$ must coincide with the restriction to D of $V(t,A)$ for all $t \in \mathbb{R}$. That can be seen, e.g., by the differentiation of $s \to V_\infty(t-s)V(s,A)u$ for each fixed $t \in \mathbb{R}$ and $u \in D$. This concludes the proof of (a).

Suppose the assumptions in (b) are satisfied for some continuous, locally equicontinuous representation of G in E. The assumption $dV(\mathfrak{g}) \subset \mathbf{A}(D)$ implies that D is contained in the maximal space of C^∞-vectors $E^\infty(V)$.

It is then well known that the commutation relations

$$V(g)\ dV(X)\ V(g^{-1})u = dV(\mathrm{Ad}(g)X)u \tag{27}$$

hold for all $g \in G$, $X \in \mathfrak{g}$, and $u \in D$. Indeed, by the homomorphism property of V we have

$$V(g)V(\exp(tX))V(g^{-1})u = V(\exp(t\ \mathrm{Ad}(g)X))u \text{ for } t \in \mathbb{R}.$$

Differentiation with respect to t at t=0 yields (27). Now an argument precisely analogous to the one given in part (a) above, in particular (24) and (25), gives the desired conclusion of part (b). E.O.P.

If V is a continuous representation of a Lie group G in a Banach space E, it follows from part (b) of the lemma that the restriction of V to the space of C^∞-vectors $E^\infty(V)$ is a differentiable representation of G in the Fréchet space $E^\infty(V)$. The following example shows that this class of differentiable group representations is not exhaustive. It is possible that a finite-dimensional Lie algebra $\mathfrak{L} \subset \mathbf{A}(D)$ is integrable to a differentiable representation V_∞ of G in (D,τ_∞) with the property that the operators $V_\infty(g)$ for $g \neq e$ are unbounded in the Banach space E. The problem is addressed in the next section. It is asked in [JM] if the smeared operators $V(\varphi) = \int\varphi(g)V_\infty(g)dg$ for $\varphi \in \mathcal{D}(G)$ have extensions to bounded operators in E. That may or may not be the case.

8.5. Example

Let E be the Hilbert space $\ell^2(\mathbb{Z})$ of square summable sequences, and let D be the space of finite linear combinations of the canonical basis vectors $\{e_n : n \in \mathbb{Z}\}$ in E. Let A be an endomorphism in D given by $Ae_{2n} = 0$, and $Ae_{2n+1} = ne_{2n}$ for $n \in \mathbb{Z}$. Let $\mathfrak{L} \subset \mathbf{A}(D)$ be the one-dimensional Lie algebra generated by A. There is a differentiable representation V_∞ of the additive group $\mathbb{R}$ in (D,τ_∞) such that V_∞ has no extension to a continuous representation of $\mathbb{R}$ in E. The operator A is given by an infinite matrix with $\begin{bmatrix} 0 & n \\ 0 & 0 \end{bmatrix}$ in the diagonal.

It is clear then that $A^2 = 0$, and that a differentiable representation V_∞ with the desired properties can be defined by $V_\infty(t) = I + tA$ where I denotes the identity mapping of D. Since A is unbounded in E, the operator $V_\infty(t)$ does not have a continuous extension to E for $t \neq 0$. In fact the resolvent set of A is empty, as can be seen from the fact that the formal inverse $(\lambda I - A)^{-1} = \lambda^{-1}(1 - \lambda^{-1}A)^{-1} = \lambda^{-1}(1 + \lambda^{-1}A)$ $= \lambda^{-1} + \lambda^{-2}A$ is also unbounded in $\ell^2(\mathbb{Z})$ and must agree with the resolvent $R(\lambda,A)$ of $\bar{A}$ wherever the latter might exist.

Pertaining to the remark above as to the possibility of smearing a differentiable representation to obtain bounded operators $V(\varphi)$ where φ is a test function, we record the following observation. The smeared operators $V(\varphi) = \int\varphi(t)V_\infty(t)dt$ for $\varphi \in \mathcal{D}(\mathbb{R})$ do not extend to the bounded operators in E. For φ positive with $\int\varphi(t)tdt \neq 0$ the operator $V(\varphi)$ is easily seen to be unbounded. E.O.E.

We conclude the section with a theorem on the integrability of Lie algebras of unbounded operators in a Banach space. As above, let D be a dense linear subspace of a Banach space E, and let $\mathfrak{L} \subset \mathbf{A}(D)$ be a finite-dimensional real Lie algebra.

Suppose $\mathcal{L}$ integrates to a differentiable group representation V_∞. It is clear that a necessary condition for extendability of the operators $V_\infty(g)$ to bounded operators on E is that the elements in $\mathcal{L}$ are infinitesimal pregenerators of strongly continuous one-parameter groups on E.

It is shown below that if D is complete in its projective C^∞-topology and invariant under the one-parameter groups, then $\mathcal{L}$ is integrable to a differentiable group representation V_∞. Moreover, V_∞ extends to a continuous group representation in E.

8.6. Theorem

Let D be a dense linear subspace of a Banach space E, and let $\mathcal{L} \subset \mathbf{A}(D)$ be a finite-dimensional real Lie algebra.

Suppose $\mathcal{L}$ is generated as a Lie algebra by a subset S of elements consisting of infinitesimal pregenerators for continuous one-parameter groups of bounded operators on E.

Furthermore, suppose that D is complete in its projective C^∞-topology and invariant under the one-parameter groups generated by the operator in S.

Then $\mathcal{L}$ is smooth and therefore integrable. Moreover the integrated differentiable group representation in D extends to a continuous group representation in E.

Proof: In Chapter 1, the same problem was discussed. The basic question is whether or not the commutation relations

$$V(t,A)BV(-t,A) = \exp(t \text{ ad } A)B \qquad (28)$$

hold on D for all A, B $\in S$. Since local boundedness of the functions $t \to BV(t,A)u$ for A, B $\in S$ and $u \in D$ is not assumed, Theorem 3.2 does not apply directly. However the local boundedness can be shown to follow from Proposition 3.6. In fact, it is shown in 3.6 and the remark after the proof that for each pair A,B $\in S$ there is a non-trivial interval J such that the set of vectors $\{BV(t,A)u: t\in J\}$ is bounded in E for all $u \in D$. (Recall that completeness of D enables one to apply a Baire category argument to the complete metric space $\mathbb{R} \times D$.)

Consequently, Theorem 3.2 does give the commutation relations (28) on D for all A, B $\in S$.

A standard argument using the fact that conjugation by $V(t,A)\big|_D$ respects the multiplicative structure of $\mathbf{A}(D)$, (see proof of Lemma 8.4(a)), then shows that the commutation relations (28) do, in fact, hold for all B $\in E(\mathcal{L})$ and A $\in S$.

8.7. Lemma

Let $\mathfrak{h}$ be a Lie generating subset for a finite-dimensional Lie algebra $\mathfrak{g}$.

(a) Then a linear basis for $\mathfrak{g}$ may be chosen from among elements of the form $\exp(t_1 \text{ ad } X_1)\cdots\exp(t_k \text{ ad } X_k)(X_{k+1})$ where $X_j \in \mathfrak{h}$, $t_j \in \mathbb{R}$, for $j=1,\ldots,k+1$.

(b) Equivalently, let $\mathfrak{g}$ be the Lie algebra of a Lie group G, and let G_0 be the subgroup of G generated be the elements $\exp(t_jX_j)$ for $t_j \in \mathbb{R}$, $X_j \in \mathfrak{h}$. (That is, the connected component of e if $\mathfrak{h}$ is Lie generating.) Then a basis for $\mathfrak{g}$ may be chosen from elements of the form $\text{Ad}(g_i)(X_i)$ where $X_i \in \mathfrak{h}$ and $g_i \in G_0$.

Proof: (a) If we introduce the function $F(t_1,\ldots,t_k)$ defined in eqn. (21), only with the A_j's replaced by X_j's, then the argument from the proof of 8.1 carries over directly, and (a) follows.

(b) By a well-known property of the adjoint representation we have the formula

$\exp(t_1 \text{ ad } X_1)\cdots\exp(t_k \text{ ad } X_k) = \text{Ad}(\exp t_1X_1)\cdots\text{Ad}(\exp t_kX_k)$ $= \text{Ad}(g_0)$ for $g_0 = \exp(t_1X_1)\cdots\exp(t_kX_k) \in G_0$. Since every g in G_0 is of this form, the conclusion in (b) follows from (a).

Remark: When the perturbation theory for operator Lie algebras is taken up in the next chapter, the lemma will be used again to carry analytic information from the operators in a Lie generating subset to a linear basis for a given Lie algebra of unbounded operators.

Finally, Lemma 8.4 (a), combined with Theorem 8.1, shows that $\mathfrak{L}$ can be integrated to a differentiable group representation V_∞ of G in D. Recall that G is a simply connected Lie group whose Lie algebra $\mathfrak{g}$ is isomorphic to $\mathfrak{L}$.

Extendability of V_∞ to a continuous representation of G in E is a consequence of the following.

8.8. Lemma

Let D be a dense linear subspace of a l.c.s. E and let $\mathfrak{L} \subset \mathbf{A}(D)$ be a finite-dimensional real Lie algebra. Let G be a connected Lie group with Lie algebra isomorphic to $\mathcal{L}$, and let V_∞ be a differentiable representation of G in D equipped with its projective C^∞-topology.

Suppose $\mathcal{L}$ is generated as a Lie algebra by a set S of elements consisting of infinitesimal pregenerators for c.l.e. one-parameter groups of bounded operators in E.

Then V_∞ extends to a continuous locally equicontinuous representation of G in E, and each operator in $\mathcal{L}$ is the infinitesimal pregenerator of a c.l.e. group in L(E).

Proof: Writing $V_\infty(g)$, for g in a coordinate neighborhood Ω of the identity, in terms of canonical coordinates of the second kind one sees that $V_\infty(g)$ for $g \in \Omega$ has a bounded extension, which we denote by V(g). Since G is connected, it is actually true that $V_\infty(g)$ extends for all $g \in G$.

Now the group law $V(ag)u = V(a)V(g)u$ for $u \in D$ and $a, g \in G$ extends to all of E by continuity of V(ag), V(a), a·.d V(b).

The last problem concerns strong continuity and local equicontinuity of the mapping $g \to V(g)$ of G into L(E). It is enough to check strong continuity on the neighborhood Ω. Now, the operator V(g) for $g \in \Omega$ can be expressed in local coordinates as $V(t_1(g),B_1)\ldots V(t_d(g),B_d)$. That is a product of c.l.e. groups. Hence, the proof of strong continuity and local equicontinuity goes over word for word from the corresponding part of the proof of Theorem 8.1 . Lemma 8.7 is needed also at this point.

We finally check that for each X in the Lie algebra of G the closure in $E \times E$ of $dV_\infty(X)$ is equal to the infinitesimal generator of the c.l.e. group $\{V(\exp tX): -\infty<t<\infty\} \in L(E)$. Let H be the infinitesimal generator of the latter group. Then D is a core for H by Theorem 1.3 of [Ps 1]. Since the restriction of V(exptX) to D is equal to $V_\infty(\exp tX)$ the desired conclusion $H=\overline{dV_\infty(X)}$ follows. E.O.P.

Chapter 9

EXPONENTIATION AND BOUNDED PERTURBATION OF OPERATOR LIE ALGEBRAS

The original application of the C^∞ commutation methods of this book was exponentiation of Lie algebras of unbounded operators in a Banach space. The present chapter contains two exponentiation theorems which are improvements upon results due to the co-authors. It also contains theorems on perturbations of Lie algebras of unbounded operators. These results are éntirely new.

The formal mathematical framework is described in the introduction (Chapter 1). Let D be a dense linear subspace of a l.c.s. E which, for the moment, we shall assume is a Banach space. Let $\mathfrak{L}$ be a finite-dimensional real Lie algebra contained in the algebra $\mathbf{A}(D)$ of linear endomorphisms in D. Let G be a connected, simply connected Lie group with Lie algebra $\mathfrak{g}$ isomorphic to $\mathfrak{L}$. Let $\exp : \mathfrak{g} \to G$ be the exponential mapping.

We say that an operator Lie algebra $\mathfrak{L} \subseteq \mathbf{A}(D)$ _can be exponentiated_ if it is the infinitesimal Lie algebra of some strongly continuous representation V of G in E. That is, $\mathfrak{L}$ is the infinitesimal Lie algebra of a group V(G) of bounded linear operators in the Banach space E.

More precisely, let ρ be a Lie algebra isomorphism of $\mathfrak{g}$ onto $\mathfrak{L}$. We say that $\mathfrak{L}$ can be exponentiated if, (1) the operators $\rho(X)$ for all $X \in \mathfrak{g}$ have closed extensions; and (2) there is a strongly continuous representation V of G in E such that for all $X \in \mathfrak{g}$ the closure of $\rho(X)$ coincides with the infinitesimal generator of the one-parameter group $\{V(\exp tX): -\infty < t < \infty\}$.

This definition gives uniqueness of the exponentiated group representation V. (The definition of exponentiability given in the introduction (Chapter 1) does not imply uniqueness of the exponential, as can be seen by an easy example (d/dx on $L^2(0, 1)$, say).)

9A. Discussion of exponentiation theorems and applications

We mention below some important exponentiation problems and discuss alternative methods for solving these problems.

Frequently in applications a finite set of operators

$A_1,\ldots,A_r$ in $\mathbf{A}(D)$ is given. One checks that the Lie subalgebra of $\mathbf{A}(D)$ generated by those elements is finite-dimensional. A necessary condition for exponentiability is: (a) that each of the operators A_j have closures $\bar{A}_j$ which are the infinitesimal generators of strongly continuous one-parameter groups.

The condition is not sufficient however, even for the case where the operators A_j are skew symmetric in a Hilbert space. This is most dramatically demonstrated by a well-known counterexample, due to Nelson [Nℓ 1, § 10, p. 606]. There is a two-dimensional manifold M; and two skew adjoint differential operators A_1 and A_2 in $L^2(M)$ such that $A_1A_2u = A_2A_1u$ for all $u \in C_c^\infty(M)$. The operators A_1 and A_2 are infinitesimal pregenerators of unitary one-parameter groups in $L^2(M)$ which do not commute. The Abelian Lie algebra spanned by A_1 and A_2 cannot be exponentiated.

One can check that there is no domain D with $C_c^\infty(M) \subseteq D \subseteq C^\infty(M)$ such that D is invariant under the unitary groups generated by A_1 and A_2. Note that any invariant domain D has to be contained in $C^\infty(M)$ by the Sobolev Lemma.

It follows that, in the general case, the condition (a) is not sufficient.

In our first theorem, we shall assume in addition to (a) that: (b) there is a C^∞-domain E_∞ for the Lie algebra which is invariant under the one-parameter groups $V(t,A_j)$, i.e., $V(t,A_j)E_\infty \subseteq E_\infty$.

It is still an open question whether the assumptions (a) and (b) together are sufficient. The problem (which has been open for ten years) is discussed in Chapter 1 and at the end of Chapters 3 and 8. We are convinced that there are no counterexamples in the literature that settle this question.

Our second theorem (9.2) does not assume the existence of such a domain. It has instead a density condition on the resolvents of the operators $A_1,\ldots,A_r$. In the process of the proof of Theorem 9.2 it is verified that the density condition implies the existence of a group invariant C^∞-domain.

The construction of such a domain is usually the hardest part of the exponentiation process. This is true also for the analytic vector constructions due to Nelson [Nℓ 1], Poulsen [Ps 2], and the "French school" [FSSS, and S]. A unified discussion of these techniques can be found in [Jo 1]. It is assumed in those treatments that $\mathfrak{L}$ consists of skew symmetric operators, and (i) that a certain quadratic expression $\Sigma_{k=1}^{r} A_k^2$ is essentially self adjoint on D, or (ii) that the operators $A_1,\ldots,A_r$ have a common dense set of analytic vectors. In both situations we get unitary one-parameter groups $V(t,A_j)$. If N is the self adjoint closure of $\Sigma_{k=1}^{r} A_k^2$ then $D_\infty(N)$ and $D_\omega(N)$ are candidates for a group invariant C^∞-domain, [Nℓ 1],[Ps 2].

Under assumption (ii) it is shown in [FSSS,S] that the maximal domain E_∞ of C^∞-vectors is invariant under the unitary one-parameter groups.

By the above remarks we hope to have motivated the formulation of our two exponentiation theorems. In order to recall the precise notation for these, let D be a dense linear subspace of a locally convex space E, and let $\mathfrak{L} \subset \mathfrak{A}(D)$ be a C^∞ operator Lie algebra where $\mathfrak{A}(D)$ denotes the associative algebra of endomorphisms of D. An element $A \in \mathfrak{L}$ is said to be an infinitesimal <u>pregenerator</u> (Chapter 3) if it has a closure $\bar{A}$ which is the infinitesimal generator of a strongly continuous locally equicontinuous one-parameter group, (in short $\{V(t,A): t \in \mathbb{R}\}$ is a c.l.e. group in L(E).)

<u>9.1. Theorem</u>

Let E be a l.c.s. and let $\mathfrak{L} \subset \mathfrak{A}(D)$ be a Lie algebra of operators in E. Suppose $\mathfrak{L}$ is generated as a Lie algebra by a subset S with the property that each $A \in S$ is a pregenerator of a c.l.e. group $\{V(t,A): t \in \mathbb{R}\}$.

If the following two conditions are satisfied then $\mathfrak{L}$ exponentiates to a continuous Lie group representation.

(1) The domain D is invariant under the operators $\{V(t,A): t \in \mathbb{R},\ A \in S\}$, (i.e., D is a group invariant domain).

(2) For each vector $u \in D$ and each pair of elements $A, B \in S$ there is an interval I (which may depend on u, A and B) such that the function $t \to BV(t,A)u$ is bounded in I.

The second theorem is stated only for operators in a Banach space E. Let $\|\cdot\|$ be the norm on E and let $\mathfrak{L} \subset \mathfrak{A}(D)$ be a C^∞ Lie algebra for a dense linear subspace D of E. Let τ_1 be the topology on D defined by the seminorms $u \to \|u\|$, and $u \to \|Bu\|$ for $B \in \mathfrak{L}$.

Alternatively, the τ_1 is determined by the norm $\|u\|_1 = \max\{\|B_k u\|;\ 0 \le k \le d\}$ where $B_1,\ldots,B_d$ is a basis for $\mathfrak{L}$ and $B_0 = I$. Two bases clearly give equivalent $\|\cdot\|_1$ norms.

9.2. Theorem

Let $\mathfrak{L} \subset \mathbf{A}(D)$ be a real finite-dimensional operator Lie algebra in a Banach space. Suppose $\mathfrak{L}$ is Lie generated by a subset S with the property that each $A \in S$ is the infinitesimal pregenerator of a C_0 group $\{V(t,A): t \in \mathbb{R}\}$, and denote by ω_A the type of this group (i.e., $\sup\{\|\exp(-\omega_A|t|)V(t,A)\|: t \in \mathbb{R}\} < \infty$).

Suppose
(GD): for each $A \in S$ there are two points $\lambda_\pm = \lambda_\pm(A)$ such that $\operatorname{Re} \lambda_+ > \omega_A + |\operatorname{ad} A|$, $\operatorname{Re} \lambda_- < -\omega_A - |\operatorname{ad} A|$, and both of the ranges $(\lambda_\pm - A)D$ are dense in (D,τ_1). Then $\mathfrak{L}$ exponentiates.

Remarks on the density condition GD: Let D_1 be the completion of (D,τ_1). Then D_1 can be realized as a linear space of equivalence classes of Cauchy sequences (u_n), and $(B_j u_n)$ with $j = 1,2,\ldots,d$ where (u_n) is a sequence of vectors from D. Since E is complete there corresponds to each equivalence class a d+1 tuple of points $(x_0,x_1,\ldots,x_d) \in E \times \cdots \times E$ such that

$$\lim_{n\to\infty} u_n = x_0, \text{ and } \lim_{n\to\infty} B_j u_n = x_j \text{ for } 1 \leq j \leq d.$$

If the operators $B_1,\ldots,B_d$ have closed extensions then $\bar{B}_j x_0 = x_j$ for $j = 1,2,\ldots,d$, such that the point x_0 determines the equivalence class completely. In that case, D_1 is viewed as a linear subspace of E which contains D.

In general, D_1 is not contained in E (Chapter 6). In order to check density conditions, and closability we need a description of the dual $(D,\tau_1)^*$ of continuous linear functionals on (D,τ_1). Let ΠE^* be the product $E^* \times \cdots \times E^*$ of E^* with itself d+1 times. Let N be the linear subspace of ΠE^* consisting of d+1 tuples $(f_0,f_1,\ldots,f_d)$ such that

$$f_0(u) + \sum_{k=1}^{d} f_k(B_k u) = 0 \text{ for all } u \in D. \tag{1}$$

Then the dual $(D,\tau_1)^*$ can be realized as the quotient $\Pi E^*/N$, as one easily verifies by the usual Hahn-Banach argument from the theory of Sobolev spaces of negative integer exponent. Our theory generalizes simple facts from the Sobolev spaces $H^{\pm 1}$.

If E is a Hilbert space and the operators in $\mathfrak{L}$ are skew symmetric we introduce the operator $\Delta = \Sigma_{k=1}^{d} B_k^2$. Let S be the square root of the Friedrichs-extension of $I - \Delta$. Then the norm $\|\cdot\|_1$ is most conveniently chosen to be

$$\|u\|_1 = \|Su\| = ((I-\Delta)u,u)^{\frac{1}{2}} \quad \text{for all } u \in D \qquad (2)$$

where $(.,.)$ is the inner product of E. In this way, D_1 becomes a Hilbert space, which is contained in E, of course. In fact D_1 can be shown to be equal to the domain of the operator S, [Jo 1]. The characterization of the C^1-vectors for group representations as the domain of the square root of the Laplace operator is well known [Gd 1].

We indicate below what density of D means in terms of the above structure of (D,τ_1).

Let $D_\lambda = (\lambda - A)D$ be the set of vectors $\lambda u - Au$ with $u \in D$ for fixed real λ with $|\lambda| > \omega$. By the Hahn-Banach theorem, condition (GD) can be expressed in terms of the dual $(D,\tau_1)^*$. The condition means that for every d+1 tuple of elements $f = (f_0, f_1, \ldots, f_d)$ in E^* we have the implication

$$f_0(\lambda u - Au) + \sum_{k=0}^{d} f_k(\lambda B_k u - B_k A u) = 0$$

for all $u \in D$, implies that f belongs to the space N defined by (1), when $B_1, \ldots, B_d$ is a fixed basis for the Lie algebra $\mathfrak{L}$.

The condition is conceptually much simpler for a Lie algebra $\mathfrak{L}$ of skew symmetric operators in a Hilbert space, and relates to Nelson's well-known exponentiation condition [Nℓ 1, theorem 5]. The latter aspect of condition (GD) is discussed in part in Chapter 11. Now (D,τ_1) becomes a pre-Hilbert space with inner product $(u,v)_1 = ((I-\Delta)u,v)$ for $u, v \in D$ where $\Delta = \Sigma_{k=1}^{d} B_k^2$, cf. (2). Condition (GD) is then equivalent to the implication: For each $x \in D_1$ the condition $(\lambda u - Au, x)_1 = 0$ for all $u \in D$ implies that $x = 0$.

The theorems are improvements upon known results due to both of the co-authors. The improvement lies mainly in the fact that the present theorems only place restrictions on the elements in a generating set for the Lie algebra. These restrictions are, in turn, weakened. The improved theorems are shown to apply to concrete operator Lie algebras in Chapter 12.

9B. Proof of the theorems

Proof of Theorem 9.1: The proof has been prepared in Chapters 3 and 8. Since the one-parameter groups V(t,A) leave D invariant their restrictions to D may be viewed as elements of $\mathbf{A}$(D), also denoted by V(t,A). When D is equipped with its projective C^∞-topology we intend to show that the restrictions form differentiable locally equicontinuous one-parameter groups on D. This follows easily once the commutation relations (3.3) are verified. The commutation relations are in turn obtained from Theorem 3.2. But we have to note that condition (2) of the present theorem is apparently much weaker than the regularity condition (R) of Theorem 3.2. It is shown in Lemma 3.3, however that the present condition (2) implies the condition (R) such that Theorem 3.2 does in fact apply (see the appendix to Part IV following this chapter).

Hence the commutation relations

$$V(t,A)\, B\, V(-t,A) = \exp(t \text{ ad } A)B$$

hold for all A, B $\in S$, and by Lemma 8.4(a) the restriction of $\{V(t,A): -\infty < t < \infty\}$ to D is actually a differentiable representation of the additive group of the line for all A $\in S$. In other words, each A $\in S$ is integrable.

It follows that the Lie algebra $\mathfrak{L} \subseteq L(D)$ is smooth, and by the Integration Theorem for Smooth Lie Algebras (8.2) the Lie algebra $\mathfrak{L}$ integrates to a differentiable representation V_∞ of G in D. We have $dV_\infty(\mathfrak{g}) = \mathfrak{L}$. Now, by the assumption of the theorem all the operators in S are infinitesimal pre-generators of continuous locally equicontinuous one-parameter groups in L(E). Hence, it follows from Lemma 8.7 that V_∞ extends to a continuous locally equicontinuous representation V of G in E. Moreover, Lemma 8.7 implies that for each X $\in \mathfrak{g}$ the operator $dV_\infty(X)$ is the infinitesimal pregenerator of the c.l.e. one-parameter group $\{V(\exp tX): -\infty < t < \infty\} \subseteq L(E)$.

This completes the proof of Theorem 9.1.

Corollary

Let D be a dense linear subspace of a Hilbert space H. Let A and B be elements in the operator algebra $\mathbf{A}$(D). Suppose the following conditions are satisfied:

(i) A and B are contained in a finite-dimensional Lie algebra $\mathfrak{L} \subseteq \mathbf{A}(D)$.

(ii) A and B are essentially skew-adjoint as operators in H.

(iii) The domain D is invariant under the unitary groups U(t,A) and U(t,B) generated by the closures of A and B respectively.

(iv) For each $u \in D$ there are non-empty intervals I_1 and I_2 such that the set of vectors $\{AU(t,B)u: t \in I_1\}$ and $\{BU(t,A)u: t \in I_2\}$ are bounded in H.

Then every operator belonging to the commutator Lie algebra generated by A and B is essentially skew-adjoint.

Proof: Let $\mathcal{L}$ be the commutator Lie algebra generated by A and B, i.e. the smallest Lie subalgebra of $\mathbf{A}(D)$ which contains A and B. Then $\mathcal{L}$ is finite-dimensional, and the subset S consisting of the two elements A and B satisfies the conditions of Theorem 9.1. Consequently, there is a unitary Lie group representation whose infinitesimal Lie algebra is equal to $\mathcal{L}$. In particular, every element $C \in \mathcal{L}$ is the infinitesimal pregenerator of a strongly continuous unitary one-parameter group in $L(H)$, and the closure of C is skew-adjoint by Stone's Theorem. E.O.P.

The following example shows that condition (i) in the Corollary cannot be relaxed. We give an example where the commutator Lie algebra $\mathcal{L}$ generated by A and B is infinite-dimensional, but where both of the ad-orbits $\mathcal{O}_A(B)$ and $\mathcal{O}_B(A)$ are finite-dimensional. Conditions (ii), (iii), and (iv) are satisfied, but the sum A+B is not essentially skew-adjoint.

Example 1: Let H be the Hilbert space $L^2(\mathbb{R})$ and let D be the Schwartz space of rapidly decreasing functions on the line. That is: $f \in D$ if and only if $\sup\{|x^n(d/dx)^m f(x)|: x \in \mathbb{R}\}$ is finite for all positive integers n and m. Let $A, B \in \mathbf{A}(D)$ be the operators $i\, d^2/dx^2$, and multiplication by ix^4, respectively. The reader can easily check that A and B satisfy the conditions (ii), (iii), and (iv) of the Corollary. (To see that D is invariant under $\exp(it(d/dx)^2)$ one can pass to the Fourier-transformed functions.) Furthermore $(\mathrm{ad}\, A)^5(B) = 0$, and $(\mathrm{ad}\, B)^3(A) = 0$. This means that $\dim \mathcal{O}_A(B) \leq 5$, and $\dim \mathcal{O}_B(A) \leq 3$. (The two nilpotency conditions above can be checked directly. But they are also consequences of a general commutator formula (23) in Chapter 4. Indeed, formula (4.23) applied to $X = d/dx$, and $Y =$ multiplication by x^4 yields $(\mathrm{ad}\, A)^5(B) = 0$; and the same formula applied to $X =$ multiplication by x, and $Y = d^2/dx^2$ yields $(\mathrm{ad}\, B)^3(A) = 0$.)

As a consequence of Theorem 3.2 we then have the following commutation identities

$$U(t,A)BU(-t,A) = \sum_{k=0}^{4} t^k/k!\ (\mathrm{ad}\, A)^k(B),$$

and

$$U(t,B)AU(-t,B) = \sum_{k=0}^{2} t^k/k!\ (\text{ad } B)^k(A)$$

in $\mathbf{A}(D)$ for $t \in \mathbb{R}$. The commutation relations imply that there are bases for the ad-orbits $\mathcal{O}_A(B)$ and $\mathcal{O}_B(A)$ consisting of essentially skew-adjoint operators.

However, the operator $A+B = i(d^2/dx^2 + x^4)$ is not essentially skew-adjoint, as is well known [Wm, Ex. 2]. So, by the Corollary, the Lie algebra generated by A and B must be infinite-dimensional. This last fact can also be checked directly.

Example 2: Let H be $L^2(\mathbb{R})$ and D the Schwartz space on the line, as above. Let $\mathcal{L} \subseteq \mathbf{A}(D)$ be the real linear space of all (formally) skew-symmetric polynomials of degree less than or equal to 2 in the operators P and Q. (That is, for every $H \in \mathcal{L}$ the Hilbert space adjoint H^* of H contains minus H.) Here P and Q denote the quantum operators $(1/i)d/dx$, and multiplication by x, respectively. The symbol i denotes the multiplication operator $f \to if$, and $\{,\}$ denotes the anti-commutator bracket. Since the commutator $[P,Q]$ is scalar it follows that $\mathcal{L}$ is, in fact, a Lie algebra. The reader can easily show, using the algebraic identity $AB = \frac{1}{2}(\{A,B\}+[A,B])$ that every element in $\mathcal{L}$ can be written as a real linear combination of the following six elements iP, iQ, i, $A_0 = i/4(P^2 + Q^2)$, $A_1 = i/4(PQ + QP)$, and $A_2 = i/4(P^2 - Q^2)$.

It is not hard to see that $\mathcal{L}$ is the semi-direct product of the Lie algebra of the Heisenberg group and $s\ell(2,\mathbb{R})$. Let H denote the Heisenberg group of upper triangular 3×3 real matrices with ones in the diagonal. It is well known that $\widetilde{SL}(2,\mathbb{R})$, a metaplectic double-sheeted covering group, is contained in the automorphism group Aut(H). The corresponding semi-direct product of $\widetilde{SL}(2,\mathbb{R})$ and H is denoted by G.

It is clear that the unitary group in $L(H)$ generated by the closure of iQ^2 leaves D invariant. As a consequence, one gets, by Fourier transform, that the unitary group generated by the closure of iP^2 also maps D into itself.

Let us check that $\mathcal{L}$ is generated as a Lie algebra by the set of elements $S = \{iP,iQ,iP^2,iQ^2\}$. Indeed, the above second-degree polynomials A_k for $k = 0,1,2$ satisfy the identity $[A_2,A_0] = A_1$. Hence, the smallest Lie algebra containing S must contain A_2, A_0, and A_1, and is therefore equal to $\mathcal{L}$.

By the defining properties of the Schwartz space D it is easily seen that invariance of D under the unitary groups generated by the closures of the operators in S implies the local boundedness conditions (2) of Theorem 9.1. Hence, Theorem 9.1 shows that there is a unique unitary representation U of G in H such that $dU(\mathfrak{g}) = \mathfrak{L}$ where $\mathfrak{g}$ is the Lie algebra of G. Moreover, each element in $\mathfrak{L}$ is essentially skew-adjoint being the infinitesimal pregenerator of a strongly continuous unitary one-parameter group, $t \to U(\exp tX)$ for $X \in \mathfrak{g}$.

Remarks: The last result in Example 2 was obtained independently in [Ps 2, Lemma 1]. Poulsen showed that the harmonic oscillator Hamiltonian $P^2 + Q^2$ analytically dominates every operator belonging to $\mathfrak{L}$. Our treatment in Example 2 is meant to be expository. There are other ways of exponentiating the Lie algebra $\mathfrak{L}$, and for the case of Lie algebras of skew-symmetric operators our Theorem 9.1 does not seem to yield exponentiability in cases where Nelson's theorem could not have been applied just as well; (cf. [Ps 2].)

Proof of Theorem 9.2: All the details of the proof have, in fact, been worked out in Chapters 5, 6, 7, and 8. Under the assumptions of Theorem 9.2 it follows directly from Corollary 6.4 that the Lie algebra $\mathfrak{L}$ has a closable basis. Let $B_1,\ldots,B_d$ be such a basis. The closure in $E \times E$ of the graph of B_j is thus the graph of an operator $\bar{B}_j$ for each j. The C^∞-topology will from now on be defined by the requirement that all the monomials $\bar{B}_{i_1}\cdot\bar{B}_{i_2}\cdots\bar{B}_{i_n}$, $1 \leq i_j \leq d$ be continuous (see Chapter 7).

Having shown the existence of a closable basis it follows from the remarks after Theorem 9.1 that D_1 is contained in E. By Theorem 6.1 and the density condition (GD) each $V(t,A)$ for $A \in S$ leaves D_1 invariant, and the restriction to D_1 of $\{V(t,A): t \in \mathbb{R}\}$ is, in fact, a strongly continuous group on D_1. Conclusion (c) of the same theorem gives a quantitative estimate of the growth of the D_1-operator norm of the induced one-parameter group on D_1.

From this point there are at least two alternative ways of finishing the proof. The first alternative also suggested in Chapter 1, is a D_1-variant of the proof of Chapter 8.1. Choose local coordinates of the second kind in a neighborhood of the identity in G, $g = \exp(t_1B_1)\cdots\exp(t_dB_d)$ and define

$$V(g) = V(t_1,B_1)\cdots V(t_d,B_d) \tag{3}$$

where $B_1,\ldots,B_d$ is a pre-closed basis of the type (8.10). (See the proof of Theorem 8.1. Elements in $\mathfrak{L}$ and in $\mathfrak{g}$ have been identified.) By the differentiation argument of Chapter 8 (sketched in the present setting in Chapter 1) one can show that (8.3) defines a local homomorphism.

The second, more elegant (and instructive), alternative is based on the construction of a group invariant C^∞-domain E_∞. Once such a domain is available, this result can be reduced to the previous one (9.1), and we get exponentiation directly.

The construction of E_∞ is carried out in Chapter 7. It is shown in Theorems 7.2 and 4 that E_∞ can be taken to be the closure in the C^∞-topology of $D(S)$. Recall that the C^∞ topology is now defined in terms of the closed operators $\bar{B}_1,\ldots,\bar{B}_d$.

The space $D(S)$ is, in turn, defined by purely algebraic conditions. It is the smallest linear subspace of E, contained in the space D_1 of C^1-vectors for $\mathfrak{L}$, which is invariant under all the operators $V(t,A)$ for $t \in \mathbb{R}$ and $A \in S$, and which contains D. The local boundedness condition (2) of Theorem 9.1 is trivially satisfied because $\|B\ V(t,A)u\| \leq \|V(t,A)u\|_1 |B|_1$ for all $B \in \mathfrak{L}$ and all $u \in E_\infty$. This completes the proof of Theorem 9.2.

<u>Remarks on C^∞-vectors</u>: Because of the commutation relations

$$\bar{B}_i V(t,A)u = V(t,A)e^{-t\ \mathrm{ad}\ A}(B_i)u \tag{*}$$

for $i=1,2,\ldots,d$; $A \in S$; and $u \in D$, (the conclusion (ii) of Theorem 6.1), it follows that $D(S)$ is contained in the maximal domain of C^∞-vectors (the intersection of the domains of all the monomials $\bar{B}_{i_1}\cdot\bar{B}_{i_2}\cdots\bar{B}_{i_n}$), and furthermore that $D(S)$ is invariant under each monomial $\bar{B}_{i_1}\cdot\bar{B}_{i_2}\cdots\bar{B}_{i_n}$.
The closure in the C^∞-topology of $D(S)$ is, in fact, equal to the maximal domain of C^∞-vectors. The Lie algebra $\mathfrak{L}$ has already been exponentiated, so the last statement can be obtained as a consequence of Theorem 1.3 of [Ps 1].

<u>Remark 1</u>: The fact that we obtain 9.2 by reduction to 9.1 might indicate that 9.1 is the strongest of the two theorems. This is not so, because if D and S satisfy conditions (1) and (2) of 9.1, it is easily shown by 6.2 that condition (GD) of 9.2 holds as well.

However, it is shown in Chapter 12 that for a particular D

(the space of K-finite vectors) condition (GD) holds and condition (1) of 9.1 does not.

In summary, the assumption that a particular C^∞-domain D be invariant under the one-parameter groups generated by the closures of the operators in S is very restrictive. In fact, Theorem 8.6 shows that under the additional assumption that the group invariant D is a Fréchet space it follows that $\mathcal{L}$ exponentiates to a continuous group representation.

One can show, using the commutation relations (*) above and the methods from the proof of Theorem 1.3 in [Ps 1], that D is complete in the C^∞-topology defined by the enveloping algebra if and only if D is equal to the maximal domain of C^∞-vectors E_∞ (defined above). Also, a given continuous group representation in a Banach space has several dense group invariant domains, i.e. the space of Gårding vectors, the hyper Schwartz space, the space of analytic vectors, and the space of C^∞-vectors [Gå 2], [Nℓ 1], [Mr 5], [Ps 1], and [Jo 2]. But of these domains only the space of C^∞-vectors is complete in the projective C^∞-topology.

Remark 2: In the proof of 9.2 one uses that for all points λ with $|\mathrm{Re}\,\lambda|$ sufficiently large the range $(\lambda - A)D$ is $\|\cdot\|_1$ dense in the space of C^1-vectors D_1. But it is enough to know this for two points λ, one with positive real part and one with negative real part, provided that the absolute value of $\mathrm{Re}\,\lambda$ is sufficiently large. Precisely how large depends on the growth exponent ω in the norm estimate $\|V(t,A)\| \leqq M \exp(\omega|t|)$, and on the norm of ad A as a linear endomorphism in $\mathcal{L}$. More precisely: it is enough to have two points λ_+ and λ_- with $\mathrm{Re}\lambda_+ > \omega + |\mathrm{ad}\,A|$ such that $(\lambda_+ - A)D$ is $\|\ \|_1$ dense in D_1, and $\mathrm{Re}\lambda_- < -\omega - |\mathrm{ad}\,A|$ such that $(\lambda_- - A)D$ is $\|\ \|_1$ dense in D_1.

We state (briefly) this strengthening of 9.2 partly in order to stress the analogy between (GD) and the conventional deficiency index conditions for (skew-) Hermitian operators in a Hilbert space, and partly because the strengthened version is needed in our Application-Chapter 12.

The plane region $\{\lambda \in \mathbb{C} : |\mathrm{Re}\,\lambda| > \omega + |\mathrm{ad}\,A|\}$ has two connected components Ω^+ and Ω^-. A technical refinement to the proof of 9.2 shows that the set of points $\lambda \in \Omega^+$ for which $D_\lambda = (\lambda - A)D$ is $\|\ \|_1$-dense in D_1 is open and closed. (cf. Remark (1) to Corollary 6.2). For the moment, let this set be denoted by N^+.

The closure of the graph of A in $D_1 \times D_1$ is the graph of an operator which we denote by $\bar{A}_1$. Then $\bar{A}_1$ is a closed operator in the Banach space D_1, and D is a core for $\bar{A}_1$. We now apply Lemma 5.7' to $T = \bar{A}_1$ and $F = D_1$. The function h defined on Ω_+ by $h(\lambda) = M(\operatorname{Re}\lambda - \omega - |\operatorname{ad} A|)^{-1}$ for $\lambda \in \Omega_+$ is locally bounded. Hence, by Lemma 5.7', N_+ is relatively open and closed in Ω_+. So, by connectedness, N_+ must be equal to Ω_+ if it is known to contain only one point. The same proof works for Ω_- of course. E.O.R.

Although Theorem 9.1 is valid for linear spaces E more general than Banach spaces, there is a Corollary which requires the following restricted setting:

9.1'. Corollary

Let D be a dense subspace of a Banach space E, and let $\mathcal{L} \subset \mathbf{A}(D)$ be a real finite-dimensional operator Lie algebra. Let $S \subset \mathcal{L}$ be a finite set of Lie generators for $\mathcal{L}$ and assume that each element A in S pregenerates a C_0 one-parameter group V(t,A). Finally assume that the space

$$D_\infty = \bigcap_{n=1}^{\infty} \bigcap_{A_i \in S} D(\bar{A}_1 \bar{A}_2 \cdots \bar{A}_n)$$

is invariant under the group V(t,A) for all A in S.

Then there is a C_0-representation V of the group G (associated to $\mathcal{L}$) by bounded operators on E such that, for every X in $\mathcal{L}$, the operator dV(X) is an extension of X, i.e., $dV(X)u = Xu$ for all $u \in D$.

Proof: Since each A in S is closable, it follows that D_∞ is a Fréchet space in the topology determined by the seminorms

$u \to \|\bar{A}_1 \cdots \bar{A}_n u\|$, for $A_i \in S$, $1 \le i \le n$, $n = 1\ldots$

Moreover, any pair of operators $\bar{A}_1, \bar{A}_2$ (with $A_i \in S$) satisfies the assumptions in Proposition 3.6 relative to D_∞. Therefore, $t \to \bar{A}_1 V(t,A_2)u$, is differentiable as an E-valued function for all $u \in D_\infty$.

By the argument in the Proof of Theorem 3.2, it follows now that the commutation relation

$$\bar{A}_1 V(t,A_2)u = V(t,A_2)\exp(-t \text{ ad } A_2)(A_1)u \qquad (*)$$

is valid for all $u \in D$. Consequently, the operator $V(t,A_2)\bar{A}_1 V(-t,A_2)$ with domain D_∞ is an extension of $\exp(t \text{ ad } A_2)(A_1)$. It is now clear, by Lemma 8.7, that $\mathcal{L}$ has a closable basis, and by Theorems 7.2 and 7.4, that the restriction operators $\{\bar{X}|_{D_\infty} : X \in \mathcal{L}\}$ form a Lie algebra, which we shall denote $\mathcal{L}_\infty$, under the commutator bracket. Moreover, the mapping, $X \to \bar{X}|_{D_\infty}$ is a Lie isomorphism of $\mathcal{L}$ onto $\mathcal{L}_\infty$.

An easy completion argument shows that the commutation relation (*) is valid also for vectors u in D_∞. Theorem 9.1 now applies to the Lie algebra $(\mathcal{L}_\infty, D_\infty)$. It follows that $(\mathcal{L}_\infty, D_\infty)$ is exponentiable, and it is easy to see that the resulting C_0-representation V satisfies $dV(X)u = Xu$ for $X \in \mathcal{L}$, and for $u \in D$. E.O.P.

9C. Phillips-perturbations of operator Lie algebras

In this section, we discuss various ways of generalizing, to the setting of operator Lie algebras and group representations, a theorem of R.S. Phillips concerning bounded additive perturbations of semigroup generators (cf. Appendix B): if A generates a C_0 semigroup and U is bounded, then A+U generates a semigroup that depends norm-analytically upon U. Here, such bounded perturbations are imposed upon some or all elements of an exponentiable operator Lie algebra $\mathcal{L}_0$ in such a way that a new "perturbed operator Lie algebra" $\mathcal{L}$ is obtained. We consider two main questions: (1) When is the perturbed $\mathcal{L}$ also exponentiable? and (2) In what senses does such a perturbed exponential depend analytically upon the perturbing operators?

The results obtained here are of interest for several reasons: There exist operator Lie algebras $\mathcal{L}$, of a general nature (such as those discussed in Chapter 12), for which the hypothesis of our exponentiation theorems 9.1 and 9.2 are very difficult to check directly, but which can be described as Phillips perturbations of much simpler "base-point" Lie algebras where the needed exponentiability is easier to obtain. Our applications of this machinery in Chapter 12 could be regarded as supplying some of the first non-trivial examples of Phillips perturbations. Moreover, for the matrix groups G (semisimple) that we have checked, the perturbations provide an embedding of $\hat{G}$ into the bounded operators on a Hilbert space. Finally, some questions of stability and analyticity properties for the corresponding smooth representations on C^∞-vector spaces are new and not completely settled even in the classical semigroup context studied by Phillips.

The following notation will be employed. Let $\mathfrak{L}_0 \subset \mathfrak{A}(D)$ be a finite-dimensional real operator Lie algebra on a normed space D, and let E be the norm-completion. The operators in $\mathfrak{L}_0$ are assumed closable, and, for each $n = 0,1,\ldots,\| \|_n$ denotes the multinomial graph norm of order n, defined from a fixed basis for $\mathfrak{L}_0$. Note that different bases give rise to equivalent graph norms. We refer to Section 9A for details. The completions $(D,\| \|_n)^\sim$ are denoted by D_n. The corresponding algebra of continuous operators on D_n is denoted by $L(D_n)$, resp., $L(E)$ for $n = 0$. An element in $L(D_n)$ will frequently be identified with its restriction to D. The subalgebra $L^u(D_\infty) = \bigcap\{L(D_n): 0 \le n < \infty\}$ of $L(D_\infty)$ will play an important role in the sequel.

We set $\mathfrak{A}_n(D) = \mathfrak{A}(D) \cap \bigcap\{L(D_k): 0 \le k \le n\}$, and $\mathfrak{A}_\infty(D) = \mathfrak{A}(D) \cap L^u(D_\infty)$, that is to say, the operators in $\mathfrak{A}(D)$ which are $\| \|_k$-bounded for all k, $0 \le k \le n$.

Our first results concern perturbations of a given continuous group representation $V_0: G_0 \to \mathrm{Aut}(E)$ of a given group G_0, where the base-point Lie algebra $\mathfrak{L}_0$ to be perturbed is the infinitesimal representation $\mathfrak{L}_0 = dV_0(\mathfrak{g}_0)$ of the Lie algebra $\mathfrak{g}_0$ of G_0, defined on its maximal natural domain, the C^∞-vectors for V_0:

$$D = C^\infty(V_0) = \{u \in E: \tilde{u}(g) = V_0(g)u \in C^\infty(G_0,E)\}.$$

The C^n-vectors will also be used:

$$C^n(V_0) = \{u \in E: \tilde{u} \in C^n(G_0,E)\},$$

and the following two facts are noted [Gd 1]: (a) $C^n(V_0) = D_n$; and, (b) each D_n is $V_0(G_0)$-invariant, and V_0 restricts to a continuous representation on D_n.

Analyticity of the perturbed global representations V_z is considered. In fact, we establish analyticity of V_z for corresponding analytic families of infinitesimal perturbations P_z. Then, for a more restricted class of perturbations, we show stability of the C^∞-spaces, i.e., $C^\infty(V_z) = C^\infty(V_0) = D_\infty$. Finally, we establish analyticity of the mapping, $z \to V_z(\cdot)|_{D_\infty}$, and in fact analyticity with respect to each of the norms $\|\cdot\|_n$ for $n = 1,2,\dots$. The proof of this fact seems to require a certain local boundedness of P_z (which is quite similar to assumption (2) in Theorem 9.1.)

The hypotheses in the perturbation theorems are in one sense less restrictive than those of the analytic continuation results, in that they concern only the perturbations of a Lie generating subset S_0 of $\mathfrak{L}_0$. In another sense they are more restrictive: the perturbations must lie in the smallest algebra $\mathfrak{A}_\infty(D)$.

A special instance of $\mathfrak{A}_\infty(D)$-perturbations occurs when the perturbations belong to a commutator subalgebra P of $\mathfrak{A}_0(D)$, such that $[\mathfrak{L}_0, P] \subset P$. This is the last case considered.

9.3. Theorem

Let V_0 be a strongly continuous (i.e., C_0) representation, and let $\mathfrak{L}_0$ be the corresponding operator Lie algebra on $D = C^\infty(V_0)$. Let $S_0 \subset \mathfrak{L}_0$ be a Lie-generating subset, and let $f: S_0 \to \mathfrak{A}_\infty(D)$ be a function such that the subset $S = \{A + f(A): A \in S_0\}$ Lie-generates a finite-dimensional Lie subalgebra $\mathfrak{L}$ of $\mathfrak{A}(D)$. (Note that $\|\ \|_n$-boundedness of $f(A)$ is assumed for all $n=0,1,\dots$, and $A \in S_0$.)

Then $\mathfrak{L}$ exponentiates to a C_0-representation of the connected simply connected Lie group G with Lie algebra isomorphic to $\mathfrak{L}$.

Proof: The proof is based on Theorem 9.1. By [Gd 3, Proposition 1.2], V_0 restricts to a C_0-representation on each $D_n = C^n(V_0)$. Since, for $A \in S_0$, the operator $f(A)$ is $\|\ \|_n$-bounded, it follows that $V_n(t, A + f(A))$ leaves D_n invariant and acts continuously there. This is a direct application of the Phillips perturbation theorem (B.7) to the induced C_0-group, $V_n(t,A) = V_0(\exp tA)|_{D_n}$. Applying the argument from the proof of the present Theorem 6.1 inductively, one gets

$$V_{n+1}(t, A + f(A)) = V_n(t, A + f(A))|_{D_{n+1}} = V(t, A + f(A))|_{D_{n+1}},$$

so the latter operator leaves invariant $D = \bigcap_1^\infty D_n = C^\infty(V_0)$, and acts smoothly on this space.

We have, for $B \in \mathcal{S}_0$, and $u \in D$, the estimate

$$\begin{aligned}\|(B+f(B))V(t,A+f(A))u\| &\leq \|V(t,A+f(A)u\|_1 + \text{const.}\|u\| \\ &\leq \text{const.}(\|u\|_1 + \|u\|),\end{aligned}$$

where the constants depend on the operator norm of the $\| \ \|_0$-bounded (resp., $\| \ \|_1$-bounded) operator $f(B)V(t,A+f(A))$ (resp., $V_1(t,A+f(A))$. Moreover, the estimate is uniform for t in compact sub-intervals. Hence, both of the conditions, (1) and (2), in Theorem 9.1 have been verified for the infinitesimally perturbed Lie algebra $\mathcal{L}$, and exponentiability follows.

It is implicit in the proof of Theorem 9.3 above that the perturbed representation $V:G \to \text{Aut}(E)$ leaves invariant the Banach spaces $C^n(V_0)$, as well as the Fréchet space $C^\infty(V_0)$, and acts continuously (resp., smoothly) there. For some applications, it is useful to know more than this: For example, we have the identity $C^n(V_0) = C^n(V)$ for all n, so that operator perturbations do not change the analytic environment of the theory. This seems to require conditions on a basis for $\mathcal{L}_0$, rather than just a Lie generating set.

9.4. Proposition

Let V_0 be a base-point representation with infinitesimal operator Lie algebra $\mathcal{L}_0$ on $D = C^\infty(V_0)$ as in Theorem 9.3. Let $\mathcal{B}$ be a basis for $\mathcal{L}_0$, and $f:\mathcal{B} \to \mathbf{A}_\infty(D)$ a function such that $\{A+f(A): A \in \mathcal{B}\}$ is a set of Lie generators for a finite-dimensional commutator Lie subalgebra $\mathcal{L}$ in $\mathbf{A}(D)$. Then $\mathcal{L}$ exponentiates to a C_0-representation V, and

$$C^n(V) = C^n(V_0) \quad \text{for } n = 1,2,\ldots,\infty. \tag{4}$$

Proof: The exponentiation claim follows as a special case of Theorem 9.3. In order to prove that $C^n(V_0) = C^n(V)$ for all n, it suffices to prove that $\mathcal{L}_0$ and $\mathcal{L}$ induce equivalent C^n-norms, call them $\|.\|_n$ and $\|.\|_n'$, respectively, on the domain $D = C^\infty(V_0)$. For if the norms are equivalent, note that D is invariant under V_0 by construction, while by the proofs of 9.1 and 9.3, it is invariant under V as well. (Inspection of the proof of 9.1 shows that D is actually a space of C^∞-

vectors for V, contained in $C^\infty(V)$.) For $n = \infty$, Theorem 1.3 of [Ps 1] directly shows that $D = C^\infty(V_0)$ is dense in $C^\infty(V)$, hence coincides with it since both spaces carry the same complete C^∞-topology. But Poulsen's proof applies _mutatis mutandis_ when the Fréchet space $C^\infty(V)$ is replaced by the Banach spaces $C^n(V_0)$ and $C^n(V)$ for finite n, whence $C^n(V_0)$ and $C^n(V)$ are completions of D with respect to equivalent norms, and must coincide.

It remains to check that $\|\cdot\|_n$ and $\|\cdot\|_n'$ are in fact equivalent. If $\mathfrak{h}$ is the generating set for abstract $\mathfrak{g} \simeq \mathfrak{L}$ that corresponds to $S \subset \mathfrak{L}$, we have by Lemma 8.7 that a basis for $\mathfrak{g}$ can be chosen from elements of the form $\mathrm{Ad}(g_r)(X_k)$ for $X_k \in \mathfrak{h}$ and g_r in the group G_0 generated by $\{\exp(tX_i): t \in \mathbb{R},\ X_i \in \mathfrak{h}\}$. But $dV(\mathrm{Ad}(g_r)(X_k)) = V(g_r)dV(X_k)V(g_r^{-1}) = V_rB_kV_r^{-1}$, where $B_k = A_k + f(A_k)$ in S and V_r is an automorphism of $C^n(V_0)$ for all n, since it is a product of operators $V(\exp(tX_k)) = V(t,B_k) = V(t,A_k+f(A_k))$ whose bounded action on $C^n(V_0)$ was established in the proof of 9.3. This, in combination with $\|\cdot\|_n$-boundedness of the $U_k = f(A_k)$, allows us to prove the equivalence of $\|\cdot\|_n$ and $\|\cdot\|_n'$ by induction, beginning with the trivial case $\|\cdot\|_0 = \|\cdot\| = \|\cdot\|_0'$. First, for $C_{kr} = V_rB_kV_r^{-1}$ a typical basis element for $\mathfrak{L}$, and $u \in D$, we have by equivalence of $\|\cdot\|_n$ and $\|\cdot\|_n'$ that the contribution of C_{kr} to $\|\cdot\|_{n+1}'$ satisfies

$$\begin{aligned}
\|C_{kr}u\|_n' &\le K_1\|V_rB_kV_r^{-1}u\|_n \\
&\le K_1\|V_r\|_n\|(A_k+U_k)V_r^{-1}u\|_n \\
&\le K_2(\|A_kV_r^{-1}u\|_n+\|U_kV_r^{-1}u\|_n) \\
&\le K_2(\|V_r^{-1}u\|_{n+1}+\|U_kV_r^{-1}\|_n\|u\|_n) \\
&\le K_2(\|V_r^{-1}\|_{n+1}\|u\|_{n+1}+\|U_kV_r^{-1}\|_n\|u\|_{n+1}) \le K_3\|u\|_{n+1}
\end{aligned}$$

where K_1, K_2, and K_3 are constants possibly depending upon k and r. Thus

$$\|u\|_{n+1}' = \max\{\|C_{kr}u\|_n' : C_{kr} \text{ in the basis, or } C_{kr} = I\} \le K_4\|u\|_{n+1}.$$

On the other hand, if $|B_i|$ denotes the norm of $B_i \in \mathfrak{L}$ with respect to the basis $\{C_{kr}\}$, we have for every A_i in the basis for $\mathfrak{L}_0$:

$$\|A_i u\|_n = \|(B_i - U_i)u\|_n \le \|B_i u\|_n + \|U_i\|_n \|u\|_n$$
$$\le K_5(\|B_i u\|_n' + \|u\|_n' \le K_5(|B_i|\ \|u\|_{n+1}' + \|u\|_{n+1}')$$
$$\le K_6 \|u\|_{n+1}',$$

with K_6 possibly depending upon i. Hence, as before

$$\|u\|_{n+1} = \max\{\|A_i u\|_n : 0 \le i \le d\} \le K_7 \|u\|_{n+1}'.$$

This establishes equivalence of $\|\cdot\|_{n+1}$ and $\|\cdot\|_{n+1}'$, completing the induction and the proof.

Theorem 9.3 is (implicitly) a result concerning stability of the hypotheses of our domain-regularity exponentiation Theorem 9.1. As might be expected, there is a corresponding result concerning stability of the hypotheses of our graph-density exponentiation Theorem 9.2. Here, we must continue the placement of restrictions upon a basis for $\mathcal{L}_0$, but we are able to relax the boundedness assumptions upon perturbing operators, at the cost of less perturbation-stability of smooth vectors. For completeness, we include the fact that domain invariance hypotheses, similar to these implicit in 9.3, will also suffice here.

9.5. Theorem

Let $\mathcal{L}_0$ and $\mathcal{L}$ be finite-dimensional Lie sub-algebras of $\mathbf{A}(D)$ for a vector space D. Let E be the completion of D with respect to a norm $\|\ \|$ which makes the operators in $\mathcal{L}_0$ and $\mathcal{L}$ pre-closed, and let $f:\mathcal{L}_0 \to \mathbf{A}_1(D)$ be a function such that, $A \to A+f(A)$, is a linear isomorphism of $\mathcal{L}_0$ onto $\mathcal{L}$. (Hence, f(A) is assumed bounded both with respect to $\|\ \|$ and $\|\ \|_1$ where $\|\ \|_1$ is the 1-norm defined by $\mathcal{L}_0$.)

Then each of the following two conditions implies exponentiability in E of $\mathcal{L}$:

(i) Condition GD is satisfied for each operator in a linear basis for $\mathcal{L}_0$.

(ii) $\mathcal{L}_0$ has a C_0-exponential in E which leaves D invariant.

Proof: Assume (i) and let $\mathcal{B}$ be the corresponding basis. Then $\mathcal{L}_0$ exponentiates by 9.2 to a C_0-representation in E, and $C^1(V_0)$ = the $\|\ \|_1$-completion of D, viz., D_1. For $B \in \mathcal{B}$ set

$U = f(B)$. Then for $|\mathrm{Re}\,\lambda|$ sufficiently large the operator $U(\lambda - B)^{-1}$ is $\|\ \|_1$-bounded with arbitrarily small norm. Hence the usual perturbation argument (of Kato and others) obtains $\|\ \|_1$-density of $(B+U-\lambda)D = -(1-U(\lambda-B)^{-1})(\lambda-B)D$ from that of $(\lambda-B)D$ when applied to operators induced on D_1 with domain D. Since $h = \mathrm{id} + f$ is an isomorphism, it follows that GD is satisfied for a basis for $\mathfrak{L}$. Hence 9.2 gives exponentiability of $\mathfrak{L}$ as well, and

$$C^1(V_0) = \bigcap_{\mathbf{B}} D(\bar{B}) = \bigcap D(\bar{B}+f(B)) = C^1(V).$$

Finally, we reduce case (ii) to case (i) by the pregenerator theorem on D_1 [Gd 3, Corollary 1.2] (which is just the $C^1(V)$-variant of Poulsen's theorem [Ps 1, 1.3] discussed in the proof of 9.4 above): That is, $V_1(\exp tB) = V_1(t,B)$ leaves $D \subset D_1$ invariant, so D is a $\|\ \|_1$-core for $\bar{B}^{(1)}$, and, whenever $\lambda \in \rho(\bar{B}^{(1)})$, $D_\lambda = (\lambda-B)D$ is $\|\ \|_1$-dense.

<u>Remark</u>: We imposed condition (ii) on an arbitrary $V_0(G_0)$-invariant domain, while Theorem 9.3 considered $D = C^\infty(V_0)$ in particular. In fact, no generality was lost in 9.3 by doing so, since if D is $V_0(G)$-invariant and $U \in \mathbf{A}_\infty(D)$, Poulsen's theorem 1.3 [op.cit.] ensures that D is dense in $C^\infty(V_0)$ and U extends to $\tilde{U} \in \mathbf{A}_\infty(C^\infty(V_0))$ preserving the perturbation relationships involved in 9.3. Similarly, (ii) holds for an arbitrary D iff it holds for $D = C^\infty(V_0)$.

One useful feature of Phillips' perturbations in the theory of one-parameter semigroups is analytic dependence of the perturbed semigroup $V(t,A+U)$ upon the perturbing operator. The analog of this property for group representations is "analytic series of representations", with possible singular behavior at exceptional points. It is to this subject that we turn next.

Let Ω be a complex domain (in one or several dimensions). As in 9.3 (where $D = C^\infty(V_0)$) we consider a base-point Lie algebra $\mathfrak{L}_0 \subset \mathbf{A}(D)$ in a fixed $\|\ \|$-completion E of D. If $S_0 \subset \mathfrak{L}_0$ is a fixed Lie-generating subset we consider functions $f:\Omega \times S_0 \to \mathbf{A}_0(D)$. We say that f is <u>admissible</u> if for every $z \in \Omega$ the mapping, $h_z:A \to A+f(z,A)$, extends from S_0 to $\mathfrak{L}_0$ as an isomorphism of Lie algebras (denoted $h_z = id+f_z$). (Hence $\mathcal{L}_z = h_z(\mathcal{L}_0)$ is finite-dimensional.)

We say that f is continuous (resp., analytic) as a L(E)-valued mapping if, for all $A \in S_0$, $z \to f(z,A)$, is continuous (resp., analytic) into L(E). (The operator $U_z = f(z,A)$ is $\| \ \|$-bounded and extends therefore uniquely to a bounded operator on E, also denoted by U_z.) Continuity is taken in the uniform (operator) norm of L(E). As is well known ([HP]) the topology on L(E) need not be specified in the analyticity definition.

9.6. Theorem

Let $\mathcal{L}_0 \subset \mathbf{A}(D)$ be a base-point operator Lie algebra, and $S_0 \subset \mathcal{L}_0$ a Lie-generating subset. Let $f: \Omega \times S_0 \to \mathbf{A}_0(D)$ be an admissible function where Ω and D are as above. Let G be the connected and simply connected Lie group with Lie algebra isomorphic to $\mathcal{L}_0$. Assume that each of Lie algebras $\mathcal{L}_0$ and $\mathcal{L}_z = h_z(\mathcal{L}_0)$ have C_0-exponentials V_0 (resp., V_z).

If $z \to f_z$ is continuous (resp., analytic), then it follows that the mapping, $z \to V_z(g)$, is L(E)-norm continuous (resp., analytic) for all $g \in G$.

Proof: For $A \in S_0$ consider the Dyson-Phillips expansion for $V(t,A+U_z)$ where $U_z = f_z(A) \in L(E)$. It is analytic as a function of U in the sense which is specified in Appendix B7. Hence $z \to V(t,A+U_z)$ is norm-continuous (resp., analytic) by composition if the perturbing operator function $z \to U_z$ has these properties.

By Lemma 8.7 we may pick a basis for $\mathcal{L}_0$ from among elements of the form

$$B = e^{s_k \operatorname{ad} A_k} \cdots e^{s_1 \operatorname{ad} A_1}(A_0) \text{ for } s_k,\ldots,s_1 \in \mathbb{R} \text{ and } A_k,\ldots,A_0 \in S_0.$$

Then

$$h_z(B) = e^{s_k \operatorname{ad} h_z(A_k)} \cdots e^{s_1 \operatorname{ad} h_z(A_1)} h_z(A_0)$$

and

$$V(t,h_z(B)) = V(s_k,h_z(A_k))\ldots V(s_1,h_z(A_1))V(t,h_z(A_0))V(s_1,h_z(A_1))^{-1}\ldots V(s_k,h_z(A_k))^{-1}.$$

Each term in this product expression is of the form $V(t,A+U_z)$ with $A \in S_0$ and $U_z = f_z(A)$. Hence the analyticity and norm-continuity conclusions follow for $z \to V(t,h_z(B))$. Since the elements $h_z(B)$ form a basis for $\mathcal{L}_z$ for each z we note that these conclusions carry over to $z \to V_z(g)$ for all g in a coordinate neighborhood of the identity in G, and hence for all $g \in G$, using canonical coordinates of the second kind as usual.

We turn next to certain admissible functions defined everywhere on the unperturbed Lie algebra $\mathcal{L}_0$, $f:\Omega \times \mathcal{L}_0 \to \mathbf{A}_0(D)$, which we shall call admissible. The standing assumptions which are thereby added are:

(i) boundedness of each operator $f_z(A)$; and

(ii) analyticity of, $z \to f_z(A)$, for all $A \in \mathcal{L}_0$ (and not just, as above, for A in a Lie generating subset).

For such f we clearly have stability of C^1-vectors in the sense that $C^1(V_z) = C^1(V_0)$ for all $z \in \Omega$.

9.7. Lemma

Let $f:\Omega \times \mathcal{L}_0 \to \mathbf{A}_0(D)$ be an analytic admissible function where $\mathcal{L}_0 \subset \mathbf{A}(D)$ is a base-point Lie-algebra. Assume exponentiability of each of the Lie algebras $\mathcal{L}_0$ and $h_z(\mathcal{L}_0)$ where $h_z = id+f_z$.

Then the mapping $z \to V_z(g)|_{D_1}$ is $\|\ \|_1$-analytic.

Proof: Note that stability of the C^1-vectors is implicit in the statement of the lemma, $C^1(V_z) = D_1$ for all $z \in \Omega$. By 9.6 we know that $z \to V_z(g)$ is already analytic into $L(E)$. In this proof we identify $\mathcal{L}_0$ with the Lie algebra of $G_0 = G$. We then have the identity

$$AV_z(g)u = V_z(g)h_z(Ad(g^{-1})A)u - f_z(A)V_z(g)u \qquad (5)$$

for all $A \in \mathcal{L}_0$, $z \in \Omega$, $g \in G$, and $u \in D_1$. Here we note that both terms on the right-hand side are z-analytic into E by composition of analytic functions; each of the functions $h_z(.)$ and $f_z(.)$ is analytic by assumption, and $V_z(.)$ is analytic by the conclusion in 9.6. It follows that $z \to V_z(g)u$ is analytic into D_1 which is equivalent to the conclusion of the lemma.

9.8. Theorem

Let $f: \Omega \times \mathfrak{L}_0 \to \mathbf{A}_\infty(D)$ be an admissible function where $\mathfrak{L}_0 \subset \mathbf{A}(D)$ is a base-point Lie algebra of operators. Assume

(i) that, $z \to f_z$, is analytic into $\mathbf{A}_\rho(D)$;

(ii) that the range of $z \to f(z,A)$ is locally bounded in $L(D_n)$, i.e., for all compacts $K \subset \Omega$, $A \in \mathfrak{L}_0$, and $n = 0,1,\ldots$

$$\sup_{z\in K} \|f(z,A)\|_n < \infty; \tag{6}$$

and

(iii) each of the Lie algebras $\mathfrak{L}_0$, and $h_z(\mathfrak{L}_0)$ for $z \in \Omega$, have C_0-exponentials.

Then the mapping $z \to V_z(g)|_{D_n}$ is analytic into $L(D_n)$ for all $n = 0,1,\ldots,\infty$ and $g \in G$.

Proof: The conclusion holds for $n = 1$ by Lemma 9.7 above. If we could show that $z \to f(z,A)$ were analytic into $L(D_n)$ for all n, the theorem would then follow from the Lemma by induction. Since $\mathfrak{L}_0$ may be considered as an operator Lie algebra in D_n, and the properties from E carry over, it is clearly enough to show analyticity of f_z into $L(D_1)$. Note that this interesting implication is, in itself, an instance of the converse of the analyticity implication which is established in the Phillips perturbation theorem (B7). It would be of independent interest to know if, in general, analyticity of the infinitesimal perturbation follows from analyticity of the perturbed one-parameter (semi) group.

We first establish local boundedness of $z \to \|V_z(g)u\|_1$, for $u \in D_1$ and $g \in G$, under the assumptions in the theorem. We may consider Lie group elements in g on the form $g = \exp(tB)$ only ("exp" denoting the exponential map of Lie theory), by connectedness of G. Set $V(t) = V_0(t,B), V_z(t) = V_z(\exp(tB))$ $= V(t,h_z(B)) = V(t,B + U_z)$ where $U_z = f(z,B)$. If the unperturbed generator B is of the type $C_0(M,\omega)$, it follows from (B7) that the perturbed generator $h_z(B) = B + U_z$ is of type $C_0(M,\omega+M\|U_z\|)$. We claim that the restricted one-parameter group $V_z(t)|_{D_1}$ is of generator type $C_0(M_1,\omega_1)$, where

$$M_1 = M(|e^{-t \operatorname{ad} B}| + \max_i \|f_z(e^{-t \operatorname{ad} B}(A_i))\| + \|U_z\|),$$

and $\omega_1 = \omega + M\|U_z\|$. The expression for M_1 contains the finite-dimensional norm of the transformation $e^{-t \text{ ad } B}$, and $\{A_i\}$ is a fixed chosen basis containing A. (Note that the stated local boundedness follows from this, together with assumption (6) for n = 0.)

The proof of the type-conclusion for $V_z(t)|_{D_1}$ is based on the following commutation relation, $A,B \in \mathcal{L}_0$:

$$\begin{aligned} AV_z(t) &= h_z(A)V_z(t) - U_zV_z(t) \\ &= V_z(t)h_z(e^{-t \text{ ad } B}(A)) - U_zV_z(t). \end{aligned}$$

The resulting estimate (see below) then establishes the claim:

$$\begin{aligned} \|AV_z(t)u\| &\le Me^{\omega_1|t|}(|e^{-t \text{ ad } B}| + \max_i \|f_z(e^{-t \text{ ad } B}(A_i))\|)\|u\|_1 + \|U_z\|Me^{\omega_1|t|}\|u\| \\ &\le M_1e^{\omega_1|t|}\|u\|_1 \end{aligned}$$

where M_1 is defined as above.

To show analyticity of $z \to U_z|_{D_1}$, we consider the approximation

$$\lim_{\substack{h\to 0 \\ h\neq 0}} \|U_z^h u - U_z u\|_1 = 0 \quad \text{for } u \in D_2, \tag{7}$$

where

$$U_z^h = h^{-1}(V_z(h) - V(h)) = h^{-1}\int_0^h V(h - s)U_zV_z(s)ds \tag{8}$$

(given by the Duhamel formula) is clearly an analytic $L(D_1)$-operator function.

Before beginning the proof we note that the approximation (7) must necessarily be valid for all $u \in D_1$ if it can be established for $u \in D_2$. This is a consequence of the usual $3\,\varepsilon$ argument, combined with local boundedness of

$z \to \|U_z^h\|_{L(D_1)}$, - uniform in h.

Denoting the $L(D_1)$-norm also by $\|\ \|_1$, we have

$$\begin{aligned} \|U_z^h\|_1 &\le h^{-1}\int_0^h \|V(h - s)\|_1\|U_z\|_1\|V_z(s)\|_1 ds \\ &\le M_1^2\|U_z\|_1 h^{-1}\int_0^h e^{\omega_1|h-s|}e^{\omega_1|s|}ds \\ &\le M_1^2\|U_z\|_1 e^{\omega_1} \end{aligned}$$

for all $h \in \mathbb{R}\setminus\{0\}$, $|h| < 1$. Hence local boundedness of $\|U_z^h\|_1$ follows from (6), n = 1.

The following calculation is needed in performing the norm estimates involved in the proof of (7):

$$U_z^h u - U_z u = h^{-1}\int_0^h (V(h-s)U_z V_z(s) - U_z)u\, ds. \tag{9}$$

The integrand decomposes further:

$$V(h-s)U_z V_z(s)u - U_z u = V(h-s)U_z(V_z(s)u-u)+(V(h-s)-I)U_z u$$

$$= T_1(h,s)+\int_0^{h-s} V(t)BU_z u\, dt.$$

Here, the first term $T_1(h,s)$ on the right-hand side is dominated in $\|\ \|_1$-norm by

$$2M_1^2 e^{\omega_1|h-s|}\|U_z\|_1 e^{\omega_1|s|}$$

So, by continuity of the integral, the conclusion $\lim_{h\to 0}\|h^{-1}\int_0^h T_1(h,s)ds\|_1 = 0$ follows.

The second term $T_2(h,s)$ is dominated in $\|\ \|_1$-norm (for $h \in \mathbb{R}\setminus\{0\}$, $|h| < 1$) by $|h-s|\|U_z u\|_2 M_1 e^{\omega_1}$. Since clearly $h^{-1}\int_0^h |h-s|ds = h/2 \to 0$, the integral $h^{-1}\int_0^h T_2(h,s)ds$ may be estimated in $\|\ \|_1$-norm, and the conclusion (7) follows. Hence, (7) is valid uniformly for z in compact subsets of Ω. (Note that assumption (6) is needed for all values of n when the induction is carried out!)

We have shown that $U_z u$ is the limit of $U_z^h u$ as $h \to 0$, and the limit is uniform for z in compact subsets of Ω. This limit holds in $\|\ \|_1$-norm, so analyticity of $z \to U_z|_{D_1}$ into $L(D_1)$ follows.

As stated in the beginning of the proof, the full conclusion of the theorem may now be derived by an inductive application of (B7) to each of the spaces D_n.

9D. Semidirect-product perturbations

In this section we consider a special class of Phillips perturbations for operator Lie algebras whose principal feature might best be described as commutator-boundedness: for every A, B in the unperturbed Lie algebra $\mathcal{L}_0$, the commutator of A with every perturbation P of B that occurs in the perturbed Lie algebra $\mathcal{L}$ must be bounded. (That is: $A,B\in\mathcal{L}_0$ and $B+P \in \mathcal{L}$ must imply that $[A,P]$ is bounded. The example where $\mathcal{L}_0 = \mathbb{R}\, d/dx$ and $P = -\cos(e^x)$ on $L^2(\mathbb{R})$ shows that this condition is not automatic - P is bounded but $[d/dx,P] = e^x\sin(e^x)$ is not.) This class of perturbations offers three advantages when compared to the more general classes considered in our "perturbation of exponentiability" results 9.3 and 9.5 above.

First, our rather technical conditions of $\|\cdot\|_n$-boundedness there are replaced by the more natural and familiar condition that commutators be bounded. Second, for many applications the process of checking commutator-boundedness is very simple and direct. (There are exceptions where the more general conditions are preferable: see Chapter 12.) Third, it becomes possible in this setting to draw useful conclusions concerning the changes in Lie structure that can occur under such perturbations: how different can $\mathcal{L}_0$ and $\mathcal{L}$ be as abstract Lie algebras? For example, one result below shows that if $\mathcal{L}_0$ and $\mathcal{L}$ both have unitary exponentials and $\mathcal{L}_0$ is non-compact simple, then when $\mathcal{L}$ is such a commutator-bounded perturbation of $\mathcal{L}_0$ it can differ only "trivially" in structure: $\mathcal{L} = \mathcal{L}_0' \oplus \mathcal{B}$, where $\mathcal{L}_0' \cong \mathcal{L}_0$ and the direct summand $\mathcal{B}$ is a Lie algebra of bounded operators that commutes with $\mathcal{L}_0'$, and $\mathcal{L}$ itself has the "Singer-trivial" structure $\mathcal{B} = \mathfrak{k} \oplus \mathfrak{z}$, where $\mathfrak{k}$ is compact semisimple and $\mathfrak{z}$ is central in $\mathcal{L}$.

In practice, commutator-boundedness is not the most economical way to formulate our condition on Phillips perturbations, because it is useful to retain the general setting in which $\mathcal{L}$ is related to $\mathcal{L}_0$ via a map $f: \mathcal{L}_0 \to \mathcal{A}_0(D)$ of $\mathcal{L}_0$ into bounded operators and $\mathcal{L}$ is merely <u>Lie-generated</u> by the set $\{A+f(A): A \in \mathcal{L}_0\}$. It is then better to consider <u>semidirect-product perturbations</u>: we assume that there exists a perturbing class $\mathcal{P} \subset \mathcal{A}(D)$ such that $f: \mathcal{L}_0 \to \mathcal{P}$, satisfying three conditions

(a) Every $P \in \mathcal{P}$ is $\|\cdot\|_0 = \|\cdot\|$ <u>bounded</u> on the dense domain D (so $\mathcal{P} \subset \mathcal{A}_0(D)$ as in Section 9C).

(b) The set $\mathcal{P}$ is itself a <u>commutator Lie algebra</u>, generally infinite-dimensional.

(c) Also $\mathcal{P}$ is normalized in $\mathcal{A}(D)$ by $\mathcal{L}_0$: $[\mathcal{L}_0,\mathcal{P}] \subset \mathcal{P}$, so, in particular, commutator-boundedness holds: $[A,P] \in \mathcal{P} \subset \mathcal{A}_0(D)$ for all $A \in \mathcal{L}_0$, $P \in \mathcal{P}$.

Thus $\mathcal{L} \subset \mathcal{L}_0 \,+\!)\, \mathcal{P}$, the (generally infinite-dimensional) semidirect product of $\mathcal{L}_0$ with P, hence the terminology is appropriate. (See Proposition 9.10(i) below for details.)

9.9. Theorem

Let D be a dense subspace of a Banach space E, and let $\mathcal{L}_0 \subset \mathbf{A}(D)$ be a finite-dimensional Lie algebra of closable operators.

(a) Every perturbing class P as in (a)-(c) above is contained in $\mathbf{A}_n(D)$ for all $n = 1,2,\ldots,\infty$.

(b) If $\mathcal{L} \subset \mathcal{L}_0 +) P$ is generated by $S = \{A + f(A): A \in \mathcal{L}_0\}$ for some $f:\mathcal{L}_0 \to P$ and $\mathcal{L}$ is finite-dimensional, then every $B \in \mathcal{L}$ is closable and $D_n(\mathcal{L}_0) = D_n(\mathcal{L})$ for $n = 1,2,\ldots,\infty$.

(c) If $\mathcal{L}$ is as in (b), then it has an exponential V that leaves the common C^∞-domain, $D_\infty = D_\infty(\mathcal{L}) = D_\infty(\mathcal{L}_0)$, invariant if and only if $\mathcal{L}_0$ has an exponential V_0 with this property. Then $C^n(V) = D_n(\mathcal{L}) = D_n(\mathcal{L}_0) = C^n(V_0)$, $n = 1,2,\ldots,\infty$.

Proof: We prove (a) and (b) together. First, we show in a single induction on n that $P \subset \mathbf{A}_n(D)$ and in (b) that if $\|\cdot\|_n'$ is the C^n-norm determined by $\mathcal{L}$, then $\|\cdot\|_n$ is stronger than $\|\cdot\|_n'$: $\|u\|_n' \leq M_n\|u\|_n$, for all finite n. The claims are true by assumption for n = 0. But if $P \in \mathbf{A}_n(D)$ as in (a), then $\|Pu\|_{n+1} = \max\{\|A_kPu\|_n: 0 \leq k \leq d\}$ by definition, and since

$$\begin{aligned}\|A_kPu\|_n &\leq \|PA_ku\|_n+\|[A_k,P]u\|_n \leq \|P\|_n\|A_ku\|_n+\|[A_k,P]\|_n\|u\|_n\\ &\leq (\|P\|_n+\|[A_k,P]\|_n)\|u\|_{n+1}\end{aligned}$$

it follows that $P \in \mathbf{A}_{n+1}(D)$. Similarly if $\|\cdot\|_n$ is stronger than $\|\cdot\|_n'$ and $\{B_k = A_k+P_k: 0 \leq k \leq f\}$, $\|u\|_{n+1}' = \max\{\|B_ku\|_n': 0 \leq k \leq f\}$ and

$$\begin{aligned}\|B_ku\|_n' &\leq M_n(\|A_ku\|_n+\|P_ku\|_n) \leq M_n(|A_k|\|u\|_{n+1}+\|P_k\|_n\|u\|_n)\\ &\leq M_n(|A_k|+\|P_k\|_n)\|u\|_{n+1}\end{aligned}$$

implies that $\|\cdot\|_{n+1}$ is stronger than $\|\cdot\|_{n+1}'$. This proves (a), and half of (b). If we show that $\mathcal{L}_0 \subset \mathcal{L} + P$ and that $[\mathcal{L},P]\subset P$, then the argument above will apply with $\mathcal{L}$ and $\mathcal{L}_0$ interchanged, so every $\|\cdot\|_n'$ will be stronger than $\|\cdot\|_n$, and hence equivalent to it. Since every $B \in \mathcal{L}$ is a bounded perturbation of a closable $A \in \mathcal{L}$, it must be closable as well, so it follows that the completion $D_n(\mathcal{L})$ is well-defined and equal to the $D_n(\mathcal{L}_0)$ for all n. But every $A \in \mathcal{L}_0$ is of the form $A = (A + f(A)) - f(A) \in \mathcal{L}+P$, and it is easy to see that $\mathcal{L}' = \{B \in \mathcal{L}:[B,P]\subset P\}$ is a subalgebra of $\mathcal{L}$ containing the generating set S, so $[\mathcal{L},P] \subset P$.

($[A+f(A),P] = [A,P]+[f(A),P] \subset P$ by conditions (b) and (c) for P so $[S,P]\subset P$, and since P is a vector space, so is $\mathcal{L}'$. But if $B_1,B_2 \in \mathcal{L}'$ then by Jacobi's identity $[[B_1,B_2],P] = [B_1,[B_2,P]] - [B_2,[B_1,P]] \subset P$ by a double application of condition (c), so $\mathcal{L}'$ is a Lie algebra.) Thus (b) follows.

To obtain (c), we observe first that the usual application of Poulsen's density theorem shows that if exponentials V_0 (respectively V) exist and leave D_∞ invariant then $C^n(V_0) = D_n (=C^n(V)$ respectively) for all n. If $\mathcal{L}_0$ has such an exponential V_0, then 9.3 and 9.4 apply to produce V satisfying the same conditions. If $\mathcal{L}$ has an exponential V leaving D_∞ invariant, we must apply the argument in the proof of 9.3 instead, to conclude that every $A \in \mathcal{L}_0$, qua $B-f(A)$ for $B = A+f(A)$ and $f(A) \in \mathbf{A}_\infty(D)$, generates a smooth group on $D_\infty(\mathcal{L}_0) = C^\infty(V)$, whence as before we can apply 9.1 to conclude that $\mathcal{L}_0$ has an exponential V_0 leaving D_∞ invariant. This completes the proof.

Our discussion of the possible changes in Lie structure that can result from semidirect-product perturbations makes use, at several points, of the following algebraic remarks:

Let $\mathbf{A}$ be an associative algebra over a field F, and let $\mathbf{A}_L$ be the corresponding commutator Lie algebra. Let $\mathcal{L}_0,P$ be a pair of Lie subalgebras of $\mathbf{A}_L$ satisfying $[\mathcal{L}_0,P] \subset P$, and finally let $f:\mathcal{L}_0 \to P$ be a function. We say that the Lie algebra $\mathcal{L} \subset \mathbf{A}_L$ which is Lie-generated by the set $\{A+f(A):A \subset \mathcal{L}_0\}$ is an f-perturbation of $\mathcal{L}_0$.

9.10. Proposition

(i) Every f-perturbation of $\mathcal{L}_0$ is contained in $\mathcal{L}_0 + P$.
(ii) Let $\mathcal{L}$ be an f-perturbation of $\mathcal{L}_0$. Then there is a Lie algebra isomorphism $\mathcal{L}/\mathcal{L} \cap P \simeq \mathcal{L}_0/\mathcal{L}_0 \cap P$ which is determined by the coset mapping: $(A + P) \to (A)$, $A \in \mathcal{L}_0$, $P \in P$.
(iii) The isomorphism in (ii) is induced by a Lie homomorphism $\psi:\mathcal{L} \to \mathcal{L}_0$ defined as follows.
$\psi(A+P) := A$, $A \in \mathcal{L}_0$, $P \in P$, $A+P \in \mathcal{L}$. Moreover $\ker(\psi) = \mathcal{L} \cap P$.

(iv) If $\mathcal{L} \cap P = \mathcal{L}_0 \cap P = 0$, then id+f(.) is onto $\mathcal{L}$. The canonical homomorphism ψ is an isomorphism and $\psi^{-1} = \text{id}+f$.

Remark: Let $\mathcal{L}$ be an arbitrary Lie subalgebra of $\mathcal{L}_0 + P$. Then the set $S_0 = \{A \in \mathcal{L}_0 : \exists P \in P \ni A+P \in \mathcal{L}\}$ is a Lie subalgebra of $\mathcal{L}_0$. If $\mathcal{L}$ is, in fact, an f-perturbation of $\mathcal{L}_0$, then $S_0 = \mathcal{L}_0$. In the general case, (ii) generalizes to the isomorphism $\mathcal{L}/\mathcal{L} \cap P \simeq S_0/S_0 \cap P$.

Proof: (i) Let $\mathcal{L}$ be an f-perturbation of $\mathcal{L}_0$. Then the following implications clearly hold:
$\{A+f(A) : A \in \mathcal{L}_0\} \subset \mathcal{L} \cap (\mathcal{L}_0+P) \subset \mathbf{A}_L$. By definition, $\mathcal{L}$ is Lie-generated by the set on the left in the chain, and the middle term is a Lie subalgebra. Hence, the inclusion $\mathcal{L} \subset \mathcal{L}_0+P$ follows.

Let I be an ideal in a Lie algebra $\mathfrak{g}$. For $X \in \mathfrak{g}$ the coset X + I in $\mathfrak{g}/I$ will be denoted (X), and we note that $\mathfrak{g}/I$ inherits a Lie algebra structure via the canonical mapping $X \to (X)$. We shall apply this simple algebraic fact to both sides $\mathcal{L}/\mathcal{L} \cap P$ and $\mathcal{L}_0/\mathcal{L}_0 \cap P$ in (ii).

We now show that the coset mapping $(A+P) \overset{\psi}{\to} (A)$ defines a Lie isomorphism $\mathcal{L}/\mathcal{L} \cap P \simeq \mathcal{L}_0/\mathcal{L}_0 \cap P$.

To see that ψ is well defined we note the following two implications:

(a) $X = A+P \in \mathcal{L} \cap P$, $A \in \mathcal{L}_0$, $P \in P \Rightarrow A \in \mathcal{L}_0 \cap P$

(b) $X = A+P = A'+P' \in \mathcal{L}$, $A,A' \in \mathcal{L}_0$, $P,P' \in P \Rightarrow A-A' \in \mathcal{L}_0 \cap P$.

Suppose $X = A+P$, and $X' = A'+P'$, are elements in $\mathcal{L}$, $A,A' \in \mathcal{L}_0$, $P,P' \in P$. Then the identity

$$[X,X'] = [A,A'] + [P,A'] + [A,P'] + [P,P'] \tag{10}$$

shows that $\psi([(X),(X')]) = [\psi((X)),\psi((X'))]$, by the assumptions (b) and (c) on P. Since linearity is clear we have shown that ψ is a Lie homomorphism.

To see that ψ is one-to-one we note the implication: $X = A+P \in \mathcal{L}$, $A \in \mathcal{L}_0 \cap P$, $P \in P \Leftrightarrow X \in \mathcal{L} \cap P$. Finally, we note that ψ is onto by the very definition of $\mathcal{L}$ in terms of the given subalgebra $\mathcal{L}_0$.

Suppose the vanishing condition in (iv) is satisfied for a given f-perturbation $\mathcal{L}$ of $\mathcal{L}_0$. Then ψ in (iii) reduces to an isomorphism of $\mathcal{L}$ onto $\mathcal{L}_0$, and a simple calculation shows that $\text{id}+f = \psi^{-1}$.

Remark: We note in particular that the vanishing condition (iv) provides a setting for analytic perturbations where the admissibility/Lie isomorphism assumption in Theorem 9.8 is automatically satisfied. By that theorem we see that the semidirect-product perturbations form a natural framework for the theory.

In Chapter 10, a special class of semidirect-product perturbations with P Abelian is considered, where P is generally infinite-dimensional. Here $\mathfrak{L}_0 \cap P$ is trivial, while $\mathfrak{L} \cap P$ is not. Chapter 12, in turn, shows that the representation theory of $SL(L,\mathbb{R})$ may be obtained as an application of the perturbation theory with P non-commutative and infinite-dimensional. In these examples $\mathfrak{L}_0 \cap P = \mathfrak{L} \cap P = 0$ so that by 9.10 (iv) the Lie algebras $\mathfrak{L}_0$ and $\mathfrak{L}$ are isomorphic. Finally, we note that induced representations, realized on a fixed Hilbert space as multiplier representations with operator-valued smooth multipliers [KS] over a compact homogeneous space can be shown to provide yet a third class of semidirect-product perturbations. It turns out that both the base-point Lie algebra $\mathfrak{L}_0$, as well as the perturbed one $\mathfrak{L}$, satisfy a condition (property U which is defined below) which is stronger than $\mathfrak{L}_0 \cap P = 0$, resp., $\mathfrak{L} \cap P = 0$.

An operator Lie algebra $\mathfrak{L} \subset \mathfrak{A}(D)$ is said to have property U if $\mathfrak{L} \cap \mathfrak{A}_0(D) = (0)$, i.e., only the zero-operator in $\mathfrak{L}$ is bounded. We say that $\mathfrak{L}$ is exponentiable if it is the infinitesimal Lie algebra of a (unique) strongly continuous representation V in $E = (D,\| \|)^\sim$. We say that $\mathfrak{L}$ has property UBE if the representation V is uniformly bounded. Obviously, a unitary representation V has property UBE, and $\mathfrak{L}$ consists of essentially skew-adjoint operators in a Hilbert space completion E.

Property U enters in the applications mentioned above through the following results from Appendix G:

(1) If $\mathfrak{L}$ is simple and exponentiable, then it has property U, or else $\mathfrak{L} \subset \mathfrak{A}_0(D)$. (Corollary G2.)

(2) If $\mathfrak{L}$ is totally non-compact semisimple (TNC, a direct sum of non-compact simple Lie ideals) and has property UBE, then $\mathfrak{L}$ has property U. (Theorem G3.)

We now turn to the structural changes and the algebraic, as well as analytic, properties of $\mathfrak{L}$ and $\mathfrak{L}_0$ in the special case of a type U basepoint Lie algebra $\mathfrak{L}_0$.

Let $\mathfrak{g}$ be a Lie algebra. We then denote by $\mathfrak{g}^n$ (resp., $\mathfrak{g}^{(n)}$) the terms in the lower central series (resp., the derived series), $\mathfrak{g}^{n+1} = [\mathfrak{g}^n,\mathfrak{g}]$, (resp., $\mathfrak{g}^{(n+1)} = [\mathfrak{g}^{(n)},\mathfrak{g}^{(n)}]$).

9.11. Proposition

Let $\mathfrak{L}_0$ have property U, and assume $\mathfrak{z}(\mathfrak{L}_0) = 0$, trivial center. Let $\mathfrak{L}$ be a f-perturbation of $\mathfrak{L}_0 \neq 0$. Then $\mathfrak{L}$ cannot be nilpotent; more generally for no n is $\mathfrak{L}^n \subset \mathfrak{A}_0(D)$.

9.12. Proposition

Let $\mathfrak{L}$ be a f-perturbation of $\mathfrak{L}_0$, and suppose $\mathfrak{L}_0$ has property U. If $\mathfrak{L}$ is solvable modulo P, i.e., $\mathfrak{L}^{(n)} \subset \mathfrak{A}_0(D)$, then it follows that $\mathfrak{L}_0$ is in fact solvable.

9.13. Corollary

Let $\mathfrak{L}_0$ and $\mathfrak{L}$ be as in 9.11 but assume $\mathfrak{L}_0$ is simple instead of just centerfree. Then $\mathfrak{L}$ cannot be solvable.

The proof of 9.12 is based on the following.

9.14. Lemma

Let $\mathfrak{L}$ be a f-perturbation of $\mathfrak{L}_0$, and assume that $\mathfrak{L}_0$ has property U and that $\mathfrak{z}(\mathfrak{L}_0) = 0$. If $X \in \mathfrak{L}$ is such that $[X,Y] = XY - YX$ is bounded for all $Y \in \mathfrak{L}$, then X is in P and hence bounded.

Proof of 9.14: Suppose $X = A+P$, $Y = B+Q$, $A,B \in \mathfrak{L}_0$, $P,Q \in P$. Since $\mathfrak{L}_0$ normalizes $P \subset \mathfrak{A}_0(D)$ we conclude by eqn. (10) that $[A,B] \in \mathfrak{L}_0 \cap \mathfrak{A}_0(D) = (0)$. Since this holds for all B in $\mathfrak{L}_0$ by 9.10 (ii), A is in the center of $\mathfrak{L}_0$, and hence $A = 0$.

Proof of 9.11: Suppose for some n, $\mathfrak{L}^n \subset \mathfrak{A}_0(D)$. Then every $X \in \mathfrak{L}^{n-1}$ satisfies the condition in Lemma 9.14. Hence $\mathfrak{L}^{n-1} \subset P \subset \mathfrak{A}_0(D)$, and by induction we get $\mathfrak{L} \subset P$. This is a contradiction by 9.10 (iii).

Proof of 9.12: Let $\mathfrak{L}$ be as in the statement of 9.12 and assume $\mathfrak{L}^{(n)} \subset \mathfrak{A}_0(D)$. We claim that $\mathbf{B}(\mathfrak{L}) = \mathfrak{L} \cap P$. For if $X = A+P \in \mathbf{B}(\mathfrak{L})$, $A \in \mathfrak{L}_0$, $P \in P$, it follows from property U that $A = 0$, and so $X \in P$. Hence $\mathfrak{L}^{(n)} \subset \mathfrak{L} \cap P$, and $(\mathfrak{L}/\mathfrak{L} \cap P)^{(n)}$ $= \mathfrak{L}^{(n)}/\mathfrak{L} \cap P = (0)$. But $\mathfrak{L}/\mathfrak{L} \cap P$ is isomorphic to $\mathfrak{L}_0$ by 9.10(iii)

We finally turn to our most definitive structure theorem for semidirect-product perturbations, which contains as a special case the "triviality" theorem for unitary perturbations of non-compact simple representations mentioned at the beginning of the section: it goes further by concluding that the exponentials V of such "trivial" Lie algebra perturbations factor as a product of a very special pair of commuting group representations U and W, where W arises from an isomorphism-producing perturbation of $\mathcal{L}_0$ and U (hence V) is the result of an especially trivial further perturbation. We view the perturbation W as the "interesting" part; the examples in Chapter 12 show that it can, in general, be quite nontrivial.

9.15. Theorem

Let $\mathcal{L}$ be a semidirect-product perturbation of a totally non-compact semisimple (TNC - cf. Appendix G) operator Lie algebra $\mathcal{L}_0 \subset \mathfrak{A}(D)$, in a Banach space E. Suppose that $\mathcal{L}$ has a uniformly bounded exponential V and that $\mathcal{L}_0 \cap P = \{0\}$.

(a) Then $\mathcal{L} = \mathcal{L}_0' \oplus \mathfrak{B}$ as a direct sum of commuting Lie ideals. Here $\mathcal{L}_0'$ has property U, and $\psi|_{\mathcal{L}_0'}: \mathcal{L}_0' \to \mathcal{L}_0$ is a Lie algebra isomorphism, while $\mathfrak{B} = \mathfrak{k} \oplus \mathfrak{z} = \mathcal{L} \cap P$ is itself the direct sum of a compact semisimple ideal $\mathfrak{k}$ with the center $\mathfrak{z}$ of $\mathcal{L}$, and consists of bounded operators.

(b) Moreover, if $f: L_0 \to P$ determines this perturbation, there exist semidirect-product perturbations $f': \mathcal{L}_0 \to P$ and $f'': \mathcal{L}_0' \to P$ such that $\mathrm{id}+f': \mathcal{L}_0 \to \mathcal{L}_0'$ is the Lie algebra isomorphism inverse to $\psi|_{\mathcal{L}_0'}$, id+f and id+f" have the same range in $\mathcal{L}$, and the diagram

$$\begin{array}{ccc} \mathcal{L}_0 & \xrightarrow{\ \mathrm{id}+f'\ } & \mathcal{L}_0' \\ & {\scriptstyle \mathrm{id}+f} \searrow \quad \swarrow {\scriptstyle \mathrm{id}+f''} & \\ & \mathcal{L} & \end{array}$$

commutes. Moreover, $f''(\mathcal{L}_0')$ generates $\mathfrak{B}$ as a Lie algebra.

(c) If G_0 and H are the Lie groups associated with $\mathcal{L}_0$ and $\mathfrak{B}$ respectively, then $G = G_0 \times H$ as a commuting direct product, and there exist uniformly bounded exponentials $W: G_0 \to L(E)$ and $T: H \to L(E)$ such that $V = WT$ (i.e., $V(g_0 h) = W(g_0)T(h)$ for all $g = g_0 h \in G = G_0 \times H$). Moreover, T is operator norm-analytic.

Remarks: (1) It is easy to see that $[\mathcal{L}_0', P] \subset P$ and that if $f'''(X) = -f'(\psi(X))$ then $\psi = \mathrm{id}+f''': \mathcal{L}_0' \to \mathcal{L}_0$. Thus $\mathcal{L}_0$ is a semidirect-product perturbation of $\mathcal{L}_0'$, and hence is exponentiable. Moreover, since $\mathcal{L}_0'$ has property U, so must $\mathcal{L}_0$, although $\mathcal{L}_0$ need not have a uniformly bounded exponential in general.

(2) Conclusion (c) can be summarized as: (multiplicatively) modulo a trivial representation, V is a representation of the same TNC Lie group G_0 that is associated with $\mathcal{L}_0$ (and its exponential).

Proof: (a). Let $\mathcal{L} = S +) R$ be a (semidirect) Levi decomposition of $\mathcal{L}$ into a semisimple subalgebra S and a solvable radical ideal R, and let $\psi: \mathcal{L} \to \mathcal{L}_0$ be the canonical Lie homomorphism of 9.10 (iii). Since $\mathcal{L}_0 \cap P = 0$, ψ induces an isomorphism $\mathcal{L}/\mathcal{L} \cap P \simeq \mathcal{L}_0$.

Let $E \subset S$ be the direct sum of all non-compact simple ideals S. Then of course $S = E \oplus \mathbf{k}$ where $\mathbf{k}$ is compact semisimple. The rest of the proof turns on the observation that both of the ideals $\psi(\mathbf{k})$ and $\psi(R)$ in $\mathcal{L}_0$ are trivial:

(i) For $\psi(\mathbf{k})$ this is a consequence of the TNC property of $\mathcal{L}_0$, since $\psi(\mathbf{k})$ is compact.

(ii) For $\psi(R)$ the triviality follows from semisimplicity of $\mathcal{L}_0$.

Indeed $\psi(R) \subset$ radical $(\mathcal{L}_0) = 0$. Since $\ker(\psi) = \mathcal{L} \cap P$, we note that both R and $\mathbf{k}$ consist of bounded operators. A direct application of The Generalized Singer Theorem ((b)-(d)), (G.3), yields the direct sum decomposition $\mathcal{L} = E \oplus \mathbf{k} \oplus \mathfrak{z}$ where both of the operator Lie algebra components $\mathbf{k}$ and $\mathfrak{z}$ $(= R)$ are bounded, $\mathfrak{z}$ is central (and hence Abelian), and the property $[E, \mathbf{B}] = 0$ is contained in conclusion (b) of (G.3). We note that $\ker(\psi) = \mathcal{L} \cap P = \mathbf{B}(\mathcal{L}) = \mathbf{k} \oplus \mathfrak{z}$ so that ψ induces a Lie isomorphism $E \simeq \mathcal{L}/\mathcal{L} \cap P \simeq \mathcal{L}_0$.

For (b), let φ denote the restriction of ψ to E. We have shown that $E +) \ker(\psi) = \mathcal{L}$ and it follows that φ is a Lie isomorphism of E onto $\mathcal{L}_0$. Hence if we define $f'(X) = \varphi^{-1}(X) - X$ for $X \in \mathcal{L}_0$ it follows that $\mathrm{id}+f' = \varphi^{-1}$ is a Lie isomorphism of $\mathcal{L}_0$ onto $E = \mathcal{L}_0'$. If we define $f''(A) = \varphi(A)+f(\varphi(A)) - A$ for $A \in \mathcal{L}_0' \subset \mathcal{L}$, it follows that since $\varphi(A) + f(\varphi(A)) \in \mathcal{L}$, so is $f''(A)$. But since $f(\varphi(A)) \in P$ and $\varphi(A) - A = \varphi(A) - (\mathrm{id} + f')\varphi(A) = -f'(\varphi(A)) \in P$, $f''(A) = f(\varphi(A)) - f'(\varphi(A)) \in P$, and $f''(A) \in \mathcal{L} \cap P = \mathbf{B}$. The exponential is V. It follows that $W = V|_{G_0 \times \{e\}}$ and $T = V|_{\{e\}\times H}$ are uniformly bounded exponentials for G_0 and H, respectively, that these commute, and that $V = WT$ as suggested. Moreover, since $\mathbf{B}$ consists of bounded operators (extendable to all of E), it follows that T is norm-C^∞ into L(E), and the proof is complete.

Appendix to Part IV

The theorems in Part IV depend, via the proofs, on technical results from other chapters in the monograph. Of particular relevance here are Chapters 3, and 5 through 7:

3. Operator semigroups: domain regularity and commutation relations.

5. The graph density condition (GD): Resolvent- and operational calculus-commutation relations from (GD).

6. Semigroup commutation relations from (GD).

7. Construction of globally semigroup-invariant C^∞-domains.

Of the seven appendices A through G presented at the end of this monograph, only A, B, E and G, are used in this part.

In Chapter 3 we consider the algebra of linear endomorphisms End(D) on a given vector space D. It will be assumed that a locally convex vector topology is given on D, and we denote by E the corresponding completion. If, for example, D is a normed space, then E is a Banach space containing D as a dense linear subspace. The elements A in End(D) may then be viewed as operators in the completed space E. As such they are only defined on the dense subspace D. In case E is a Banach space, A will typically be unbounded operator on E. We use the notation $\mathbf{A}(D)$ for End(D) to stress the unbounded operator point of view.

The following four conditions are imposed on a pair of elements (alias operators) in $A,B \in \mathbf{A}(D)$:

(E) global semi-exponentiability of A, in the sense that the closure $\bar{A}$ is an operator which generates a continuous locally equicontinuous (c.l.e.) semigroup $\{V(t,A): t \in [0,\infty)\}$;

(D) global domain invariance, in the sense that $V(t,A)D \subset D$ for all $t \in [0,\infty)$;

(F) finite-dimensionality of the ad-orbit, in the sense that $\mathcal{O}_A(B)$ = real span $\{(\text{ad } A)^k (B): k \in \mathbb{N}\}$ is finite-dimensional;

(R) regularity: for all $u \in D$, the function $G_u(t) = BV(t,A)u$ is t-differentiable. (Frequently, boundedness in some non-empty interval will be enough.)

3.2. Theorem

Let A,B be a pair of operators in $\mathbf{A}(D)$. Suppose that conditions (E), (D), (F), and (R) hold. Then the power series $e(t) = \exp(-t \text{ ad } A)(B) = \Sigma_0^\infty (n!)^{-1}(-t \text{ ad } A)^n(B)$ converges in $\mathbf{A}(D)$ with the strong-operator-topology induced by D, and we have

$$BV(t,A) = V(r,A)\, e(t) \quad \text{for all } t \in [0,\infty). \tag{1}$$

A special case of the theorem (viz., Proposition 3.1) is used in the proof of Lemma 8.2. In 3.1 it is assumed that A and B are contained in a finite-dimensional Lie sub-algebra $\mathbf{C}$ in $\mathbf{A}(D)$, and moreover that both operators are continuous. Then we get the commutation relation

$$V(t,A)BV(-t,A) = \exp(t \text{ ad } A)(B) \tag{2}$$

with both sides analytic in $t \in \mathbb{R}$.

In Lemma 3.3, two additional regularity properties are introduced for the operator pair A,B:

(CWC) the operator B is closable, and $G(t) = \bar{B}\, V(t,A)$ is weakly continuous on $[0,\infty)$, in the sense that for all $u \in D$, and $f \in E^*$, the scalar function $t \to f(G(t)u)$ is continuous.

(ELB) the closure $\bar{B}$ generates a c.l.e. semigroup, and for all $u \in D$ and compact $K \subset [0,\infty)$, the subset $\{G(t)u\colon t \in K\}$ is bounded in E.

(If $\bar{A}$ is known to generate a group, then the conditions on, $t \to G(t)u$, are only required to hold at $t = 0$, - boundedness in some neighborhood $(-\varepsilon,\varepsilon) \subset \mathbb{R}$, $\varepsilon = \varepsilon_u$.)

3.3. Lemma

Suppose that the pair $A,B \in \mathbf{A}(D)$ satisfies the two conditions (E) and (D). Then either of the conditions (CWC) or (ELB), ensure that condition (R) is also satisfied.

This lemma then leads to a number of formal improvements upon Theorem 3.2. The consequence for the setting of Chapters 8 and 9 is the observation that condition (R) may be replaced by the boundedness requirement of $G_u(t) = BV(t,A)u$ in some non-empty t-interval which may possibly depend on the vector u in D.

In Proposition 3.6, we derive the conclusions in Theorem 3.2 without the assumption of local boundedness. But 3.6 is stated in a different topological context. It serves as a technical lemma for the exponentiation theorems in Chapter 9.

3.6. Proposition

Suppose that D is equipped with two topologies τ_0, and τ_∞, such that D is a topological vector space in both topologies. The following conditions are assumed:

(i) τ_∞ is finer than τ_0.

(ii) τ_0 is given by a norm on D, and the corresponding completion is the Banach space E.

(iii) D is a Fréchet space in the τ_∞ topology.

Suppose that $A,B \in \mathbf{A}(D)$ are closable as operators in E such that

(iv) $\bar{A}$ generates a c.l.e. group $\{V(t,A): t \in \mathbb{R}\} \subset L(E)$ with $V(t,A)D = D$; and

(v) $\bar{B}$ generates a c.l.e. semigroup $\{V(r,B): r \in [0,\infty)\} \subset L(E)$, then $G_u(t) = BV(t,A)u$ is differentiable on $\mathbb{R}$ for all $u \in D$.

This proposition is tailored to the various C^n-topologies that are introduced for the analysis of Lie algebras of unbounded operators in Banach spaces. The following simple lemma is used towards the end of the proof of Theorem 9.2.

5.7'. Lemma

Let T be a closed operator in a Banach space F. Let D be a core for T, and let h be a locally bounded function defined on a subset $\Omega \subset \mathbb{C}$. Suppose that $h(\lambda)\|(\lambda-T)u\| \geq \|u\|$ for all $u \in D$ and $\lambda \in \Omega$. Then the set $\{\lambda \in \Omega: (\lambda-T)D \text{ is dense in } F\}$ is relatively closed in Ω.

We now consider the case where the operator pair A,B is contained in a finite-dimensional real Lie algebra $\mathcal{L} \subset \mathbf{A}(D)$ with D dense in the Banach space E. We fix a linear basis

$B_1,\dots,B_d$ for $\mathfrak{L}$, and define

$$\|u\|_1 = \max_{k=0}^{d} \|B_k u\| , \quad u \in D, \quad B_0 = I.$$

We let D_1 be the completion of D with respect to this norm, $\|\ \|_1$. We assume that $\bar{A}$ exists and generates a strongly continuous semigroup $\{V(t,A): t \in [0,\infty)\} \subset L(E)$ of type ω, i.e., $\|V(t,A)\| \le Me^{\omega t}$ for some constant M. Consider the spectral radius $\nu(\text{ad } A)$ of the endomorphism ad A: $\mathfrak{L}_{\mathbb{C}} \to \mathfrak{L}_{\mathbb{C}}$, i.e., $\nu(\text{ad } A) = \lim_{k\to\infty} |(\text{ad } A)^k|^{1/k}$.

6.1. Theorem

Let A and $\mathfrak{L}$ be as above; and assume that it is possible to choose the basis such that each B_k is a closable operator in E. Suppose that for some λ with $\text{Re}(\lambda) > \omega+\nu(\text{ad } A)$ the subspace $D_\lambda = (\lambda-A)D$ is $\|\cdot\|_1$-dense in D.

(i) Then each of the operators $V(t,A)$, $t \in [0,\infty)$, leaves D_1 invariant and restricts to a strongly continuous semigroup $\{V_1(t,A): t \in [0,\infty)\}$ on D_1. If A is regarded as an operator A_1 say, in D_1, then A_1 is closable and $\bar{A}_1$ is the generator of the semigroup $V_1(t,A)$.

(ii) We have the estimate, $\|V_1(t,A)\|_1 \le M \exp((\omega+\nu(\text{ad } A))t)$, and hence $V_1(t,A)$ is of type $\omega_1 = \omega+\nu(\text{ad } A)$.

(iii) Every B in $\mathfrak{L}$ extends by limits to a bounded operator, $B^{(1)}: D_1 \to E$, and we have the relation

$$B^{(1)}V_1(t,A) = V(t,A)[\exp(-t \text{ ad } A)(B)]^{(1)} \tag{3}$$

as an identity in $L(D_1,E)$.

6.2. Corollary

Let A and $\mathfrak{L}$ be as in Theorem 6.1. Then the following are equivalent:

(a) For some $\lambda > \omega+\nu(\text{ad } A)$, $D_\lambda = (\lambda-A)D$ is $\|\cdot\|_1$-dense in D.

(b) For all $\lambda > \omega+\nu(\text{ad } A)$, the resolvent $R(\lambda,A) = (\lambda-\bar{A})^{-1}$ leaves D_1 invariant, and restricts to a ($\|\cdot\|_1$-bounded) two-sided inverse $R_1(\lambda,A)$ for $\lambda - \bar{A}_1$.

(c) For all $t \in [0,\infty)$, the semigroup operators $V(t,A)$ leave D_1 invariant, and their restrictions $V_1(t,A)$ define a strongly continuous semigroup on D_1 with infinitesimal generator $\bar{A}_1$.

If it is not assumed at the outset that $\mathcal{L}$ possesses a closable basis we still have a result which runs parallel to Corollary 6.2 above, but a technical complication arises from the fact that the completion D_1 of D in the $\|\cdot\|_1$-norm is no longer imbedded as a subspace of E. Instead there is a bounded linear mapping $J_1: D_1 \to E$ which extends the natural injection of $(D,\|\cdot\|_1)$ into $(E,\|\cdot\|)$. If $\mathcal{L}$ has a closable basis, then J_1 is injective, but in general it is not. In this generality we have the following,

6.3. Lemma

(a) For each $B \in \mathcal{L}$ there exists a unique bounded extension-by-limits to an operator $B^{(1)}: D_1 \to E$.

(b) If B is closable, then $\text{kernel}(J_1) \subset \text{kernel}\ (B^{(1)})$ and $J_1 B \subset \bar{B} J_1 \subset B^{(1)}$. Since D_1 is a completion of D we may regard B also as an operator B_1 in D_1. As such it is closable, and $J_1\bar{A}_1 \subset \bar{A}J_1$, so that in particular $\bar{A}_1(\ker(J_1) \cap D(\bar{A}_1)) \subset \ker(J_1)$. I.e., $\ker(J_1)$ is invariant under $\bar{A}_1$.

6.4. Theorem

Let A and $\mathcal{L}$ be as in Theorem 6.1 excepting the provision that $\mathcal{L}$ has a closable basis. Then the following relations hold:

(a) For all large $\lambda > \omega_1$, there exists on D_1 a bounded resolvent $R_1(\lambda,A)$ for $\bar{A}_1$ satisfying $J_1R_1(\lambda,A) = R(\lambda,A)J_1$, and

$$B^{(1)}R_1(\lambda,A) = \Sigma_0^\infty R(\lambda,A)^{k+1}[(-\text{ad}\ A)^k(B)]^{(1)}. \tag{4}$$

(b) The operator $\bar{A}_1$ is the generator of a strongly continuous semigroup $\{V_1(t,A): t \in [0,\infty)\} \subset L(D_1)$ which satisfies $J_1V_1(t,A) = V(t,A)J_1$, and

$$B^{(1)}V_1(t,A) = V(t,A)[\exp(-t\ \text{ad}\ A)B]^{(1)}. \tag{5}$$

7.1. Definition

Let $\mathcal{F}$ be a family of closed operators on a Banach space E.

(a) A subspace $F \subset E$ is said to be a C^∞-domain for $\mathcal{F}$ iff $F \subset D(T)$, and $TF \subset F$, for all $T \in \mathcal{F}$.

(b) The space $E_\infty(\mathcal{F}) = \cup\{F: F \text{ is a } C^\infty\text{-domain for } \mathcal{F}\}$ is the space of C^∞-vectors for $\mathcal{F}$. (Clearly, $E_\infty(\mathcal{F})$ is the maximal C^∞-domain. We are interested in conditions sufficient to ensure that $E_\infty(\mathcal{F})$ is "big enough".)

7.2. Theorem

Let $A,\mathcal{L}$ be a pair satisfying the conditions in Theorem 6.1 excepting the existence of a closable basis. Let $B \in \mathcal{L}$ be a closable operator. Then the subspace $D_A = \text{span}\{V(t,A)u: u \in D, t \in [0,\infty)\}$ has the following properties:

(a) it is invariant under $V(t,A)$ for all $t \in [0,\infty)$;

(b) it is a C^∞-domain for each of the following three sets $\mathcal{F}_1 = \{\bar{B}\}$, $\mathcal{F}_2 = \{\bar{A},\bar{B}\}$, and $\mathcal{F}_3 = \{\bar{C}: C \in \mathcal{O}_A(B) \text{ closable}\}$;

(c) if $\tilde{A}$ is the restriction of $\bar{A}$ to D_A, $\tilde{B}$ the comparable restriction of $\bar{B}$, and $\tilde{\mathcal{O}}_A(B) = \text{real span}\{(\text{ad}\,\tilde{A})^k(\tilde{B}): k \in \mathbb{N}\}$, then $\sim$ extends to a natural isomorphism of $\mathcal{O}_A(B)$ as an ad A - module onto $\tilde{\mathcal{O}}_A(B)$ as an ad $\tilde{A}$ - module, and the identity $\tilde{C}V(t,A) = V(t,A)\widetilde{e(t)}$ is valid on the dense subspace D_A for all $t \in [0,\infty)$, and $\tilde{C} \in \tilde{\mathcal{O}}_A(B)$.

PART V

LIE ALGEBRAS OF VECTOR FIELDS ON MANIFOLDS

"He thought he saw a Garden-Door
  That opened with a key:
He looked again, and found it was
  A Double Rule of Three:
'And all its mystery', he said,
  'Is clear as day to me!'"

Sylvie and Bruno
A Musical Gardener
LEWIS CARROLL

... Born wanted a theory which would generalize these matrices or grids of numbers into something with a continuity comparable to that of the continuous part of the spectrum. The job was a highly technical one, and he counted on me for aid ... I had the generalization of matrices already at hand in the form of what is known as operators. Born had a good many qualms about the soundness of my method and kept wondering if Hilbert would approve of my mathematics. Hilbert did, in fact, approve of it, and operators have since remained an essential part of quantum theory. (Wiener, 1956, pp. 108-109, about the paper, Born-Wiener, 1926.)

INTRODUCTION TO PART V

The exponentiation problem for Lie algebras of unbounded operators is most conveniently stated in terms of representations ρ of finite-dimensional real Lie algebras $\mathfrak{g}$. If E is a Banach space, a representation ρ on E maps into unbounded operators in the Banach space E. We then say that ρ is exact if there is a strongly continuous representation V of the simply connected Lie group G, with Lie algebra $\approx \mathfrak{g}$, such that $\rho = dV$, where the infinitesimal representation dV is the one defined in Gårding's paper [Gå]. The operator Lie algebra $\mathfrak{L} = \rho(\mathfrak{g})$ is said to be exponentiable in E if $\rho = dV$ with strongly continuous representation $V: G \to L(E)$ in the Banach space E.

The simplest instance of the exponentiation problem is perhaps the problem of integrability of Lie algebras of vector fields which was formulated and solved in Palais' Memoir [Pℓ]. Palais' solution was based on the theory of foliations and it is interesting that much later (in fact recently) the foliations have found far reaching applications in operator theory [Co]. The exponentiation problem for Lie algebras of unbounded operators was first formulated in the thesis [Sr] of I. Singer. Here we give a solution to this problem which reduces in a special case to the theorem of Palais. We note that our theorem is more general, since it applies, for instance, to a variety of Lie algebras of second-order differential operators, and moreover, that our proof is based only on operator theory. (Hence we have reversed the "traffic" from foliations into operator theory).

Palais considered in [Pℓ] general Lie algebras of smooth vector fields. A C^∞-manifold M is given, and $\mathfrak{L}$ is a finite-dimensional real Lie algebra of vector fields on M. Let S be a subset of $\mathfrak{L}$ such that $\mathfrak{L}$ is spanned by S and commutators $[X_1,[X_2,[\ldots[X_{r-1},X_r]\ldots]$ with $X_i \in S$, $r = 2,3,\ldots$. Palais showed that if the elements in S generate global flows $t \to \gamma(t,\cdot)$, $-\infty < t < \infty$, on M, then it follows that there is a Lie group action on M which yields $\mathfrak{L}$ as the infinitesimal vector field Lie algebra. The assumption is that for every $X \in S$ there is a flow $\gamma(t)$ on M with $(d/dt)\gamma(t)|_{t=0} = X$.

If G is the simply connected Lie group (which exist by Ado's theorem) with Lie algebra isomorphic to $\mathcal{L}$, then Palais' theorem gives a smooth action of G on M, such that the corresponding infinitesimal action by vector fields is identical to the original action of $\mathcal{L}$ on M.

The classical exponentiation problem for single operators considered first by von Neumann for symmetric operators in Hilbert space, and by Hille-Yosida for unbounded operators in Banach space, may naturally be viewed as a generalization of the integration problem for vector fields. (The applications are to partial differential equations, probability theory, and ergodic theory, etc.). We shall be particularly concerned with strongly continuous one-parameter groups of bounded operators. The terminology C_0 will be used for strong continuity. A C_0 one-parameter group is a homomorphism V of $\mathbb{R}$ into the algebra of all bounded operators on a given Banach space E: $V : \mathbb{R} \to L(E)$, satisfying $V(s+t) = V(s)V(t)$, and $\lim_{t\to 0} \| V(t)u - u \| = 0$, for all $u \in E$.

The infinitesimal generator A is a closed operator with dense domain D(A) given by $Au = \lim_{t\to 0} t^{-1}(V(t)u-u)$, $u \in D(A)$. It is well known that V(t) is determined uniquely by A, and we shall employ the notation V(t,A) for the one-parameter group with generator A.

Now let A be an unbounded operator with dense domain D in a given Banach space E. Suppose that the closure $\bar{A}$ of A exists as a closed operator, and that $\bar{A}$ is the infinitesimal generator of a C_0 one-parameter group $V(t,\bar{A})$. Then we shall say that A is a pregenerator.

We now consider the algebra End(D) of all linear endomorphisms of D. Note that End(D) naturally carries the structure of a Lie algebra with commutator Lie product $[A,B] = AB - BA$ for $A,B \in \text{End}(D)$. Hence we may consider finite-dimensional real Lie subalgebras $\mathcal{L}$ of End(D), and the problem of Palais now has an obvious generalization (to this much more general, and applicable, setting of operator theory.) Suppose $\mathcal{L}$ is spanned by a subset S, and commutators of elements in S, then one might expect that $\mathcal{L}$ is the infinitesimal operator Lie algebra associated with a C_0 representation of the Lie group G on E, viz., $V : G \to L(E)$, provided only that the individual elements A in S are pregenerators for C_0 one-parameter groups. We now state the problem more precisely: A representation V maps the group G into the bounded invertible operators on E : We have:
$V(g_1 g_2) = V(g_1)V(g_2)$, $g_1,g_2 \in G$, and $\lim_{g\to e} \| V(g)u-u \| = 0$, for $u \in E$.

Let $A \in \mathcal{L}$, and let $g(t)$ be the corresponding one-parameter group in G which is given by the exponential mapping of Lie theory. Then the operator $\tilde{A} = (d/dt)V(g(t))|_{t=0}$ is the infinitesimal generator of a C_0 one-parameter group. If for all A in $\mathcal{L}$, the operator $\tilde{A}$ is the least closed extension of A, then we say that the Lie algebra $\mathcal{L}$ exponentiates on E. If $\tilde{A}$ is only an extension of A, then we say that $\mathcal{L}$ is a sub-generator ([Ra]). The extension requirement is that $D(A) \subset D(\tilde{A})$, and $Au = \tilde{A}u$ for all $u \in D(A)$.

The Palais-theorem does not generalize directly to the setting of operator theory as stated. (The counterexample is due to Nelson [Nℓ 1].) It turns out that the naive transformation of the problem into operator theory loses too much of the original regularity structure. However the present authors have shown that, in fact, very little is missing [Mr 1, Mr 2, Jo 1, Jo 3]. It turns out that the implication does hold true for operator Lie algebras when only a relatively mild compatibility assumption is imposed on the elements in the operator Lie algebra $\mathcal{L}$ considered above. Moreover this compatibility condition can easily be verified for the vector field Lie algebras such that Palais' theorem drops out from "pure" operator theory.

In Chapter 9, we prove two exponentiation theorems (9.1 and 9.2) for Lie algebras of unbounded operators. (The second theorem will not be discussed here). The first theorem (9.1) is based on a certain regularity assumption on the invariant domain for the infinitesimal operators in the Lie algebras. It is this theorem which, in the case of vector fields, specializes to the Palais' theorem. In applications we shall need the generality of locally convex spaces (ℓ.c.s.), and the reader is referred to [Yo] for background.

We recall that strongly continuous one-parameter groups, infinitesimal generators, exponentiability etc., are defined, mutatis mutandis, for ℓ.c.s. E, as in the case of Banach spaces (real or complex). In addition to the usual strong continuity condition on the group representations, a mild local equicontinuity assumption is imposed. We consider a given ℓ.c.s. E and a fixed dense linear subspace $D \subset E$. An operator Lie algebra $\mathcal{L}$ is a finite-dimensional real Lie subalgebra of End(D) = the linear endomorphisms of D.

9.1. Theorem

Let S be a Lie generating subset of a given operator Lie algebra $\mathfrak{L} \subset \mathrm{End}(D)$ in a ℓ.c.s. E. Suppose the one-parameter exponentials $\{V(t,A) : t \in \mathbb{R}\} \subset L(E)$ exist as C_0 groups (locally equicontinuous in t) for all elements A in S. If the following two conditions are satisfied then $\mathfrak{L}$ exponentiates to a C_0 Lie group representation in E:

(i) The domain D is invariant under the operator family $\{V(t,A) : t \in \mathbb{R}\ ,\ A \in S\}$, i.e., $V(t,A)u \in D$ for all $u \in D$, $t \in \mathbb{R}$, and $A \in S$.

(ii) For each $A,B \in S$ and $u \in D$ there is a positive interval I (which may depend on u,A,B) such that the vector function $t \to B\ V(t,A)u$ is bounded in I.

The proof of Theorem 9.1 is based on techniques developed by the co-authors in Chapter 8 of this monograph. Hence a smooth (i.e. C^∞ -) version of the operator Lie algebras is introduced. The C^∞-vector method transforms the exponentiability problem for Lie algebras $\mathfrak{L}$ of unbounded operators into an easier problem concerning continuous everywhere defined operators, - but now operators in different spaces, viz., the spaces of C^∞-vectors for the original Lie algebra $\mathfrak{L}$ of not everywhere defined (unbounded) operators. In Chapter 8, the resulting smooth operator Lie algebras, $\mathfrak{L}_\infty$, are defined, and our exactness (exponentiability, or integrability) theorem takes a form which , in fact, generalizes the classical Lie - Ado theorem, to the effect that every finite-dimensional Lie algebra is the infinitesimal Lie algebra of some Lie group.

The setting of our infinite-dimensional theorem 8.1 is abstract: a locally convex space D, and a smooth Lie algebra $\mathfrak{L}_\infty \subset L(D)$ of continuous linear endomorphisms upon it, are given at the outset. (Typically, in applications, D is a space of C^∞-vectors for a Lie algebra of unbounded operators in a Banach space B containing D as a dense linear subspace). We say that $\mathfrak{L}_\infty$ is <u>integrable</u> (exact, with a smooth exponential) if there is a differentiable representation V of G (the simply connected Lie group with associated Lie algebra $\mathfrak{g}$ isomorphic to $\mathfrak{L}_\infty$) such that $dV(\mathfrak{g}) = \mathfrak{L}_\infty$.

An operator $A \in L(D)$ is said to be a <u>smooth infinitesimal generator</u> if the one-dimensional Lie algebra $\mathfrak{L}_A = \mathbb{R} \cdot A$ is integrable.

8.1. Theorem

Let $\mathcal{L}_\infty$ be a finite-dimensional real Lie algebra of operators contained in L(D). Suppose $\mathcal{L}_\infty$ is generated as a Lie algebra by a subset (commutators and linear span) which consists of smooth infinitesimal generators. Then $\mathcal{L}_\infty$ is integrable; (i.e., is the differential of a smooth representation V in the ℓ.c.s. D).

Chapter 10

APPLICATIONS OF COMMUTATION THEORY TO VECTOR-FIELD LIE ALGEBRAS AND SUB-LAPLACIANS ON MANIFOLDS

The results in this chapter illustrate typical applications of the invariant-domain exponentiation theory for operator Lie algebras treated in Theorem 9.1, as well as direct applications of the integration theory for smooth operator Lie algebras developed in Chapter 8. We also discuss some secondary applications to the solution of certain higher-order differential equations on manifolds that are determined by second-order elements of the enveloping algebra of a vector field Lie algebra; these operators are generally known as "sub-Laplacians" [Fo], [Jo 2], and can be viewed as abstract models for an interesting special class of hypoelliptic operators. We show that a number of qualitative properties of the solutions of these equations can be dropped out quite quickly via a suitable combination of exponentiation theorems and results concerning representation of differential operators on Lie groups [Jo 2].

The discussion is arranged as follows. In Section 10A, we consider Lie algebras $\mathcal{L}$ of smooth vector fields on a manifold M, viewed as derivations acting on the pointwise-multiplication algebra $D = C_0^\infty(M)$ of compactly-supported (complex) infinitely-differentiable functions. Here, D is viewed as a dense subalgebra of one of three natural locally convex algebras of functions: the sup-norm Banach algebra $C_\infty(M)$ of continuous functions vanishing at infinity on M, the locally m-convex Fréchet algebra $\mathcal{E}(M)$ of C^∞ functions with compact convergence of derivatives, and $\mathcal{D}(M) = D$ itself as the usual Schwartz LF test function algebra. We begin there by observing that a vector field X is a pregenerator in any of these settings iff it is "complete" in the geometrical sense: it is integrable to a smooth transformation group action (one-parameter flow) $\gamma:\mathbb{R} \times M \to M$ connected to the automorphism group $\{V(t,X): t \in \mathbb{R}\}$ via the usual composition action $[V(t,X)u](x) = u(\gamma(t,x))$, $x \in M$. From this, we obtain three variations upon the same basic functional-analytic result: if a vector-field Lie algebra $\mathcal{L}$ has a Lie-generating set $S \subset \mathcal{L}$ of complete vector fields, then it exponentiates to a group representation on $C_\infty(M)$, $\mathcal{E}(M)$ and $\mathcal{D}(M)$, respectively, with range in the multiplicative automorphisms of these algebras. As a corollary, we are then able to read off Palais' global version of a theorem of Lie on geometric integration of vector field Lie algebras [Pℓ]: if a generating set $S \subset \mathfrak{L}$ consists of complete vector fields, then $\mathfrak{L}$ "integrates" to a global action of the simply-connected Lie group G with Lie algebra $\mathfrak{g} \simeq \mathfrak{L}$, as a smooth transformation group on M. We discuss the relationship of this transformation group with the contragredient representation of G on the respective dual spaces of measures, compactly-supported distributions, and arbitrary distributions on M, where M is viewed as embedded qua point-evaluation functionals.

Exponentials on L^p spaces are then considered in Section 10B, with integration performed in terms of Riemannian densities on M. We also broaden the context to consider Lie algebras of first-order differential operators $A = X + b$, where X is a vector field on M and $b \in C^\infty(M)$ is an "infinitesimal multiplier". We first establish a technical lemma which characterizes those pregenerators in L^p whose vector field part is complete, as those for which the growth of Re(b) at infinity is cancelled by that of a suitable expression in the Riemannian divergence of a vector field determined by X. This characterization permits us to give an analog of the exponentiation theorem for $C_\infty(M)$ using the Banach space L^p: $\mathfrak{L}$ is assumed to be Lie-generated by S consisting of operators $A = X + b$ satisfying the divergence condition mentioned above. We also briefly discuss the attractive possibility of extending this theorem to the case where the vector-field parts X of the operators $A \in S$ are assumed only to be "almost complete" in the sense of Nelson [Nℓ 4]: points in a set of measure 0 may "propagate to infinity" in finite time under the local flow that integrates X.

Section 10C is then devoted to a general theorem in which second-order equations are solved. Specifically, if $S = \{X_1,\dots,X_r\}$ generates a vector-field Lie algebra, and

$$S_0 = \Sigma X_i^{\,2} \tag{1}$$

is the second-order sub-Laplacian determined by these vector fields, we show that S_0 is a pregenerator of a semigroup on $C_\infty(M)$ and on $L^p(M)$ whenever S satisfies the hypotheses of our exponentiation theorems. Moreover, the method of proof from [Jo 2] permits the resolvents $R(\lambda,S_0)$ and the semigroup elements $W(t,S_0)$, respectively, to be expressed in terms of the integrated forms of suitable Green's and Gauss' kernels on G, thereby representing them as a generalized type of integral operator.

Then, in Section 10D, we use these kernel representations from the group setting to give rather explicit formulae for the semigroups and resolvents of certain sub-Laplacians on manifolds M, in terms of integral kernels defined on M × M that are obtained by averaging the kernels on G with respect to suitable isotropy subgroups when G acts transitively. Various positivity and smoothness properties of these kernels can then be read off from these formulae, providing a simple account of several typical results in hypoelliptic regularity via global group theory and representation theory. The flavor of the examples can briefly be suggested by the following special cases. For the first, $M = \mathbb{R}^2$,

$$S_0 = X_1^2 + X_2^2 = (\partial/\partial x_1)^2 + (x_1\partial/\partial x_2)^2 \tag{2}$$

is a degenerate hypoelliptic operator of a type considered by Bony [By], and $L^p(\mathbb{R}^2)$ is formed with respect to Lebesgue measure. The vector field Lie algebra is $\mathcal{L} = \mathrm{span}_{\mathbb{R}}\{\partial/\partial x_1, \partial/\partial x_2, x_1\partial/\partial x_2\}$, which is isomorphic to the three-dimensional nilpotent Heisenberg Lie algebra $\mathfrak{h}_3$ (Chapters 2 and 11), so the Gauss' and Green's kernels live on the Heisenberg group. In the second example, M is iteself a Lie group: the two-dimensional solvable "ax + b group",

globally parametrized as $\left\{\begin{pmatrix} e^{x_1} & x_2 \\ 0 & 1 \end{pmatrix} : (x_1, x_2) \in \mathbb{R}^2\right\}$,

$$S_0 = X_1^2 + X_2^2 = (\partial/\partial x_1 + x_2\partial/\partial x_2)^2 + ((1-x_2)\partial/\partial x_2)^2 \tag{3}$$

is also degenerate hypoelliptic, and $L^p(M)$ is formed with respect to the natural left Haar measure $\exp(-x_1)dx_1dx_2$ on M. Here $\mathcal{L} = \mathrm{span}_{\mathbb{R}}\{\partial/\partial x_1, \partial/\partial x_2, x_2\partial/\partial x_2\}$ is three-dimensional solvable, and our Green's and Gauss' kernels come from the corresponding group. In Section 10D, (2) is treated as a special case of a more general class of operators on $\mathbb{R}^{n+1}$ considered by Bony [By] for arbitrary n, with an associated nilpotent Lie group of dimension $d(n) = (n+1)(n+2)/2$. Similarly, (3) is discussed in terms of a sub-Laplacian for a Lie generating set $\mathcal{S} \subset \mathcal{L}$, where $\mathcal{L}$ integrates to a transitive action of a Lie group G = KH on M with H the isotropy group of a point $x_0 \in M$, $K \cap H = \{e\}$, and K is normal in G. (K is the ax + b group in (3).) We suggest a unifying framework for these examples, but much of the development is carried out using the special structure of the examples, in order to obtain sharper and more explicit results.

10A. Exponentials versus geometric integrals of vector-field Lie algebras

In this section, we examine the interplay between two kinds of "exponentiation" or "integration" of a finite-dimensional real Lie algebra $\mathcal{L}$ of C^∞ vector fields $X \in \mathcal{L}$ on a real C^∞ manifold M: functional-analytic exponentials as automorphism groups of various algebras of continuous and test-functions on the one hand, and geometric integrals as transformation groups on the other. Our functional-analytic results serve in part as tools for subsequent sections: exponentiation in L^p spaces and solution of higher-order differential equations. But our main objective, as described in the introduction above, is to show how the existence of a geometric integral for such a vector field Lie algebra can be recovered in a natural way from the easily-obtained functional-analytic exponentials.

As indicated in the introduction, three different functional-analytic settings are technically useful, for a variety of reasons. The most classical variant works with the Banach algebra $C_\infty(M)$ of complex continuous functions u vanishing at infinity on M, equipped with the uniform norm $\|u\|_\infty = \sup\{|u(x)|:x\in M\}$, viewing $\mathcal{L}$ as a family of (closable) operators on the dense C^∞-domain and subalgebra $D = C_0^\infty(M)$ of compactly-supported C^∞-functions. When $\mathcal{L}$ has a strongly continuous exponential acting upon this Banach algebra, essentially classical methods lead immediately to a zero-order approximation to the Lie-Palais integration theorem: $\mathcal{L}$ integrates to a continuous transformation group of homeomorphisms of M. In order to recover the smooth action of the transformation group, it is necessary instead to exponentiate $\mathcal{L}$ on the locally convex algebra $E(M)$ of C^∞-functions, equipped with the topology of uniform convergence of derivatives on compacta. For our purposes later on in the discussion, it is most economical to view this topology as defined by a slightly nonstandard collection of sub-multiplicative "algebra semi-norms" which exhibit the structure of $E(M)$ as a locally multiplicatively convex algebra in the sense of Michael [Mi]. For each compact subset K of a coordinate neighborhood $U \subset M$, let $\{X_j:1 \le j \le d\}$ be the vector fields on U represented by $\{\partial/\partial x_j:1 \le j \le d\}$ in local coordinates, and let $p_K^0(u) = \sup\{|u(x)|:x \in K\}$ be the familiar submultiplicative "sup-norm": $p_K^0(uv) \le p_K^0(u)p_K^0(v)$ for all $u,v \in E(M)$. Then for $\ell = 0,1,2,\ldots,$

$$p_K^{\ell+1}(u) = p_K^{\ell}(u) + \Sigma\{p_K^{\ell}(X_j u) : 1 \le j \le d\} \tag{4}$$

inductively defines a sequence of seminorms which are easily seen by induction to be submultiplicative because the X_j are derivations:

$$\begin{aligned} p_K^{\ell+1}(uv) &= p_K^{\ell}(uv) + \Sigma\, p_K^{\ell}((X_j u)v + u(X_j v)) \\ &\le p_K^{\ell}(u)p_K^{\ell}(v) + [\Sigma\, p_K^{\ell}(X_j u)]p_K^{\ell}(v) + p_K^{\ell}(u)\,\Sigma\, p_K^{\ell}(X_j v) \\ &\le p_K^{\ell+1}(u)p_K^{\ell+1}(v) \end{aligned} \tag{5}$$

Since every vector field X on M can be described on K by a linear combination of the X_j with C^∞ coefficients, one easily checks that $p_K^{\ell}(Xu) \le C_X p_K^{\ell+1}(u)$ for all ℓ, and a suitable $C_X < \infty$ possibly depending upon ℓ. Hence every such X maps

$\mathcal{E}(M)$ continuously into itself. (The same is true for any differential operator Q of "order $\le q$": $p_K^{\ell}(Qu) \le C_Q p_K^{\ell+q}(u)$. We leave details to the reader.) This means that an exponential for a vector-field Lie algebra $\mathfrak{L}$ acts smoothly on $\mathcal{E}(M)$ (cf. Chapter 8) with a smooth contragredient (acting upon the dual space $\mathcal{E}'(M)$ of compactly-supported distributions) from which the full "smooth transformation group" action of Lie-Palais can be recovered. For completeness, we also bring $\mathcal{D}(M) = D$ on stage, equipped with the "Schwartz topology" obtained as a (strict sequential) inductive limit from the (Fréchet) subspaces $\mathcal{D}_K(M) = \{u \in \mathcal{E}(M)\colon \text{support}(u) \subset K\}$ when these are equipped with the relative topology from $\mathcal{E}(M)$. While results in this setting add little that is new to the information obtainable via $C_\infty(M)$ and $\mathcal{E}(M)$, "exponentials" for vector field Lie algebras $\mathfrak{L} \subset \mathbf{A}(D)$ directly become "integrals" on $\mathcal{D}(M)$ in the sense of Chapter 8; their contragredients contain all of our other exponentials as subrepresentations on the appropriate invariant subspaces, thereby unifying the entire discussion within one framework.

The hypotheses of our exponentiation-integration theorems are couched in terms of the geometrical notion of a <u>complete vector field</u> X; which integrates to a <u>global</u> C^∞ <u>flow</u> (or <u>$\mathbb{R}$-transformation group</u>) $\gamma\colon \mathbb{R} \times M \to M$. Recall that for any Lie group G, a C^∞ mapping $\gamma\colon G \times M \to M$ is a G-transformation group iff $g \to \gamma(g,\cdot)$ is a homomorphism of G into the composition-group Diff(M) of diffeomorphisms of M. When $G = \mathbb{R}$ (additive) γ integrates a vector field X iff for all C^∞-functions u and points $x \in M$

$$d/dt\; u(\gamma(t,x)) = (Xu)(\gamma(t,x)).$$

(We recall that compactly-supported vector fields are always complete in this sense: if $Xu = 0$ for all u supported outside a compact K, X integrates to a flow on K which fixes every point in the complement of K. Various other sufficient conditions for completeness can be given, essentially in terms of "uniform" Lipschitz conditions. See Section 2.1 of Abraham and Marsden [AM], particularly Propositions 2.1.20 and 2.1.21.) It is convenient to adopt the suggestive notation $\gamma(t,\cdot) = V_*(t,X)$ when γ integrates X, and then to put

$$[V(t,X)u](x) = u(V_*(t,X)x) \tag{6}$$

for u a function chosen from $C_\infty(M)$, $\mathcal{E}(M)$ or $\mathcal{D}(M)$. The following lemma is the easy half of a characterization of those vector fields X which are pregenerators on D of C_0 (or smooth) operator groups on these spaces; we later obtain its converse from our version of the Lie-Palais theorem.

10.1. Lemma

Let X be a complete C^∞-vector field on a C^∞-manifold M, acting upon $D = C_0^\infty(M)$ as a dense subalgebra of E, where E is any one of the locally convex algebras $C_\infty(M)$, $\mathcal{E}(M)$ or $\mathcal{D}(M)$. Then X is the pregenerator of a group $\{V(t,X): t \in \mathbb{R}\}$ of automorphisms of E, given by (6), that is C_0 on E (smooth for $E = \mathcal{E}(M)$ or $\mathcal{D}(M)$). Moreover, for any vector field Y and any $u \in D$, the function $G(t) = YV(t,X)u$ is C^∞ from $\mathbb{R}$ into E.

Proof: Composition mappings like (6) clearly respect pointwise multiplication of functions, which yields the automorphism claim for the invertible operators V(t,X). The fact that $\gamma: \mathbb{R} \times M \to M$ is C^∞ immediately entails that D is invariant under each V(t,X). It is also clear from uniform continuity of continuous functions on compacta that for each $u \in D$ V(t,X)u must be uniformly C^∞ on compacta, so $t \to V(t,X)u$ is C^∞ into $E = C_\infty(M)$. Similarly, for X_j a coordinate-vector field on U and K compact in U, $X_jV(t,X)u$ on compact $K \subset U$ can be estimated in terms of its values on $K' \supset K$ in U when t is small, and uniform continuity consideration again yield sup-norm-C^{∞} properties of $t \to X_j V(t,X)u$ for t near 0. Hence $p_{K'}$ sees $V(t,X)u$ as C^{∞}, and our usual kind of induction argument extends this to all p_K^{ℓ}. This establishes smoothness of $V(t,X)$ as its acts upon $\mathcal{E}(M)$ and $\mathcal{D}(M)$; continuity of action upon $C^{\infty}(M)$ follows by the usual 3-ε argument since the $V(t,X)$ are sup-norm isometries. Then the fact that X is a pregenerator follow from invariance of D, by the pregenerator core-theorem (Appendix B, Corollary B.4 for n = 1).

Smoothness of G(t) also follows from the remarks above, since Y and QY are differential operators on M whenever Q is.

It is most economical to unify our three exponentiation/integration theorems in one result, in the same style as the lemma above.

10.2. Theorem

Let E be one of the locally convex algebras $C_{\infty}(M)$, $\mathcal{E}(M)$ or $\mathcal{D}(M)$, with $D = C_0^{\infty}(M)$ a dense subalgebra. Suppose that a finite-dimensional real Lie algebra $\mathfrak{L}$ of vector fields, acting upon D, has a Lie generating set $S \subset \mathfrak{L}$ consisting of complete vector fields. Then $\mathfrak{L}$ is exponentiable to a representation, with range in the multiplicative automorphisms of E, of the simply-connected Lie group G with Lie algebra isomorphic to $\mathfrak{L}$. When $E = \mathcal{E}(M)$ or $\mathcal{D}(M)$, this exponential is smooth. (That is, it is an integral in the sense of Chapter 8.)

Proof: For all three algebras $C_{\infty}(M)$, $\mathcal{E}(M)$ and $\mathcal{D}(M)$, viewed simply as locally convex spaces, the existence of exponentials follows immediately from the Lemma and Theorem 9.1: every $X = A \in S$ is a differential operator, whence the local boundedness condition is certainly satisfied. Since the vector fields $X \in \mathfrak{L}$ are differential operators, it follows from the construction of the topologies of $\mathcal{E}(M)$ and $\mathcal{D}(M)$ that these are continuous operators, and the Lemma ensures that the $X \in S$ are smooth infinitesimal generators in the sense of Chapter 8 whence Theorem 8.1 applies to ensure that the exponential is smooth (or an integral) in these cases.

It remains only to check that every V(g) is a multiplicative automorphism of E. But this follows easily from the well-known property of Lie-generating sets $\mathfrak{h}$ for Lie algebras $\mathfrak{g}$ that we have recorded in Sublemma 8.7: $\exp(\mathbb{R}\mathfrak{h})$ generates the (connected) group G. Here, for each $X \in S \cong \mathfrak{h}$, $V(t,X) = V(\exp(t X))$ is multiplicative, products of multiplicative automorphisms are multiplicative, and the representation $V:G \to L(E)$ respects products , so $V(\exp(\mathbb{R}\mathfrak{h})) = \{V(t,X):X \in S\}$ generates $V(G)$ and the latter consists of multiplicative operators as claimed.

The following corollary on contragredient representations is then the second major step toward the Lie-Palais integration theorem. We recall that if $V:G \to L(E)$ is a representation, then the contragredient $V^+:G \to L(E^*)$ on the dual space E^* is defined by

$$V^+(g)f = V^*(g^{-1})f , \quad f \in E^* \tag{7}$$

where $(V^*f)(u) = f(Vu)$ defines the adjoint V^* of $V \in L(E)$ when f runs over E^* and u runs over E.

10.3. Corollary

Let M, D, $\mathcal{L}$ and S be as in Theorem 10.2.
(a) If the dual $C^*_\infty(M) = E^*$ is equipped with the topology of uniform convergence on $\|\cdot\|_\infty$-precompact subsets of E, then $V^+:G \to L(E^*)$ supplies a strongly continuous, locally equicontinuous representation of G which leaves the set of nonzero bounded multiplicative functionals invariant.
(b) For $E = \mathcal{E}(M)$ or $\mathcal{D}(M)$, $V^+:G \to L(E^*)$ is a strongly smooth locally equicontinuous representation which leaves the set of multiplicative functionals on E invariant. (Here, the strong topology $\beta(E^*,E)$ is the topology of uniform convergence on bounded sets in E.)

Proof: For (a), the contragredient representation is trivially weak-* or $\sigma(E^*,E)$-continuous, and by the Mackey-Arens theorem [Sch, Section IV.3] E is the dual of E^* with the topology of norm-precompact convergence, so by Theorem 3.3 of [Mr 6, p.24] it suffices to check local equicontinuity of V^+ in order to obtain continuity for this topology. But for each compact $K \subset G$, joint continuity of $(g,u) \to V(g)u$ from $G \times E$ to E ensures that $P_K = V(K)P$ is precompact whenever $P \subset E$ is precompact, so local equicontinuity follows by polarity. (The neighborhood P_K^0 is contained in $(V^+(g))^{-1}(P^0)$ for all $g \in K$.) Clearly, since each $V(g)$ is multiplicative, $V^+(g)m = m \circ V(g)$ is a composite of multiplicative mapping, hence multiplicative. Moreover, the adjoint $V^*(g^{-1})$ of an automorphism $V(g^{-1})$ is an automorphism, so $V^+(g)m = 0$ iff $m = 0$.

The claims in (b) follow by the same kind of argument, simplified by the fact that $\mathcal{E}(M)$ and $\mathcal{D}(M)$ are reflexive (indeed, Montel). (The proof that bounded sets in these spaces are precompact (which implies reflexivity) is essentially the usual one for $M = \mathbb{R}^d$, modulo book-keeping. In fact, the usual Ascoli argument using boundedness of derivatives of the next higher order, shows that $p_K^{\ell+1}$-bounded sets are p_K^{ℓ}-precompact, and the rest follows by Tychonoff's theorem with K and ℓ as indices. We leave these details to the reader.)

Given reflexivity, the desired result can be read of directly from Theorems 5.4 and 5.5 of [Mr 6, pp. 43-44].

E.O.P.

The final ingredient in our Lie-Palais proof is the following characterization of the multiplicative functionals on these three algebras as the obvious ones: the point-evaluations on M and "at ∞". The $E = C_\infty(M)$ case is well-known, but we have not seen the $\mathcal{E}(M)$ and $\mathcal{D}(M)$ cases written down anywhere so a proof is recorded here for the reader's convenience.

For $C_\infty(M)$, $C(M_\infty) = \mathbb{C} \oplus C_\infty(M)$ is the uniform algebra of continuous functions on the compact Hausdorff space M_∞, so its spectrum (multiplicative functionals alias maximal ideals) consists of point-evaluations (cf. Rudin [Rd, p. 271]). As remarked before, $C_\infty(M)$ itself is just the ideal of functions vanishing at ∞: δ_∞ restricts to the zero functional there, so the δ_x for $x \in M$ exhaust the nonzero multiplicative functionals.

The case of multiplicative continuous functionals on $\mathcal{E}(M)$ can be reduced immediately to the following:

10.4. Lemma

A nonzero distribution m on a manifold M, $m \in \mathcal{D}'(M)$, is multiplicative iff it is point-evaluation $m = \delta_x$ for some point $x \in M$. (Here $\delta_x u = u(x)$ for all $u \in \mathcal{D}(M)$.)

Proof: Since the lemma has an elementary proof we shall limit ourselves here to an indication of the details. It is an advantage to reduce the problem early to Euclidean space $\mathbb{R}^d$ so that the Fourier transform can be applied. We do this by showing that the support of m must necessarily be a single point. If not there would be a pair of test-functions u, $v \in \mathcal{D}(M)$ such that $m(u) \neq 0$, $m(v) \neq 0$, and $uv \equiv 0$. But then $m(u)m(v) = m(uv) = m(0) = 0$, and we have a contradiction. By choosing an open coordinate neighborhood around the single-point-support of m, we may reduce the problem to a distribution, u say, in $\mathbb{R}^d$ with support at the origin. By a standard theorem in distribution theory [Yo, Theorem I.13.3] we then have $\hat{u}(\xi) = P(\xi)$ for the Fourier transform $\hat{u}$ of u, where P is a polynomial in the variable $\xi = (\xi_1,\ldots,\xi_d) \in \mathbb{R}^d$. The multiplicative property of u is equivalent to the equation $P(\xi+\eta) = P(\xi)P(\eta)$, for all $\xi,\eta \in \mathbb{R}^d$. Indeed

$$\begin{aligned} P(\xi+\eta) &= \hat{u}(\xi+\eta) = u(e^{i(\xi+\eta)\cdot x}) \\ &= u(e^{i\xi\cdot x})u(e^{i\eta\cdot x}) = P(\xi)P(\eta). \end{aligned}$$

We have $u \not\equiv 0$, and therefore $P \not\equiv 0$. The conclusions $P(\xi) \equiv 1$, and $u = \delta_0$, now follow since $P \equiv 1$ is the only non-zero polynomial solution to the functional equation above.

E.O.P.

The ground is now completely prepared for the statement and proof of our version of the Lie-Palais integration theorem [Pℓ, Theorem IV.III, p. 95]. For emphasis, we explicitly display some of our functional-analytic machinery in the statement of the result.

<u>10.5. Corollary</u>

Let M, $\mathcal{S}$ and $\mathcal{L}$ be as in Theorem 10.2. Then there exists a global C^∞ transformation group action $\gamma: G \times M \to M$ on M of the simply-connected Lie group G whose Lie algebra is isomorphic to $\mathcal{L}$, such that for each $X \in \mathcal{L}, \gamma(\exp(tX),\cdot)$ defines a global integral for X. Indeed, the map $x \to \delta_x$ embeds M homeomorphically into the dual E^* of $E = C_\infty(M)$, $\mathcal{E}(M)$ or $\mathcal{D}(M)$ in such a way for each $x \in M$ and $g \in G$, $\gamma(g,x)$ is represented by the contragredient action on δ_x of the exponential $V: G \to \mathrm{Aut}(E)$ of $\mathcal{L}$ obtained in Theorem 10.2:

$$\gamma(g,x) = V^+(g)\delta_x. \tag{8}$$

Thus for each $u \in E$ we have

$$(V^+(g)\delta_x)(u) = \delta_x(V(g)u) = u(\gamma(g,x)) = u(V_*(g)x). \tag{9}$$

<u>Proof</u>: Corollary 10.3 yields a continuous action of G on $C_\infty^*(M)$ and a smooth action of it on $\mathcal{E}^*(M)$ (with respect to group parameters); these leave the multiplicative functionals invariant and consequently preserve the natural copy $\tilde{M} = \{\delta_x : x \in M\}$ of M in E^* by virtue of Lemma 10.4. Considering open subsets of M with compact closures, or the one-point

compactification M_∞, it is clear that $x \to \delta_x$ is a continuous (indeed, C^∞) map of M into E* that is injective, hence a homeomorphism. (Notice that in both cases, the topology used on E* involves uniform convergence on equicontinuous families of functions, since p_K^1-boundedness implies equicontinuity, so the map $x \to \delta_x$ is continuous into E*: a net $x_\alpha \to x$ implies that $|u(x_\alpha) - u(x)| \to 0$ uniformly for the u in such a family.)

If we choose $E = C_\infty(M)$, only continuity of the map $G \times M \to M$ can easily be read off from the representation theory. Smoothness can then be obtained by the Bochner-Montgomery theorem, once we use V(G)-invariance of D to see that every $V^+(g)$ restricts to a diffeomorphism of M (see Montgomery and Zippin [MZ, Theorem on pp. 212-213]. Chernoff and Marsden [CM] give a more representation-theoretic line of argument.) Joint continuity can be obtained here from local equicontinuity of V^+, which entails that $(g,f) \to V^+(g)f$ is jointly continuous on $K \times E^*$ for any compact neighborhood of $e \in G$.

The C^∞ structure of $E = \mathcal{E}(M)$ gives the full strength of the Lie-Palais theorem without the intervention of Bochner-Montgomery. Here, continuity of differential operators on M (in particular, operators in the enveloping algebra $\mathcal{E}(\mathcal{L})$ of $\mathcal{L}$ as endomorphisms of E combines with local equicontinuity of $V^+(G)$ on the strong dual E^*_β to produce joint continuity of the maps $(g,f) \to V^+(g)Q^*f$ for any operator Q of "mixed" derivatives in group and manifold parameters since β-continuity of Q* implies equicontinuity of $V^+(K)Q^*$ on compact $K \subset G$. Hence, joint continuity of $(g,x) \to [V^+(g)Q^*\delta_x](u) = [QV(g)u](x)$ follows for all $u \in \mathcal{E}(M)$, so it correspondingly holds for d-tuples of coordinate functions as maps into $\mathbb{R}^d$, and the jointly-C^∞ characteristics of γ follow. E.O.P.

We conclude by extracting the promised converse to Lemma 10.1. This shows that completeness of the vector fields in S in Theorem 10.2 is necessary as well as sufficient for its conclusions.

10.6. Proposition

Let M, D and E be as in Lemma 10.1. Then a vector field X is a pregenerator on D of a locally equicontinuous group of multiplicative automorphisms iff it is complete.

Proof: Lemma 10.1 and the argument of Corollary 10.5 with $G = \mathbb{R}$. (Notice that our proofs in 10.3, 10.4, and 10.5 really show that the conclusions of 10.2 entail the existence of a transformation group action.) E.O.P.

10B. Exponentiation on L^p spaces

In the preceding section, we discussed the exponentiation of finite-dimensional Lie algebras of vector fields on function-spaces of sup-norm type, where the functional-analytic exponential is intimately tied to the geometric integral of the Lie algebra as a Lie transformation group. Here, we take up measure-theoretic matters, considering exponentials on L^p spaces whose connection with transformation groups is, in principle, more tenuous, as we discuss in detail at the end of the section. Here, the L^p spaces are defined with respect to a (pseudo-) Riemannian measure on M, and it is natural to consider as generators not just vector fields but general first-order differential operators. We proceed at once to recall the necessary terminology and to make these remarks precise.

We recall that a pseudo-Riemannian structure on a manifold M is a smooth assignment $x \to g_x$ of a nondegenerate bilinear form on the tangent space TM_x; g is Riemannian iff each g_x is an inner product (cf. [Hℓ, Chapter 1]). Let $\varphi(x) = (\xi_1(x),\ldots,\xi_n(x))$ be a coordinate system valid on an open subset Ω of M. The corresponding coordinate functions of g are defined as follows

$$g_{ij} = g\left(\frac{\partial}{\partial\xi_i}, \frac{\partial}{\partial\xi_j}\right) \quad \text{for } 1 \le i,j \le n. \tag{10}$$

The notation $\bar{g} = |\text{determinant } (g_{ij})|$ is also used.

Let X be a vector field which on Ω is given in local coordinates by

$$X = \sum_{i=1}^{n} a_i \frac{\partial}{\partial\xi_i} \tag{11}$$

The divergence of X is then given in local coordinates by

$$\text{div } X = \bar{g}^{-\frac{1}{2}} \sum_{i=1}^{n} (\partial/\partial\xi_i)(\bar{g}^{\frac{1}{2}} a_i) \tag{12}$$

It can be shown [Hℓ] that the right-hand side does not in fact depend on the particular coordinate system $(\xi_1,\ldots,\xi_n)$

chosen. Note that the Riemannian measure dx on M is given in coordinates by

$$\int_M u(x)dx = \int_{\mathbb{R}^n} u(\xi_1,\ldots,\xi_n)\bar{g}^{\frac{1}{2}}(\xi)d\xi_1\ldots d\xi_n, \tag{13}$$

where u is supported in the given coordinate neighborhood. This integral extends in the usual way (via partitions of unity) to every $u \in C_0^\infty(M)$ as a sup-norm bounded (positive) linear functional, hence via the Riesz theorem to a measure on M with respect to which we form the $L^p(M)$ Banach spaces treated here.

The operators A on the domain $D = C_0^\infty(M)$ of compactly supported smooth functions that we consider as elements of our operator Lie algebras are first-order differential operators $A = X + b$, where X is a complete C^∞ vector field on M and $b \in C^\infty(M,\mathbb{C})$ is a smooth complex-valued infinitesimal multiplier. (This terminology is explained later on.) We let I denote the set of first-order operators, and I_0 the set of vector fields on M, as operators on D. The following simple algebraic result is basic.

10.7. Lemma

(a) The decomposition $A = X + b$ of a first-order differential operator is unique, so the mapping $\psi: I \to I_0$ defined by $\psi(X+b) = X$ is well-defined.
(b) Both I and I_0 are commutator Lie subalgebras of $\mathbf{A}(D)$, and the mapping ψ is a Lie algebra homomorphism with kernel $C^\infty(M,\mathbb{C}) = P$.
(c) If $S = \{A_i = X_i + b_i : 1 \le i \le n\} \subset I$ Lie-generates a finite-dimensional operator Lie algebra $\mathfrak{L} \subset \mathbf{A}(D)$, then $\mathfrak{L} \subset I$, and $\mathfrak{L}_0 = \psi(\mathfrak{L}) \subset I_0$ is a finite-dimensional vector-field Lie algebra generated by $S_0 = \psi(S)$.

Proof: (a) Clearly uniqueness could fail iff for some $X \in I_0$ and $b \in C^\infty(M)$, $X + b = 0$. But then for every $u \in D$ we would have

$$2u(Xu)+bu^2=(X+b)(u^2)=0=2u(X+b)(u)=2u(Xu)+2bu^2$$

so $bu^2 = 0$ for all such u, b=0, and X=0. Thus ψ is well-defined.
(b) Since I_0 is well-known to be a real Lie algebra,

$I = I_0 \oplus C^\infty(M,\mathbb{C})$ is clearly a vector space and ψ is clearly linear by (a), with kernel $C^\infty(M,\mathbb{C})$. Then, since the derivation law for vector fields $X \in I_0$ shows that for every $d \in C^\infty(M,\mathbb{C})$,

$[X,d]u = (Xd)u + (Xu)d - d(Xu) = (Xd)u$ for all $u \in D$, it follows that for all $A = X + b$ and $B = Y + d$ in $\mathcal{I}$

$$[A,B]u = [X+b,Y+d]u = [X,Y]u+[X,d]u+[b,Y]u+[b,d]u$$
$$= [X,Y]u+(Xd-Yb)u. \tag{14}$$

Hence, $\mathcal{I}$ is also a Lie algebra and we have

$$\psi([A,B]) = \psi([X+b,Y+c]) = [X,Y] = [\psi(A),\psi(B)] \tag{15}$$

so ψ is a Lie homomorphism.

(c) It is immediate from (a) and (b) that ψ is a Lie homomorphism from $\mathfrak{L} \subset \mathcal{I}$ onto $\mathfrak{L}_0 = \psi(\mathfrak{L}) \subset \mathcal{I}_0$. But every $A \in \mathfrak{L}$ is a finite linear combination of iterated commutators formed from S (see Lemma 8.7), whence $\psi(A)$ is the corresponding linear combination of iterated commutators from $S_0 = \psi(S)$ and S_0 generates $\mathfrak{L}_0$. E.O.P.

<u>Remark</u>: In fact, $P = C^\infty(M)$ is an Abelian Lie ideal in $\mathcal{I}$, and so by the proof alone, $\mathcal{I} = \mathcal{I}_0 \dotplus P$ as a semidirect product of Lie algebras. Thus on a purely formal level, $\mathfrak{L}$ and $\mathfrak{L}_0$ in (c) are related via the "semidirect product perturbation" machinery of Section 9D. Analytically, there is less contact with 9D, since the $b \in P$ that we consider below do not in general define bounded operators on $L^p(M)$, and the vector fields $X \in \mathcal{I}_0$ are not, in general, pregenerators in L^p. We return to this matter later.

Turning from infinitesimal to global considerations, the following lemma serves primarily to characterize the L^p pregenerators among the first-order differential operators, as a first step toward an exponentiation theorem for the Lie algebra $\mathfrak{L}$ of these described in Lemma 10.7(c) above. Its secondary purpose is to give an explicit description of the operator groups that are obtained, as multiplier representations of the Lie group $G = \mathbb{R}$. For the latter purpose, and to facilitate the extension of this explicit description to general G, it is useful to obtain the corresponding smooth integral of a general first-order operator, as a smooth multiplier representation on D qua test-function space ($D = \mathcal{D}$), thereby extending Lemma 10.1. Since the smooth integral exists in greater generality than the exponential, and since it serves as a tool in obtaining that exponential, it is discussed first in the lemma.

10.8. Lemma

Let $A = X + b \in \mathcal{I}$, and suppose that $X = \psi(A)$ is complete.
(a) Then A is the infinitesimal generator of a smooth locally equicontinuous one-parameter operator group $\{V_\infty(t,A)\}$ on $\mathcal{D}$ given for all $u \in \mathcal{D}$, $t \in \mathbb{R}$ and $x \in M$ by the multiplier formula

$$[V_\infty(t,A)u](x) = \mu(t,x)[V_\infty(t,X)u](x), \tag{16}$$

where $[V_\infty(t,X)u](x) = u(\gamma(t,x))$ for $\gamma:\mathbb{R} \times M \to M$ the integral flow determined by $X = \psi(A)$, and

$$\mu(t,x)=\exp(\int_0^t[V_\infty(s,x)b](x)ds)=\exp(\int_0^t b(\gamma(s,x))ds). \tag{17}$$

Here, the multiplier operator μ satisfies the usual "cocycle" identity for all $x \in M$, t_1, $t_2 \in \mathbb{R}$:

$$\mu(t_1,x)\mu(t_2,\gamma(t_1,x)) = \mu(t_1 + t_2,x). \tag{18}$$

(b) Moreover, for $1 < p < \infty$, the operator A is the pre-generator on $\mathcal{D}$ of a strongly continuous one-parameter group $\{V(t,A)\}$ on $L^p(M)$ if and only if it satisfies the divergence condition

$$\|Re(b) - p^{-1}div(X)\|_\infty < \infty \tag{19}$$

where $\|\cdot\|_\infty$ denotes the sup-norm. Then (16) extends in the obvious way to $\{V(t,A)\}$ for almost every $x \in M$.

Remarks: (1) Although the result in (b) can be extended to the extreme values $p = 1$ and (suitably interpreted) $p = \infty$, we have not included either case in the statement of the result above for the following reasons. For $p = \infty$, the result must be interpreted as applying to $E_\infty = C_0(M)$ with the sup-norm, and (19) must be interpreted as $\|Re(b)\|_\infty < \infty$. In that form, the result becomes a trivial consequence of (a), which we have omitted only because it is devoid of any measure-theoretic or geometric content. (Indeed, it extends further in a reasonable way to continuous b, and X the generator of a continuous flow on any locally compact Hausdorff space M.) For $p = 1$, certain technicalities intrude in the proof that we use below (as we point out there); these make it impractical to use this functional analytic proof scheme. We sketch an

alternative approach to the p = 1 case following the conclusion of the present proof.
(2) For all $1 \leq p < \infty$, our result in (b) then extends easily by limits to all measurable b with $Re(b) - p^{-1}div(X) \in L^\infty(M)$. We leave details to the reader.

Proof: (a) We begin by checking (18) and recalling how the group represetnation property for $\{V_\infty(t,A)\}$ in (16) follows from it. First, by the group property for γ and change-of-variables in (17) we have

$$\begin{aligned}\mu(t_2,\gamma(t_1,x)) &= \exp(\int_0^{t_2} b(\gamma(s,\gamma(t_1,x))ds)\\ &= \exp(\int_0^{t_2} b(\gamma(s+t_1,x))ds)\\ &= \exp(\int_{t_1}^{t_1+t_2} b(\gamma(s,x))ds).\end{aligned} \tag{20}$$

Thus, the product on the left in (18) adds exponents, and the integrals combine to yield $\exp(\int_0^{t_1+t_2} b(\gamma(s,x))ds = \mu(t_1+t_2,x)$ as on the right there. Then as usual

$$\begin{aligned}[V_\infty(t_1,A)V_\infty(t_2,A)u](x) &= \mu(t_1,x)[V_\infty(t_2,A)u](\gamma(t_1,x))\\ &= \mu(t_1,x)\mu(t_2,\gamma(t_1,x))u(\gamma(t_1,\gamma(t_2,x)))\\ &= \mu(t_1+t_2,x)u(\gamma(t_1+t_2,x))\\ &= [V_\infty(t_1+t_2,A)u](x).\end{aligned} \tag{21}$$

As described in detail below, Lemma 10.1 can be used to check that for $u \in \mathcal{D}$, $V_\infty(t,A)u \in \mathcal{D}$ and is differentiable there, with derivative

$$\frac{d}{dt} V_\infty(t,A)u = V_\infty(t,A)Au = AV_\infty(t,A)u. \tag{22}$$

Since $Au \in \mathcal{D}$, this will show that $\{V_\infty(t,A)\}$ is a C^∞ group (necessarily locally equicontinuous, since $\mathcal{D}$ is barreled) with A as its infinitesimal generator.

The details for (22) can be supplied as follows: we check that $\mu(t,\cdot)$ is a C^∞-function with values in $\mathcal{E} = \mathcal{E}(M)$ using the $\mathcal{E}$ case of Lemma 10.1, and that $V_\infty(t,X)u$ is a C^∞ function with values in $\mathcal{D}(M) = \mathcal{D}$ by the $\mathcal{D}$ case in 10.1, while for $t \in [-T,T]$ for $T < \infty$, $V_\infty(t,X)u \in \mathcal{D}_K$ for some fixed compact set $K \subset M$. Moreover, the product mapping $E \times E \to E$ is bilinear and jointly continuous since the p_K^n are sub-multiplicative, $\mathcal{D}_K$ is injected homeomorphically into E, and $E\,\mathcal{D}_K \subset \mathcal{D}_K$ (products preserve or shrink support), whence $t \to V_\infty(t,A)u = \mu(t,\cdot)V_\infty(t,X)u$ is C^∞ with first derivative

$$\frac{d}{dt} V_\infty(t,A)u = [\frac{d}{dt}\mu(t,\cdot)]V_\infty(t,x)u + \mu(t,\cdot)V_\infty(t,X)(Xu).$$

Thus if we verify that $\mu(t,\cdot)$ is C^∞ into E and $(d/dt)\mu(t,\cdot) = \mu(t,\cdot)V_\infty(t,X)b$ we will obtain $(d/dt)V_\infty(t,A)u = \mu(t,\cdot)V_\infty(t,X)Au$ and the first identity in (22) will follow, with the second identity following automatically from the group property. To obtain the claimed properties of $\mu(t,\cdot)$, we interpret b as an element of E, so that by Lemma 10.1 (E case) $s \to V_\infty(s,X)b$ is C^∞ into E and $e(t) = \int_0^t V_\infty(s,X)b\,ds$ is a well-defined C^∞ E-valued function with $e'(t) = V_\infty(t,X)b$. Uniform differentiability of exp on compacta then easily shows that $\mu(t,\cdot) = \exp(e(t))$ is sup-norm (or $p_K^{(0)}$)-differentiable for each compact $K \subset M$, with the obvious chain rule derivative $\exp(e(t))e'(t)$. The obvious product-rule argument then shows by induction that since $e'(t)$ is $p_K^{(0)}$-C^∞, so is $\mu(t,\cdot)$. The claim for the higher-order seminorm $p_K^{(n)}$ follows by a similar induction. That is, for any vector field X_j on a coordinate-neighborhood U, we have $X_j\exp(e(t)) = \exp(e(t))(X_je)(t)$ by chain rule. Hence, if $\exp(e(t))$ is already known to be $p_K^{(n)}$- C^∞ into E, continuity of $X_j\colon E \to E$ entails that $(X_je)(t)$ is C^∞ into E, whence $X_j\exp(e(t))$ is $p_K^{(n)}$- C^∞ and by the inductive definition (4), $\exp(e(t))$ is $p_K^{(n+1)}$- C^∞ as well. Thus $\mu(t,\cdot)$ is indeed C^∞ as a function into E.

Our proof of (b) uses (22) in combination with another identity that must be derived by smooth analysis and geometry. If $u \in D$, $1 < p < \infty$ we abbreviate $u_t = V_\infty(t,A)u$ and $\ell(t) = \|u_t\|_p^p = \int_M |u_t|^p(x)dx$. Then the required result asserts that $\ell(t)$ is differentiable and

$$\ell'(t) = p\int_M [\mathrm{Re}(b) - p^{-1}\mathrm{div}(X)]\,|u_t|^p dx. \tag{23}$$

We defer the verification of this fact for the moment, first showing that it leads to the operator-theoretic characterization of L^p-pregenerators given in (b).

First, we check that if (23) holds and $\omega = \|\mathrm{Re}(b)-p^{-1}\mathrm{div}(X)\|_\infty$, then $V_\infty(t,A)$ is $\|\cdot\|_p$-bounded on D and satisfies the estimate

$$\|V_\infty(t,A)\|_p \leq \exp(\omega|t|),\ t \in \mathbb{R} \tag{24}$$

We verify (24) from (23) as follows. First, for $t \geq 0$ we observe that by (23)

$$\frac{d}{dt} e^{-p\omega t}\ell(t) = -p\omega e^{-p\omega t}\ell(t) + e^{-p\omega t}\ell'(t)$$
$$= p\int_M[\mathrm{Re}(b) - p^{-1}\mathrm{div}(X) - \omega]\,|u_t|^p dx \tag{25}$$

Hence, if $\omega = \|\mathrm{Re}(b) - p^{-1}\mathrm{div}(X)\|_\infty$, the integrand is everywhere non-positive, so $e^{-p\omega t}\ell(t)$ has non-positive derivative and must be maximized at $t = 0$: $\|u\|_p^p = e^0\ell(0) \geq e^{-p\omega t}\|u_t\|_p^p$ for all $t \geq 0$, so $\|V_\infty(t,A)u\|_p = \|u_t\|_p \leq e^{\omega t}\|u\|_p$ and (24) follows when $t \geq 0$. For $t < 0$, a similar argument with $s = -t = |t|$ shows that

$$(d/ds)e^{-p\omega s}\ell(-s) = p\int_M[p^{-1}\mathrm{div}(X)-\mathrm{Re}(b)-\omega]\,|u_t|^p ds \tag{25'}$$

and thus that $\|V_\infty(t,A)u\|_p = \ell(-s) \leq e^{\omega s}\|u\|_p \leq e^{\omega|t|}\|u\|_p$, since (23) also shows that the integrand in (25') is non-positive. Thus (24) is proved for all $t \in \mathbb{R}$ and $\{V_\infty(t,A)\}$ extends as claimed to all of L^p. By the Riesz-Fisher theorem, the extending sequences for each $t \in \mathbb{R}$ can be assumed to converge pointwise a.e., thus extending (16) a.e. to all $u \in L^p$. Moreover, (24) permits us to extend the algebraic and analytic properties of $\{V_\infty(t,A)\}$ to L^p. That is, since $T < \infty$ and $t \in [-T,T]$ implies that $V_\infty(t,A)u \in \mathcal{D}_K$ (i.e. $\mathrm{supp}(u) \subset K$) for a suitable compact $K \subset M$, and is a uniformly C^∞ function on K, the fact that $\|u\|_p \leq \mathrm{vol}(K)^{1/p}\sup\{|u(x)|: x \in K\}$ implies that $V_\infty(t,A)$ is $\|\cdot\|_p$-C^∞ and that (22) holds as a $\|\cdot\|_p$-derivative. Hence, $\{V_\infty(t,A)\}$ extends by limits to a strongly continuous one-parameter group $\{V(t,A)\}$ on all of L^p. (As remarked above, algebraic properties are preserved, and local L^p-norm boundedness of the $V_\infty(t,A)$ extends continuity by uniform limits, while the norm estimate also extends.) The pregenerator theorem then ensures that $D = \mathcal{D}$ is a core for the generator of this group, which agrees with A on D by (22), so A is a pregenerator.

Turning to the converse in (b), suppose A pregenerates a strongly continuous group $\{V(t,A)\}$ on L^p. Then, since our argument on L^p-differentiability of $V_\infty(t,A)u$ for $u \in D = \mathcal{D}$ did not depend upon (23), we can use the L^p version of (22) and $V_\infty(t,A)D \subset D$ to check that $V_\infty(t,A)$ and $V(t,A)$ agree on D. (The argument is the same as that used in the proof of Theorem 6.1 to show that $V(t,A) = V_1(t,A)$. We refer the reader there for details.)

Now, $\|V(t,A)\| \le Me^{\omega|t|}$ for some $M,\omega < \infty$ (Appendix B), and this in combination with the defining equations (16) and (17) for $V_\infty = V|_D$ can be used to check that in fact $\|\mathrm{Re}(b) - p^{-1}\mathrm{div}(X)\|_\infty \le \omega\log(M)$. We use the sufficiency part of (b) proved above as one step in this proof. Indeed, let $d = p^{-1}\mathrm{div}(X)$ and $c = b - d$, so we need to prove that $\|\mathrm{Re}(c)\|_\infty \le \omega\log(M)$. Let $\tilde{A} = X + d$, and observe that by construction $\tilde{A}$ has $\omega = \|\mathrm{Re}(c)-p^{-1}\mathrm{div}(X)\|_\infty = 0$, so it exponentiates on L^p to an isometry group: $\|V(t,\tilde{A})u\|_p = \|u\|_p$ for all $t \in \mathbb{R}$, $u \in L^p$ (since $V(\pm t,\tilde{A})$ are contractions ...). But if $\mu_1(t,x) = \exp(\int_0^t V_\infty(s,X)(c)ds)$ and $\mu_2(t,x) = \exp(\int_0^t V_\infty(s,X)(d)ds)$, it is easy to see that $\mu_1(t,x)\mu_2(t,x) = \exp(\int_0^t V_\infty(s,X)(c+d)ds) = \mu(t,x)$. Consequently, we get for all $t \in \mathbb{R}$ and $u \in D$

$$\begin{aligned} Me^{\omega|t|}\|u\|_p \ge \|V(t,A)u\|_p &= \|V_\infty(t,A)u\|_p = \|\mu(t,\cdot)V_\infty(t,X)u\|_p \\ &= \|\mu_1(t,\cdot)[\mu_2(t,\cdot)V_\infty(t,X)u]\|_p = \|\mu_1(t,\cdot)V(t,\tilde{A})u\|_p. \end{aligned} \tag{26}$$

Putting $v = V(t,\tilde{A})u$, and using the fact that the invertible isometry $V(t,\tilde{A})$ must map D onto D, we get $Me^{\omega|t|}\|v\|_p \ge \|\mu_1(t,\cdot)v\|_p$ for all $v \in D$, hence by limits for all $v \in L^p$. Hence, since the norm of a bounded multiplication operator on L^p is its sup-norm, we obtain $Me^{\omega|t|} \ge \|\mu_1(t,\cdot)\|_\infty$. But

$$|\mu_1(t,x)| = \exp(\mathrm{Re}(\int_0^t c(\gamma(s,x))ds) = \exp(\int_0^t \mathrm{Re}(c)(\gamma(s,x))ds)$$

so we obtain $Me^{\omega|t|} \ge \exp(\int_0^t \mathrm{Re}(c)(\gamma(s,x))ds$ and by taking logarithms

$$|t|\omega \log(M) \geq \int_0^t \mathrm{Re}(c)(\gamma(s,x))ds \tag{27}$$

In particular, for $t > 0$ we have

$$\omega\log(M) \geq \lim_{t\to 0} t^{-1}\int_0^t \mathrm{Re}(c)(\gamma(s,x))ds = \mathrm{Re}(c)\gamma(0,x) = \mathrm{Re}(c)(x). \tag{28}$$

For $t < 0$

$$\omega\log(M) \geq \lim_{|t|\to 0} |t|^{-1}\int_0^{-|t|} \mathrm{Re}(c)(\gamma(s,x))ds = -\mathrm{Re}(c)(x) \tag{28'}$$

so $\|\mathrm{Re}(c)\|_\infty \leq \omega\log(M)$ and (19) follows.

Thus (b) is proved modulo the geometric identity (23).

We observe that the integrand in $\ell(t) = \int_M |u_t|^p dx$ satisfies

$$|u_t|^p = |\mu(t,x)u(\gamma(t,x))|^p = \exp\left(\int_0^t (p\mathrm{Re}(b))(\gamma(s,x))ds\right)|u|^p(\gamma(t,x))$$
$$= \mu_p(t,x)[V_\infty(t,X)|u|^p](x) = [V_\infty(t,A_p)|u|^p](x) \tag{29}$$

where $A_p = X + p\,\mathrm{Re}(b)$ is another first-order differential operator. Here, if $p > 1$ and $u \in D$, it is not difficult to see that $|u|^p$ is at least of class C^1, and thus that (29) is sup-norm-differentiable in t on a compact $K \supset \text{support}(u_t)$ for $t \in [-T,T]$, with derivative satisfying the extension of (22) to C^1-functions:

$$\frac{d}{dt}|u_t|^p = A_p V_\infty(t,A_p)|u|^p = (p\,\mathrm{Re}(b) + X)V_\infty(t,A_p)|u|^p$$
$$= p\,\mathrm{Re}(b)|u_t|^p + X(|u_t|^p). \tag{30}$$

(We elaborate the details below.) Thus $\ell(t)$ is differentiable under the integral and

$$\ell'(t) = p\int_M \mathrm{Re}(b)|u_t|^p dx + \int_M X(|u_t|^p)dx. \tag{31}$$

But then a routine application of the Gauss-Stokes-divergence theorem yields

$$\int_M X(|u_t|^p)dx = -\int_M \mathrm{div}(X)|u_t|^p dx = p\int_M -p^{-1}\mathrm{div}(X)|u_t|^p dx \tag{32}$$

and when this is substituted into (31), (23) follows. (See Abraham and Marsden [AM], Sections 2.6 and 2.7 for details concerning Stokes Theorem and its pseudo-Riemannian applications, noting that compact support of $|u_t|^p$ eliminates boundary terms. The development in [AM] normally carries C^∞ assumptions, but Lang [Lg 2, see Chapter XVIII] shows that C^1 suffices.) Identity (30) can be justified by direct computation, extending Lemma 10.1 to C^1-functions and pushing through each step in the derivation of (22) in that generality. It is more economical to observe that $\mathcal{D}$ is dense in $C^1(M)$ with respect to the C^1-topology and that (22) itself can be extended by limits to an identity relating continuous maps from $C^1(M)$ to $C^0(M)$. We omit the routine details.

The claim that $|u|^p$ is C^1 follows by composition from the fact that $|z|^p$ is a C^1 function of the components (x,y) of $z = x + iy$: $|z|^p = (x^2 + y^2)^{p/2}$. The only difficulty occurs at $(0,0)$ when $p < 2$, for then the partial derivatives contain a singular factor $S(x,y) = (x^2 + y^2)^{p/2-1}$. But for $p > 1$, the partial derivatives $\partial|z|^p/\partial x = 2xS(x,y)$ and $\partial|z|^p/\partial y = 2yS(x,y)$ still approach 0 at $(0,0)$, hence are continuous everywhere. (For $p = 1$, $|z|^1 = |z|$ is not differentiable at $(0,0)$ and the partial derivatives fail to have unique limits there, as the reader may check.)

Thus (30) holds and the proof is complete. E.O.P.

Remarks on the case $p = 1$: (1) It is possible to fix up the proof given above when $p = 1$, since the singularity of $|z|$ at $(0,0)$ is mild and produces difficulties only at transversal zeros of u (where the differential $du \neq 0$): these form a set of measure 0 so $X(|u_t|^p)$ can be shown to be continuous a.e. and in $L^\infty(M)$. But detailed inspection of behavior near these transversal zeros is needed in order to justify both (30) and (32) for this case.

(2) There is an alternative approach to the case $p = 1$ that can be used to obtain a superficially different solution to the pregenerator problem on $L^p(M)$ for all $1 \leq p < \infty$. One begins with the observation that the diffeomorphism $\gamma(t,\cdot)$ defines a new positive measure ν_t on M via

$$\int_M u(x)d\nu_t(x) = \int_M u(\gamma(t,x))dx \tag{33}$$

and that $d\nu_t(x)$ and dx are mutually absolutely continuous, so $J(t,x) = (d\nu_t/dx)(x)$ is a positive nonvanishing function. The chain rule for Radon-Nikodym derivatives and the group property of γ in fact show that $J(t,x)$ is a multiplier. Moreover, so is $J^{-1/p}(t,x)$ and the formula $[\tilde{V}(t)u](x) = J^{-1/p}(t,x)u(\gamma(t,x))$ is easily seen to define an isometry group on $L^p(M)$ since

$$\int|\tilde{V}(t)u|^p(x)dx = \int_M J^{-1}(t,x)|u|^p(\gamma(t,x))dx = \int_M|u|^p(x)dx.$$

To obtain strong continuity, and differentiability for $u \in D = C_0^\infty(M)$, one notices that $J(t,x)$ is in fact the Jacobian determinant of $d_x\gamma(t,\cdot)$, computed with respect to the parallel translation of tangent spaces induced by the pseudo-Riemannian connection that defines the measure dx. Hence $J(t,x)$ and $\mu_p(t,x) = J^{-1/p}(t,x)$ are smooth functions, so one can argue as in the proof above that $t \to \tilde{V}(t)u$ is C^∞ when $u \in D$. Thus if $d_p(x) = d/dt\ \mu_p(0,x)$, then

$$(d/dt)\tilde{V}(t)u = (X + d_p)\tilde{V}(t)u = \tilde{V}(t)(X + d_p)u. \tag{34}$$

Hence, one can conclude that $\{\tilde{V}(t):t \in \mathbb{R}\}$ is a strongly continuous isometry group pregenerated by $\tilde{A} = X + d_p$. Then, arguing exactly as in the proof of the pregenerator characterization in (b) above, one can show that $A = X + b$ pregenerates a strongly continuous group on L^p if $\|\mathrm{Re}(b) - d_p\|_\infty < \infty$. In fact, a complicated geometrical argument shows that $d_p = p^{-1}\mathrm{div}(X)$, but we do not know a reference for this fact. Consequently, a fully infinitesimal characterization as in (23) for pregenerators on L^p is difficult to obtain from the line of development discussed here. (Notice, however, that both d_p and $p^{-1}\mathrm{div}(X)$ are real valued, so if both are known to generate isometry groups and either (b) of the Lemma or the corresponding result above is known, then

$$0 = \mathrm{Re}(d_p)-p^{-1}\mathrm{div}(X) = d_p-\mathrm{Re}(p^{-1}\mathrm{div}(X)) = d_p-p^{-1}\mathrm{div}(X)$$

so $d_p = p^{-1}\mathrm{div}(X)$ follows from our functional-analytic development.)

Both the proof of Lemma 10.8 and the remarks directly above call attention to the following special cases of the lemma, which we record because of contact with the prior literature.

10.9. Corollary

Let $1 < p < \infty$, let M be pseudo-Riemannian, and let X be complete.

(a) For every complete vector field X on M, $X+p^{-1}\mathrm{div}(X)$ pregenerates an isometry group on $L^p(M)$.

(b) In particular, X pregenerates an isometry group if and only if it is incompressible ($\mathrm{div}(X) = 0$).
(c) If X is incompressible, $A = X + b$ pregenerates a strongly continuous group iff $\mathrm{Re}(b)$ is bounded.

10.10. Corollary

Let $p = 2$, let M be pseudo-Riemannian; and let X be complete.
(a) Then $A = i(X + b)$ is essentially self-adjoint on D iff $b - 1/2\mathrm{div}(X)$ is pure imaginary.
(b) In particular, iX itself is essentially self-adjoint iff X is incompressible.

In these cases, the groups involved are unitary.

With the pregenerator question settled (for first-order differential operators in $L^p(M)$) by Lemma 10.8, we turn to the exponentiation problem for Lie algebras $\mathfrak{L}$ of such operators. The principal conclusion of our main result below, which characterizes the L^p-exponentiable $\mathfrak{L}$, is an easy consequence of the lemma and our exponentiation theorem 9.1. Because of its classical interest (and its relationship to the corresponding one-parameter result) we also exhibit the explicit form of the exponential as a multiplier representation.

10.11. Theorem

Let $\mathfrak{L}$ be a finite-dimensional real Lie algebra of first-order differential operators on a (pseudo-) Riemannian manifold M, and let $1 < p < \infty$. Suppose that a Lie generating subset $S \subset \mathfrak{L}$ consists of operators with complete vector field part ($\psi(A) = X$ is complete for all $A \in S$.)
(a) Then $\mathfrak{L}$ is exponentiable in $L^p(M)$ if and only if for all $A = X + b$ in S,

$$\|\mathrm{Re}(b) - p^{-1}\mathrm{div}(X)\|_\infty < \infty. \tag{35}$$

Then for all $A = X + b \in \mathfrak{L}$, (35) holds.
(b) Moreover, if G is the Lie group associated with $\mathfrak{L}$ there exists an action $\gamma: G \times M \to M$ of G on M and a function $\mu \in C^\infty(G \times M, \mathbb{C})$ such that for all $u \in L^p(M)$ and $g \in G$, the exponential $V: G \to L(L^p(M))$ satisfies

$$[V(g)](x) = \mu(g,x)u(\gamma(g,x)) \quad \text{a.e. } x. \tag{36}$$

Here, μ satisfies the multiplier identity

$$\mu(g_1g_2,x) = \mu(g_1,x)\mu(g_2,\gamma(g_1,x)). \tag{37}$$

Proof: For (a), the condition (35) for every $A \in S$ implies that $\mathcal{L}$ is Lie-generated by pregenerators satisfying the hypotheses and conclusions of Lemma 10.8. Consequently, for $u \in D_\infty = \mathcal{D}$, the one-parameter exponentials $V(t,A)u = V_\infty(t)u$ are C^∞ into $\mathcal{D}$. Consequently, if $B = Y + c \in S$, the fact that both vector fields Y and smooth multipliers c map $\mathcal{D}$ continuously into $\mathcal{D}$ entails that $BV_\infty(t)u = YV_\infty(t)u+cV_\infty(t)u$ is C_∞ into $\mathcal{D}$ and is certainly therefore C^∞ into $L^p(M)$. Thus, the hypotheses of Theorem 9.1 have been checked, and the existence of the exponential follows.

In order to finish the proof of (a) and to obtain the group action γ for (b), we observe that by Lemma 10.7(c), $\mathcal{L}_0 = \psi\mathcal{L}$ is a finite-dimensional Lie algebra of vector fields generated by $S_0 = \psi(S)$, and every $X \in S_0$ is complete. Hence the theory of Section 10A applies, and Corollary 10.5 supplies a transformation group action $\gamma_0:G_0 \times M \to M$ for the connected, simply connected Lie group G_0 associated with $\mathcal{L}_0$. It also implies that every $X \in \mathcal{L}_0$ is complete. Lemma 10.8 now asserts that a general element $A \in \mathcal{L}$ is a pregenerator iff (35) holds, while by definition if $\mathcal{L}$ is exponentiable every $A \in \mathcal{L}$ is a pregenerator, so (a) is proved. Moreover, the Lie algebra homomorphism $\psi:\mathcal{L} \to \mathcal{L}_0$ lifts to a C^∞ group homomorphism $\psi:G \to G_0$ of the connected, simply connected group G associated with $\mathcal{L}$, and then $\gamma(g,x) = \gamma_0(\psi(g),x)$ defines the action $\gamma:G \times M \to M$ required for (b).

Next, we observe that for any $A \in \mathcal{L}$, the one-parameter group $\{V(\exp(tA)):t \in \mathbb{R}\}$ determined by A (qua member of the Lie algebra $\mathfrak{g} \simeq \mathcal{L}$ of G) must be the multiplier group $V(t,A) = \mu_A(t,\cdot)V_\infty^A(t)$ determined by A via Lemma 10.8. Moreover the flow $\gamma(\exp(tA),\cdot)$ must be the integral $\gamma(t,\cdot)$ of the complete vector field $X = \psi(A)$, so it follows that if we now define $\mu(\exp(tA),\cdot) = \mu^A(t,\cdot)$ for all $A \in \mathcal{L}$ and $t \in \mathbb{R}$, we have

$$[V(\exp(tA))u](x) = \mu(\exp(tA),x)u(\gamma(\exp(tA),x)). \tag{38}$$

This is just a local version of (36). But then if g_1 and g_2 are elements of $\exp(\mathcal{L}) \subset G$ such that $g_1g_2 \in \exp(\mathcal{L})$ as well, (37) follows by the G-version of the calculation (21) that led to the $\mathbb{R}$-version (18) in the proof of Lemma 10.8. (We leave these details to the reader.) Since G is connected, we can then use (36) inductively to extend the definition of $\mu(g,\cdot)$ to all $g \in G$, and to check that (35) extends in the same way, proving (b). E.O.P.

The following little result shows that we can also generalize to Lie algebras $\mathcal{L}$ of first-order differential operators the result 10.8(a) on "smooth integrals" acting on the Schwartz test-function space $\mathcal{D}$. Since the proof is the same as that given above, we leave it to the reader.

10.12. Proposition

Let $\mathcal{L}$ be a Lie algebra of first-order differential operators on a C^∞ manifold M, and suppose that for a Lie-generating set $S \subset \mathcal{L}$, the vector fields in $S_0 = \psi(S)$ are complete. Then $\mathcal{L}$ has a smooth integral $V = V_\infty : G \to L(\mathcal{D})$ in the sense of Chapter 8, and this integral satisfies (36) and (37).

The interest of this result lies in the fact that it is independent of the Riemannian structure of M and the L^p structure on functions. Theorem 10.11 can then be viewed as a set of necessary and sufficient conditions on a choice of pseudo-Riemannian structure and L^p structure such that this canonical integral for $\mathcal{L}$ extends to an exponential in L^p. Considerations similar to those in Remark (2) following Lemma 10.8 then permit Theorem 10.11 to be generalized to $p = 1$. In that case, $\mu(g,x) = \mu_1(g,x)\mu_2(g,x)$ and $\mu_2(g,x) = J^{1/p}(g,x)$ where $J(g,x)$ is the Jacobian of $\gamma(g,x)$.

We close this section with some remarks concerning the completeness assumptions that we have placed upon the vector-field parts $X = \psi(A)$ of the differential operators considered here. It is well-known in the theory of dynamical systems (at least in the special cases $\mathrm{div}(X) = 0$, $p = 2$ and $G = \mathbb{R}$) that the condition on X equivalent to "X is a pregenerator on $D = C_0^\infty(M)$ of a unitary group" is the (weaker) condition "X is almost complete". That is, the local integral flow $\gamma(t,x)$ need not always be well-defined, so long as the exceptional set $E \subset M$ of points x "carried to infinity by γ in finite time" is of Riemannian measure 0. (See Abraham and Marsden [AM], Section 1.2, for an informative discussion and further references. The first proof of sufficiency known to us is in Nelson's work [Nℓ 1].) There are several reasons why that assumption is inappropriate here. First, it is not clear how to formulate conditions on individual vector fields X which will ensure that the finite-dimensional Lie algebras $\mathcal{L}_0$ generated by sets of these as in Lemma 10.7(c) will be "jointly almost-complete" (integrating to almost-everywhere-defined global Lie flows.) More important, it is clear that the connection between almost-completeness and group-generation is not so close in the case of general first-order operators and multiplier groups, as the following example indicates.

10.13. Example

Let $[V(t)u](x) = (1 - xt)^{-1}u(x(1 - xt)^{-1})$ for $x \neq t^{-1}$, 0 for $x = t^{-1}$. It is not difficult to check that $V(t)$ defines a one-parameter unitary group on $L^2(\mathbb{R})$, in fact a one-parameter subrepresentation of a well-known induced unitary representation of $SL(2,\mathbb{R})$ on $L^2(\mathbb{R})$ in the "noncompact picture" ([St 2], [Sa], 12A). The corresponding flow $\gamma(t,x) = x(1-xt)^{-1}$ sends the entire complement of $(-\infty,t^{-1})$ to infinity in time $\leq t$, according to standard conventions. It is the integral of the vector field $x^2\, d/dx$, and $V(t)$ is associated with the differential operator $A = x^2\, d/dx + x$. But the skew-symmetric operator obtained by restricting iA to $D = C_0^\infty(\mathbb{R})$ is not essentially self-adjoint on that domain. (One can check that the function $v_+(x) = \chi_{[0,\infty)}(x)x^{-1}\exp(-x^{-1})$ is in $L^2(\mathbb{R})$ and is orthogonal to $D_+ = (1-A)D$ while $v_-(x) = \chi_{(-\infty,0]}(x)x^{-1}\exp(x^{-1})$ is in L^2 and orthogonal to $D_- = (-1-A)D$. But A <u>is</u> essentially self-adjoint on its own Gårding domain, for example, and fails to have an almost-complete vector-field part. Thus, the problem of describing appropriate pregenerator domains for this kind of first-order operator remains open, and is clearly unrelated to almost-completeness. What the example shows is that the equivalence "pregenerator iff almost-complete" is an artifact of the particular choice $D = C_0^\infty(M)$.

10C. Sub-Laplacians on manifolds

In this section we consider second-order differential operators of the form $S_0 = \Sigma X_i^2$ where the X_i are complete vector fields on a pseudo-Riemannian manifold M, with pseudo-Riemannian measure dx. We will use the exponentiation theorems from Sections 10A and 10B to derive semigroup generation properties of the so called sub-Laplacian S_0.

Rather than considering $S_1 = \Sigma(X_i+b_i)^2$ (cf. Section 10B), we restrict ourselves to the case $b_i=0$. This has the advantage of allowing unity with Section 11A where the perturbations b_i are not considered.

Our method makes use of the finite-dimensional Lie algebra $\mathcal{L}$ as follows. $\mathcal{L}$ will always be the Lie algebra of a real Lie group G which we may assume is simply-connected and connected. This means that the elements in $\mathcal{L}$, (and in the enveloping algebra of $\mathcal{L}$), may be identified with right invariant vector fields (resp., differential operators). In particular, this means that S_0 is associated with a second-order differential operator, Σ_0 say, on G. But it is important for the solution of the problem that this operator Σ_0 is right-invariant, for, as is shown in [Jo 2], the theory of right-invariant partial differential operators on Lie groups parallels that of constant coefficient p.d.o. on Euclidean space.

10.14. Theorem

Let $X_1,\ldots,X_n$ be complete vector fields on a pseudo-Riemannian manifold M, and suppose that the commutator Lie algebra $\mathcal{L}$ generated in the Lie sense by the X_i is finite-dimensional.
(i) Then $S_0 = \Sigma X_i^2$ is the pregenerator of a positive semigroup $\{W(t): 0 \leq t < \infty\}$ on the Banach space $C_\infty(M)$. Moreover, W(t) has a holomorphic extension $W(\zeta)$ to the open half-plane $\mathrm{Re}(\zeta) > 0$. There is a kernel $p_t(g)$ on $\mathbb{R}^+ \times G$ which is $C^\infty(\mathbb{R}^+ \times G)$, and has analytic extension $\zeta \to p_\zeta$ as a $L^2(G)$-valued function, such that

$$W(\zeta)u(x) = \int_G p_\zeta(g)u(\gamma(g,x))dg,$$

where γ is a C^∞-action of G on M and dg is a left-invariant Haar measure on G.
(ii) Let $1 \leq p < \infty$, and suppose that $\mathrm{div}(X_j)$ is bounded. Then S_0 has the same pregenerator properties on the Banach spaces $L^p(M,dx)$ as listed above in (i) for $C_\infty(M)$.

Proof: We first invoke Corollary 10.5 to get a global action $\gamma\colon G \times M \to M$ which exponentiates $\mathcal{L}$. That is, for every $X \in \mathcal{L}$ with corresponding G-exponential one-parameter group $\{a(t): -\infty < t < \infty\} \subset G$, the identity $(d/dt)u(\gamma(a(t),x))|_{t=0} = (Xu)(x)$ holds for $u \in C_0^\infty(M)$. In the L^p-setting of (ii), Theorem 10.11 of Section 10B is applied instead.

In view of [Hö, Theorem 1.1] and [Jo 2, Section 3] there is a convolution semigroup $p_t(.)$ of C^∞-densities on G such that $(t,g) \to p_t(g)\colon \mathbb{R}_+ \times G \to \mathbb{R}_+$ is jointly C^∞, and the map $t \to p_t$ from $\mathbb{R}_+$ into $L^2(G)$ has a holomorphic extension $\zeta \to p_\zeta$, of the half-plane $\mathrm{Re}(\zeta) > 0$ into $L^2(G)$. If we put

$$(W(\zeta)u)(x) = \int_G p_\zeta(g)u(\gamma(g,x))dg \tag{39}$$

($u \in C_\infty(M)$, dg = a fixed left-invariant Haar measure on G), then $\{W(\zeta):\mathrm{Re}(\zeta) > 0\}$ defines a holomorphic semigroup with the properties stated in (i).

If the assumptions in (ii) are satisfied, then the action γ induces a C_0 representation of G on $L^p(M,dx)$ by Theorem 10.11. Therefore, the integral in (39) converges in the $L^p(M,dx)$ norm, which one can show implies that $W(\zeta)$ extends to a holomorphic semigroup on $L^p(M,dx)$. It is a consequence of [Jo 2, Theorem 3.1] that the restriction to $C_0^\infty(M)$ of the infinitesimal generator S for W is equal to the differential operator S_0. The identity $\bar{S}_0 = S$ (the infinitesimal generator) expresses that S_0 is a pregenerator, as claimed. Said identity is in turn a consequence of [Jo 2, Prop. 3.2]. See also [Ps 1, Theorem 1.4].

For the sake of completeness, we indicate briefly how the results of [Jo 2] lead to the above conclusions regarding the kernel p_ζ.

The operator S_0 is first regarded as a symmetric operator in $L_2(G)$. Measure theoretic Dirichlet methods [Jo 2, Lemma 3.1] show that S_0 is self-adjoint. Moreover $(\lambda-S_0)^{-1}$ exists as a convolution operator for $\lambda > 0$.

Since $\exp(tS_0) = \lim_{n\to\infty} ((n/t)((n/t)-S_0)^{-1})^n$ it follows that $\exp tS_0$ is also a convolution operator. Let p_t be the corresponding semigroup of measures, $0 < t < \infty$. Hörmander's result implies hypoellipticity of $S_0-\partial/\partial t$ on $\mathbb{R}^+ \times G$.

Since $(S_0-\partial/\partial t)p_. = 0$, it follows that $p_. \in C^\infty(\mathbb{R}^+ \times G)$. Note that this argument requires S_0 to be homogeneous of second order (i.e., no first-order term). We check in [Jo 2, Proposition 3.1] that $p_t(x)$ decays exponentially in the x-variable. This result allows us in turn to define for $\zeta = t + is \in \mathbb{C}$, $t > 0$, $s \in \mathbb{R}$, $p_\zeta = \exp(isS_0)(p_t)$ such that $\zeta \to p_\zeta$ is analytic with values in $L_2(G)$. If V denotes the representation obtained in Theorem 10.2 and Corollary 10.5, then

$$W(\zeta) = \int_G p_\zeta(g)V(g)dg \tag{40}$$

defines a holomorphic semigroup of operators on $C_\infty(M)$ by [Jo 2, Theorem 3.1], and moreover $W(\zeta)u(x)$ is given by (39) for $u \in C_\infty(M)$, $x \in M$.

The formula (40) is also valid if V denotes the $L^p(M,dx)$-exponential in (ii) obtained through application of Theorem 10.11, cf. the proof of Corollary 10.15 below.

10.15. Corollary

Under the assumptions of the theorem, the closure of $\lambda - S_0$ in $C_0(M)$, and in $L^p(M,dx)$ is invertible for $\text{Re}(\lambda) > 0$. The inverse $R(\lambda)$ is an integral operator given by

$$(R(\lambda)u)(x) = \int_G r_\lambda(g)u(\gamma(g,x))dg \tag{41}$$

where

$$r_\lambda(g) = \int_0^\infty p_t(g)e^{-\lambda t}dt. \tag{42}$$

Proof: Each X_j is a pregenerator on $C_0^\infty(M)$ for a C_0-one-parameter group of bounded operators $V(t,X_j)$ on $L^p(M,dx)$ by Lemma 10.8. If $w_j = \|p^{-1}\text{div}(X_j)\|_\infty$ then eqn. (24) in the proof of the lemma shows that $\|V(t,X_j)\|_p \leq \exp(w_j|t|)$. Hence the Lie group representation V generated by the X_j is also of pure exponential growth. Let dg denote Haar measure on the Lie group G from Theorems 10.5 and 10.11. Then the semigroup W(t) is given by the integral $\int_G p_t(g)V(g)dg$ which is convergent in the $L^p = L^p(M,dx)$-operator norm as a Bochner integral. As a consequence, note that W(t) is positive, and that the resolvent $R(\lambda)$ exists for $\text{Re}(\lambda)$ sufficiently large, and is given by (41) and (42). The verification of (42) requires the Fubini theorem since

$$R(\lambda) = \int_0^\infty e^{-\lambda t}W(t)dt = \int_0^\infty \int_G e^{-\ t}p_t(g)V(g)dg\,dt.$$

Finally, note that $r_\lambda(.) \in L^\infty(G)$, and that, as a consequence, the validity of (42) extends to all λ, $\text{Re}(\lambda) > 0$.

10D. Solution kernels on manifolds

In the preceding section, we considered sub-Laplacian operators $S_0 = \Sigma X_j^2$ on function spaces $C_\infty(M)$ and $L^p(M)$, representing their resolvents $(\lambda - S_0)^{-1}$ and evolution semigroups $W(\zeta) = $ "$\exp(\zeta S_0)$" in terms of integral kernels carried by a Lie transformation group G acting upon M. (Recall that M is pseudo-Riemannian with pseudo-Riemannian measure dx, and $X_1,\ldots,X_r$ is a set of complete vector fields which Lie-generates a finite-dimensional Lie algebra $\mathfrak{L}$.) Motivated by applications to different partial differential equations on M we here examine special cases in which these kernels can be dropped down to the manifold M, by an averaging process, to obtain integral representations of the operators on M itself. This approach derives useful properties of the operators $(\lambda - S_0)^{-1}$ and $\exp(\zeta S_0)$, and of their kernels $E(x,y)$ on $M \times M$, from comparable qualitative properties of the kernels on G.

Our development focusses primarily upon two examples, the first of which was considered earlier by Bony [By] from the point of view of degenerate elliptic equations. We choose for expository reasons to present first the manifolds M, Riemannian measures μ, and vector fields $\{X_i\}$ involved in these examples, stating without proof several facts concerning the group-theoretic contexts of these. The conclusions concerning the integral kernels $E(x,y)$ will then be summarized in Theorem 10.16, followed by details of proof. The examples will be unified in the proof, which is, in fact, carried out for the most part in a more general Lie group setting.

Example 1: M is Euclidean space $\mathbb{R}^{n+1}$ with Lebesgue measure dx. The notation $x = (x_0,\ldots,x_n)$ will be convenient for the points in M, so $dx = dx_0\ldots dx_n$. The vector fields $X_1 = \partial/\partial x_0$ and $X_2 = \sum_{j=1}^{n} x_0^{\,j}(\partial/\partial x_j)$ are both pregenerators in $L^p(\mathbb{R}^{n+1})$ and $C_\infty(\mathbb{R}^{n+1})$, as explained below. The Lie algebra $\mathfrak{L}$ generated by X_1, X_2 is in fact spanned by the basis of vector fields $x_0^{\,k}\,\partial/\partial x_j$, $0 \le k \le j \le n$. Hence its dimension $d(n) = (n+1)(n+2)/2$ is certainly finite, so the general theory applies. Let $\mathfrak{L}_K$ [resp., $\mathfrak{L}_H$] be the Lie subalgebras of $\mathfrak{L}$ spanned by $\partial/\partial x_j$, $0 \le j \le n$, [resp., $x_0^{\,k}\,\partial/\partial x_j$, $1 \le k \le j \le n$]. Then the mapping $X,Y \to X+Y$ of $\mathfrak{L}_K \times \mathfrak{L}_H$ into $\mathfrak{L}$ is an isomorphism onto. If K,H, and G, respectively, denote the connected simply-connected Lie groups with Lie algebras $\mathfrak{L}_K$, $\mathfrak{L}_H$, and $\mathfrak{L}$, then the product mapping $k,h \to kh$ is a diffeomorphism of $K \times H$ onto G. Here, we use multiplicative notation for the Lie group product, and we summarize the results in the form

$$\mathfrak{L} = \mathfrak{L}_K + \mathfrak{L}_H\,, \quad \mathfrak{L}_K \cap \mathfrak{L}_H = (0) \tag{43}$$

$$G = KH, \quad K \cap H = (e). \tag{44}$$

Each vector field in $\mathfrak{L}$ turns out to be complete on M, and in fact, by Corollary 10.5 in Section 10A, G acts as a transitive transformation group on M with infinitesimal vector field Lie algebra equal to $\mathfrak{L}$. This group action can be written explicitly in terms of coordinates $k = (k_0,\ldots,k_n)$, $h = (h_{ij})$, $1 \le i \le j \le n$, as follows. If the action of a group element $g = kh$ on a point x in M is written $\gamma(g,x) = y$ then

$$\begin{aligned} y_0 &= h_0 + x_0, \\ y_j &= k_j + \sum_{i=1}^{j} x_0^i h_{ij} + x_j. \end{aligned} \tag{45}$$

Example 2: Here we let M be the (ax+b)-group of 2×2 matrices $\begin{pmatrix} a & b \\ 0 & 1 \end{pmatrix}$ with $a > 0$, $b \in \mathbb{R}$, with the left-invariant Riemann structure for this group. Then the Riemannian measure becomes left Haar measure : $dx = a^{-1}\, da\, db = d\mu(a,b)$. (The reasons for this choice of measure will become evident as the discussion proceeds.) Here, we consider the following pair X_1, X_2 of vector fields on M : $X_1 = a\,\partial/\partial a + b\,\partial/\partial b$ and $X_2 = (1-b)\partial/\partial b$. They generate a 3-dimensional commutator Lie algebra $\mathfrak{L}$, spanned, f.ex., by basis elements $a\,\partial/\partial a$, $\partial/\partial b$ and $b\,\partial/\partial b$.

It turns out here that $\mathfrak{L}$ is the Lie algebra of a particular 3-dimensional matrix Lie group G which acts transitively on M such that the corresponding infinitesimal vector field Lie algebra is also equal to $\mathfrak{L}$. As in Example 1, we mean by this that $\mathfrak{L}$ is related to the pair G, M as described in the Lie-Palais result, cf. Corollary 10.5.

Indeed, we can realize G as a 2×2 matrix group with elements $\begin{pmatrix} a & b \\ 0 & h \end{pmatrix}$ with $a, h > 0$ and $b \in \mathbb{R}$. We identify M with the subgroup K determined by $h \equiv 1$, and let H be the subgroup of matrices $\begin{pmatrix} 1 & 0 \\ 0 & h \end{pmatrix}$, $h > 0$. Then, clearly identities (44) are satisfied, and H normalizes K. It follows from this that a group action γ of G on M may be defined as follows: For $g \in G$, $g = kh$, $k \in K$, $h \in H$, and $x \in M$, define $y = \gamma(g,x) \in M$ by

$$y = khxh^{-1} \tag{46}$$

where the product on the right-hand side of (46) is the matrix product in G. Note that for h = e (the identity matrix) this action reduces to the left-action of the group K on itself. Hence, the corresponding infinitesimal vector field Lie algebra $\mathcal{L}_K$ is the Lie algebra of all right invariant vector fields on the (ax + b)-group.

In both examples, reduction of the kernels to M × M uses the fact that the action of G on M is transitive, and that H is the isotropy group of a suitable point $x_0 \in M$: the point (0,...,0) in Example 1, and $e = \begin{pmatrix} 1 & 0 \\ 0 & 1 \end{pmatrix}$ in Example 2. In both cases it follows that for all $x,y \in K \sim M$, and $h \in H$, the equation $\gamma(zh,x) = y$ has a unique "change-of-variables-solution" z, and moreover that z depends differentiably on y. That is, there is a diffeomorphism T_x^h of M such that $z = T_x^h(y)$. The solution of eqns. (45) yields

$$z_j = y_j - \sum_i x_0^i h_{ij} - x_j \tag{47}$$

in Example 1, while in Example 2

$$z = T_x^h(y) = y\, h\, x^{-1} h^{-1}. \tag{48}$$

10.16. Theorem

Let $S_0 = X_1^2 + X_2^2$ be the sub-Laplacian, where the vector fields X_1, X_2 are given as in either of the examples 1 and 2 above. In both cases S_0 is degenerate elliptic. Let B be any one of the operators $R(\lambda) = (\lambda - S_0)^{-1}$ or $\exp(\zeta S_0)$ in $L^p(M,\mu)$ considered in Section 10C, and let A be the corresponding kernel function on G, (the Green's function in the case of $R(\lambda)$ and the Gauss' kernel in the case of $\exp(\zeta S_0)$). Then for each case, there exists a corresponding integral kernel $E(\cdot,\cdot)$ on M × M such that $(Bu)(x) = \int_M E(x,y)u(y)d\mu(y)$, $(M \sim K)$ where

$$E(x,y) = \int_H A(T_x^h(y)h)dh \tag{49}$$

and $T_x^h(y)$ is defined by (47) for the operators in Example 1. In Example 2, the corresponding formula is

$$E(a_0,b_0;a,b) = \int_0^\infty A\begin{pmatrix} aa_0^{-1} & bh-ab_0a_0^{-1} \\ 0 & h \end{pmatrix} \frac{dh}{a_0} \tag{50}$$

where $\begin{pmatrix} aa_0^{-1} & bh-ab_0a_0^{-1} \\ 0 & h \end{pmatrix}$ is the matrix expression for $T^h_{(a_0,b_0)}(a,b)$.

Proofs: The claims regarding Example 1 which are stated before the theorem are fairly routine to verify, so we turn directly to the second example.

The group action (46) of Example 2 may be calculated in local coordinates as follows: $x = (\alpha,\beta) \sim \begin{pmatrix} \alpha & \beta \\ 0 & 1 \end{pmatrix} \in K$, $g = kh = \begin{pmatrix} a & b \\ 0 & h \end{pmatrix}$, and $\gamma(g,x) = (a\alpha, a\beta h^{-1}+b)$. Let Z_1, Z_2, Z_3 be the respective vector fields on M corresponding to the action of the respective one-parameter subgroup $\begin{pmatrix} a & 0 \\ 0 & 1 \end{pmatrix}$, $\begin{pmatrix} 1 & b \\ 0 & 1 \end{pmatrix}$, $\begin{pmatrix} 1 & 0 \\ 0 & h \end{pmatrix}$ of G. Then $Z_1 = a\, \partial/\partial a + b\, \partial/\partial b$, $Z_2 = \partial/\partial b$, and $Z_3 = b\, \partial/\partial b$. Hence, X_1 and X_2 belong to the infinitesimal Lie algebra of the Lie-Palais pair G,M. To see that X_1, X_2 are pregenerators for C_0 one-parameter groups of bounded operators on each of the function spaces under consideration, we appeal to Lemma 10.8(b). Let g denote the Riemannian structure on M (in Example 2) associated to left-Haar measure μ, and let (x_1,x_2) be coordinates on M given by $\exp x_1 = a$, $x_2 = b$. Then since $\mu = da\, db/a^2 = (\bar{g})^{\frac{1}{2}} dx_1\, dx_2$, it follows by eqn. (10) that $(\bar{g})^{\frac{1}{2}} = \exp(-x_1)$, and $X_1 = \partial/\partial x_1 + x_2\, \partial/\partial x_2$, $X_2 = (1-x_2)\partial/\partial x_2$. Hence, the corresponding divergence calculates by formula (12) to $\mathrm{div}(X_1) = 0$ and $\mathrm{div}(X_2) = -1$. Consequently, Lemma 10.8 and Theorem 10.11(b) apply and yield exponentiability in each of the function spaces under consideration.

We now turn to general considerations which apply to a pair of Lie sub-groups $K, H \subset G$ such that $G = KH = \{kh : k \in K, h \in H\}$ satisfies the product property (44) and the corresponding infinitesimal relations (43). We let dk and dh be fixed left-invariant Haar measures on K, respectively, H. The following function $\rho(h) = \det \mathrm{Ad}_H(h)(\det \mathrm{Ad}_G(h))^{-1}$ is discussed in [Hℓ, p. 372] where it is shown that the product measure $f \to \int_{K\times H} f(kh)\rho(h)\,dk\,dh$ is left-invariant on G. It will be denoted dg.

In the transformation of the integrals (40) and (41) in Section 10C the following family of diffeomorphisms T^h_x of

$K \sim M$ enter. We assume throughout that G acts transitively on M, and note that this assumption, together with (44), imply the diffeomorphism $M \sim K$. Indeed, if $x_0 \in M$ is suitable chosen such that M is the isotropy group of x_0, then it follows that $k \to \gamma(k,x_0)$ is a diffeomorphism of K onto M. For fixed $h \in H$, $x \in K$, the transformation $T_x^h:K \to K$ is determined by $T_x^h(y) = z$ where z is the unique solution in K to the equation $\gamma(zh,x) = y$. Note that H is also the isotropy group for the point e in K under the induced action of G on K. (This action will also be denoted by γ.) Hence, the action of G induces a transitive and effective action of K on itself. For the pair of points x_1 and y, where $x_1 = \gamma(h,x)$ the equation $\gamma(z,x_1) = y$ therefore has a unique solution z in K. It can be verified in a general setting that T_x^h is a diffeomorphism, and that $h,y \to T_x^h(y)$ is C^∞. It is certainly clear, and easy to check directly, in the examples. In the present discussion the C^∞ properties will simply be assumed.

If $\mu = dk$ is the left-Haar measure, we therefore have $(T_x^h)^*(\mu) = J(h,x,\cdot)\mu$ where $J(h,x,\cdot)$ is the Jacobian of T_x^h calculated with respect to the left-invariant Riemannian connection associated with μ. This means that J is calculated in local coordinates of points y, z which are related through right group multiplication by $y^{-1}z$. Since for $f \in C_0(K)$ we have

$$\int f \, d(T^*\mu) = \int (f \circ T^{-1}) d\mu \tag{51}$$

by [Hℓ, p. 363), the integrals (40) and (41) in Section 10C may be transformed as follows: $u \in C_0(K)$,

$$\begin{aligned}\int_G A(g)u(\gamma(g,x))dg &= \iint A(kh)u(\gamma(kh,x))\rho(h)dk\, dh \\ &= \iint A(T_x^h(y)h)u(y)J(h,x,y)\rho(h)dy\, dh \\ &= \iint A(T_x^h(y)h)J(h,x,y)\rho(h)dh\, u(y)\, dy\end{aligned}$$

where the second step makes use of the transformation formula (51) for $f \circ T^{-1}(z) = A(zh)u(\gamma(zh,x))$ and $f(y) = A(T(y)h)u(y)$, and the third step is a simple application of Fubini's theorem, once the suitable decay properties of the kernel function $A(T_x^h(y)h)$ in the h-variable, which are needed, are verified. Indeed, it is shown in Section 10C and [Jo 2] that the G-kernel A is of rapid decay at ∞, and the terms $T_x^h(y)h$ and

$J(h,x,y)$ can readily be calculated in the two examples (cf. (47) and (48)), and hence the existence of the double integrals can be verified directly. For each G-kernel A, the corresponding operator B, given by $Bu(x) = \int_G A(g)u(\gamma(g,x))dg$, is in fact an integral operator on K with kernel

$$E(x,y) = \int_H A(T_x^h(y)h)J(h,x,y)\rho(h)dh \tag{52}$$

and formulas (49) and (50) in Theorem 10.16 follow by direct substitution into (52).

Completion of the argument requires the calculation of the functions J and ρ in each of the two examples. In Example 1, we have $J = \rho \equiv 1$. For J this follows by direct calculation of the Jacobian of the transformation T_x^h given in (Riemannian) coordinates in (47). For ρ it follows by unimodularity of G and H. In Example 2, we have $\rho(h) = \det \mathrm{Ad}_H(h)/\det \mathrm{Ad}_G(h) = 1$. Indeed

$$\begin{pmatrix}1 & 0\\ 0 & h\end{pmatrix}\cdot\begin{pmatrix}a & b\\ 0 & 1\end{pmatrix}\cdot\begin{pmatrix}1 & 0\\ 0 & h\end{pmatrix}^{-1} = \begin{pmatrix}a & bh^{-1}\\ 0 & 1\end{pmatrix}.$$

Since the differential of $h\cdot h^{-1}$ is independent of x, both numerator and denominator is equal to h^{-1}. The calculation of J may be performed by use of the modular function Δ_K of K. Since right Haar measure is $da\,db/a$, $\Delta_K(a,b) = 1/a$. Formula (48) then shows that $J(h,x,y) = J(h;(a_0,b_0);(a,b)) = \Delta_K(h\ x\ h^{-1}) = 1/a_0$.

PART VI

DERIVATIONS ON MODULES OF UNBOUNDED OPERATORS WITH APPLICATIONS TO PARTIAL DIFFERENTIAL OPERATORS ON RIEMANN SURFACES

Come, listen, my men,
  while I tell you again,
The five unmistakable marks
By which you may know,
  wheresoever you go,
The warranted genuine Snarks.

The Hunting of the Snarks
Fit the Second
LEWIS CARROLL

Otherwise, during my student years I was not much interested in higher mathematics. Mistakenly it seemed to me that it was a field with so many branches that one could easily squander one's entire energy in one of its remote parts. Furthermore, I thought in my innocence that it sufficed for a physicist to understand elementary mathematical concepts clearly in order to be able to apply them, and that the remainder consisted of subtleties that were fruitless for a physicist. (Einstein, 1956, p. 11.)

On 30 August 1925 Niels Bohr delivered an address on 'Atomic Theory and Mechanics' to the sixth Scandinavian Mathematical Congress in Copenhagen:

It will interest mathematical circles that the mathematical instruments created by the higher algebra play an essential part in the rational formulation of the new quantum mechanics. Thus the general proofs of the conservation theorems in Heisenberg's theory carried out by Born and Jordan are based on the use of the theory of matrices, which go back to Cayley and were developed by Hermite. It is to be hoped that a new era of mutual stimulation of mechanics and mathematics has commenced. To the physicist it will seem first deplorable that in atomic problems we have apparently met with such a limitation of our usual means of visualisation. This regret will, however, have to give way to thankfulness that mathematics, in this field too, presents us with the tools to prepare the way for further progress. (Bohr, 1925, p. 852.)

INTRODUCTION TO PART VI

For the present chapter we shall need the wider generality of modules of unbounded operators.

The operator modules are defined in the preliminary section below, and the main theorem 11.4 is proved in Section 11A. The two application sections 11B and 11C treat different algebraic settings corresponding to pairs of commuting operators; respectively nilpotent Heisenberg-Weyl relations.

In Chapter 9 we worked with Lie algebras of unbounded operators $\mathfrak{L}$, and with individual integrability assumptions on the elements (alias, operators) A in $\mathfrak{L}$. A classical example of Nelson [Nℓ 1] demonstrates that $\mathfrak{L}$ may <u>not</u> be exponentiable, even though the individual operators A in $\mathfrak{L}$ are pregenerators for strongly continuous one-parameter groups of bounded operators (C_o one-parameter exponentials).

In Theorems 9.1 and 9.2, we work with the abstract algebraic and analytic setting of a real finite-dimensional Lie sub-algebra $\mathfrak{L}$ of the endomorphism algebra End(D), for some fixed dense linear subspace D of a Banach space E. Existence of a Lie generating subset $S \subset \mathfrak{L}$ is assumed, as well as one-parameter C_o exponentials $\{V(t,A) : t \in \mathbb{R}\}$ for each $A \in S$. The requirement is that the infinitesimal generator for V(t,A) coincides with $\bar{A}$, and in addition that each A in S satisfies one of two regularity-compatibility conditions vis à vis the other operators in S. To formulate the conclusion of the theorems we let G be a simply-connected Lie group with Lie algebra isomorphic to $\mathfrak{L}$ as a real Lie algebra (cf. Ado's theorem). Then the exponentiability conclusion in Theorems 9.1 and 9.2 yields the existence of a C_o representation V of G in E such that the derived Lie algebra of V is identical to $\mathfrak{L}$, now regarded as an operator Lie algebra on E. (We recall that the infinitesimal operators dV(X) are defined on the Gårding space of V. The identity dV(X) = A, for some A in $\mathfrak{L}$, is understood here in the sense of closures, $dV(X)^{-} = \bar{A}$, and so the Gårding space may, in fact, differ from the initially given common core domain D for $\mathfrak{L}$.)

Nelson's example has, over the years, inspired several papers: [Pw 1,2], [Jo 7], [Fu 2], as well as the present part, (which, in time precedes both [Jo 7], and [Fu 2], by two years). The different authors point out a variety of interesting connections between, on the one hand, abstract operator theory, and, on the other, density questions for partial differential operators on the Riemann surfaces occurring in Nelson's example. We shall focus here on two cases, the 2-dimensional Abelian Lie algebra, and the 3-dimensional Heisenberg algebra, which both allow natural Lie generating two-element subsets S. Inspired by a paper of Kato [Kt 1] we consider the operator $T = (I-\bar{A})(I-\bar{B})$. First we show, in the abstract, that the integrated commutation relations for A and B follow from various (equivalent) core-conditions for the operator T. We then proceed with a number of examples involving partial differential operators on Riemann surfaces where the core-condition is not satisfied. In particular, we consider in Example 11.1 a certain compact surface with mixed boundary condition for $A = \partial/\partial x$ and $B = \partial/\partial y$: In each variable we mix periodic conditions, and phase-inversion. (This work was initiated by Powers in [Pw 1]).

An integral formula of an algebraic nature is available for the example, and for similar ones, but with different geometric constants. We show that the basic properties of the example reflect themselves through the geometry of the surface. The integral formula is a useful tool at the different steps of the analysis. Let V be the surface described above, and $E = L^p(V)$ the corresponding L^p-space with respect to 2-dimensional Lebesgue measure. Our mixed boundary conditions define a dense subspace $D \subset E$, and D serves as a common core domain for $A, B \in \mathbf{A}(D)$, with $[A,B] = 0$ on D. Nonetheless, the space $D_{\lambda\mu} = (\lambda-\bar{A})(\mu-\bar{B})D$ has positive co-dimension in E (co-dimension 3, in fact) for $(\lambda,\mu) \in \rho(\bar{A}) \times \rho(\bar{B})$, and strong commutativity fails. Indeed, when $p = 2$, we check that the algebra generated by $\bar{A},\bar{B}$ is irreducible.

Fuglede [Fu 2] has just shown that the domain space D <u>is</u> a core domain for the product operator $\bar{A}\,\bar{B}$ in the log z-Riemann surface version of Nelson's example. (In fact, Fuglede considered the operator $S = p(iA,iB)$ on L^2 for real polynomials of degree 2, and established the equivalence between essential selfadjointness on D of S, on the one hand, and non-ellipticity of p, on the other.) The core property carries over to our example 11.10, and, in fact, the calculations are considerably easier here because of the simplifications in the geometry, and the integral formulas.

Chapter 11

RIGOROUS ANALYSIS OF SOME COMMUTATOR IDENTITIES FOR PHYSICAL OBSERVABLES

This chapter serves as the primary "applications" sequel to Part III. Here, we connect the general "graph-density" sufficient condition for rigorous commutation theory with several equivalent conditions introduced by Kato [Kt 1] in his discussion of the canonical commutation relations. We indicate that generalizations of Kato's conditions can be applied to a number of other commutation-theoretic matters that play an important role in mathematical physics: strong commutativity, dynamics of the harmonic oscillator, Cartan subalgebra methods (alias "raising and lowering operators") in the study of the rotation group and Lorentz groups for two and three space dimensions, etc.

In addition to their physical relevance, the results and examples in this chapter serve to illustrate a number of other purely mathematical aspects of the theory that are at present understood only for these selected low-dimensional ad-orbits and Lie algebras. First, we exhibit commutation relations connecting the resolvents $(\mu - A)^{-1}$ and $(\lambda - B)^{-1}$ of two operators; these relations turn out to imply that the operators, their (semi-) group exponentials and their spectral projections also satisfy appropriate commutation relations. Second, commutation relations between resolvents and (group) exponentials are used to derive results on spectral structure and uniform multiplicity. Third, we take a few tentative steps toward direct use of graph-density data for operators (observables) A,B to obtain commutation relations for the resolvents and exponentials of polynomials in these operators; some of these identities exhibit the sort of "infinite-dimensional ad-orbit" behavior discussed in Sections 2D and 3D. In these three respects, the present chapter serves to indicate the principal open problems, and the potential directions for further development, of the purely infinitesimal spectral-theoretic commutation theory of Part III. It is for these mathematical reasons that we carry the generality of (1) Banach spaces rather than Hilbert spaces, and (2) semigroup-pregenerators rather than essentially self- (or skew-) adjoint operators, despite the lack of convincing physical applications for much of this generality.

The discussion of these matters in this chapter is organized as follows. In Section 11A, we consider operators $A \in \mathbf{A}(D)$ and finite-dimensional complex subspaces $M \subset \mathbf{A}(D)$ (for D a dense C^∞ domain in a Banach space E), with M invariant under ad A as in Chapter 5-7. But here we impose the additional condition that M be dominated by some closable $C \in M$: for each $B \in M$ there exist bounds a,b such that $\|Bu\| \leq a\|Cu\| + b\|u\|$. This has the effect of ensuring that the Banach space $(D_1, \|\cdot\|_1)$ of Chapters 5-7 reduces to $D(\bar{C})$ when the latter carries the topologically equivalent graph-norm for $\bar{C}: \|u\|_C = \|u\| + \|Cu\|$. For the case where $\mu \in \rho(C)$ exists, this enables us to reduce the graph-density (GD) condition for $\lambda \in \mathbb{C}$ of Chapters 5-7 ($D_\lambda = (\lambda - A)D$ is $\|\cdot\|_1$-dense in D) to the generalized Kato condition: that $D_{\mu\lambda} = (\mu - C)(\lambda - A)D$ be dense in E. For a special case that arises repeatedly in the present chapter and in Chapter 12, $\mu - C$ maps D onto D (equivalently, $R(\mu,C) = (\mu - \bar{C})^{-1}$ maps D into D), so that this reduction enables us to obtain (GD) automatically where A is a (semi-) group pregenerator on D.

The first set of applications for this machinery is supplied in Section 11B, where we treat commuting operators $[A,B] = 0$, with $M = \mathbb{C}B$ and ad $A = 0$. This results in an analysis of "strong commutativity" (commutation of spectral projections) for the special case where A and B are essentially self-adjoint. We also study a variant of the counter-examples of Nelson [Nℓ 1] and Powers [Pw 1], exhibiting commuting group pregenerators A,B whose groups and resolvents fail to commute, and explaining (by direct computation of annihilating functionals) precisely why the appropriate Kato and (GD) conditions fail. (A flaw in Powers' example is also pointed out and repaired.)

Then in Section 11C, we look at nilpotent commutation phenomena of step 1, motivated in large part by the canonical commutation relations $i(PQ - QP) = 1$. Our Kato condition/(GD) analysis there reduces in essence to Kato's earlier work [Kt 1]. However, our methods apply directly and usefully to $[A,B] = Z$ for any bounded Z commuting with A and B ($M = \mathbb{C}B + \mathbb{C}Z$, $C = B$). Useful commutation relations are obtained which connect resolvents, operational images, and semigroups for A and B with each other; these include the historically important Weyl (semi-)group commutation relations. If A and B are essentially skew-adjoint, these results can be extended to unbounded Z, showing that Kato's conditions can be sufficient in the absence of domination (e.g. the "regular" representation of the Heisenberg Lie algebra in $L^2(\mathbb{R}^3)$ or the Bony example in Chapter 10.)

(It is possible to adapt the machinery of Section 11A to the solvable commutation relation $[A,B] = \alpha B$ for $0 \neq \alpha \in \mathbb{C}$ and $M = \mathbb{C}B$, $C = B$. Again, tractable commutation relations are obtained for bounded operator pairs consisting of resolvents, operational images, or (semi-)groups. For essentially self adjoint A,B with $\alpha \in \mathbb{R}$, the operator ad B induces shifts (by distance α) in the spectral projections for A, creating effects that are familiar for the harmonic oscillator and for group representations where B is a "creation/annihilation" or "ladder" operator. Details are omitted.)

The machinery also applies to the "abstract" harmonic oscillator on an arbitrary Banach space with $A = 1/2(P^2 + Q^2)$ $(i[P,Q] = 1)$, and B chosen from among P, Q, iP, iQ, P^2 and Q^2, obtaining commutation relations for the holomorphic semigroup $\{V(\zeta,A): \mathrm{Re}(\zeta) > 0\}$ and its possible boundary group $\{V(t,iA): t \in \mathbb{R}\}$ for P,Q and A essentially self-adjoint. (Here every M contains A, which dominates.) One obtains sufficient conditions that $\mathfrak{L}$ = real span $\{iP^2, iQ^2, [P^2,Q^2]\}$ exponentiates to a representation of the two-fold metaplectic covering group of $SL(2,\mathbb{R})$, in terms of Kato conditions for A, etc.

While omitting details, we note that the techniques also yield known results on the irreducible, self-adjoint representations of the canonical commutation relations in terms of "spectral shift" ad A-eigenoperators for $A = 1/2(P^2 + Q^2)$ acting in the sequence space $\ell^2(\omega)$, where $\omega = \{0,1,2,...\}$. Then the exponentiation theory for the "almost-diagonal" Heisenberg matrix representation of $i(PQ - QP) = 1$ on the various $\ell^p(\omega)$ results using $M = \mathbb{C}P + \mathbb{C}Q + \mathbb{C}P^2 + \mathbb{C}Q^2 + \mathbb{C}A$, with $A \in M$ as the dominating operator, assumed to have nonempty resolvent set.

Finally, we take the opportunity to briefly indicate that the dominated-M and Kato-condition machinery can be brought into play in the commutation theory for standard basis operators in the (complexifications of) various "infinitesimal symmetry" operator Lie algebras of physical interest: the "angular momentum" Lie algebra $o(3)$ for the rotation group and its central dominating Casimir "square of the total angular momentum", the Lorentz Lie algebra $so(2,1) \simeq s\ell(2,\mathbb{R})$ in two space dimensions and its three-dimensional analog $so(3,1) \cong s\ell(2,\mathbb{C})$, and the inhomogeneous Lorentz (Poincaré) Lie algebra for two and three space-dimensions.

11A. Variations upon the graph-density and Kato conditions

As indicated above, the purpose of this section is to show that if an ad A-invariant $M \subset \mathbf{A}(D)$ is dominated by a suitable $C \in M$, then the usual graph-density sufficient condition for commutation relations can be replaced by other conditions that are more familiar or easily checked. First, we dispense with some routine preliminaries connecting $\|u\|_C = \max\{\|\bar{C}u\|, \|u\|\}$ on $D(\bar{C})$ with the M-graph norm on $D: \|u\|_1 = \max\{\|B_i u\|: 0 \leq i \leq d\}$ (where $B_o = 1$ and $B_1 \ldots B_d$ form a complex basis for M).

11.1. Lemma

Suppose the closable operator $C \in M$ dominates M. Then the C-graph norm $\|\cdot\|_C$ is topologically equivalent to $\|\cdot\|_1$ on D and $(D(\bar{C}), \|\cdot\|_C)$ is topologically isomorphic to the completion $(D_1, \|\cdot\|_1^{\sim})$ of $(D, \|\cdot\|_1)$.

Remark: If M has a closable basis $\mathbf{B} = \{B_1, \ldots, B_d\}$, then this reduces to the claim that $D(\bar{C})$ is identical to the $\|\cdot\|_1$-closure of D in $\cap\{D(\bar{B}_i): 1 \leq i \leq d\} = E(\mathbf{B})$. In general, (as in Lemma 6.3), it is to be interpreted as saying that the mapping J_1 sending the abstract completion $(D_1, \|\cdot\|_1)$ into E is a topological isomorphism of D_1 onto $D(\bar{C})$. We separate the discussion of these matters in the proof, since in most applications below a closable basis is available. (The relevant portions of Chapters 5 and 6 have been collected in the Appendix to Part VI at the end of the chapter.)

Proof: From Section 5B, we trivially obtain $\|Cu\| \leq |C| \, \|u\|_1$ on D_1 so that $\|u\|_C \leq \max(1, |C|)\|u\|_1$. But if

$$\|B_i u\| \leq a_i\|Cu\| + b_i\|u\| \text{ for } 0 \leq i \leq d \text{ (with } a_o = 0,\ b_o = 1)$$

then $\|u\|_1 \leq \max\{a_i + b_i\}\|u\|_C$, so the norms are equivalent as claimed.

In the closable basis case, it is immediate that $D_1 \subset D(\bar{C})$ by the discussion in Section 5B, since C acts boundedly from $(D_1, \|\cdot\|_1)$ to E. But any $u \in D(\bar{C})$ is a limit $u = \lim u_n$ of $u_n \in D$ with $Cu_n \to \bar{C}u$, whence $\{u_n\}$ is $\|\cdot\|_C$ (alias $\|\cdot\|_1$)-convergent to u, and $u \in D_1$. Thus, the estimates used above extend to $D(\bar{C})$ (i.e., $\|\bar{B}_i u\| \leq a\|\bar{C}u\| + b\|u\|$) to extend the norm equivalence.

In general, the same argument works, except for bookkeeping. First, C extends to a bounded operator $C^{(1)}$ from $(D_1, \|\cdot\|_1)$ to $(E, \|\cdot\|)$ with $C^{(1)} = \bar{C}J_1$ and J_1 maps D_1 into $D(\bar{C})$ by Lemma 6.3(b) and (c). It is easy to check that J_1 is bijective, since u_n is $\|\cdot\|_1$-Cauchy (or convergent) iff u_n and $\bar{C}u_n$ are $\|\cdot\|$-Cauchy (or convergent). For example, $J_1 u = 0$ implies that for $\|u_n - u\|_1 \to 0$, $u_n \to 0$ in E and $\bar{C}u_n$ converges, so $\bar{C}u_n \to 0$ and $u = 0$. Surjectivity follows similarly.

Our basic result then combines Proposition 5.5 and Lemma 5.6 to provide several new equivalents to the sufficient condition (GD) in the two main commutation theorems 5.4 and 6.4.

11.2. Proposition

Suppose that A, M, and C are as above. Let $\lambda \in \rho(\bar{A})$ and $\mu \in \rho(\bar{C})$. Then the following are equivalent.

(a) (Kato's condition) the space $D_{\mu\lambda} = (\mu - C)(\lambda - A)D$ is dense in E.

(b) The space $D_\lambda = (\lambda - A)D$ is a core for $\bar{C}$.

(c) (GD) The space D_λ is $\|\cdot\|_1$-(alias $\|\cdot\|_C$-)dense in D.

(d) The resolvent $R(\lambda, A) = (\lambda - \bar{A})^{-1}$ leaves $D(\bar{C})$ invariant, and supplies a two-sided inverse for the $\|\cdot\|_1$-closure $\lambda - \bar{A}_1$ of $\lambda - A$ on $D(\bar{C})$.

Proof: The equivalence of (a), (b) and (c) is just Lemma 5.6 applied to the domain D_λ, using $\|\cdot\|_C$ as the norm. Then 11.1 translates the Lemma 5.6 version of (c) into (GD), whence 5.5 establishes the equivalence of (c) and (d) modulo the identification of $D(\bar{C})$ with D_1 in Lemma 11.1.

Remark: Notice that since conditions (b)-(d) are independent of $\mu \in \rho(C)$, (a) holds for one such μ iff it holds for all of these. It is also clear from (d) that (a) holds for all λ in an open subset of $\rho(A)$: $\rho(\bar{A}_1)$ in fact. Detailed analysis of the specific commutation relations (as in Lemma 5.4) is generally needed in order to determine how this open λ-set is related to $\rho(A)$. However, the following purely analytic result is useful both in weakening (GD) still further and in establishing "reversed commutation relations" (Sections 11B-C below).

11.3. Lemma

Let $A, C \in \mathbf{A}(D)$ be closable.

(a) Suppose that $[A,C]$ is dominated by C (i.e., relatively C-bounded), so that $P(\mu) = [A,C]R(\mu,C)$ is bounded for each $\mu \in \rho(C)$ (e.g. suppose C dominates an ad A-invariant M with $C \in M$). Suppose further that $D_\mu = (\mu - C)D$ is a core for A. Then $D_{\mu\lambda} = (\mu - C)(\lambda - A)D$ is dense in E for all $\lambda \in \rho(A)$ which satisfy $\|R(\lambda,A)\| \, \|P(\mu)\| < 1$.

(b) In particular, if $[A,C]$ is bounded on D, then one of $D_{\mu\lambda} = (\mu - C)(\lambda - A)D$ and $D_{\lambda\mu} = (\lambda - A)(\mu - C)D$ is dense for some pair $(\lambda,\mu) \in \rho(A) \times \rho(C)$ such that $\|R(\lambda,A)\| \, \|R(\mu,C)\| \, \|[A,C]\| < 1$ iff both $D_{\lambda\mu}$ and $D_{\mu\lambda}$ are dense for all such pairs.

Proof: (a) First, recall that if $\|[A,C]u\| \le c\|Cu\| + d\|u\|$ for all $u \in D$, we have for $v \in D$ (dense) that $\|[A,C]R(\mu,C)v\| \le c\|CR(\mu,C)v\| + d\|R(\mu,C)v\|$ and since $CR(\mu,C) = \mu R(\mu,C) - 1$ the two operators on the right are bounded and $\|P(\mu)v\| \le M\|v\|$ for suitable M. By density, $P(\mu)$ is bounded.

Now, for $u \in D$, we put $v = (\mu - C)u \in D_\mu$ to get
$(\mu - C)(\lambda - A)u = (\mu - C)(\lambda - A)R(\mu,C)v$
$= (\lambda - A)v - [A,C]R(\mu,C)v = (\lambda - (A + P(\mu))v$. Hence, density of $D_{\mu\lambda}$ is equivalent to density of $(\lambda - (A + P(\mu))D_\mu$. Since $(\lambda - A)D_\mu$ is assumed dense (D_μ is a core for $\bar{A}$, using Lemma 5.6), this reduces the proof to a perturbation argument. In fact, $A + P(\mu)$ is closable with $(A + P(\mu))^- = \bar{A} + P(\mu)$, and if $\|R(\mu,A)\| < \|P(\mu)\|^{-1}$ then $\|R(\lambda,A)P(\mu)\| < 1$, so a standard Neumann Series argument shows that $\lambda \in \rho(\bar{A} + P(\mu))$. Since $\bar{A} + P(\mu)$ is the closure of its restriction to D_μ, this means by Lemma 5.6 that $(\lambda - (A + P(\mu))D_\mu$ is dense as desired. (Here $D(\bar{A} + P(\mu)) = D(\bar{A})$, whence for $u \in D(\bar{A} + P(\mu))$ there is a sequence $u_n \in D_\mu$ with $u_n \to u$ and $Au_n \to \bar{A}u$, whence $(A + P(\mu))u_n \to (\bar{A} + P(\mu))u$).

In (b), if $D_{\lambda\mu}$ is dense then D_μ is a core for A by 5.6, while $[A,C]$ is trivially relatively bounded and $\|P(\mu)\| \le \|[A,C]\| \, \|R(\mu,C)\|$ so $\|R(\lambda,A)\| \le \|[A,C]\|^{-1}\|R(\mu,C)\|^{-1} \le \|P(\mu)\|^{-1}$ under the conditions in (b) and (a) applies to yield density of $D_{\mu\lambda}$ for all such (λ,μ). But A and C enter symmetrically into (b), so that by interchanging A with C and λ with μ, the remaining implication follows.

The following result exhibits a common instance for dominated M in which $D_{\lambda\mu}$ is trivially dense.

11.4. Theorem

Let $A \in \mathbf{A}(D)$ be a (semi-)group pregenerator, and M a finite-dimensional complex ad A-invariant subspace of $\mathbf{A}(D)$ that is dominated by $C \in M$. Suppose there exists a point $\mu \in \rho(\bar{C})$ such that $D_\mu = (\mu - C)D$ is a core for A. (For example, suppose $D_\mu = D$.) Then for all complex values of λ with $|\mathrm{Re}(\lambda)|$ sufficiently large (with $\mathrm{Re}(\lambda) > 0$ for the semigroup case) the range $D_\lambda = (\lambda - A)D$ is $\|\cdot\|_1$-dense in D.

Proof: Here (c) in Lemma 11.3(a) applies for $|\mathrm{Re}(\lambda)|$ large, since $[A,C] \in M$ is dominated by C, D_μ is a core by hypothesis, and $\|R(\lambda,A)\| \le M(|\mathrm{Re}(\lambda)| - \omega)^{-1}$ by a standard estimate from semigroup theory. Hence Kato's condition follows for $|\mathrm{Re}(\lambda)|$ large, ($\mathrm{Re}(\lambda) > 0$ for semigroups) by Lemma 11.3, and Proposition 11.2 converts that condition into (GD) as claimed.

Remark: Notice that $D_\mu = D$ is equivalent to $R(\mu,C)D \subset D$, a "domain-invariance" condition for resolvents.

Turning finally to generalizations, we mention that these involve relaxing the condition $C \in M$ to permit the dominating operator to be "stronger" than those in M. For E a Hilbert space, the "second-order" operator $C = \Sigma\{B_i^*B_i : 1 \le i \le d\}$ on D is a typical choice. If the B_i are skew-symmetric, $C = -\Sigma\{B_i^2 : 1 \le i \le d\} = -\Delta$, where Δ is the "Laplacian" for M. (Recall that for each B_j, $\|B_ju\|^2 \le \Sigma\|B_ju\|^2 = \Sigma(B_ju,B_ju)$ $= |(u, (\Sigma B_j^*B_j)u)| \le \|u\| \, \|Cu\| \le \|u\|_C^2$ so that $\|B_ju\| \le \|u\|_C \le \|Cu\| + \|u\|$. Such estimates then extend easily to a general linear combination $B = \Sigma a_jB_j \in M$.) If $B_1,\ldots,B_d$ are a basis for a real Lie algebra $\mathcal{L}$ of skew-symmetric operators, with $M = \mathcal{L} \oplus i\mathcal{L}$, then Lemma 6.1 in Nelson [Nℓ 1] shows that $D(\bar{C}) = D_2$ and $\|\cdot\|_C = \|\cdot\|_2$, the second-order Sobolev space of C^2-vectors for M. But it is obviously true that if D_λ in 11.2 is dense in D with respect to a possibly stronger norm $\|\cdot\|_C$, it must be dense with respect to $\|\cdot\|_1$, whence the only change in the formulation of 11.2 that is needed in order to remove the $C \in M$ condition is to make conditions (a) and (b) sufficient for (c) and (d) but possibly no longer necessary.

Carrying through to Lemma 11.3 and Theorem 11.4, we observe that domination by Δ often still occurs. Indeed, for $M = \mathfrak{L} \oplus i\mathfrak{L}$ and $C = -\Delta$ as above, Nelson's calculations [Nℓ, 6.1] indicate that when $A \in M$ as well, $[A,C] = -[A,\Delta]$ is a complex linear combination of elements in $E_2(\mathfrak{L})$ (second-order polynomials in elements of M) and is also dominated by $C = -\Delta$. Moreover, the condition that $(1 - \Delta)D$ be a core for all $A \in M$ is also natural in that setting. In fact, our relatively innocent-looking assumption that $\rho(\bar{C})$ is nonempty reduces to the full strength of the single hypothesis in Nelson's famous Exponentiation Theorem 5 [Nℓ 1]: Δ is essentially self-adjoint. When this occurs, Nelson's preliminary development in Lemma 5.2 [Nℓ 1] shows that $\mathfrak{L}$ (hence E) can be extended to the domain $D_\infty(\bar{\Delta})$ of C^∞-vectors for $\bar{\Delta}$ to supply a new set of operators in $\mathbf{A}(D_\infty(\bar{\Delta}))$ such that $R(\mu,\Delta)$ leaves the new domain invariant and (c) = GD holds in Lemma 11.3. In fact, although it is only of academic interest, we note that our present methods supply a second proof for a weakened form of Nelson's exponentiation theorem: if $\mathfrak{L}$ is a Lie algebra of <u>essentially skew-adjoint</u> operators and Δ is essentially self-adjoint, then the extension of Theorem 11.4 to operators like Δ shows (via passage to $D_\infty(\bar{\Delta})$) that every $A \in \mathfrak{L}$ satisfies the (GD) hypothesis of our "C^∞-vectors" exponentiation theorem 9.2. It follows that $\mathfrak{L}$ exponentiates. (Nelson's analytic domination technique shows that the condition of essential skew-adjointness of the skew-symmetric $A \in \mathfrak{L}$ is redundant; this we cannot yet replace by a "C^∞-vectors" argument.)

Perhaps more to the point, we observe that for the quasi-simple skew-symmetric representations $\mathfrak{L}$ of semisimple Lie algebras on K-finite vector domains D, the trivially essentially self-adjoint "Casimir Laplacian" $\Delta = Q + 2\Delta_K = q + 2\Delta_K$ has resolvents which are necessarily reduced by the K-isotypic components, whence $R(\lambda,\Delta)(D) \subset D$ for these domains. Here again, then, essential skew-adjointness of the Lie algebra elements becomes the only issue. For $\mathfrak{L} \cong s\ell(2,\mathbb{R})$, we show in Chapter 12 how to solve that "pregenerator problem" directly (Lemma 12.1), using difference-equation methods. In fact, for the Lie algebras $\mathfrak{L}$ of Chapter 12 (and for various others for other rank-1 groups at least) the dominator can be chosen to be a regular element in the compact subalgebra and Theorem 11.4 will apply as given, even to many non-unitary representations, without recourse to "analytic domination". Further development of this C^1-domination technique, in the nonunitary theory of group representations, seems highly desirable. We regard it as one of the more promising open problem areas. The applications of the techniques below should suffice to illustrate the possibilities.

11B. Various forms of strong commutativity

The simplest possible commutation relation is, of course, $[A,B] = 0$. Our methods permit a simple, direct analysis of situations in which this commutativity implies stronger forms of commutativity, up to and including the usual sense of strong commutativity for (essentially-) self-adjoint operators on Hilbert spaces: commutativity of the projections in the spectral resolutions for A and B. We recall that the example of Nelson [Nℓ 1] shows that strong commutativity of essentially self-adjoint operators need not follow from their "infinitesimal" commutativity. A related example of independent interest is constructed below, borrowing ideas of Powers [Pw] to illustrate the sort of failure of condition (GD) that occurs in situations of this sort.

Before proceeding to the main result of the section, we note informally that if $[A,B] = 0$ in $\mathbf{A}(D)$, then $D_{\mu\lambda} = (\mu-B)(\lambda-A)D = (\lambda-A)(\mu-B)D = D_{\lambda\mu}$, so that A and B enter symmetrically into any Kato conditions which may hold, and the slight asymmetry in the following result is only apparent.

11.5. Theorem

Let $A,B \in \mathbf{A}(D)$ be commuting closable operators in a Banach or Hilbert space as appropriate below. Suppose that for some $\lambda \in \rho(A)$ and $\mu \in \rho(B)$, $D_{\mu\lambda} = (\mu-B)(\lambda-A)D$ is dense in E. Then the following hold.

(a) For every $\lambda \in \rho(A)$ and $\mu \in \rho(B)$, the Kato condition holds and (GD) holds for A and $M = \mathbb{C}B$. Thus every resolvent $R(\lambda,A)$ and every operational image $\varphi(A)$ (φ holomorphic in a neighborhood of $\sigma^*(A) = \sigma(A) \cup \{\infty : \text{if } A \text{ is unbounded}\}$) leaves $D(\bar{B})$ invariant, and $\bar{B}$ commutes with these: $\bar{B}R(\lambda,A) \supset R(\lambda,A)\bar{B}$, $\bar{B}\varphi(A) \supset \varphi(A)\bar{B}$.

(b) Indeed, all resolvents $R(\lambda,A)$, and operational images $\varphi(A)$, commute with all resolvents $R(\mu,B)$, and operational images $\psi(B)$.

(c) If $\bar{A}$ generates a (semi-)group $\{V(s,A): s \in [0,\infty) \text{ or } \mathbb{R}\}$ then every $V(s,A)$ leaves $D(\bar{B})$ invariant and commutes with $\bar{B}$, every $R(\mu,B)$, and every $\psi(B)$. Similarly, if $\bar{B}$ generates $\{V(t,B): t \in [0,\infty) \text{ or } \mathbb{R}\}$, then $V(s,A)$ commutes with every $V(t,B)$, as does $R(\lambda,A)$ and $\varphi(A)$.

(d) If A is essentially self-adjoint, then its spectral

projections leave $D(\bar{B})$ invariant, and these commute as above with $\bar{B}$, $R(\mu,B)$, $\psi(B)$ and $V(t,B)$ for all appropriate t. If B is also essentially self-adjoint, the projections for A commute with those for B: A and B commute strongly.

Remark: In (c) and (d), the results concerning $\varphi(A)$ and $\psi(B)$ extend to the most general operational calculi appropriate for the operators A and B under discussion. We omit details.

Proof: (a) Since $D_{\mu\lambda} = D_{\lambda\mu}$ as observed above, we first use the reduction to "$D_\lambda = (\lambda-A)D$ is a core for B" as in Lemma 11.3 to conclude that $D_{\mu\lambda} = D_{\lambda\mu}$ is dense for all $\mu \in \rho(B)$. We then note that $D_{\lambda\mu}$ being dense for all μ yields that D_μ is a core for A and $D_{\lambda\mu} = D_{\mu\lambda}$ is dense for all $(\lambda,\mu) \in \rho(A) \times \rho(B)$. Consequently, Proposition 11.2 ensures that D_λ is $\|\cdot\|_1$-dense for all $\lambda \in \rho(A)$ when $M = \mathbb{C}B$, $\|u\|_1 = \max\{\|u\|,\|\bar{B}u\|\}$ on $D_1 = D(\bar{B})$. The rest of (a) is then immediate from Theorems 5.1-5.4 and Theorem 5.8, once we observe that ad A is the zero operator on M so that $\sigma(A,M) = \sigma(A)$.

(b) Commutativity of resolvents is an easy consequence of Corollary 5.3(b), since

$$[R(\lambda,A),R(\mu,B)] \supseteq R(\mu,B)R(\lambda,A)[A,B]R(\lambda,A)R(\mu,B) = 0$$

on a dense domain on the right, whence the bounded left-hand side vanishes. But then $R(\lambda,A)R(\mu,B) = R(\mu,B)R(\lambda,A)$ implies that the integrands in the contour integral formulae for $\varphi(A)R(\mu,B)$ and $R(\mu,B)\varphi(A)$ agree when the resolvent factors are taken inside the operator-norm-convergent integrals $\int_\Gamma \varphi(\lambda)R(\lambda,A)d\lambda$, whence $\varphi(A)R(\mu,B) = R(\mu,B)\varphi(A)$. The same argument yields commutativity of $\psi(B)$ with $\varphi(A)$ and with $R(\lambda,A)$.

(c) Here, Theorem 6.1 yields semigroup or group invariance of $D(\bar{B}) = D_1$ and $\bar{B}V(s,A) \,\dot{\supseteq}\, B^{(1)}V_1(s,A) = V(s,A)B^{(1)} = V(s,A)\bar{B}$. Hence for $\mu \in \rho(B)$, $(\mu - \bar{B})V(s,A) \supseteq V(s,A)(\mu - \bar{B})$ and since $R(\mu,B)E = D(\bar{B})$, $(\mu-\bar{B})V(s,A)R(\mu,B) = V(s,A)(\mu-\bar{B})R(\mu,B)=V(s,A)$, whence $V(s,A)R(\mu,B) = R(\mu,B)(\mu-\bar{B})V(s,A)R(\mu,B) = R(\mu,B)V(s,A)$. The quickest route to $V(s,A)V(t,B) = V(t,B)V(s,A)$ is by uniqueness of Laplace transforms in t, since the transform of the left-hand side, $V(s,A)R(\mu,B)$, agrees with that of the right, $R(\mu,B)V(s,A)$. (Notice that $V(s,A)R(\mu,B) = R(\mu,B)V(s,A)$

can be obtained from commutation of resolvents by a similar argument, while a direct proof of semigroup commutation is quickly obtained from $V(t,B) = \lim\{[n/tR(n/t,B)]^n : n \to \infty\}$ since the finite powers clearly commute with $V(s,A)$. Reversing the argument to obtain $R(\lambda,A)$ as a Laplace transform of $V(s,A)$, we derive the commutation of $R(\lambda,A)$ with $V(t,B)$ (or, using uniqueness, this follows from commutation for resolvents) whence commutation of $\varphi(A)$ with $V(t,B)$ follows as usual by contour integration.

(d) Perhaps the simplest argument uses the Cayley transform $C(A) = -(i-\bar{A})(i+\bar{A})^{-1} = (i-\bar{A})R(-i,A) = -2iR(-i,A) - 1$, which commutes with $\bar{B}$ and all bounded operators derived from it by (a)-(c) above. Thus a standard result on bounded normal operators such as unitary $C(A)$ shows that the spectral projections for $C(A)$ (on the circle $T = \{z \in \mathbb{C} : |z| = 1\}$) commute with all of these bounded operators; these projections correspond with those for A via the mapping $\lambda \to (\lambda-i)(\lambda+i)^{-1}$ from $\mathbb{R}$ to $\mathbb{C}$ in the usual way. But if a projection P commutes with $R(\lambda,A)$, it preserves $D(\bar{A}) = R(\lambda,A)E$ and commutes with $\bar{A}$ there ($P(\lambda-\bar{A})u = P(\lambda-\bar{A})R(\lambda,A)v = Pv = (\lambda-\bar{A})R(\lambda,A)Pv = (\lambda-\bar{A})PR(\lambda,A)v = (\lambda-\bar{A})Pu$ for $u = R(\lambda,A)v \in D(\bar{A})$, whence $P\bar{A}u = \bar{A}Pu$ follows by cancellation and sign-reversal.) When $\bar{B}$ is also self-adjoint, the same argument shows that $C(B)$ and its projections commute with all bounded operators associated with A, including P, and the commutation lifts back to $C(B)$ as before. E.O.P.

The following result is then a partial converse for Theorem 11.5, as well as a corollary for some of its conclusions.

11.6. Proposition

(a) If the resolvents of closed operators $\bar{A}$ and $\bar{B}$, $R(\lambda,A)$ and $R(\mu,B)$, respectively, commute for some $\lambda \in \rho(A)$ and $\mu \in \rho(B)$, then $D = D(\bar{A}\bar{B}) \cap D(\bar{B}\bar{A})$ is a dense core domain for $\bar{A}$ and $\bar{B}$ such that $D_{\mu\lambda} = (\mu-\bar{B})(\lambda-\bar{A})D = D_{\lambda\mu}$ is dense in E, and for $u \in D$, $\bar{A}\bar{B}u = \bar{B}\bar{A}u$. Indeed, if $\lambda = 0$ and $\mu = 0$ in the above, then $D = D(\bar{A}\bar{B}) = D(\bar{B}\bar{A})$ and $\bar{A}\bar{B} = \bar{B}\bar{A}$ in the sense of unbounded operators.

(b) If $\bar{A}$ and $\bar{B}$ generate commuting (semi-)groups, then $D = E_\infty(\{\bar{A},\bar{B}\})$ is a dense core domain for $\bar{A}$ and $\bar{B}$ such that $\bar{A}$ and $\bar{B}$ restrict to commuting members of $\mathbf{A}(D)$ and $D_{\mu\lambda}$ is dense in E.

Proof: (a) We show that $D(\bar{A}\bar{B}) \cap D(\bar{B}\bar{A}) = R(\lambda,A)R(\mu,B)E = R(\mu,B)R(\lambda,A)E$. (If $\lambda = 0 = \mu$ then $D(\bar{A}\bar{B}) = R(0,B)R(0,A)E = R(0,A)R(0,B)E = D(\bar{B}\bar{A})$ as well.) Indeed, if $u = R(\lambda,A)R(\mu,B)v$ for $v \in E$, then $u \in D(\bar{A})$ and $\bar{A}u = (\lambda R(\lambda,A) - 1)R(\mu,B)v = R(\mu,B)(\lambda R(\lambda,A) - 1)v \in D(\bar{B})$ by commutativity, so $u \in D(\bar{B}\bar{A})$ and $\bar{B}\bar{A}u = (\mu R(\mu,B) - 1)(\lambda R(\lambda,A) - 1)v$. But by commutativity $u = R(\mu,B)R(\lambda,A)v \in D(\bar{B})$ as well, and a similar argument shows that $u \in D(\bar{A}\bar{B})$ with $\bar{A}\bar{B}u = (\lambda R(\lambda,A)-1)(\mu R(\mu,B)-1)v$, so by commutativity, $\bar{A}\bar{B}u = \bar{B}\bar{A}u$. Thus $D = R(\mu,B)R(\lambda,A)E = R(\lambda,A)R(\mu,B)E \subset D(\bar{A}\bar{B}) \cap D(\bar{B}\bar{A})$ as well. But if $u \in D(\bar{A}\bar{B}) \cap D(\bar{B}\bar{A}) \subset D(\bar{A}) \cap D(\bar{B})$ then $\bar{B}u$, μu and $(\mu-\bar{B})u \in D(\bar{A})$, whence $(\mu-\bar{B})u = R(\lambda,A)v$ for some $v \in E$ and $u = R(\mu,B)R(\lambda,A)v \in D$, reversing the inclusion.

Now, $R(\mu,B)E = D(\bar{B}) = (\lambda-\bar{A})D$ is dense in E, and $D_{\mu\lambda} = (\mu-\bar{B})(\lambda-\bar{A})D = E$. Also, $D = R(\lambda,A)D(\bar{B})$ is the image of a dense subspace, hence is a core for $\lambda - \bar{A}$ and must be dense. (For any $u = R(\lambda,A)v \in D(\lambda-\bar{A})$ there exists a sequence $v_n \in D(\bar{B})$ with $v_n \to v$ and $R(\lambda,A)v_n \to R(\lambda,A)v = u$.)

For (b), the obvious adaptation of Gårding's smoothing argument (Appendix B) shows that $D = E_\infty(\{\bar{A},\bar{B}\})$ is dense: for $\varphi \in C_c^\infty((0,\infty) \times (0,\infty))$ we put

$V(\varphi)u = \int_0^\infty\int_0^\infty \varphi(s,t)V(s,A)V(t,B)u\,ds\,dt$, noting that (by commutativity) we have $\bar{A}V(\varphi)u = -V(\partial\varphi/\partial s)u$, $\bar{B}V(\varphi)u = -V(\partial\varphi/\partial t)u$, etc, so the span of the $V(\varphi)u$ is a dense (by approximate identities) C^∞ domain for $\{\bar{A},\bar{B}\}$. But since $R(\lambda,A)D(\bar{B}) = R(\lambda,A)R(\mu,B)E = R(\mu,B)R(\lambda,A)E \subset D(\bar{B})$ and $R(\mu,B)D(\bar{A}) \subset D(\bar{A})$ similarly, an easy induction shows that $R(\lambda,A)E_n(\{\bar{A},\bar{B}\}) \subset E_n(\{\bar{A},\bar{B}\})$ for all n, and $R(\lambda,A)D \subset D$, while $R(\mu,B)D \subset D$ as well. Hence all of $R(\lambda,A)$, $R(\mu,B)$, $\lambda - \bar{A}$ and $\mu - \bar{B}$ map D bijectively onto itself, so D is a core for $\bar{A}$ and $\bar{B}$ by Lemma 5.6 and $D_{\mu\lambda} = D$ is dense, etc. E.O.P.

Turning next to negative results, where commuting group-pregenerators $A,B \in \mathbf{A}(D)$ violate both the hypotheses and the conclusions of Theorem 11.5, we first review the three simplest examples which exhibit these properties. In all three, D is a suitable space of compactly-supported C^∞ functions on a two-dimensional manifold M which can be charted so that $A = \partial/\partial x$ and $B = \partial/\partial y$ in local coordinates, while the Banach space $E = L^p(M)$, $1 \le p < \infty$, formed with respect to the two-form that agrees locally with Lebesgue measure in the chart-domains. The examples differ only in the way in which the topology in M or boundary conditions on the functions in D are used to produce noncommutativity of the isometry groups

V(s,A) and V(t,B) generated by A and B. We confine ourselves at first to quick qualitative descriptions.

11.7. Example

(Reed and Simon [RS 1]). Let M be the Riemann surface for $z^{\frac{1}{2}}$. For $u \in C_c^\infty(M)$, the group-actions V(s,A)u and V(t,B)u are well-defined as "translation in the x (respectively, y) direction on the same sheet" for small s,t. (If the support does not lie above the real or imaginary axis, the remark holds for all s,t.) Easy applications of Lemma 11.1, or the pregenerator theorem (cf. Example 11.10) show that $\bar{A}$ and $\bar{B}$ generate strongly continuous isometry groups. Global non-commutativity arises from the fact that for large s,t the functions V(s,A)V(t,B)u and V(t,B)V(s,A)u are supported on different sheets. (Taking the cut along Im (z) $\geq$ 0, u supported near (1,-1) on the bottom sheet, and s = 2 = -t, V(s,A)V(t,B)u lies over (-1,1) on the top sheet, while V(t,B)V(s,A)u does not cross the cut and lies over (-1,1) on the bottom sheet.) Poulson described a variant of this example to one of the present authors (Jo) about 1971, taking M to be the Riemann surface for log(z): the covering space of that for $z^{\frac{1}{2}}$.

11.8. Example (The subscript c refers to vanishing conditions in a neighborhood of deleted points.)
(Simplification of Nelson [Nℓ 1])

(a) Geometric Model. Let V be the L-shaped region in $\mathbb{R}^2$ formed from the union of the unit squares with lower left-hand corners at (0,0), (0,1) and (1,0). Let V_0 be V with all integer-coordinate points deleted, and let M be the manifold obtained by identifying "opposite sides". (The line joining (0,0) to (0,1) is identified with that joining (2,0) to (2,1) while that joining (0,1) to (0,2) is identified with that joining (1,1) to (1,2) etc.) If u is supported near $(\frac{1}{2},\frac{1}{2})$ then V(1,A)V(1,B)u is supported near $(\frac{1}{2},\frac{3}{2})$ while V(1,B)V(1,A)u is supported near $(\frac{3}{2},\frac{1}{2})$ instead.

(b) Analytic Model. Instead, take M = V_0 and D the subspace of $C_c^\infty(V_0)$ all of whose partial derivatives satisfy "periodic boundary conditions" in x and y of the form u(y,0) = u(y,2), 0 < y < 1; u(y,0) = u(y,1), 1 < y < 2; u(0,x) = u(2,x), 0 < x < 1; and u(0,x) = u(1,x), 1 < x < 2. Here, the groups V(s,A) and V(t,B) are easily seen to act by "translation mod 1" or "translation mod 2" as appropriate: [V(s,A)u](x,y) = u(x + s[mod 2],y) for 0 < y < 1, but

$[V(s,A)u](x,y) = u(x + s(\text{mod } 1),y)$ for $1 < y < 2$, etc. As in (a), a function supported near $(\frac{1}{2},\frac{1}{2})$ ends up near $(\frac{3}{2},\frac{1}{2})$ or $(\frac{1}{2},\frac{3}{2})$ depending upon the order in which $V(1,A)$ and $V(1,B)$ are applied.

In both (a) and (b), either Lemma 11.1 or the pregenerator theorem can be used to check that $\bar{A}$ and $\bar{B}$ generate isometry groups. Nelson's original example [Nℓ 1] differs only by using a more complicated initial region V: the square annulus obtained by deleting the open unit square with lower left-hand corner (1,1) from the closed 3×3 square cornered at (0,0). Nelson's formulation is geometrical, but admits a (lengthy!) analytic version as in (b).

11.9. Example (Repair of Powers [Pw 1].)

Let M be the 2×2 square cornered at (0,0) with all corners and the two midpoints (1,0) and (0,1) deleted. Let D be the space of functions in $C_c^\infty(M)$ all of whose partial derivatives satisfy periodic boundary conditions in x of the form $u(0,y) = u(2,y)$, $0 < y < 2$; and in y to the left of the midline $u(x,0) = u(x,2)$, $0 < x < 1$; but "phase-inverted" to the right $u(x,0) = -u(x,2)$, $1 < x < 2$.

It turns out (as explained in detail for the comparable example 11.10 below) that $A = \partial/\partial x$ and $B = \partial/\partial y$ have closures which generate isometry groups: $V(s,A)$ acts by translation mod 2, while $V(t,B)$ acts by translation mod 2 modified by a $(-1)^{[t/2]}$ phase factor when $1 < x < 2$. The noncommutativity that this induces is most easily seen for $s = 1$, $t = 2$ and u supported close to $(\frac{1}{2},\frac{1}{2})$: then $V(1,A)V(2,B)u = -V(2,B)V(1,A)u$. Powers has pointed out to us (personal communication) that a "subrepresentation" of the Reed-Simon example 12.7 works in much the same way: take D to be the subspace of $C_c^\infty(M)$ (M the Riemann surface for $z^{\frac{1}{2}}$) consisting of functions that are odd under exchange of sheets (rotation by 2π). We note that the difficulty with Power's original example [Pw 1] lies in the fact that $B \notin \mathbf{A}(D)$ for the domain D described there, while the C^∞ domain $E_\infty(\{\bar{A},\bar{B}\})$ (Section 7A) for generators $\bar{A},\bar{B}$ of the unitary groups discussed there is not a core for $\bar{B}$. (Powers imposes the variable phase-shift condition $u(0,y) = \exp(iy)u(2,y)$, $0 < y < 2$; in order for $\partial u/\partial y$ to satisfy the same condition it is necessary that $u(0,y) = 0 = u(2,y)$. We omit further details.)

In all of these examples, it is possible to exhibit the failure of both the hypotheses (GD and the Kato condition) and the conclusions of Theorem 11.5. However, the details can most economically be presented for the following symmetrized variant of Example 11.9. We also record a verification that this example exhibits the same extreme noncommutativity (trivial commutant) correctly established by Powers for the unitary groups in his example 5.3.

Remark: We proceed with an explicit calculation of the "deficiency vectors" associated to the 2-dimensional non-commutative examples. The analytic details of the calculation are of special interest in relation to the geometrical exponentiation problem which, in turn, is central to the commutation theory developed in Chapters 3, and 5 through 9. In the abstract setting, non-exponentiability arises at two levels:
(i) The individual elements in $\mathfrak{L}$ may not generate C_0 one-parameter groups.
(ii) If the one-parameter groups exist, they may not satisfy the relevant commutation relations.

The second type of obstruction has received relatively little attention in the literature, and is yet quite poorly understood, in spite of its interesting connections to geometry of boundary manifolds, and to the boundary conditions for partial differential operators [Jo 7]. In fact, the two types of deficiency obstructions occur naturally in a more general setting in von Neumann algebras (as in [Pw 1,2], and [Jo 6]). The present "deficiency vector" calculations for $\partial/\partial x$ and $\partial/\partial y$, although elementary, seem to be the first of their kind. Recent calculations (1980) of Fuglede [Fu 2] are related, but go in a different direction.

In our examples, the analysis of obstruction (ii) is isolated, since the individual one-parameter exponentials $V(s,A)$, and $V(t,B)$, can easily be shown to exist as C_0 groups on the Banach spaces $L^p(M)$ under consideration. For $p = 2$, Nelson's theorem [Nℓ 1, theorem 5] shows that the non-commutativity is reflected in the deficiency of the Laplace operator $\Delta = (\partial/\partial x)^2 + (\partial/\partial y)^2$ on $D = C_c^\infty(M)$, where M is one of the manifolds in Examples 11.7 through 11.9, or the Riemann surface of $z^{\frac{1}{2}}$, or log z. However, the present Theorem 11.5 shows that the non-commutativity information (ii) is contained in the operator $T_{\lambda,\mu} = (\lambda-A)(\mu-B)$ for $(\lambda,\mu) \in \rho(A) \times \rho(B)$, f.ex., non-purely imaginary values. Moreover, this works for $p \neq 2$, as well, and we calculate the solutions $v_{\lambda\mu} \in L^p(M)^* = L^q(M)$ to the deficiency equation, $\langle T_{\lambda,\mu} u, v_{\lambda\mu} \rangle = 0$ for all $u \in D$. In contrast to the situation for Nelson's Laplace operator it turns out that the geometry of M reflects itself explicitly, and in a surprisingly simple manner, in the solutions $v_{\lambda\mu}$ to the deficiency equations, and that the geometric symmetries in M serve to facilitate the partial integrations on the different surfaces involved. Hence, Theorem 11.5 yields new information on the different types of non-commutativity, which are reflected in the geometry of M.

11.10. Example

Let $E = L^p(V)$ $1 \le p < \infty$ with respect to Lebesgue measure in the planar set $V \subset \mathbb{R}^2$ consisting of the 2×2 square $\{(x,y): 0 \le x,y \le 2\}$ with all 8 integer-coordinate boundary points deleted. Let $A = \partial/\partial x$ and $B = \partial/\partial y$ on the domain D of all C^∞ functions in V that vanish in a neighborhood of the deleted points and satisfy the boundary conditions

(a) $[(\partial/\partial x)^n u](0,y) = [(\partial/\partial x)^n u](2,y)$, $0 \le y \le 1$ (periodic),
$[(\partial/\partial x)^n u](0,y) = -[(\partial/\partial x)^n u](2,y)$, $1 \le y \le 2$ (phase-inverted);

(b) $[(\partial/\partial y)^n u](x,0) = [(\partial/\partial y)^n u](x,2)$, $0 \le x \le 1$ (periodic),
$[(\partial/\partial y)^n u](x,0) = -[(\partial/\partial y)^n u](x,2)$, $1 \le x \le 2$ (phase-inverted).

Since functions in D vanish in a neighborhood of the (deleted) boundary points where the formal boundary conditions (a)-(b) exhibit jumps, it is not difficult to verify that both A and B leave D invariant: $A,B \in \mathbf{A}(D)$. It is also evident that A and B commute: $[A,B] = 0$.

We show by direct calculation that these operators violate both the hypotheses and the conclusions of Theorem 11.5 and Proposition 11.6, but that they are pregenerators with large resolvent sets. We begin with the Kato and graph-density hypotheses that are basic to the section. In Kato's condition, failure of $D_{\mu\lambda} = (\mu-B)(\lambda-A)D$ to be dense is equivalent to the existence of $v_{\mu\lambda} \in L^q(V)$ $(p^{-1} + q^{-1} = 1)$ orthogonal to $D_{\mu\lambda}$. By the density of D this must require that $[(\mu-B)(\lambda-A)]^* v_{\mu\lambda} = 0$ in $L^q(V)$, where $(\lambda-A)^*(\mu-B)^* v_{\mu\lambda} = (\bar{\lambda}+\partial/\partial x)(\bar{\mu}+\partial/\partial y)v_{\lambda\mu}$ is interpreted in the sense of distributions. Despite the peculiarities of the boundary conditions (a)-(b), it turns out that the most naive solution $v_{\mu\lambda}(x,y) = \exp[-(\bar{\lambda}x+\bar{\mu}y)]$ actually works. (Notice that the two formal solutions $\exp(-\bar{\lambda}x)$ and $\exp(-\bar{\mu}y)$ to $(\bar{\lambda}+\partial/\partial x)v = 0$ and $(\bar{\mu}+\partial/\partial y)v = 0$ do not annihilate $(\lambda-A)D$ and $(\mu-B)D$, respectively.) Abbreviating $w = (\lambda-A)u$, so $((\mu-B)(\lambda-A)u,v_{\mu\lambda}) = ((\mu-B)w,v_{\mu\lambda})$, we obtain by integration-by-parts that

$$\begin{aligned}&((\mu-B)w,v_{\mu\lambda})\\ &= \int_0^2 dx e^{-\lambda x}\{\mu\int_0^2 e^{-\mu y}w(x,y)dy - e^{-\mu y}w(x,y)\Big|_{y=0}^{y=2} - \mu\int_0^2 e^{-\mu y}w(x,y)dy\}\\ &= \int_0^2 e^{-\lambda x}(e^{-2\mu}H_1(x)+1)w(x,0)dx \qquad (1)\end{aligned}$$

where H_1 is the Heaviside unit function at $x = 1$: $H_1(x) = -1, 0 \le x \le 1$; $H_1(x) = 1$, $1 \le x \le 2$. Breaking (1) into two integrals and applying it to $w = (\lambda-A)u$, we get on $0 \le x \le 1$ that

$$\begin{aligned}&\int_0^1 e^{-\lambda x}(-e^{-2\mu}+1)(\lambda u - \frac{\partial u}{\partial x})dx\\ &= (-e^{-2\mu}+1)\{\lambda\int_0^1 e^{-\lambda x}u(x,0)dx - e^{-\lambda x}u(x,0)\Big|_{x=0}^{x=1} - \lambda\int_0^1 e^{-\lambda x}u(x,0)dx\}\\ &= 0 \qquad (2)\end{aligned}$$

since $u(x,0)$ vanishes when x approaches 0 or 1. Similarly, the corresponding integral for $1 \le x \le 2$ from (1) vanishes on $w = (\lambda-A)u$, so that $((\mu-B)(\lambda-A)u,v_{\mu\lambda}) = 0$ for all $u \in D$ as claimed. Inspection of the argument in fact reveals that if $\chi_1^{\pm}(x,y) = \frac{1}{2}(1 \pm H_1(x))$ and $\chi_2^{\pm}(x,y) = \frac{1}{2}(1 \pm H_1(y))$, then $v_{\mu\lambda}^{\pm,i}(x,y) = \exp(-\bar{\lambda}x+\bar{\mu}y))\chi_i^{\pm}(x,y)$ annihilates $D_{\mu\lambda}$ for $i = 1,2$. Hence all functions in the three-dimensional complex span of the $v_{\mu\lambda}^{\pm,i}$, $i=1,2$ annihilate $D_{\mu\lambda}$. (In fact, $D_{\mu\lambda}^{\perp} = \{v \in L^q\colon (w,v)=0$ for all $w \in D_{\mu\lambda}\}$ can be shown to be precisely the span of these four functions. We omit details.) Clearly $v_{\mu\lambda}$ is just the sum of $v_{\mu\lambda}^{+1}$ and $v_{\mu\lambda}^{-1}$, or of $v_{\mu\lambda}^{+2}$ and $v_{\mu\lambda}^{-2}$, whence any one of these is in the span of the other three.

The functions $v_{\mu\lambda}^{\frac{1}{2},i}$ can then be used to manufacture functionals that are orthogonal to $D_\lambda = (\lambda-A)D$ and bounded with respect to graph-norms involving B. (That is, these are functionals in $D_1(B)^*$ that turns out to be in E^*.) In fact, for $w \in D$ we can define

$$f_\lambda^{\pm,i}(w) = ((\mu-B)w, v_{\mu\lambda}^{\pm,i}) \tag{3}$$

so that when $w = (\lambda-A)u$ for $u \in D$ we have $f_\lambda^{\pm,i}((\lambda-A)u) = 0$ by the above. But

$$|f_\lambda^{\pm,i}(w)| \le \|f_{\mu\lambda}^{\pm,i}\|_q \|(\mu-B)w\|_p \le \|v_{\mu\lambda}^{\pm,i}\|_q \{|\mu|\,\|w\|_p + \|Bw\|_p\},$$

whence $f_\lambda^{\pm,i}$ is $\|\cdot\|_B$-bounded for $\|u\|_B = \max\{\|u\|, \|Bu\|\}$ and is certainly bounded for the $M = \mathbb{C}A + \mathbb{C}B$ graph-norm. If $\mathrm{Re}(\mu) \ne 0$ so that $e^{2\mu} \ne \pm 1$, then specializations of the calculation (1) to the four $v_{\mu\lambda}^{ij}$ cases reveals both that $f_\lambda^{\pm,i} \ne 0$, and that its (implicit) μ-dependence lies only in a scalar multiplier. (In fact, $(\lambda-A)D = D_\lambda$ has complex codimension exactly three in $D_1(B)$.) Indeed, the $f_\lambda^{\pm,i}$ are represented by measures (qua derivatives of L^q functions) supported on the lines $y = 0$, $y = 1$, and $y = 2$.

Similar calculations show that the $g_\mu^{\pm,i}(w) = ((\lambda-A)w, v_{\mu\lambda}^{\pm,i})$ are $\|\cdot\|_A$-bounded nonzero functionals orthogonal to $D_\mu = (\mu-B)D$ accounting for the failure of (GD) for B with respect to $M = \mathbb{C}A$ or $M = \mathbb{C}A + \mathbb{C}B$. Details are left to the reader.

Information about the commutation properties of groups and resolvents is derived from the analysis of the isometry group defined for $u \in L^p(V)$, and almost all x,y, by the recipe

$$[V(s,A)u](x,y) = \begin{cases} u(x+s-2[(x+s)/2],y), & 0 \le y \le 1, \\ (-1)^{[(x+s)/2]}u(x+s-2[(x+s)/2],y), & 1 \le y \le 2, \end{cases} \tag{4}$$

where $[x]$ denotes the greatest integer $\le x$. For each $u \in D$, there exists $\varepsilon(u) > 0$ such that $|s| < \varepsilon(u)$ implies that $V(s,A)u \in D$ as well, since small translates of u still vanish near the deleted boundary points in V, while the boundary conditions for D imply a smooth fit at the jumps of $[(x+s)/2]$ in x and are preserved at the boundaries by the group recipe. In fact by calculus, $d/ds\, V(s,A)u|_{s=0} = Au$ for such u, first computed with respect to the sup-norm and by inference with respect to the L^p norm. Moreover, if we let D_A be the (dense) subspace of D consisting of functions which vanish in a neighborhood of the lines $y = 0$, $y = 1$ and $y = 2$, then it is not difficult to check that $V(s,A)D_A \subset D_A$ for all $s \in \mathbb{R}$, and for $u \in D_A$, $d/ds\, V(s,A)u = AV(s,A)u = V(s,A)Au$. Hence, since each $V(s,A)$ is clearly a L^p-isometry, $\{V(s,A)\colon s \in \mathbb{R}\}$ extends from D_A to $L^p(V)$ as a strongly-continuous isometry group, and the pregenerator theorem (Appendix B) shows that D_A is a core for its generator, whence the generator is the closure of $A = \partial/\partial x$ on D_A (hence on D). This justifies our notation "$V(s,A)$". The group $V(t,B)$ is described by an analogous formula involving translation of y by $t(\mathrm{mod}\ 2)$ essentially as in (4). (The domain D_B of functions vanishing in a neighborhood of $x = 0$, $x = 1$ and $x = 2$ replaces D_A in the pregenerator argument.)

The extreme degree of noncommutative behavior exhibited by these two period-4 one-parameter groups is best analyzed in terms of qualitative information concerning the action of the multiplicative commutator $C(s,t) = V(s,A)V(t,B)V(-s,A)V(-t,B)$ on function-values. In fact, $[C(s,t)u](x,y) = \pm u(x,y)$ for almost all (with respect to Lebesgue measure) $(x,y) \in V$, with the sign chosen as follows: if the open rectangle with corners (x,y), $(x - s,y)$, $(x - s,y - t)$ and $(x,y - t)$ contains exactly one deleted midpoint, the sign is negative while otherwise it is taken positive (unless a deleted point lies in the boundary of the rectangle – we can safely ignore this measure-0 set of (x,y) for each (s,t).) This recipe is justified by a routine analysis of the following cases: no midpoint or corner in the interior, one midpoint only, one corner only, one midpoint and one corner, and two midpoints plus the adjacent corner. It is geometrically clear that since the rectangle involved is no larger than V, this exhausts all cases except those "ignored" above. If there are no deleted points inside the rectangle, it either lies entirely in V or crosses one boundary twice between the same two deleted points. The operators $V(\pm s,A)$ and $V(\pm t,B)$ (acting mod 2) in effect carry (x,y) around this rectangle, inducing either no sign-reversals in the function-value or exactly two at the boundary-crossings. If a single midpoint is interior, then one boundary line is crossed twice on opposite sides of that midpoint, once with sign-reversal and once without, while no other boundary is crossed, so the sign reverses. If a single corner is interior, then two boundaries are crossed, each once in a sign-change region and once not, so that two sign reversals occur and cancel. The cases with two and three deleted points interior are combinations of these, with the obvious sign effects.

(Visualization is aided by drawing the basic 2×2 square and all of its contiguous neighbors and assigning + and - signs to all unit boundary segments according to the boundary conditions/group formulae, consistent mod (2,2).)

This analysis clearly shows that $C(s,t) \neq 0$ for $0 < |s|,|t| \leq 2$, since a function $u \neq 0$ with small support close to a suitably chosen midpoint has its entire support carried around only that midpoint, whence $C(s,t)u = -u \neq u$. Since both groups have period 4, this establishes, by the group property, that $C(s,t)$ vanishes only when one of the groups passes through the unit.

An easy contradiction argument then shows that no resolvent $R(\lambda_o,A)$ could commute with any resolvent $R(\mu_o,B)$, or any group operator $V(t,B)$ when $t \neq 4n$. For if $R(\lambda_o,A)$ commutes with any bounded T, the Neumann expansion in powers of $R(\lambda_o,A)$ shows that $R(\lambda,A)$ commutes with T for all nearby λ, hence for all λ in the same component of $\rho(A)$. (It is easy to see that $\sigma(A) \subset i\mathbb{R}$ since $\|V(s,A)\|_p \equiv 1$, so that $\rho(A)$ has at most two half-plane components. A more detailed analysis using periodicity shows that $\sigma(A)$ consists of isolated points: $\rho(A)$ has a single component.) But if $R(\lambda,A)$ commutes with T for all λ in a half-plane, the argument for 11.5(c) shows that every $V(s,A)$ must commute with T too. If $T = R(\mu_o,B)$ a repetition of the argument shows that every $V(t,B)$ commutes with every $V(s,A)$. In either case, a contradiction results.

For $p = 2$ (where $\bar{A}$ and $\bar{B}$ are skew-adjoint) an adaptation of an argument of Powers ([Pw 1, Lemma 5.3] shows that no nontrivial spectral projection for A or B can commute with the group (or resolvents) for the other. In fact, the von Neumann algebra **A** generated by the $V(s,A)$ and $V(t,B)$ has trivial commutant and acts irreducibly upon $L^2(V)$: no non-scalar operator whatsoever commutes with both groups.

In order to check these claims, we define projections $P_A(\lambda)$ and $P_B(\mu)$ by $P_A(\lambda) = \frac{1}{2}(1-C(\lambda,1))$, $P_B(\mu) = \frac{1}{2}(1-C(1,\mu))$. It is not difficult to verify that if $1 < y < 2$,

$$[P_A(\lambda)u](x,y) = \begin{matrix} u(x,y), & 0 < x < \lambda; \\ 0 \quad , & \lambda < x < 2; \end{matrix} \tag{5}$$

while when $0 < y < 1$,

$$[P_A(\lambda)u](x,y) = \begin{matrix} u(x,y), & \lambda-1 < x < 1, \text{ or } \lambda+1 < x < 2; \\ 0 \quad , & 1 < x < \lambda+1. \end{matrix} \tag{6}$$

Thus if we put $S_1 = \int_0^2 e^{i\lambda\pi} dP_A(\lambda)$, it is apparent that

$$[S_1u](x,y) = \begin{cases} e^{i(x+1)\pi}u(x,y) = -e^{i\pi x}u(x,y), \; 0 < y < 1; \\ e^{ix\pi}u(x,y) = e^{i\pi x}u(x,y), \; 1 < y < 2; \end{cases} \tag{7}$$

or more briefly,

$$[S_1u](x,y) = H_1(y)e^{i\pi x}u(x,y) \quad \text{a.e. } [x,y]. \tag{8}$$

Similarly, if $S_2 = \int_0^2 e^{i\pi\mu}dP_B(\mu)$,

$$[S_2u](x,y) = H_1(x)e^{i\pi y}u(x,y) \text{ a.e. } [x,y] \tag{9}$$

It is evident by construction that S_1 and S_2 are commuting unitary operators on $L^2(V)$, so that their polynomial algebra $\mathbf{A}_o$, and its weak-operator closure $\mathbf{A}_1$, are self-adjoint commutative algebras. Following Lemma 5.4 in Powers [Pw 1], we show that the identity function $u_1(x,y) \equiv 1$ is a cyclic vector for $\mathbf{A}_1$, whence $\mathbf{A}_1$ is maximal commutative and its commutant $\mathbf{A}_1'$ consists of multiplication operators. Then, again as in Powers' argument, we observe that any operator in the commutant of the von Neumann algebra $\mathbf{A} \supset \mathbf{A}_1$ generated by $\{V(s,A): s \in \mathbb{R}\} \cup \{V(t,B): t \in \mathbb{R}\}$ is a multiplication operator which commutes with "small translations" and must be constant (i.e. scalar). We refer the reader to Powers' discussion for details, turning to the much more delicate issue of cyclicity.

The idea for showing that u_1 is cyclic is simply (as in [Pw 1]) to show that the modified "trigonometric polynomials with jumps" in the span $\mathbf{A}_o u_1$ of the $e_{mn}(x,y)=[S_1^m S_2^n u_1](x,y)$ $= H_1^n(x)H_1^m(y)e^{i(mx+ny)\pi}$ are dense in $L^2(V)$. In fact, let W be the locally compact space obtained by deleting the Lebesgue-measure-0 set consisting of the lines $x = 0$, $x = 1$, $x = 2$, $y = 0$, $y = 1$ and $y = 2$; notice that $L^2(V)$, and $L^2(W)$, can be identified. Then the restrictions of the functions in $\mathbf{A}_o u_1$ to W (defined for <u>all</u> $(x,y) \in W$ by the formula above) supply a self-adjoint algebra $\mathbf{A}$ of bounded continuous functions on W that separates points and contains the constants. (Notice that $e_{11}(x,y) = H_1(x)e^{i\pi x}H_1(y)e^{i\pi y}$ separates all points except the pairs $(\frac{1}{2},\frac{1}{2}),(\frac{1}{2},\frac{3}{2}),(\frac{3}{2},\frac{1}{2})$ and $(\frac{3}{2},\frac{3}{2})$, while $e_{12}(x,y)$ $= e^{i\pi x}H_1(y)e^{2i\pi y}$ and $e_{21}(x,y) = H_1(x)e^{2i\pi x}e^{i\pi y}$ separate these as well. Details are left to the reader.) Extending $\mathbf{A}$ to the Stone-Čech compactification $\beta(W)$ and then to the compact quotient $\bar{W}$ of $\beta(W)$ obtained by identifying points in $\beta(W)-W$ that are not separated by $\mathbf{A}$, we obtain by the Stone-Weierstrass theorem that $\mathbf{A}$ is uniformly dense in $C(\bar{W})$. (In fact, $\bar{W}$ is nothing but W with two copies of each deleted line $x = 1$, $y = 1$ segment restored, along with one copy each of $x = 0$, $x = 2$, $y = 0$ and $y = 2$). Consequently, each compactly-supported continuous function on W is a uniform (and L^2) limit of functions in $\mathbf{A}$, and these in turn are dense in $L^2(W)$ (alias $L^2(V)$) so u_1 is indeed cyclic.

An integral formula for the calculation of $D_{\lambda\mu}^{\perp}$

The failure of Kato's condition and of GD in Examples 11.7-11.9 is verified by calculations that are quite close in spirit to those discussed above. In Example 11.8, $D_{\lambda\mu}$ has codimension 3: $\exp(-(\bar{\lambda}x + \bar{\mu}y))\chi \in D_{\lambda\mu}^{\perp}$ if χ is the characteristic function of any one of the unit sub-squares. (Nelson's original example gives codimension 5 when all identifications are taken in account.) The Reed-Simon example 11.7 turns out to be more complicated: Its manifold M has infinite measure so that care must be taken in writing down annihilating functionals for the $D_{\lambda\mu}$. Clearly $v_{\lambda\mu} = \exp(-\bar{\lambda}x-\bar{\mu}y)$ will be in L^q of exactly one of the four quadrants in $\mathbb{C} = \mathbb{R}^2$, depending upon the signs of $\mathrm{Re}(\lambda)$ and $\mathrm{Re}(\mu)$. Since the Riemann surface $z^{\frac{1}{2}}$ contains two copies of that quadrant, if χ is the characteristic function of either of these, $\exp(-(\bar{\lambda}x + \bar{\mu}y))\chi \in D_{\lambda\mu}^{\perp}$, and it can be shown that these functionals span $D_{\lambda\mu}^{\perp}$.

Instead of giving complete details of the dimension calculations for the spaces $D_{\lambda\mu}^{\perp}$ we include here an integral formula which lies at the basis of the dimension count in the different examples. Consider the rectangle

$$F = \{(x,y) \in \mathbb{R}^2 : \alpha \le x \le \alpha + \ell \ , \ \beta \le y \le \beta + m\}.$$

The following terminology is convenient for functions u, defined on F:

$$[u]_{x=\alpha}^{\alpha+\ell}(y) = u(\alpha+\ell,y) - u(\alpha,y) \ ,$$

$$[u]_{y=\beta}^{\beta+m}(x) = u(x,\beta+m) - u(x,\beta) \ ,$$

and finally

$[[u]]_{\alpha\beta} = u(\alpha+\ell,\beta+m) - u(\alpha+\ell,\beta) - u(\alpha,\beta+m) + u(\alpha,\beta).$

The partial derivatives are denoted by subscripts: $u_x = u/\partial x$, $u_{xy} = \partial^2 u/\partial x\partial y$, etc. The (distribution) solutions to the p.d.e., $v-v_x-v_y+v_{xy} = 0$, are given by the formula

$v(x,y) = f(x)e^{-y} + g(y)e^{-x}.$

Consider again the operators A,B in the Banach space $L^p(F)$ with dual L^q, $p^{-1} + q^{-1} = 1$. Let $v \in L^q$ be a distribution solution to $(I-\partial/\partial x)(I-\partial/\partial y)v = 0$. Then

$$B_F(u,v) = \iint_F (u-u_x-u_y+u_{xy})v\ dx\ dy$$

$$= -\int_\beta^{\beta+m}[u(v+v_y)]_{x=\alpha}^{\alpha+\ell}(y)dy - \int_\alpha^{\alpha+\ell}[u(v+v_x)]_{y=\beta}^{\beta+m}(x)dx+[[uv]]_{\alpha,\beta}.$$

Here, we have restricted attention to the case $\lambda = \mu = 1$, since the general case $(\lambda,\mu) \in \rho(A) \times \rho(B)$ is completely parallel. If, in Example 11.10, we apply this integral formula for $v = v_{\lambda\mu}$, or $v = v^{\pm,i}$, on the 2 × 2 square, or each of the four sub-rectangles, then it follows that span $\{v^{\pm,i} : i = 1,2\}$ is contained in $D_{\lambda\mu}^{\perp}$. Hence, dim $D_{\lambda\mu}^{\perp} \geq 3$. To see that the dimension is precisely 3, we note that every $v \in D_{\lambda\mu}^{\perp}$ is a (distribution) solution to $(\bar{\lambda} - \partial/\partial x)(\bar{\mu} -\partial/\partial y)v = 0$, in the open 2 × 2 square, as well as in each of the four open sub-rectangles. Using the integral formula on the identity

$$B_V^{\lambda,\mu}(u,v) = \Sigma_{\alpha=0}^{1}\Sigma_{\beta=0}^{1}\ B_{\alpha,\beta}^{\lambda,\mu}(u,v) = 0 \text{ for all } u \in D,$$

(where D is given in the beginning of Example 11.10), we then conclude that v necessarily belongs to span $\{v_{\lambda\mu}^{\pm,i}\}$.

In fact, if $v = f(x)\exp(-\bar{\mu}y)+g(y)\exp(-\bar{\lambda}x)$, then, by considering the boundary integrals, we arrive at the differential equations $\bar{\lambda}f+f' = 0$, resp., $\bar{\mu}g+g' = 0$. (Apply the identity $B_V^{\lambda,\mu}(u,v) = 0$, for carefully chosen u in D!)

A second interesting consequence of the integral formula is the fact that strong commutativity may hold in examples where $D_{\lambda,\mu}$ is not dense. Hence, the implication in Theorem 11.5 cannot be reversed. However, Proposition 11.6 does serve as a partial converse to Theorem 11.5.

Let F be the unit square with lower left corner at the origin (0,0), and let E be the Hilbert space $L^2(F)$. Let $A = \partial/\partial x$ and $B = \partial/\partial y$ on the domain $D \subset E$ of all C^∞ functions in F that vanish in a neighborhood of each of the four corners, and together with all derivatives satisfy periodic boundary conditions in both variables. Then A, B $\in$ **A**(D), and D is a core for each of the unitary group generators $\bar{A}, \bar{B}$. Finally, $\bar{A}, \bar{B}$ commute strongly, but the dimension of $D_{\lambda\mu}^{\perp}$ is one, as follows from a direct substitution into the integral formula with $v = \exp(-\bar{\lambda}x - \bar{\mu}y)$, and $(\lambda,\mu) \in \rho(A) \times \rho(B)$.

Indeed, the identity

$$B^{1,1}(u,v) = -\int_0^1 [u(\bar{v}+\bar{v}_x)]_{y=0}^1(x)dx - \int_0^1 [u(\bar{v}+\bar{v}_y)]_{x=0}^1(y)dy + [[u\bar{v}]]$$

$$= u(0,0)\ [[\bar{v}]]$$

is valid for all u that satisfy the periodic boundary conditions.

A final consequence of the integral formula is an interesting short proof of the known essential selfadjointness of the product operator AB on D, in Example 11.10 ($E = L^2(V)$.) While essential self-adjointness on D of AB, in Example 11.7, is due to Fuglede [Fu 2], and quite difficult, it is of some interest that our 11.10 version of Nelson's example yields a relatively quick proof of the essential selfadjointness. For u in the domain D in Example 11.10, (mixed boundary conditions on the 2 × 2-square V), and $v \in L^2(V)$, we have

$$\iint_V u_{xy}\ v\ dxdy$$

$$= -\Sigma\int_\alpha^{\alpha+1} [uv_x]_{y=\beta}^{\beta+1}(x)dx - \Sigma\int_\beta^{\beta+1} [uv_y]_{x=\alpha}^{\alpha+1}(y)dy + \iint_V uv_{xy}\ dxdy$$

as an identity in $u \in D$. (Recall that the [[uv]] term vanishes!) Hence, the deficiency equations for $v \in ((AB \pm iI)D)^{\perp}$, $v \in L^2(V)$, take a particularly simple form. We leave details to the reader.

11C. Nilpotent commutation relations of generalized Heisenberg-Weyl type

The discussion in this section serves three purposes. First, we show how the general machinery of Chapters 5-7 (based upon the graph-density condition (GD) as in Section 11A above) can be used to recover and generalize the work of Kato [Kt 1] on rigorous analysis of the canonical (or Heisenberg) commutation relation

$$PQ - QP = ci, \quad c \geq 0, \tag{10}$$

from quantum physics. (Here, P in that application is a self-adjoint momentum observable, while Q is a coordinate observable.) Classically, one sought to replace the formal identity (10) in unbounded operators by the more tractable Weyl commutation relations for their unitary group-exponentials

$$\exp(isP)\exp(itQ) = \exp(istc)\exp(itQ)\exp(isP) \tag{11_u}$$

Kato [Kt 1] pointed out that if $A = iP$, $B = iQ$ and $Z = -ic = [A,B]$ then density of $D_{\mu\lambda} = (\mu-B)(\lambda-A)D$ was sufficient to imply the reformulated Weyl identity

$$V(s,A)V(t,B) = \exp(-stZ)V(t,B)V(s,A) \tag{11}$$

even in the general case where A,B are merely semigroup generators. In the process, he obtained two other important commutation relations: one for resolvents

$$\begin{aligned}[R(\lambda,A), R(\mu,B)] &= ZR(\lambda,A)R(\mu,B)^2R(\lambda,A) \\ &= ZR(\mu,B)R(\lambda,A)^2R(\mu,B)\end{aligned} \tag{12}$$

and another connecting resolvents and semigroups

$$R(\mu,B)V(s,A) = V(s,A)R(\mu,B-sZ) = V(s,A)R(\mu-sZ,B) \tag{13}$$

for suitable values of λ,μ and s. Our development permits us to recover all three identities easily for $Z = [A,B]$ any bounded operator commuting with A and B (rather than a scalar) and to extend (12) and (13) to cases where B (and A in (12)) need not be semigroup generators. With more difficulty, even unbounded Z can be accommodated, yielding examples where Kato's condition is sufficient in the absence of the dominating operators for ad A-invariant M which are needed in Section 11A to obtain the graph-density condition.

Second, we examine the (basically known) ways in which identities like (11)-(13) can be used to establish restrictions upon the structure of the spectrum of $\sigma(B)$ if A is a group-generator, deriving the operational calculus shift-identity

$$\psi(B)V(s,A) = V(s,A)\psi_s(B); \quad \psi_s(\mu) = \psi(\mu+sz), \tag{14}$$

and the projection-intertwining identity

$$P_B(S)V(t,A) = V(t,A)P_B(S-tZ), \tag{15}$$

$P_B(S)$ = the projection supported on the Borel subset $S \subset \sigma^*(B)$ for B self-adjoint. (Think of $P_B(S)$ as $\chi_S(B)$ where χ_S is the characteristic function of S; the operational calculus having been extended by limits to bounded measurable functions.)

Third, examples are given that serve to illustrate interesting applications of our sufficient conditions of (12)-(15) where A and B are not group-generators. These also serve to refute possible conjectures on the extendability of spectral structure results to that setting. (Our examples take A and B to be linear combinations of d/dx and the multiplication operator ix in $L^p[a,b]$, $-\infty < a < b < \infty$ with various boundary conditions at a.) We also recall the example of Reed and Simon [RS] extending Nelson's counterexample of Section 11B to the present setting, with a sketch of the reasons why Kato's condition fails in this example.

As in Section 11B, the various results stated and proved below are encyclopedic in character: we group together in lengthy multipart theorems an assortment of smaller propositions that hold under a fixed set of assumptions. The following conventions will shorten the formulations:

(1) by "$Z = [A,B]$ is bounded" we shall mean that the commutator $[A,B] \in \mathbf{A}(D)$ is the restriction to D of a bounded operator Z, and we shall not distinguish notationally between $Z \in L(E)$ and its restrictions;

(2) for A, B supplied by the context, $\rho_K(A)$ will denote the "Kato resolvent set"

$\rho_K(A) = \{\lambda \in \rho(A)\colon \exists\, \mu \in \rho(B)$ with $D_{\mu\lambda} = {}_|(\mu-B)(\lambda-A)D$ dense in E$\}$, while $\sigma_K^*(A) = \mathbb{C} \sim \rho_K(A) \cup \{\infty$ if A is unbounded$\}$.

<u>11.11. Theorem</u>

Let $A,B \in \mathbf{A}(D)$ be closable operators in a Banach space E such that $Z = [A,B]$ is bounded, and $[A,Z] = 0 = [B,Z]$ in the obvious sense.

(a) If $\rho_K(A) \neq \emptyset$, it is a union of components of $\rho(A)$, and if in addition there exists a sequence $\mu_n \in \rho(B)$ with $\|R(\mu_n,B)\| \to 0$, then $\rho_K(A) = \rho(A)$ and $\rho_K(B) \neq \emptyset$. Hence if there also exists a sequence $\lambda_n \in \rho(A)$ with $\|R(\lambda_n,A)\| \to 0$, $\rho(B) = \rho_K(B)$ as well.

(b) Let $\lambda \in \rho_K(A)$ and let φ be holomorphic in a neighborhood of $\sigma_K^*(A)$. Then $R(\lambda,A)$ and $\varphi(A)$ leave $D(\bar{B})$ invariant, commute with Z, and satisfy the unbounded commutation relations

$$\bar{B}R(\lambda,A) = R(\lambda,A)\bar{B} - ZR(\lambda,A)^2 = R(\lambda,A)(\bar{B}-ZR(\lambda,A)); \quad (16)$$

$$\bar{B}\varphi(A) = \varphi(A)\bar{B} - Z\varphi'(A) \quad (17)$$

(c) Suppose $\lambda \in \rho_K(A)$, $\mu \in \rho_K(B)$, φ is holomorphic in a neighborhood of $\sigma_K^*(A)$ and ψ is holomorphic in a neighborhood of $\sigma_K^*(B)$. Then the bounded operator commutation relations (12) and

$$\begin{aligned} [R(\lambda,A), \psi(B)] &= ZR(\lambda,A)\, \psi'(B)R(\lambda,A) \\ [\varphi(A), R(\mu,B)] &= ZR(\mu,B)\, \varphi'(A)R(\mu,B) \end{aligned} \quad (18)$$

hold.

Proof: (a) Let $M = \mathbb{C}B + \mathbb{C}Z$, with $B = C$ dominating since Z is bounded. Then Proposition 11.2 ensures that $\lambda \in \rho_K(A)$ iff $(\lambda-A)D$ is $\|.\|_1$-dense (GD), whence Corollary 5.7 ensures that $\rho_K(A)$ is a union of components of the M-diminished resolvent set $\rho(A;M)$. But $(\text{ad } A)^2(M) = \{0\}$, so $\sigma(\text{ad } A) = \{0\}$ and $\rho(A;M) = \rho(A)$ for such nilpotent commutation relations (cf. the remarks at the end of Section 5C). Thus, the first claim follows. If $\|R(\mu_n,B)\| \to 0$, let $\lambda \in \rho(A)$ and $\nu \in \rho_K(A)$, and choose $\mu = \mu_n$ so that both $\|R(\lambda,A)\|\,\|R(\mu,B)\| < \|Z\|^{-1}$ and $\|R(\nu,A)\|\,\|R(\mu,B)\| < \|Z\|^{-1}$. By 11.2, $D_{\mu\nu}$ is dense in E for this particular μ, since it is dense for some μ, whence by 11.3, $D_{\nu\mu}$ is dense as well, and by 11.2 again, $D_{\lambda\mu}$ must be dense. Invoking 11.3 again in reverse, we obtain that $D_{\mu\lambda}$ is dense and $\lambda \in \rho(A)$. In the process, we found that $\mu = \mu_n \in \rho(B)$ as well, so the same argument shows that $\rho_K(B) = \rho(B)$ if $\|R(\lambda_n,A)\| \to 0$ for a suitable sequence.
In (b) (which holds vacuously if $\rho_K(A)$ is empty) we have (GD) for all $\lambda \in \rho_K(A)$ by the proof of (a) above, while $D_1 = D(\bar{B})$ by 11.1. Consequently, Theorems 5.1 (or 5.4) and 5.8 apply to yield finite-sum commutation relations with unique ad A-eigenvalue $\alpha = 0$ and step $s = 1$: $C^{(1)}R_1(\lambda,A) = R(\lambda,A)C^{(1)} - R(\lambda,A)^2[A,C]^{(1)}$ for all $C \in M$ with a corresponding identity for $\varphi(A)$. Here, if $C = Z$, $[A,C] = 0$ and

Z commutes on $D(\bar{B}) = D_1$ with $R(\lambda,A)$, whence it commutes on all of E by boundedness. If $C = B$, then $C^{(1)} = \bar{B}$ and (16) follows from $R_1(\lambda,A) = R(\lambda,A)|_{D(\bar{B})}$. Details for (17) are left to the reader.

For (c)(also vacuous if either $\rho_K(A)$ or $\rho_K(B)$ is empty) we obtain (10) easily by a double application of Corollary 5.3, which directly asserts here that $[R(\lambda,A), R(\mu,B)] = R(\mu,B)R(\lambda,A)[\bar{A},\bar{B}]R(\lambda,A)R(\mu,B)$. But by (b), $R(\lambda,A)$ maps $R(\mu,B)E=D(\bar{B})$ back into $D(\bar{B})$ while producing vectors in $D(\bar{A})$ so the right-hand side is everywhere-defined, while on dense $D_{\mu\lambda} = (\mu-B)(\lambda-A)D$, $[\bar{A},\bar{B}]R(\lambda,A)R(\mu,B)$ reduces to $ZR(\lambda,A)R(\mu,B)$, so the right-hand side is in fact $R(\mu,B)R(\lambda,A)ZR(\lambda,A)R(\mu,B)$ on all of E, and Z can be commuted outside by (b). Since A and B enter symmetrically into the hypotheses here, (b) also holds with A and B interchanged, (and λ replaced by μ, etc.) and the argument above applies to yield $[R(\mu,B), R(\lambda,A)] = -ZR(\lambda,A)R(\mu,B)^2R(\lambda,A)$ whence the rest of (10) follows by reversing signs. The identities in (18) follow routinely from (10) by contour integration in μ and λ, respectively:

$$[R(\lambda,A),\int_\Gamma\psi(\mu)R(\mu,B)d\mu] = \int_\Gamma\psi(\mu)[R(\lambda,A), R(\mu,B)]d\mu$$

$$= ZR(\lambda,A)\ \{\int_\Gamma\psi(\mu)R(\mu,B)^2d\mu\}\ R(\lambda,A)$$

$$= ZR(\lambda,A)\{\int_\Gamma\psi'(\mu)R(\mu,B)d\mu\}R(\lambda,A)$$

as in Section 5D, whence the first identity follows. (The second uses the other form of (10).)

We next tabulate the additional identities which become available when A and B are semigroup-pregenerators. (For some of these, it suffices for A alone to pregenerate, but the Kato hypothesis is simplified if both are assumed pregenerators.) Our result contains the key implication of Kato's Theorem [Kt 1].

11.12. Theorem

Let $A, B \in \mathbf{A}(D)$ be pregenerators satisfying the commutation conditions of Theorem 11.11. Suppose that for some $\lambda \in \rho(A)$ and $\mu \in \rho(B)$, $D_{\mu\lambda} = (\mu-B)(\lambda-A)D$ is dense. Then all conclusions of Theorem 11.11 hold for all $\lambda \in \rho(B)$, etc., and we have the following as well.

(d) The semigroup operators $V(s,A)$ leave $D(\bar{B})$ invariant for all $s \in [0,\infty)$ and

$$\bar{B}V(s,A) \supseteq V(s,A)(\bar{B} - sZ) \tag{19}$$

(e) For each $\mu \in \rho(B)$ and $s \in [0,\infty)$

$$[V(s,A), R(\mu,B)] = sZ\, R(\mu,B)V(s,A)R(\mu,B), \tag{20}$$

and if $\|V(t,B)\| \leq M_B \exp(\omega_B t)$ for all $t \in [0,\infty)$, then for $Re(\mu) > \omega_B$ and $0 \leq s < M_B^{-1} \|Z\|$ $(Re(\mu) - \omega_B)$, $\mu \in \rho(B - sZ)$ and

$$R(\mu,B)V(s,A) = V(s,A)R(\mu,B - sZ) \tag{13}$$

(f) For all s, $t \in [0,\infty)$, the Weyl commutation relation

$$V(s,A)V(t,B) = \exp(stZ)V(t,B)V(s,A) \tag{11}$$

holds.

Remark: In (d) and (e), the obvious variants with A and B exchanged are also true.

Proof: First, recall that for suitable $M_A, M_B < \infty$ and $0 \leq \omega_A$, ω_B, $Re(\lambda) > \omega_A$ implies that $\lambda \in \rho(A)$ and $\|R(\lambda,A)\| \leq M(Re(\lambda)-\omega_A)^{-1}$, while $Re(\mu) > \omega_B$ implies $\mu \in \rho(B)$ and $\|R(\mu,B)\| \leq M(Re(\mu)-\omega_B)^{-1}$. By hypothesis, $\rho_K(A) \neq 0$ as in Theorem 11.11(a), and these estimates imply that $\|R(\lambda,A)\| \to 0$ as $Re(\lambda) \to \infty$ and $\|R(\mu,B)\| \to \infty$, so $\rho_K(A) = \rho(A)$ and $\rho_K(B) = \rho(B)$.
Thus (b) and (c) of Theorem 11.11 are simplified as claimed, and (GD) holds for all $\lambda \in \rho(A)$ with respect to $M = \mathbb{C}B + \mathbb{C}Z$.

Applying Theorem 6.1, noting that $\Gamma_1(M) = D(\bar{B})$ as in Lemma 11.1, we immediately obtain (d), since $(ad\ A)(B) = Z$ and $(ad\ A)^k(B) = 0$ for $k > 1$. Both claims in (e) are then easy consequences of the following variant of (19):

$$(\mu-\bar{B})V(s,A) \supseteq V(s,A)(\mu-(\bar{B}-sZ)) = V(s,A)(\mu-\bar{B})+sV(s,A)Z \tag{19'}$$

In fact (2) is the immediate result of pre-and-post multiplying both sides of (19') by $R(\mu,B)$. To get (13), recall that by Neumann expansions, $\mu \in \rho(\bar{B}-sZ)$ whenever $s\|Z\|\ \|R(\mu,B)\| < 1$, so that by the resolvent estimate above, $s\|Z\|M_B(\mathrm{Re}(\mu)-\omega_B)^{-1}<1$ is sufficient, or $s < M_B^{-1}\|Z\|^{-1}(\mathrm{Re}(\mu)-\omega_B)$. Then one premultiplies (19') by $R(\mu,B)$ and postmultiplies by $R(\mu,B-sZ)$ to get (13) (Note in both arguments that $R(\mu,B)$ and $R(\mu,B-sZ)$ map E onto $D(\bar{B})$, so that (19') is an equality on the range of either, and both (20) and (13) hold on all of E.)

It is easiest to obtain (11) from (13) by uniqueness of Laplace transforms. (Kato [Kt 1] proceeds by Laplace-inversion.) That is, left-hand side of (13) is the t-Laplace transform of $B(t,B)V(s,A)$, which should agree with $\exp(-stZ)V(s,A)V(t,B)$ if (11) holds. But since Z commutes with $R(\lambda,A)$ and $R(\mu,B)$ (12.8), both Z and $\exp(-stZ)$ trivially commute with $V(s,A)$ and $V(t,B)$, and in fact $\exp(-stZ)V(t,B) = V(t,B-sZ)$, as the reader may check. Hence, for fixed s and $\mathrm{Re}(\mu)$ sufficiently large as in (e), the t-Laplace transform of $\exp(-stZ)V(s,A)V(t,B) = V(s,A)V(t,B-sZ)$ is $V(s,A)R(\mu,B-sZ)$: the right-hand side of (13). Hence, by uniqueness, $V(t,B)V(s,A) = \exp(-stZ)V(s,A)V(t,B)$, and (11)follows. E.O.P.

When A pregenerates a group, (e) can be sharpened and useful group/operational calculus relations are obtained. The formulations are clearer if Z is scalar as well.

11.13. Proposition

Suppose the hypotheses of Theorem 11.12 hold, with A a group-pregenerator, and $z \in \mathbb{C}$ is scalar.

(e') Then for every $\mu \in \rho(B)$ and $s \in \mathbb{R}$, $\mu \pm sz \in \rho(B)$ and (13) holds without restriction upon s. Equivalently,

$$R(\mu,B+sz) = V(s,A)R(\mu,B)V(-s,A),\ s \in \mathbb{R},\ \mu \in \rho(B). \tag{13'}$$

(g) Moreover, if ψ is holomorphic in a neighborhood of $\sigma^*(B)$, so is ψ_{sz} ($\psi_{sz}(\mu) = \psi(\mu+sz)$), and

$$\psi_{sz}(B) = V(s,A)\psi(B)V(-s,A) \tag{21}$$

Proof (e'): Replacing A by -A, we may apply Example 11.9 to $V(s,A) = V(-s,-A)$ for $s < 0$ to extend (19) to all $s \in \mathbb{R}$. Reversing the sign of s, and premultiplying both sides of (19) by $V(s,A)$, we get

$$V(s,A)\bar{B}V(-s,A) \supseteq \bar{B} + sz \tag{19'}$$

so

$$V(s,A)(\mu-\bar{B})V(-s,A) \supseteq \mu-(\bar{B}+sz) \qquad (19'')$$

But $V(s,A)R(\mu,B)V(-s,A)$ is clearly a bounded two-sided inverse for the left-hand side of (19), and since $V(s,A)$ leaves $D(\bar{B})$ = range $(R(\mu,B))$ invariant this candidate-inverse sends E into $D(\bar{B}) = D(\mu-(\bar{B}+sZ))$ so it right-inverts $\mu - (\bar{B}+sZ)$ on all of E. Since it is clearly a left-inverse as well, we see that $V(s,A)R(\mu,B)V(-s,A)$ is indeed the resolvent for $\bar{B}+sZ$ at μ, and (13') follows. Then (13) is obtained by sign-reversal and algebra.

The claim (g) is obtained by contour-integration from (13'), taking the bounded operators $V(\pm s,A)$ under the integral:

$$\begin{aligned} V(s,A)\psi(B)V(-s,A) &= \psi(\infty)+(2\pi i)^{-1} \int_\Gamma \psi(\mu)V(s,A)R(\mu,B)V(-s,A)d\mu \\ &= \psi(\infty)+(2\pi i)^{-1} \int_\Gamma \psi(\mu)R(\mu-sz,B)d\mu \\ &= \psi(\infty)+(2\pi i)^{-1} \int_{\Gamma+sz} \psi(\nu,B)d\nu \\ &= \psi_{sz}(B). \end{aligned}$$

Here, the translation-invariance of $\rho(B)$ (hence $\sigma^*(B)$) in (e') ensures that Γ and $\Gamma + sz$ both lie in $\rho(B)$ and have the same winding-index, while $\psi_{sz}(\infty) = \psi(\infty)$, so our change-of-variable and shift-of-contour are justified above.

11.14. Corollary

(a) Under the hypotheses of Proposition 11.13, $\sigma(B)$ is a union of lines with direction Arg(z). (b) If $\bar{A}$ is skew-adjoint and $\bar{B}$ is self-adjoint, then $z \in \mathbb{R}$ and $\sigma(B)$ has uniform spectral multiplicity: the spectral subspaces for bounded Borel $S \subset \sigma(B)$ and any translate $S + \tau$ are unitarily equivalent.

Proof: (a) is immediate from Proposition 11.13(a). To obtain (b), there are two arguments of equal interest. First, we may use the folk-fact that the holomorphic calculus "determines" the calculus for bounded measurable ψ, whence (after limits) one obtains (21) for such ψ, including characteristic functions. Recall, for example, that the algebra A of functions on $\mathbb{R}$ generated by restrictions of $\psi(\mu) = (\mu-z)^{-1}$ ($\mathrm{Re}(z) \neq 0$) on $\mathbb{R}$ is self-adjoint, separates points, and consists of functions holomorphically extendable into a neighborhood of

$\sigma^*(B) \subset \mathbb{R} \cup \{\infty\}$, whence A is uniformly dense in $C_\infty(\mathbb{R})$ by Stone-Weierstrass. But for these functions, $\psi(B) = \int_{-\infty}^{\infty} \psi(\mu)\, dP(\mu)$ and $\psi \to \psi(B)$ is uniformly operator norm bounded, so that for any $\psi \in C_\infty(\mathbb{R})$ there exists $\psi_n \in A$ with $\|\psi-\psi_n\|_\infty \to 0$ and $\psi(B) = \lim \psi_n(B)$, whence $V(s,A)\psi(B)V(-s,A) = \lim V(s,A)\psi_n(B)V(-s,A) = \lim(\psi_{sz})_n(B) = \psi_{sz}(B)$. For any bounded Borel S, the characteristic functions χ_S is then pointwise-approximable by a uniformly-bounded sequence $\psi_n \in C_\infty(\mathbb{R})$ supported in some compact $K \supset S$; the fact that $\psi_n(B) = \psi_n(P_K B)$ for all n then means by Corollary X.29(v) in [DS] ($P_K B$ is bounded) that $\chi_S(B)$ is the strong-operator limit of the $\psi_n(B)$, whence our identity extends to $V(s,A)\chi_S(B)V(-s,A) = (\chi_S)_{sz}(B) = \chi_{S-sz}(B)$. But $\chi_S(B) = P_B(S)$, etc., completing the argument: the projections are unitarily equivalent under $V(s,A)$.

The second line of proof uses the resolvent formula for $a < b$ in $\mathbb{R}$ (Theorem XII.2.10 [DS])

$$P_B((a,b)) = \lim\{(2\pi i)^{-1} \int_{a+\delta}^{b-\delta} (R(\mu-\varepsilon i,B) - R(\mu+\varepsilon i,B)d\mu : \delta \to 0_+,\ \varepsilon \to 0_+\}$$

as a strong-operator limit. Thus

$$\begin{aligned} V(s,A)P_B((a,b))V(-s,A) &= \lim(2\pi i)^{-1} \int_{a+\delta}^{b-\delta}(R(\mu-\varepsilon i-sz,B)-R(\mu+\varepsilon i-sz,B))d\mu \\ &= \lim(2\pi i)^{-1} \int_{a-sz+\delta}^{b-sz-\delta}(R(\nu-\varepsilon i,B)-R(\nu+\varepsilon i,B))d\nu \\ &= P_B((a-sz,\ b-sz)). \end{aligned}$$

Since all boundedly-supported projections for B are "determined" (via strong-operator limits) by those supported on intervals, the result extends to all boundedly-supported $P_B(S)$.

Finally, we extend Examples 11.9 and 11.10 to unbounded skew-adjoint Z.

11.15. Corollary

Let $A,B,Z \in \mathbf{A}(D)$ be essentially skew-adjoint operators in a Hilbert-space E, with $[A,B] = Z$, $[A,Z] = [B,Z] = 0$. Suppose that for some $\lambda \in \rho(A)$, $\mu \in \rho(B)$ and $\nu \in \rho(Z)$, $D_{\mu\lambda}$, $D_{\mu\nu}$ and $D_{\lambda\nu}$ are dense in E (usual notation: $D_{\mu\nu} = (\mu-B)(\nu-Z)D$.) Then the Weyl commutation relations hold, in the form

$$V(s,A)V(t,B) = V(st,Z)V(t,A)V(s,B) \tag{11'}$$

Proof: First, the theory of Section 11B applies to the pairs (A,Z) and (B,Z) to produce the conclusion that Z strongly commutes with A and B. Consequently, if P is any spectral projection for Z with bounded support, Z reduces on $E_P = PE$ to a bounded operator Z_P. Moreover, the relations $PA \subseteq AP$ and $PB \subseteq BP$ show that PD (necessarily dense in PE) is a C^∞-domain for $A_P = A|_{PE}$ and $B_P = B|_{PE}$, while $[A,B] = Z$ implies that $[A_P,B_P] = Z_P$, and Z_P commutes with A_P and B_P. Finally, $(\mu-B_P)(\lambda-A_P)D_P = PD_{\mu\lambda}$ is dense in E_P as well, so that Theorem 11.9 applies on E_P: $V(s,A_P)V(t,B_P) = \exp(stZ_P)V(t,B_P)V(s,A_P)$. But $V(s,A_P)V(t,B_P)Pu = V(s,A)V(t,B)Pu$ for all $u \in E$, while $\exp(stZ_P)V(t,B)V(s,A)u = V(st,Z)V(t,B)V(s,A)Pu$, whence since P was arbitrary, we may let $P \to 1$ strongly to obtain (11') in the limit.

Remark: Suitable versions of the other conclusions from Example 11.8, 11.9 and 11.10 can be patched together in roughly the same way. We leave details to the reader.

Before taking up the examples which illustrate the relative generality of the results 11.8-11.10, we pause briefly to consider the issue of "converses" of these results: the extent to which the validity of the various bounded-operator commutation relations ensures the existence of suitable domains upon which the unbounded-operator relations and Kato/(GD) conditions can be checked. In contrast to our procedure in Section 11B above (the commutative case, Proposition 11.6), we proceed informally here since much of the discussion already appears in [Kt 1].

As pointed out by Kato in [Kt 1], either of the two expressions for the commutator of resolvents in (12) suffices to ensure that $D = D(\bar{A}\bar{B}) \cap D(\bar{B}\bar{A})$ is a dense core and that $D_{\lambda\mu} = (\lambda-A)(\mu-B)D$ is dense, with $[A,B]u = Zu$ for all $u \in D$. We recall the calculation, using the second identity in (12) in the form with bounded (central) Z nested in the middle:

$$R(\lambda,A)R(\mu,B) = R(\mu,B)R(\lambda,A)\,[1-ZR(\lambda,A)R(\mu,B)] \qquad (12')$$

we put $D = R(\lambda,A)R(\mu,B)E$ and check that $D = D(\bar{A}\bar{B}) \cap D(\bar{B}\bar{A})$ essentially as in Proposition 11.6. Indeed, if $u = R(\lambda,A)R(\mu,B)v \in D$, then $u \in D((\mu-\bar{B})(\lambda-\bar{A}))$ immediately, so $u \in D(A)$ and $(\lambda-A)u = R(\mu,B)v \in D(\bar{B})$. But since $u = R(\mu,B)R(\lambda,A)[v+ZR(\lambda,A)R(\mu,B)v]$ also, it follows that $u \in D((\lambda-\bar{A})(\mu-\bar{B}))$ as well, whence in particular $u \in D(\bar{B})$ and $\bar{A}u = -(\lambda-\bar{A})u + \lambda u$ so $u \in D(\bar{B}\bar{A})$. Similarly, $(\mu-\bar{B})u \in D(\bar{A})$ now implies that $\bar{B}u \in D(\bar{A})$ so $u \in D(\bar{A}\bar{B})$. But any $u \in D(\bar{A}\bar{B}) \cap D(\bar{B}\bar{A})$ has form $u = R(\lambda,A)R(\mu,B)v$ for $v = (\mu-\bar{B})(\lambda-\bar{A})u$, so the domain claim follows. Density of $D_{\lambda\mu}$ (and $D_{\mu\lambda}$) follows as usual by 5.5. The commutation relation $[A,B]u = Zu$ is obtained for $u = R(\lambda,A)R(\mu,B)v$ by

$$\begin{aligned}\bar{B}\bar{A}u &= (\bar{B}-\mu+\mu)(\bar{A}-\lambda+\lambda)u = (\bar{B}-\mu)(\bar{A}-\lambda)u+\mu(\bar{A}-\lambda)u+\lambda(\bar{B}-\mu)u+\lambda\mu u\\ &= -v+\mu(\bar{A}-\lambda)u+\lambda(\bar{B}-\mu)u+\lambda\mu u;\end{aligned}$$

$$\bar{A}\bar{B}u = (\bar{A}-\lambda)(\bar{B}-\mu)u+\lambda(\bar{B}-\mu)u+\mu(\bar{A}-\lambda)u+\lambda\mu u$$

so

$$\begin{aligned}[\bar{A},\bar{B}]u &= v + (\bar{A}-\lambda)(\bar{B}-\mu)R(\mu,B)R(\lambda,A)[v-ZR(\lambda,A)R(\mu,B)v]\\ &= v - v + Zu = Zu\end{aligned}$$

by (12').

The matter of obtaining a dense C^∞ domain $D \subset D(\bar{A}\bar{B}) \cap D(\bar{B}\bar{A})$ is more technical. The work of the second author [Mr 9] on resolvent-smoothing shows (after lengthy development) that such a domain can be obtained without the assumption that $\bar{A}$ and $\bar{B}$ are semigroup-generators. At the opposite extreme, if $\bar{A}$ and $\bar{B}$ generate groups which satisfy the Weyl relations (11), then one has (implicitly) a representation V of the Heisenberg nilpotent Lie group (Chapters 1 and 2) and the usual Gårding domain D is available, and is easily checked to satisfy the various GD and Kato conditions. (Indeed, D is a core for $\bar{A},\bar{B}$ and both A and B satisfy GD conditions for the $\|\cdot\|_1$ determined by $M = \mathbb{R}A + \mathbb{R}B + \mathbb{R}Z$, hence satisfy GD and Kato conditions for the weaker graph-norms induced by each other.) The Gårding convolution-smoothing idea is also easily adapted to the semigroup situations in Theorem 11.12 using convolutions over the "Lie semigroup" $[0,\infty) \times [0,\infty) \times \mathbb{R}$ with the appropriate nilpotent Heisenberg product. (This is just the sub-semigroup of the Heisenberg group obtained by appropriately restricting the non-central one-parameter subgroups. We omit details.)

The following two-part example illustrates applications of Theorem 11.12 outside the semigroup/Weyl relation setting; it is based upon ideas that have been known at least since Singer's work [Sr 1]. It also shows that the spectral structure Corollary (11.14) is best possible: the conclusions fail under weaker hypotheses.

11.16. Example

Let $E = L^p([a,c])$ for $1 \leq p < \infty$, $-\infty < a < c < \infty$, and for $b \in (a,c)$, let $D = \{u \in C^\infty([a,c]): u^{(n)}(b) = 0,\ n = 0,1\ldots\}$.
(a) The operators $A = iP = d/dx$ and $B = iQ = ix$ are in $\mathbf{A}(D)$ and satisfy the canonical commutation relation $[A,B] = i$. Here A is closable, $\rho(A) = \mathbb{C}$, but for no $\theta \in [0,2\pi]$ is $e^{i\theta}A$ a pregenerator, while B is bounded with $\sigma(B) = i[a,c]$ and when $p = 2$, B is skew-adjoint with uniform multiplicity 1. Moreover, for all $\lambda \in \mathbb{C}$, $\mu \in \mathbb{C} \sim i[a,c]$ the domains $D_\lambda = (\lambda-A)D$, $D_\mu = (\mu-B)D$, $D_{\lambda\mu} = (\lambda-A)(\mu-B)D$ and $D_{\mu\lambda} = (\mu-B)(\lambda-A)D$ all coincide with D, whence all possible Kato and graph-density conditions hold as in 11.11. All conclusions in Theorem 11.11 hold, and can independently be checked by calculus.
(b) If, instead, $B_z = A-izB$, then $[A,B_z] = z \in \mathbb{C}$, B_z as well as A has $\sigma(B_z) = \mathbb{C}$, no $e^{i\theta}B_z$ is a pregenerator, but $D_\lambda = D_\mu = D_{\lambda\mu} = D_{\mu\lambda} = D$ and hypotheses as well as conclusions in Theorem 11.11 hold for this pair.

Verification of Example 11.16: Beginning with B, it is obvious that $(\mu-ix)^{-1}$ and ix are both bounded functions on $[a,c]$ if $\mu \notin i[a,c]$, whence these define bounded multiplication operators on $L^p([a,c])$. Since they are C^∞, a routine Leibniz rule argument shows that both preserve D. Hence (since D is dense) $(\mu-ix)$ and $(\mu-ix)^{-1}$ are closures of their restrictions to D, $\mu \in \rho(B)$, and $R(\mu,B) = (\mu-ix)^{-1}$ leaves D invariant, so $(\mu-ix)D = D$. The claims for $p = 2$ are equally obvious:

multiplication operators are the standard model for "multiplicity 1".

If we verify the claims for B_z in (b), for all $z \in \mathbb{C}$, then the choice $z = 0$ yields them for $B_o = A$. Notice that once $(\mu-B_z)D = D$ (hence $(\lambda-A)D = D$) is proved, we will have $D_{\mu\lambda} = (\mu-B_z)(\lambda-A)D = (\mu-B_z)D = D$ (similarly for B) and $D_{\lambda\mu} = (\lambda-A)(\mu-B_z)D = D$ as well, so Kato's condition is satisfied and Theorem 11.12 will apply.

In fact, the standard theory of first-order linear differential equations explicitly exhibits the fact that $R(\mu,B_z)$ exists for all $\mu \in \mathbb{C}$ and that $R(\mu,B_z)D \subset D$, whence both $(\mu-B_z)$ and $R(\mu,B_z)$, map D bijectively on itself. That is, for $u \in D$, the inhomogeneous differential equation

$$[(\mu-B_z)v](x) = \mu v(x) - v'(x) + izxv(x) = u(x) \tag{22}$$

has a unique solution $v \in D$:

$$v(x) = [R(\mu,B_z)u](x) = -\exp(izx^2/2+\mu x)\int_b^x \exp(-izt^2/2-\mu t)u(t)dt$$
$$= \int_a^b K(t,x)u(t)dt \tag{23}$$

where $K(t,x) = -\exp\{iz^2/2(x^2-t^2) + \mu(x-t)\}\chi_{[b,x]}$ is a bounded kernel on $[a,c] \times [a,c]$, so that (23) extends to all $u \in L^p$ to define a bounded operator on $L^p = E$ (Example 2.4, pp. 143-144 [Kt 2]). Consequently, B_z is closable, $R(\mu,B_z)$ maps E onto $D(B_z)$, and $R(\mu,B_z)$ is indeed the resolvent of B_z.

This explicit form of $R(\mu,B_z)$ makes it possible to verify that no multiple $e^{i\theta}B_z$ of B_z ($\theta \in [0,2\pi)$) pregenerates a C_Θ semigroup, whence Theorem 11.12 (and Chapter 5) applies to this example, but neither the subsequent results below nor the semigroup-based theories in Chapters 3 and 6 can be applied. Specifically, we show that the strong convergence of $\mu R(\mu,e^{i\theta}B_z) = (e^{-i\theta}\mu)R(e^{-i\theta}\mu,B_z)$ to the identity operator as $0 < \mu \to \infty$ (required by semigroup-pregenerator) necessarily fails here. We discuss the case $z = 0$ (i.e., $B_o = A$) first, later reducing the general case to it. In fact, putting $z = 0$, $B_z = A$, $\mu = \lambda$ and $u(x) \equiv 1$ in (23) we have

$$[\lambda R(\lambda,A)u](x) = e^{\lambda x}\int_b^x e^{-\lambda t}d(-\lambda t) = 1-e^{\lambda(x-b)} = u(x)-e^{\lambda(x-b)}; \tag{24}$$

so that for all $1 \leq p < \infty$ we have

$$\|\lambda R(\lambda,A)u - u\|_p^p = \int_a^c e^{pRe(\lambda)(x-b)}dx. \tag{25}$$

If $Re(\lambda) \neq 0$, the right-hand side becomes $(p\,Re(\lambda))^{-1}$ $(e^{pRe(\lambda)(c-b)} - e^{-pRe(\lambda)(b-c)})$ whence as $|\lambda| \to \infty$ along any ray with $\arg(\lambda) = -\theta \neq \pm\,\pi/2$, one of these terms goes to ∞ while the other goes to zero, and $\lambda R(\lambda,A)u$ fails to converge to u. If $Re(\lambda) = 0$ (i.e., $-\theta = \pm\,\pi/2$), the right-hand side is constantly $c-a \neq 0$, and again $\lambda R(\lambda,A)u$ fails to converge to u.

For general z, one replaces u by $v(x) = \exp(izx^2/2)$ to obtain by cancellation in the integrand that $[\mu R(\mu,B_z)v](x)$ $= \exp(izx^2/2)[\mu R(\mu,A)u](x) = v(x) - \exp(izx^2/2)e^{\mu(x-b)}$. Hence $\mu R(\mu,B_z)v - v = -M_z w_\mu$, where $w_\mu(x) = e^{\mu(x-b)}$ and M_z is the boundedly-invertible multiplication operator $\exp(izx^2/2)$, so since $w_\mu \not\to 0$ as $|\mu| \to \infty$ (above), $\mu R(\mu,B_z)v \not\to v$ and B_z is not a pregenerator as before.

Finally, elementary direct proofs of many of the conclusions from Example 11.16 can be obtained from the following commutation relation for $R(\lambda,A)$ with multiplication operators $(Mu)(x) = m(x)u(x)$ defined by a C^∞ function m, with $(M'u)(x)$ $= m'(x)u(x)$:

$$[M, R(\lambda,A)] = R(\lambda,A)M'R(\lambda,A). \tag{26}$$

In the applications for (a), $m(x) = ix$, $m(x) = (\mu-ix)^{-1}$, or $m(x) = \psi(ix)$ so that $M = R(\mu,B)$ or $\psi(B)$, yielding (16), (12) and (17), respectively. In (b), we can obtain only (12) from this, by noting that $B_zR(\lambda,A) = AR(\lambda,A) - izM\,R(\lambda,A)$ for $m(x) = ix$; the other follows more easilv from (12) by the proof of Theorem 11.12 than by direct calculation. To obtain (26), we apply the right-hand side to $u \in D$, say, and integrate by parts:

$$\begin{aligned}[R(\lambda,A)M'R(\lambda,A)u](x) &= e^{\lambda x}\int_b^x e^{-\lambda t}m'(t)e^{\lambda t}\int_b^t e^{-\lambda s}u(s)ds\,dt \\ &= e^{\lambda x}\{\int_b^x m'(t)\int_b^t e^{-\lambda s}u(s)ds\,dt\} \\ &= e^{\lambda x}\{m(x)\int_b^x e^{-\lambda s}u(s)ds - 0 - \int_b^x m(t)e^{-\lambda t}u(t)dt\}\end{aligned}$$

$$= [MR(\lambda,A)u](x) - [R(\lambda,A)Mu](x). \qquad (26')$$

The identity extends to all $u \in L^p$ by boundedness. The remaining details are left to the reader. E.O.E.

Remark: The absence of semigroup-generation properties in Example 11.16 above is of interest primarily in that it exhibits a case where the resolvent commutation relation (12) is available in the absence of the Weyl commutation relation (11). This implicitly establishes that Chapter 5 is non-trivially more general than Chapter 6, and that condition (GD) of Part III can come into play in settings where the invariant-domain semigroup methods of Part II are inapplicable.

Finally, we turn to a brief inspection of the Reed-Simon extension of Nelson's counter-example [RS, p. 275] to this noncommutative (nilpotent) setting.

11.17. Example

Let $P = \partial/\partial x$ and $Q = ix + \partial/\partial y$, defined on $D = C_c^\infty(M)$ for M the Riemann surface of $z^{\frac{1}{2}}$. Then $[P,Q] = i$, and for all $1 \leq p < \infty$, both P and Q are pregenerators of isometry groups on $L^p(M)$, but the Weyl relations (11) fail, as do Kato's condition and the three possible (GD) conditions.

We confine ourselves to a general sketch of these results. As before, P generates the group of "translations in the x direction" $[V(s,P)u](x,y) = u(x + s,y)$ operating on (x,y) in either sheet away from the two copies of the x axis; it leaves the C^∞-functions supported away from those axes invariant; hence P is a pregenerator on a domain smaller than D. Likewise, Q generates the multiplier-translation group $[V(t,Q)u] = \exp(itx)u(x, y+t)$ on both sheets, and is a pregenerator on the dense invariant sub-domain of functions supported away from the y-axes. It is easy to see that failure of the y-translation group to commute with the x-translation group on M causes the Weyl relations to fail as before: for u with small support $V(s,P)V(t,Q)u$ and $\exp(ist)V(t,Q)V(s,P)$ will be supported on opposite sheets.

The failure of Kato's conditions and of the various graph-density conditions can be exhibited explicitly in the same spirit as our treatment of the commutative case in Section 11B above: we compute the "deficiency functionals" orthogonal to the appropriate images of D. The simplest place to start, from a computational perspective, involves finding the L^q functions $v_{\mu\lambda}$ (for $1/q + 1/p = 1$) that are orthogonal to $D_{\mu\lambda} = (\mu-Q)(\lambda-P)D$. If we put $f_{\mu\lambda}(x,y) = \exp(-(\bar{\lambda}x+(\bar{\mu}+ix)y))$, then for each choice of λ and μ with $\mathrm{Re}(\lambda) \neq 0 \neq \mathrm{Re}(\mu)$, there will be one quadrant in $\mathbb{R}^2$ such that $v_{\mu\lambda} = f_{\mu\lambda}\,\chi \in L^q$ when χ is the characteristic function of that quadrant. (For example, when $\mathrm{Re}(\lambda) > 0 < \mathrm{Re}(\mu)$, the quadrant $\{(x,y)\colon x,y \geq 0\}$ is appropriate.) In M, there are two copies of this quadrant, one on each sheet, yielding two characteristic functions χ_u and χ_ℓ (for "upper" and "lower"), hence $v^u_{\mu\lambda} = f_{\mu\lambda}\chi_u$ and and $v^\ell_{\mu\lambda} = f_{\mu\lambda}\chi_\ell$ are distinct members of L^q. Let χ denote either of these characteristic functions, and coordinatize the appropriate sheet by $\mathbb{R}^2$ with the cut taken outside the quadrant determined by λ and μ. Then (with $\mathrm{Re}(\lambda)$, $\mathrm{Re}(\mu) > 0$ for concreteness) for all $u \in D$ we have

$$((\mu-Q)(\lambda-P)u,\ v_{\mu\lambda}) = \int_0^\infty\!\!\int_0^\infty \bar{f}_{\mu\lambda}(x,y)(\mu-\partial/\partial y-ix)(\lambda-\partial/\partial x)u(x,y)\,dx\,dy.$$

Abbreviating $w_\lambda(x,y) = (\lambda-\partial/\partial x)u(x,y)$, one obtains upon integrating with respect to y by parts

$$\begin{aligned}\int_0^\infty \bar{f}_{\mu\lambda}(x,y)(\lambda-\partial/\partial y-ix)w_\lambda(x,y)\,dy &= \int_0^\infty \bar{f}_{\mu\lambda}(x,y)(\mu-ix)w_\lambda(x,y)\,dy \\ &\quad - \bar{f}_{\mu\lambda}(x,y)w_\lambda(x,y)\Big|_0^\infty - \int_0^\infty (\mu-ix)\bar{f}_{\mu\lambda}(x,y)w_\lambda(x,y)\,dy \\ &= \bar{f}_{\mu\lambda}(x,0)w_\lambda(x,0) = \exp(-\lambda x)(\lambda-\partial/\partial x)u(x,0),\end{aligned}$$

since u and its derivatives vanish for all large $|y|$. Integration with respect to the x-variable then gives

$$\begin{aligned}&\int_0^\infty \bar{f}_{\mu\lambda}(x,0)(\lambda-\partial/\partial x)u(x,0)\,dx \\ &= \int_0^\infty \bar{f}_{\mu\lambda}(x,0)\lambda u(x,0)\,dx - \bar{f}_{\mu\lambda}(x,0)u(x,0)\Big|_0^\infty - \int_0^\infty \lambda\exp(-\lambda x)u(x,0)\,dx = 0\end{aligned}$$

since $u(x,0)$ vanishes both for large x and near the deleted branch point $(x,y) = (0,0)$.

Then as indicated in Section 11B above, the formal

functionals $(\mu-Q)^*v^u_{\mu\lambda}$ and $(\mu-Q)^*v^\ell_{\mu\lambda}$ are $\|\cdot\|_Q$-bounded on $D(\bar{Q})$ and $\|\cdot\|_1$ bounded on the completion D_1 of D with respect to $\|u\|_1 = \max\{\|Pu\|, \|Qu\|, \|u\| = \|iu\|\}$; our calculations show that these functionals are distributions supported on the "upper" (resp. "lower") copies of the x axis:

$$((\lambda-P)u, (\mu-Q)^*v_{\mu\lambda}) = \int_0^\infty \exp(-\lambda x)(\lambda-\partial/\partial x)u(x,0)dx.$$

The corresponding annihilators $v_{\lambda\mu}$ for $D_{\lambda\mu} = (\lambda-P)(\mu-Q)D$ differ slightly from the $v_{\mu\lambda}$, since P and Q do not commute. In fact, if $g_{\lambda\mu}(x,y) = \exp(-[(\bar{\lambda}-i)x + (\bar{\mu}-ix)y])$, then $v_{\lambda\mu}=g_{\lambda\mu}\chi \perp D_{\lambda\mu}$ where χ is the characteristic function of the appropriate upper or lower quadrant determined by $\mathrm{Re}(\lambda)$ and $\mathrm{Re}(\mu)$. As above, $(\lambda-D)^*v_{\lambda\mu}$ is then the $\|\cdot\|_p$ (and $\|\cdot\|_1$)-bounded annihilator of $D_\mu = (\mu-Q)D$, as the reader may check.

It can also be shown that the functionals described above span the annihilating subspaces in which they lie. We omit details.

As the reader may have noticed, our argument applies just as well to P and Q as described, acting in $L^p(\mathbb{R}^2 \setminus \{0\})$ rather than $L^p(M)$. In that setting, the corresponding group $V(s,P)$ and $V(t,Q)$ do satisfy the Weyl relations (11), and $\mathfrak{L}$ = real span $\{P,Q,i\} \subset \mathbf{A}(D)$ is the restriction to D of a representation of the Heisenberg group, but D is too small to determine that representation. $\mathfrak{L}$ is not exponentiable!

Indeed, a similar phenomenon occurs in $L^p(M)$, which splits naturally into the direct sum of two subspaces X_E and X_0 of functions which are "even" (respectively "odd") under "exchange of sheets". (If we coordinatize M instead in polar form (r,θ), for $r > 0$ and $0 \leq \theta < 4\pi$, with $0 \leq \theta < 2\pi$ on one sheet and $2\pi \leq \theta < 4\pi$ on the other, then

$X_E = \{u \in L^p(M)\colon u(r,\theta) = u(r, \theta+2\pi(\mathrm{mod}\ 4\pi))\mathrm{ae}\}$ and

$X_0 = \{u \in L^p(M)\colon u(r,\theta) = -u(r,\theta+2\pi(\mathrm{mod}\ 4\pi))\mathrm{ae}\}$. One can easily check that the groups $V(s,P)$ and $V(t,Q)$ on $L^p(M)$ restrict on X_E to groups which satisfy the Weyl relations (11), while on X_0, the Weyl relations fail. (Both groups leave X_E and X_0 invariant: the action on X_E is naturally isomorphic to that on $L^p(\mathbb{R}^2 \setminus \{0\})$.) Moreover, it is easy to see that $v^u_{\lambda\mu} - v^\ell_{\lambda\mu} \perp X_E$, while $v^u_{\lambda\mu} + v^\ell_{\lambda\mu} \perp X_0$, so $\frac{1}{2}(v^u_{\lambda\mu} + v^\ell_{\lambda\mu})$ restricts to the (unique up to multiples) Kato deficiency functional on X_E that corresponds to the one on $L^p(\mathbb{R}^2 \setminus \{0\})$.

Appendix to Part VI

In Chapter 5 we consider a normed linear space D, and the corresponding Banach space completion is denoted by E. The associative endomorphism algebra End(D) is equipped with the structure of a Lie algebra with the commutation bracket $[A,B] = AB - BA$. But the elements in End(D) may also be viewed as unbounded operators in E with dense invariant domain D. We use the symbol $\mathbf{A}(D)$ to indicate that End(D) is equipped with this additional structure. We consider a fixed element A in $\mathbf{A}(D)$, as well as a finite-dimensional complex linear subspace $M \subset M(D)$. The following two assumptions are in force throughout:

(i) When A is regarded as an operator in E it is closable, and the closure is denoted by $\bar{A}$. Hence, if $\{u_n\} \subset D$, and $v \in E$, satisfy $\lim_{n\to\infty} \|u_n\| = 0$, as well as $\lim_{n\to\infty} \|v-Au_n\| = 0$, then it follows that $v = 0$. Recall that, for example, in the theory of unbounded derivations in C*-algebras [PS], closability is an important and nontrivial issue.

(ii) When ad A (i.e., $(\text{ad } A)(B) = [A,B]$ for $B \in \mathbf{A}(D)$) is regarded as a derivation in $\mathbf{A}(D)$, the subspace M is ad A invariant. Recall that the Jordan-Wedderburn decomposition is then available for ad A: $M \to M$.

Let the corresponding spectrum be denoted $\sigma_M(\text{ad } A) = \{\alpha_1,\ldots,\alpha_p\}$, and let $M = \Sigma_j M_j$ be the associated generalized eigen-space decomposition, $M_j = \{C \in M: (\text{ad } A - \alpha_j)^s(C) = 0$ for some integer $s\}$. If $(\text{ad } A-\alpha_j)^{s_j} \neq 0 = (\text{ad } A-\alpha_j)^{s_j+1}$ on M_j, then the algebraic structure is given by the list of eigenvalues $\{\alpha_1,\ldots,\alpha_p\}$, and the corresponding eigenvalue ascents $\{s_1,\ldots,s_p\}$.

Some of the interesting applications, and the main analytic difficulties, are present already in the first steps of the list of algebraic possibilities. The first two cases are treated in detail in Chapter 11.

(a) $M = \mathbb{C}B$, $\sigma_M(\text{ad } A) = \{0\}$, $s = 0$.

Here we have $[A,B] = 0$, and the problem is strong commutativity of a pair of unbounded operators A,B. The graph density conditions, and the Kato type variations thereof, are the main technical tools in our approach to the problem of strong commutativity.

(b) $M = \mathbb{C}B + \mathbb{C}I$, (I = the identity operator).

$\sigma_M(\text{ad } A) = \{0\}$, $s = 1$. Here we have $[A,B] = cI$ where c is a scalar, and the analytic problem is that of deriving integrated Weyl relations from infinitesimal Heisenberg commutation relations.

(c) $M = \mathbb{C}B_+ + \mathbb{C}B_-$, $\sigma_M(\text{ad } A) = \{\pm 1\}$, $s_+ = s_- = 0$.

Here we have, $[A,B_\pm] = \pm B_\pm$ which are the familiar up-down shift operator relations, or the ladder operators, which are used in semisimple Lie algebras, and in spectral analysis of operators in mathematical physics, (see Chapters 4 and 12, or [Ørs].) The global exponentiability question can be attacked through the GD-condition, rather than through the analytic vector methods of Nelson [Nℓ 1].

Existence of a closable basis

For a general ad-module M, we select a basis $B_1,\ldots,B_d$, and setting $B_0 = I$, we define $\|u\|_1 = \max\{\|B_i u\| : 0 \leqq i \leqq d\}$. Clearly, then $\|u\| \leqq \|u\|_1$ for all $u \in D$. The completion of D in the norm $\|\cdot\|_1$ is, of course, again a Banach space D_1, and the norm D_1 is also denoted by $\|\cdot\|_1$.

If it is possible to choose the basis $\mathbf{B} = \{B_1,\ldots,B_d\}$ such that each B_i is a closable operator in E, then we may define $E_1 = E_1(\mathbf{B}) = \cap_{i=1}^d D(\bar{B}_i)$ where $D(\bar{B}_i)$ denotes the domain of the closed operator $\bar{B}_i$. But when we use the closures $\bar{B}_i$ in defining the C^1-norm, $\|u\|_1 = \max\{\|\bar{B}_i u\| : 0 \leq i \leq d$, then it is easy to show that D_1 is simply the $\|\cdot\|_1$-closure of D in E_1. We recall that, in general, (that is, if a closable basis is not available) the elements in the completion D_1 are equivalence classes of $\|\cdot\|_1$-Cauchy sequences. Since $\|\cdot\|_1$ is stronger than the original norm, there is a naturally defined bounded mapping, $J_1 : D_1 \to E$, extending the injection of D into E. Typically the elements of E are functions, and, in doing analysis in D_1, it is important to be able to associate functions to the elements in the abstract space D_1. The important observation is that the mapping, $J_1 : D_1 \to E$, is injective if a closable basis exists.

In general, elements B in M may be regarded as bounded operators from D_1 to E. The symbol $B^{(1)}$ is then used, and, if

$B = \Sigma_i \beta_i B_i$ then $\|B^{(1)}u\| \leq |B| \, \|u\|_1$ for $u \in D_1$, when $|B| = \Sigma_i |\beta_i|$. Similarly, we may consider A as an operator in D_1 with dense domain. As such it is a closable operator, and the closure is denoted $\bar{A}_1$. (Here, the D_1-closability of A is a consequence of the E-closability which was assumed, but the argument is not completely trivial.)

In practice, a bootstrap situation often arises, as we prove below: one does not know, at the outset, that M has a closable basis, but after a theory for general M is developed and applied, one is able to use the resulting generalized commutation relations to show that the M in question actually must have had such a basis, and the possibility of pathology is then banished after-the-fact. In other cases, the question is still open (after commutation relations are obtained) as to whether the ad A-invariant subspace M has a closable basis. Consequently in this section we carry the full generality necessary for the worst cases.

The next lemma, and the example which follows it, indicate the potential pathologies involved in the possible absence of a closable basis. They describe the substitutes for the space $(D_1, \|\cdot\|_1)$ and for the operators $B^{(1)} \in L(D_1, E)$ and $\bar{A}_1$ in D_1 that were so easily introduced prior to the statement of Theorem 5.4.

6.3. Lemma

Let $M \subset \mathbf{A}(D)$ be a finite-dimensional complex subspace. Let τ_1 be the weakest (normable) topology on D, stronger than the initial topology τ_0, which renders all $B \in M$ continuous from (D, τ_1) into E, and let D_1 be the abstract completion of this space.

(a) There exists a bounded mapping $J_1 : D_1 \to E$ which extends the natural injection of (D, τ_1) into (E, τ_0). If M has a closable basis, then J_1 is injective.

(b) For each $B \in M$, viewed as a map from $D \subset D_1$ into E, there exists a unique bounded extension-by-limits to an operator $B^{(1)}: (D_1, \|\cdot\|_1) \to (E, \|\cdot\|)$. If B is closable, then kernel $(B^{(1)}) \supset$ kernel (J_1), and $B^{(1)} \supseteq \bar{B}J_1 \supseteq J_1 B$ (where B is viewed as acting as an endomorphism of $D \subset D_1$ in the last product).

(c) Suppose that $A \in \mathbf{A}(D)$ is closable in E, and that M is ad A-invariant. Then when A is viewed as an endomorphism of $D \subset D_1$, it is closable in D_1 with closure $\bar{A}_1$, and the intertwining relation $\bar{A}J_1 \supseteq J_1\bar{A}_1$ holds, so that in particular $\bar{A}_1(\ker(J_1) \cap D(\bar{A}_1)) \subset \ker(J_1)$.

It is clear from Lemma 11.1 that the mapping $J_1 : D_1 \to E$ has trivial kernel if M is dominated by some closable C in M. However this condition seems to be independent of the existence problem for closable basis.

Resolvent commutation (Section 5B)

Even though the domination by a closable operator in M is not directly related to existence of a closable basis, those of our commutation theorems in Chapters 5 and 6 which require a closable basis carry over, mutatis mutandis, to the ad-modules M which satisfy the domain condition in Lemma 11.1.

Our main commutation theorems are stated in terms of the Jordan-Wedderburn structure of ad A on M. The list of terminology includes:

$$\sigma(A,M) = \sigma(A) \cup \{\mu-\alpha_j : \mu \in \sigma(A),\ \alpha_j \in \sigma_M(\text{ad } A)\}$$

and

$$\rho(A,M) = \mathbb{C} \setminus \sigma(A,M),$$

the M-augmented spectrum, resp., the M-diminished resolvent set.

The projections onto the generalized eigen-spaces $M_{\alpha j}$ corresponding to $\alpha_j \in \sigma_M(\text{ad } A)$ is denoted P_j.

5.4. Theorem

Let $A \in \mathbf{A}(D)$ be a closable operator, and let $M \subset \mathbf{A}(D)$ be a finite-dimensional complex ad A-invariant subspace of $\mathbf{A}(D)$. Suppose that M has a closable basis (or is closably dominated) and that $D_\lambda = (\lambda-A)D$ is τ_1-dense in D.

(1) If $\lambda \in \rho(A,M)$, then the resolvent $R(\lambda,A) = (\lambda-\bar{A})^{-1}$ leaves D_1 invariant and restricts there to a $\|.\|_1$-bounded resolvent $R_1(\lambda,A) = (\lambda-\bar{A}_1)^{-1}$ for $\bar{A}_1$ that satisfies the following commutation relation in $L(D_1,E)$ with respect to the bounded extension $B^{(1)}$ of any $B \in M$ to D_1:

$B^{(1)}R_1(\lambda,A) = \Sigma\{(-1)^k R(\lambda+\alpha_j,A)^{k+1}[(\mathrm{ad}\ A-\alpha_j)^k(P_jB)]^{(1)}: 1\leq j\leq p,\ 0\leq k\leq s_j\}$.

(2) If instead $\mathrm{dist}(\lambda,\sigma(A))) > \max\{|\alpha_j| : 1 \leq j \leq p\}$ then $B^{(1)}R_1(\lambda,A) = \Sigma\{(-1)^k R(\lambda,A)^{k+1}[(\mathrm{ad}\ A)^k(B)]^{(1)} : 0 \leq k < \infty\}$.

(3) Moreover, if $\|R(\lambda,A)\|\ |\mathrm{ad}\ A| < 1$, then the operator norm of $R_1(\lambda,A)$ in $L(D_1)$ admits the following estimate

$$\|R_1(\lambda,A)\|_1 \leq \|R(\lambda,A)\|\ (1-\|R(\lambda,A)\|\ |\mathrm{ad}\ A|)^{-1}.$$

Using this result, we are able to obtain the following necessity-sufficiency result for graph density.

5.5. Proposition

Let A and M be as in 5.4. Suppose that $\lambda \in \rho(A,M)$. Then the following are equivalent.
(a) $D_\lambda = (\lambda-A)D$ is $\|.\|_1$-dense in D.
(b) $R(\lambda,A)$ leaves D_1 invariant and restricts to a two-sided inverse for $\lambda-\bar{A}_1$ there.

Remark: Note that this does not quite imply that the graph-density condition is equivalent to (1), since (1) by itself ensures only that $R(\lambda,A)$ sends D_1 into $E_1(\mathbf{B})$, which may be properly larger.

Proof: That (a) implies (b) is precisely the first claim in 5.4(1). For the converse, we first observe that since $R(\lambda,A)$ is bounded, it is relatively $\|.\| \times \|.\|$-closed on D_1, hence its restriction $R_1(\lambda,A)$ is necessarily $\|.\|_1 \times \|.\|_1$-closed as well. It follows from the closed graph theorem that $R_1(\lambda,A)$ is bounded on D_1 and is the resolvent of $\bar{A}_1$ at λ. The rest of the argument uses the following folk-lemma, which we also need in Chapter 11.

5.6. Lemma

Let T be a closed, densely-defined operator on a Banach space F, and suppose that T has a bounded inverse. Let D be a dense subspace of D(T). Then the following are equivalent.

(1) D is a core for T (i.e., T is the closure of its restriction to D).

(2) D is dense in D(T) with respect to the graph norm $\|u\|_T = \max\{\|u\|, \|Tu\|\}$.

(3) TD is dense in F.

(4) $D = T^{-1}F_0$ for some dense subspace F_0 of F.

Proof: (1) ⇔ (2) is immediate: $\|u_n - u\|_T \to 0$ for $u \in D(T)$, $u_n \in D$ if and only if $u_n \to u$ and $Tu_n \to Tu$. If (2) holds, then the facts that $TD(T) = F$ and T is bounded from $(D(T), \|\cdot\|_T)$ to F combine to show that T sends D into a $\|\cdot\|$-dense subspace. Claim (4) is just a rephrasing of (3). Closing the circle, if (4) holds, let $u \in D(T)$, $v = Tu$ and $v_n \in F_0$ with $v_n \to v$. Then $u_n = T^{-1}v_n \to u = T^{-1}v$ by boundedness of T^{-1}, so $Tu_n = v_n \to v = Tu$ and T is the closure of its restriction to D.

Completing the proof of Proposition 5.5, we take $T = \lambda - \bar{A}_1$ in $F = D_1$. By definition, D is a core for $\bar{A}_1$, hence for T and (1) holds in Lemma 5.6, whence by (3) there, $TD = (\lambda - \bar{A}_1)D = D_\lambda$ is dense.

Commutation for the holomorphic functional calculus (Section 5D)

Let φ scalar function, holomorphic in a neighborhood U of $\sigma^*(A)$ in the Riemann sphere $\mathbb{C}^* = \mathbb{C} \cup \{\infty\}$, and let the operator system (A,M,D) satisfy the conditions in 5.4. Let the operator φ(A) be given by the Dunford holomorphic operational calculus $\varphi(A) = \varphi(\infty) + (2\pi i)^{-1} \int_\Gamma \varphi(\lambda)R(\lambda,A)d\lambda$, for any choice of Γ a finite union of Jordan arcs positively enclosing $\sigma^*(A)$ (index + 1 about each $\lambda \in \sigma^*(A)$, lying within $\rho(A) \cap U$. The following commutation relations for φ(A) drop quite quickly out of the $L(D_1,E)$ forms of the resolvent commutation relations as given in 5.4.

5.8. Theorem

Let $A \in \mathbf{A}(D)$ be closable and let $M \subset \mathbf{A}(D)$ be a finite-dimensional complex ad A-invariant subspace with closable basis. Let φ be holomorphic in a neighborhood U of $\sigma^*(A)$.

(a) If U contains $\sigma(A,M)$ and if there exists at least one Γ in U as above that is contained entirely within components of $\rho(A,M)$ where the graph-density condition holds, then φ(A) leaves D_1 invariant and restricts there to a $\|\cdot\|_1$-bounded operator $\varphi_1(A) \in L(D_1)$ which satisfies

$$B^{(1)}\varphi_1(A)=\Sigma\{(-1)^k/k!\varphi^{(k)}(A-\alpha_j)[(\text{ad } A-\alpha_j)^k(P_jB)]^{(1)}:1\leq j\leq p,\ 0\leq k\leq s_j\}$$

for every $B \in M$.

(b) If in fact U contains a Γ such that the graph-density condition holds everywhere on Γ and $\text{dist}(\Gamma,\sigma(A)) > \max\{|\alpha_j|: 1 \leq j \leq p\}$, then in addition

$$B^{(1)}\varphi_1(A) = \Sigma\{(-1)^k/k!\varphi^{(k)}(A)[(\text{ad } A)^k(B)]^{(1)}: 0\leq k < \infty\}$$

for all $B \in M$.

Semigroup commutation relations with a closable basis (Section 6A)

Unlike the more general setting above, we assume here only that $\bar{A}$ generates a C_0 semigroup $\{V(t,A): t \in [0,\infty)\}$ on the Banach space E. Recall that this semigroup is of type ω if and only if there exists a constant $M < \infty$ such that $\|V(t,A)\| \leq Me^{\omega t}$ for all $t \in [0,\infty)$.
Then $\sigma(A) \subset \{\lambda \in \mathbb{C} : \text{Re}(\lambda) \leq \omega\}$. If we write $\nu(\text{ad } A) = \max\{|\alpha_j| : \alpha_j \in \sigma(\text{ad } A)\} = \lim\{|(\text{ad } A)^k|^{1/k}: k \to \infty\}$ for the spectral radius of ad A on M, this means that the augmented spectrum $\sigma(A,M)$ is entirely contained in the half-plane $H_{\omega+\nu(\text{ad } A)} = \{\lambda \in \mathbb{C} : \text{Re}(\lambda) \leq \omega + \nu(\text{ad } A)\}$, hence that the complementary half-plane is entirely contained in a single component of the diminished resolvent set $\rho(A,M) = \mathbb{C}\setminus\sigma(A,M)$. Therefore, Corollary 5.7 ensures that if a single λ with $\text{Re}(\lambda) > \omega + \nu(\text{ad } A)$ satisfies the graph-density-condition then all λ satisfying this inequality must have the graph-density property. We shall apply this remark below (without explicit citation) at several points.

6.1. Theorem

Suppose that A and M are as described above. Suppose further that for some λ with $\text{Re}(\lambda) > \omega + \nu(\text{ad } A)$ the subspace $D_\lambda = (\lambda-A)D$ is $\|\cdot\|_1$-dense in D.

(i) Then each of the operators $V(t,A)(t \in [0,\infty))$ leaves D_1 invariant and restricts to a $\|\cdot\|_1$-bounded operator $V_1(t,A) \in L(D_1)$. Moreover, $\{V_1(t,A) : t \in [0,\infty)\}$ acts as a C_0 semigroup on $(D_1,\|\cdot\|_1)$ whose infinitesimal generator is $\bar{A}_1$.

(ii) The semigroups $\{V(t,A) : t \in [0,\infty)\}$ and $\{V_1(t,A) : t \in [0,\infty)\}$ satisfy the following commutation relations with respect to each $B \in M$, as identities in $L(D_1,E)$:

$$\begin{aligned} B^{(1)}V_1(t,A) &= V(t,A)[\exp(-t \text{ ad } A)(B)]^{(1)} \\ &= \Sigma\{(-t)^k/k! \exp(-t\alpha_j)V(t,A)[(\text{ad } A-\alpha_j)^k(P_jB)]^{(1)} : \\ &\qquad 1 \leq j \leq p;\ 0 \leq k \leq s_j\} \end{aligned}$$

for all $t \in [0,\infty)$.

(iii) On $(D_1,\|\cdot\|_1)$, $V_1(t,A)$ is bounded by $\|V_1(t,A)\|_1 \leq M \exp((\omega + \nu(\text{ad } A))\, t)$.

PART VII

LIE ALGEBRAS OF UNBOUNDED OPERATORS:

PERTURBATION THEORY, AND

ANALYTIC CONTINUATION OF $s\ell(2,\mathbb{R})$- MODULES

For the Snark's a peculiar creature,
    that won't
Be caught in a commonplace way.
Do all that you know,
    and try all that you don't:
Not a chance must be wasted to-day!

The Hunting of the Snark
Fit the Fourth
LEWIS CARROLL

Pauli emphasized that the occurrence of the half-integral values of the quantum numbers had to be related to the anomalous values of the gyromagnetic factor and the 'anomaly of the relativistic correction'. (Pauli, 1925, p. 766.)

We were [initially] entirely in Heisenberg's footsteps. He had the idea that one should take matrices, although he did not know that his dynamical quantities were matrices ... And when one had such a programme of formulating everything in matrix language, it takes some effort to get rid of matrices. Though it seemed quite natural for me to represent perturbation theory in the algebraic way, this was not a particularly new way. (Born. Conversations, p. 48, quoted in Mehra-Rechenberg, 1982, vol. 3, p. 129).

INTRODUCTION TO PART VII

The exponentiation problem for Lie algebras of unbounded operators is most conveniently stated in terms of representations ρ of finite-dimensional real Lie algebras $\mathfrak{g}$. If E is a Banach space, a representation ρ on E maps into unbounded operators in the Banach space E. We then say that ρ is exact if there is a strongly continuous representation V of the simply connected Lie group G, with Lie algebra $\approx \mathfrak{g}$, such that $\rho = dV$, where the infinitesimal representation dV is the one defined in Gårding's paper [Gå]. The operator Lie algebra $\mathcal{L} = \rho(\mathfrak{g})$ is said to be exponentiable in E if $\rho = dV$ with strongly continuous representation $V : G \to L(E)$ in the Banach space E.

Now let A be an unbounded operator with dense domain D in a given Banach space E. Suppose that the closure $\bar{A}$ of A exists as a closed operator, and that $\bar{A}$ is the infinitesimal generator of a C_o one-parameter group $V(t,\bar{A})$. Then we shall say that A is a pre-generator.

We consider the algebra End(D) of all linear endomorphisms of D. Recall that End(D) naturally carries the structure of a Lie algebra with commutator Lie product $[A,B] = AB - BA$ for $A,B \in$ End(D). Hence we may consider finite-dimensional real Lie subalgebras $\mathcal{L}$ of End(D), and the problem of Palais now has an obvious generalization to this much more general, and applicable, setting of operator theory. Suppose $\mathcal{L}$ is spanned by a subset S, and commutators of elements in S, then one might expect that $\mathcal{L}$ is the infinitesimal operator Lie algebra associated with a C_o representation of the Lie group G on E, viz., $V: G \to L(E)$, provided only that the individual elements A in S are pre-generators for C_o one-parameter groups. We state the problem more precisely: A representation V maps the group G into the bounded invertible operators on E : We have: $V(g_1 g_2) = V(g_1)V(g_2)$, $g_1,g_2 \in G$, and $\lim_{g \to e} \| V(g)u - u \| = 0$, for $u \in E$. Let $A \in \mathcal{L}$, and let $g(t)$ be the corresponding one-parameter group in G which is given by the exponential mapping of Lie theory. Then the operator $\tilde{A} = (d/dt)V(g(t))|_{t=0}$ is the infinitesimal generator of a C_o

one-parameter group. If for all A in $\mathfrak{L}$, the operator $\tilde{A}$ is the least closed extension of A, then we say that the Lie algebra $\mathfrak{L}$ <u>exponentiates</u> on E. If $\tilde{A}$ is only an extension of $\bar{A}$, then we say that $\mathfrak{L}$ is a sub-generator [Ra]. The extension requirement is that $D(A) \subset D(\tilde{A})$, and $Au = \tilde{A}u$, for all $u \in D(A)$.

The purpose of the present part is to apply the exponentiation theorems to the representation theory of $SO_e(2,1)$, and the covering groups, which include $SL(2,\mathbb{R}) \approx SU(1,1)$, and the universal covering group $\tilde{G}$. Our emphasis is on the corresponding harmonic analysis of the classical Banach spaces ℓ^p of doubly infinite sequences, $L^p(T)$, and $C(T)$, of functions on the circle $T = \{z : |z| = 1\}$.

Using perturbation theory for representations of Lie groups, cf. Chapter 9, we obtain several new results (listed below) on the intersection between the representation theory of $SL(2,\mathbb{R})$, and the associated harmonic analysis of the classical Banach spaces mentioned above.

As indicated in the title of the first section, we approach the representation theory of $s\ell(2,\mathbb{R})$ via the ∞ by ∞ matrices of the Heisenberg formalism of quantum mechanics. Our starting point is therefore the space of all doubly infinite sequences $u = \{u(n)\}_{n \in \mathbb{Z}}$. We consider at the outset the space D of "compactly supported" sequences, i.e., the complex linear span of the canonical basis vectors e_n where $e_n(m) = \delta_{nm}$ for $n,m \in \mathbb{Z}$. For every norm on D there is a corresponding completion E which is, of course, a Banach space containing D as a dense subspace. The identification correspondence $e_n \leftrightarrow e^{inx}$ allows us to regard the elements in D also as functions on the circle T. The elements in End(D) may therefore be thought of as unbounded densely defined linear operators in E. In particular, a Lie subalgebra $\mathfrak{L} \subset \mathrm{End}(D)$ will be regarded as a Lie algebra of unbounded operators in E. If $\mathfrak{L}$ is isomorphic to $s\ell(2,\mathbb{R})$ as a Lie algebra, then $\mathfrak{L}$ may also be thought of as a Lie algebra representation of $\mathfrak{g} = s\ell(2,\mathbb{R})$ by unbounded operators in the Banach space E. Since there is a functional correspondence between representations of the Lie algebra $\mathfrak{g}$ on the one hand, and representations of the corresponding universal enveloping algebra $\mathfrak{U}(\mathfrak{g})$ on the other, we shall shift between the Lie algebraic, and the associative algebra point of view, of representations, or equivalently modules.

We may also think of the triple $(\mathfrak{L},D,E)$, Lie algebra of operators ($\mathfrak{L}$), domain space (D), and Banach space (E), as a $\mathfrak{g}$- (resp., $\mathfrak{U}(\mathfrak{g})$ -) module in the given Banach space E. (In fact, the terminology "$\mathfrak{g}$-module" is used in [Ra], "$\mathfrak{U}(\mathfrak{g})$-representation" in [Pw 2], and "$\mathfrak{U}(\mathfrak{g})$-module" in [Dx 3].) The starting point ($\mathfrak{L}$,D,E) is the one which is taken in Harish-Chandra's work [H-Ch], and in Kunze - Stein [KS]. But our approach, via the exponentiation theorems, and perturbation theory for operator Lie algebras, is new, and our results do not compare directly with previous results in the area. The Banach space E is fixed at the outset, and we consider the same space E for all representations in a given complete analytic family. The following two additional requirements are placed on E:

(i) E must be one of the familiar spaces in classical harmonic analysis : ℓ^p, L^p, C(T) etc.

(ii) The elements A in $\mathfrak{L}$ have simple analytic expressions in the Heisenberg ∞ by ∞ matrix picture, and each A is a regular operator in the Banach space E.

The Lie algebras $\mathfrak{L} \subset \mathrm{End}(D)$ are parameterized by the pair $(q,\tau) \in \mathbb{C}^2$. An element A in $\mathfrak{L}(q,\tau)$ is then an unbounded (regular, i.e., closable) operator in the Banach space E. If K is a compact operator in E, then A + K is said to be a compact perturbation. The $\mathfrak{L}(q,\tau)$-modules include naturally the unitary principal series in the special case $E = \ell^2$.

Among our new results is the fact that the analytic continuations of the unitary principal series appear as compact perturbations (with respect to q for τ fixed), either as $\mathfrak{L}(q,\tau)$-modules, or as orthogonal direct summands. The bounded perturbations reach <u>all</u> other unitary series in this manner. Moreover, we unify all the topologically completely irreducible (TCI) Banach representations in the same analytic continuation-perturbation picture.

We reduce analytic questions regarding these representations, via a new Lie algebraic approach to Phillips perturbation theory, to similar questions for certain "base-point" representations with particularly transparent analytic properties. The exponentiation properties are in focus, and we show that for $E = \ell^p$, $p \neq 2$, the corresponding operator Lie algebras $\mathfrak{L}(q,\tau)$ do <u>not</u> exponentiate to strongly continuous representations of $\widetilde{G}$ for any values of q,τ. However, our exponentiation theorems produce strongly continuous representations for the other Banach spaces $L^p(T)$, C(T), etc., under consideration.

On the positive side, we show for $E = \ell^p$ that the infinite simal representation $\mathfrak{L}(q,\tau)$ may be "smeared" with a test function on $\tilde{G}$, $\varphi \in \mathcal{D}(\tilde{G})$, and we produce so-called "smeared representations" in $E = \ell^p$: i.e., the distribution representation $V_{q,\tau}(\varphi)$ maps into the bounded operators on E, and

$$V_{q,\tau}(\varphi_1 * \varphi_2) = V_{q,\tau}(\varphi_1)\, V_{q,\tau}(\varphi_2) \quad \text{for } \varphi_1, \varphi_2 \in \mathcal{D}(\tilde{G}).$$

But the L(E) - valued distribution $\varphi \to V_{q,\tau}(\varphi)$ cannot be defined as a "point-representation" of the group $\tilde{G}$. As a technical tool in the construction we introduce the C^∞-vectors associated to $(\mathfrak{L}(q,\tau),E)$ [Ps 1], and corresponding C^∞-integrals. The C^∞-integrals, in turn, are differentiable representations of $\tilde{G}$ which act on the C^∞-spaces D_∞, resp., E_∞, contained in E, and containing D.

Chapter 12

EXPONENTIATION AND ANALYTIC CONTINUATION OF HEISENBERG-MATRIX REPRESENTATIONS FOR $s\ell(2,\mathbb{R})$

This final chapter serves two purposes. First, it rounds out the spectral-theoretic approach to group representations, via the graph-density condition and Phillips perturbations, developed in Parts III and IV of this monograph. Second, it initiates an ongoing program in the representation theory of semisimple Lie groups that is specifically tailored to rigorous treatment of the matrix-operator formalism frequently employed in mathematical physics. (This program therefore differs from the investigations of harmonic analysts, notably Harish-Chandra, both in techniques and in emphasis.)

Although some aspects of our discussion in this section require familiarity with the structure theory of semisimple Lie algebras and with a few rather deep results in the Banach representation theory of the associated groups, much of what we do below can be viewed as a direct application of the methods of Parts III and IV to certain explicit operator Lie algebras. We discuss these self-contained aspects first.

The operator Lie algebras to be considered here are defined algebraically as endomorphisms of the space $D = C_c(\mathbb{Z})$ of finitely-supported complex bilateral sequences, where D is later densely embedded in various natural complex Banach spaces (e.g., $\ell^2(\mathbb{Z})$). The <u>base-point Lie algebra</u> $\mathfrak{L}_0$ is spanned by the linear extensions to D of

$$A_0 e_n = i n e_n, \quad A_1 e_n = i\frac{n}{2}(e_{n+1}+e_{n-1}), \quad A_2 e_n = \frac{n}{2}(e_{n+1}-e_{n-1}) \qquad (1)$$

where $e_n(m) = \delta_{mn}$ (= 1 if m = n, 0 if m ≠ n) is the nth canonical basis vector. These operators can easily be shown (Section 12A) to satisfy the commutation relations which abstractly define the Lie algebra $\mathfrak{g}$ known as $s\ell(2, \mathbb{R})$ (and as $su(1,1)$ and $so(2,1)$):

$$[A_0,A_1] = -A_2, \quad [A_0,A_2] = A_1, \quad [A_1,A_2] = A_0. \qquad (2)$$

This operator Lie algebra $\mathfrak{L}_0$ serves as the base-point for a perturbation-theoretic analysis of the two-parameter "analytic" family $\{\mathcal{L}(q,\tau): (q,\tau) \in \mathbb{C}^2\}$ of operator Lie algebras whose exponentials are of primary interest. These infinitesimal representations are essentially familiar from Bargmann's work [Bg]:

$$B_0(q,\tau)e_n = i(n+\tau)e_n, \qquad B_+(q,\tau)e_n = \gamma_n(q,\tau)e_{n+1},$$
$$B_-(q,\tau)e_n = \gamma_{n-1}(q,\tau)e_{n-1}, \quad B_1(q,\tau) = \tfrac{i}{2}(B_+(q,\tau)+B_-(q,\tau)) \tag{3}$$
$$B_2(q,\tau) = \tfrac{1}{2}(B_+(q,\tau)-B_-(q,\tau))$$

where the weight $\gamma_n(q,\tau)$ for the weighted shifts $B_\pm(q,\tau)$ is

$$\gamma_n(q,\tau)=\operatorname{sgn}(n)(q+(n+\tau)(n+\tau+1))^{\frac{1}{2}}=\operatorname{sgn}(n)(q-\tfrac{1}{4}+(n+\tau+\tfrac{1}{2})^2)^{\frac{1}{2}}. \tag{4}$$

(Here, we fix a branch-cut for the square root along the negative imaginary axis and choose the branch that is positive on the positive axis. The signum (n) factor is present for technical reasons.) We refer to $\mathcal{L}(q,\tau)$ = real span $\{B_0,B_1,B_2\}$ as balanced (with respect to the basis $\{e_n: n \in \mathbb{Z}\}$) because the weighted shifts $B_+: e_n \to e_{n+1}$ and $B_-: e_{n+1} \to e_n$ apply the same multiplier $\gamma_n(q,\tau)$ to the resultant vectors: $B_+e_n = \gamma_n e_{n+1}$, $B_-e_{n+1} = \gamma_n e_n$. (That is, the matrix for B_- is the transpose of the matrix for B_+.) By contrast, the corresponding $A_\pm = A_2 \pm iA_1$ for the base-point Lie algebra $\mathcal{L}_0$ are not quite balanced: $A_+e_n = ne_{n+1}$ but $A_-e_{n+1} = (n+1)e_n$. Further background concerning this notion of balance is supplied in Section 12A.

The program of the chapter proceeds in two basic steps: exponentiation of $\mathcal{L}_0$ and perturbation continuation for the $\mathcal{L}(q,\tau)$. Specifically, Section 12B supplies a check that the graph-density hypothesis of the exponentiation theorem 9.2 is satisfied whenever D carries a norm such that $\|e_n\| \leq |n|^\nu$ for some integer ν. This leads to the conclusion that $\mathcal{L}_0$ exponentiates in the natural Hilbert space $\ell^2(\mathbb{Z}) \supset D$, and also in the Banach space completions E_p of D with respect to the $L^p(T)$ norms lifted back from the circle $T = \{z \in \mathbb{C}: |z|=1\}$

via the Fourier transform when $1 \leq p \leq \infty$. (That is, we identify D with the trigonometric polynomials in $L^p(T)$ via $e_n = (e^{in\theta})^\wedge$.) In Sections 12C-D, generalized exponentials for $\mathcal{L}_0$ in $\ell^p(\mathbb{Z})$ for $p \neq 2$ are studied, both as differentiable Fréchet-space "integrals" on the C^∞ vectors and as "smeared" or "distribution" exponentials on the $\ell^p(\mathbb{Z})$ themselves. Then our perturbation-continuation Theorem 9.3 is applied in Section 12E to obtain exponentials for the $\mathcal{L}(q,\tau)$ in $\ell^2(\mathbb{Z})$ and E_p, with $\ell^p(\mathbb{Z})$ generalizations in Section 12F for $p \neq 2$. We also obtain the fact that for fixed τ, the perturbation $U_i(q,q',\tau) = B_i(q,\tau)-B_i(q',\tau)$ is compact in E_p and in $\ell^2(\mathbb{Z})$, $i = 0,1,2$, yielding the fact that the exponentiated group representations also differ by compact additive perturbations which exhibit nice operator-norm dependence upon q and upon group-parameters. Similar perturbation results are obtained in Section 12F for the differentiable and smeared $\ell^p(\mathbb{Z})$ exponentials when $p \neq 2$. Despite the fact that the defining formulae for the $\mathcal{L}(q,\tau)$ are essentially familiar from Bargmann's initial work on $SL(2,\mathbb{R})$ [Bg], the analytic series of exponentials for these Lie algebras in $\ell^2(\mathbb{Z})$, E_p and $\ell^p(\mathbb{Z})$ have not previously been discussed, the non-unitary representations among them lie outside the scope of previously-published exponentiation methods, and the compact-perturbation phenomenon seems to be new.

In order to clarify the significance of this last remark, it is necessary to relate the elementary operator theory for these Heisenberg-matrix Lie algebras to deeper issues in the theory of topologically completely irreducible (TCI) Banach representations for the Lie group $SL(2,\mathbb{R})$ and of its simply-connected covering group $\tilde{G}$ (cf. Chapter 4 of [Wr] for terminology). In Section 12A, we take up the more algebraic aspects of these issues, pointing out that all algebraically-irreducible $\mathfrak{k}$-finite modules for the Lie algebra $s\ell(2,\mathbb{R})$ (with nontrivial $\mathfrak{k}$-isotopic components) appear as direct summands in the $s\ell(2,\mathbb{R})$ modules implicitly defined by the $\mathcal{L}(q,\tau)$ on D. Consequently all TCI Banach representations turn out to be infinitesimally (and Naimark-) equivalent to sub-representations of the exponentials for the $\mathcal{L}(q,\tau)$. On $\ell^2(\mathbb{Z})$ in particular, every "infinitesimally unitary" representation is realized unitarily, with "principal" and "complementary" series appearing irreducibly and "discrete" series appearing direct-summed with non-unitary representations (cf.

Section 12G). (For τ integral or half-integral, discrete series appear in "discrete triples" consisting of the two unitary discrete series and the associated finite-dimensional non-unitary representation with the same infinitesimal character.) Moreover, these singular reducible discrete series exponentials for the $\mathfrak{L}(q,\tau)$ are reached as "compact" norm-limits of non-unitary analytic continuations (in the Casimir-character parameter q) from unitary principal and complementary series. As we point out more fully in Section 12H*, these $\mathfrak{L}(q,\tau)$ exponentials in $\ell^2(\mathbb{Z})$ provide an alternative to Harish-Chandra's subquotient construction ([Wr, Chapter 5] and [Lg]) which exhibits several quite different technical advantages.

Other matters relating to the representation theory of $SL(2,\mathbb{R})$ and $\tilde{G}$ are taken up in Sections 12A, 12G and 12H: Section 12A discusses generalizations beyond $\ell^2(\mathbb{Z})$, E_p and $\ell^p(\mathbb{Z})$, while Section 12G takes up unitary equivalences ($\tau \simeq \tau+m$ for $m \in \mathbb{Z}$) direct-sum decompositions and single-valued representations of $SO(2,1)$ and $SL(2,\mathbb{R})$, and Section 12H describes the less-tractable perturbation behavior of other familiar analytic series of representation for $\tilde{G}$.

Our reasons for devoting so much attention to the representation theory of $s\ell(2,\mathbb{R})$ are the usual ones (cf. Lang's book $SL_2(\mathbb{R})$ [Lg]). Primarily, $SL(2,\mathbb{R})$ and its covering group serve as the most accessible (and computationally simplest) prototypes for other higher-dimensional semisimple Lie groups of interest in physics and harmonic analysis. For example, $s\ell(2,\mathbb{R})$ is related in two ways to the Lorentz group $SO(3,1)$ (the invariance group of the Minkowski metric $x_0^2-x_1^2-x_2^2-x_3^2$ in $\mathbb{R}^4$) from relativistic quantum theory: $s\ell(2,\mathbb{R}) \simeq so(2,1)$ (the Lie algebra of Minkowsky space-time with two space-dimensions, still of active interest in constructive field theory) while its complexification is $s\ell(2,\mathbb{C}) \approx so(3,1)$. Much of the perturbation theory developed here extends to $SO(3,1)$ and $SL(2,\mathbb{C})$, as can be checked by inspection of the appropriate sections in Naimark [Nk]. The portions which have been checked carry over in more interesting ways at least to the corresponding unitary theory of the de Sitter group $SO(4,1)$ as treated by Dixmier [Dx 2]. (The de Sitter group has discrete series, while the Lorentz does not.) Moreover, $s\ell(2,\mathbb{R}) \simeq su(1,1)$, the Lie algebra of the

---

* Section 12H, as presented below, is a shortened version of the original version which appeared in an earlier version of the manuscript.

conformal group $SU(1,1) = \left\{\begin{pmatrix}\alpha & \bar{\beta} \\ \beta & \bar{\alpha}\end{pmatrix} : |\alpha|^2-|\beta|^2 = 1\right\}$, which serves as a prototype for $SU(2,2) \simeq SO(4,2)$ in the study of various groups which have proven important in particle physics [BB] and in the chronometric cosmology of Segal [Sg 3]. Preliminary examination of the incomplete (and apparently in part incorrect) analysis of "balanced" SU(2,2) modules by Yao [Ya] indicates that many of the perturbation phemonema that are described here persits in that setting as well. Generally, it appears likely that all semisimple Lie algebras $\mathfrak{g}$ have a series of "balanced" $\mathfrak{k}$-finite modules which exhibit compact perturbations with respect to "characters" (Casimir values such as q), while τ-continuation properties are to be expected for those whose compactly-embedded subalgebras have a center (i.e., those associated with Hermitian symmetric spaces and holomorphic discrete series). The situation for discrete series and for "all" TCI Banach representations is less clear from the existing literature. Many of our methods presently require so much detail that computations become unmanageable as the dimension of $\mathfrak{g}$ becomes large. These matters are discussed further in Sections 12A and 12H.

12A. Connections to the theory of TCI representations of semisimple groups on Banach spaces

This section relates the operator Lie algebras $\mathfrak{L}_0$ and $\{\mathfrak{L}(q,\tau): (q,\tau) \in \mathbb{C}^2\}$ to the representation theory of the appropriate classical Lie groups: those whose Lie algebra $\mathfrak{g}$ are spanned by a basis $\{X_i: i = 0,1,2\}$ which satisfies the commutation relations corresponding to (2):

$$[X_0,X_1] = -X_2, \quad [X_0,X_2] = X_1, \quad [X_1,X_2] = X_0. \qquad (2')$$

Various features of known or conjectured generalizations to higher-dimensional $\mathfrak{g}$ are also discussed. This material is largely independent of the main developmental in Sections 12B-12G, except insofar as it serves to account for some of the terminology employed there.

We begin with a quick check that the operator triples $\{A_0,A_1,A_2\}$ and $\{B_i(q,\tau): i = 0,1,2\}$ satisfy the relations corresponding to (2') (i.e., (2) and its analog for the B_i). It is simpler to compute with the weighted shifts $B_\pm$ and their analogs $A_\pm e_n = ne_{n\pm1}$, verifying the "compact Cartan root" commutation relations $[A_0,A_\pm] = \pm iA_\pm$ (respectively, $[B_0,B_\pm] = \pm iB_\pm$) and $[A_+,A_-] = 2iA_0$ (respectively, $[B_+,B_-] = 2iB_0$). Then (2) will follow by

$$[A_0,A_1] = \frac{i}{2}([A_0,A_+]+[A_0,A_-]) = -\tfrac{1}{2}A_+ + \tfrac{1}{2}A_- = -A_2,$$

with $[A_0,A_2] = A_1$ obtained similarly, and by

$$[A_1,A_2] = \frac{i}{4}([A_+,A_+]-[A_+,A_-]+[A_-,A_+]+[A_-,A_-]) = \frac{i}{4}(-4iA_0) = A_0.$$

(The corresponding calculation yields the analogous relation for the B's.) To obtain the Cartan relations, one computes

$$(A_0A_\pm-A_\pm A_0)e_n = [i(n\pm1)n-n(in)]e_{n\pm1} = \pm ine_{n\pm1} = \pm iA_\pm e_n$$

and

$$(A_+A_--A_-A_+)e_n = [(n-1)n-(n+1)n]e_n = -2ne_n = 2iA_0e_n.$$

(We leave the calculations for the B's to the reader, noting only that $[B_+,B_-]e_n = (\gamma_{n-1}^2-\gamma_n^2)e_n = -2(n+\tau)e_n$ is the only verification in which the explicit form of $\gamma_n = \mathrm{sgn}(n)(q+(n+\tau)(n+\tau+1))^{\frac{1}{2}}$ plays a role; even there, the sign choice is "squared out.")

Next, we recall some of the elementary background concerning classical matrix Lie algebras and Lie groups associated with the commutation relations (2'). (Strictly speaking, these are irrelevant to our main purposes: one can check directly that the elements of $\mathfrak{g}$ = real span$\{X_i: i = 0,1,2\}$ satisfy the Jacobi and anti-symmetry relations for [,] when the latter is bilinearly extended to $\mathfrak{g}$, and then invoke a standard general result asserting the existence of a simply-connected Lie group G with $\mathfrak{g}$ as Lie algebra [Hc, Chapter XII], thereby obtaining a Lie group whose representation theory is to be discussed. However, as indicated in the introductory paragraphs for the chapter, the significance of some results depends upon their connections with the classical groups SL(2, $\mathbb{R}$), SO(2,1) and SU(1,1). The Lie group most frequently discussed in connection with the commutation relations (2') is the conformal group of the upper half-plane,

$SL(2,\mathbb{R}) = \left\{\begin{pmatrix} a & b \\ c & d \end{pmatrix} : ad-bc = 1;\ a,b,c,d \in \mathbb{R}\right\}$ (e.g., in our section title). The Lie algebra $s\ell(2,\mathbb{R})$ of this group consists of the real 2×2 matrices of trace 0, for which a suitable basis is $X_0 = \frac{1}{2}\begin{pmatrix} 0 & -1 \\ 1 & 0 \end{pmatrix}$, $X_1 = \frac{1}{2}\begin{pmatrix} 0 & 1 \\ 1 & 0 \end{pmatrix}$ and $X_2 = \frac{1}{2}\begin{pmatrix} 1 & 0 \\ 0 & -1 \end{pmatrix}$. (Readers unfamiliar with this group can easily verify that (2') holds for these matrices. Note that if $\exp(tX) = \begin{pmatrix} a(t) & b(t) \\ c(t) & d(t) \end{pmatrix} \in SL(2,\mathbb{R})$, then $X = \begin{pmatrix} a'(0) & b'(0) \\ c'(0) & d'(0) \end{pmatrix}$ must have

$$0 = \frac{d}{dt}(a(t)d(t)-b(t)c(t))\Big|_{t=0}$$

$$= a'(0)d(0)+a(0)d'(0)-b'(0)c(0)-b(0)c'(0) = a'(0)+d'(0)$$

since $\begin{pmatrix} a(0) & b(0) \\ c(0) & d(0) \end{pmatrix} = \begin{pmatrix} 1 & 0 \\ 0 & 1 \end{pmatrix}$, the identity matrix, whence every $X \in s\ell(2,\mathbb{R})$ has trace 0. Since $SL(2,\mathbb{R})$ is clearly three-dimensional, all trace-0 matrices must appear.)

Here, $\exp(tX_0) = \begin{pmatrix} \cos t/2 & \sin t/2 \\ -\sin t/2 & \cos t/2 \end{pmatrix}$ is easily verified, whence $K = \{\exp(tX_0): 0 \le t \le 4\pi\}$ is a maximal compact subgroup of this group.

The conformal group of the unit disc, $SU(1,1)$ $= \left\{\begin{pmatrix} \alpha & \bar{\beta} \\ \beta & \bar{\alpha} \end{pmatrix} : |\alpha|^2-|\beta|^2 = 1;\ \alpha,\beta \in \mathbb{C}\right\}$ is probably of equal mathematical interest; it is isomorphic to $SL(2,\mathbb{R})$ via the inner automorphism of $SL(2,\mathbb{C})$ induced by $\Phi = \begin{pmatrix} 1 & -1 \\ -i & -i \end{pmatrix}$: $SU(1,1) = \Phi^{-1}SL(2,\mathbb{R})\Phi$ [Sℓ]. In the Lie algebra $su(1,1)$, X_0 corresponds to $\frac{1}{2}\begin{pmatrix} i & 0 \\ 0 & -i \end{pmatrix}$, X_1 to $\frac{1}{2}\begin{pmatrix} 0 & -i \\ i & 0 \end{pmatrix}$ and X_2 to $\frac{1}{2}\begin{pmatrix} 0 & -1 \\ -1 & 0 \end{pmatrix}$ under the infinitesimal version of this isomorphism.

One of the principal advantages of $SU(1,1)$ over $SL(2,\mathbb{R})$ is that the former suggests a natural parametrization which can be used to realize the covering group $\widetilde{G}$ explicitly

[Sℓ, Section II.I]: for $\begin{pmatrix}\alpha & \bar{\beta}\\ \beta & \bar{\alpha}\end{pmatrix} \in SU(1,1)$ one puts $\gamma = \beta/\alpha \in \Delta = \{z \in \mathbb{C}: |z| < 1$ and $\omega = \arg(\alpha) \in [0,2\pi)$ so that $\alpha = e^{i\omega}(1-|\gamma|^2)^{-\frac{1}{2}}$ and $\beta = e^{i\omega}\gamma(1-|\gamma|^2)^{-\frac{1}{2}}$, expressing SU(1,1) topologically as $\Delta\times T$, so that topologically $\tilde{G} = \Delta\times\mathbb{R}$. In fact, the group-products in this coordinatization can be described by $(\gamma,\omega)(\gamma',\omega') = (\gamma'',\omega'')$ where

$$\begin{aligned}\gamma'' &= (\gamma+\gamma' e^{-2i\omega})(1-\bar{\gamma}\gamma' e^{-2i\omega})^{-1},\\ \omega'' &= \omega+\omega'+(2i)^{-1}\log\{(1+\bar{\gamma}\gamma' e^{-2i\omega})(1+\overline{\gamma\gamma'}e^{2i\omega})^{-1}\},\end{aligned} \tag{5}$$

with the second identity interpreted Mod 2π in SU(1,1) and literally in $\tilde{G}$. We shall have no direct need for this explicit realization of $\tilde{G}$; it simply provides an elementary alternative to the general machinery metnioned above when supplying a simply-connected $\tilde{G}$ with $\mathfrak{g}$ as Lie algebra.

At the opposite extreme to $\tilde{G}$, we have the physics-related "little Lorentz group" SO(2,1), whose identity component $SO_e(2,1)$ is the "smallest" connected Lie group with $\mathfrak{g}$ as Lie algebra, in the sense that it is a homomorphic image of every connected Lie group G with $\mathfrak{g}$ as Lie algebra. This observation results from the fact that $SO_0(2,1)$ can be realized as the identity component in the automorphism group $\mathrm{Aut}(\mathfrak{g})$ of $\mathfrak{g}$, hence as the range of the adjoint representation of any such G on its Lie algebra $\mathfrak{g}$. Here, the Ad G-invariant Cartan-Killing form $B(X,Y) = \mathrm{trace}(\mathrm{ad}\,X\ \mathrm{ad}\,Y)$ plays the role of the Lorentz metric: if $X = x_0X_0 + x_1X_1 + x_2X_2$, then $B(X,X) = x_1^2+x_2^2-x_0^2$. It is this identification that motivates the convention that $X \in \mathfrak{g}$ is <u>spacelike</u> if and only if $B(X,X) > 0$, <u>lightlike</u> (or in the <u>light cone</u>) if and only if $B(X,X) = 0$ and <u>timelike</u> if and only if $B(X,X) < 0$. Note that X_0 is timelike and X_1, X_2 are spacelike. We also use the fact [Bg] that $SL(2,\mathbb{R})$ and SU(1,1) are double-covers for $SO_0(2,1)$: in SU(1,1) the element $\begin{pmatrix}-1 & 0\\ 0 & -1\end{pmatrix} = \begin{pmatrix}e^{\pi i} & 0\\ 0 & e^{-\pi i}\end{pmatrix}$ $= \exp(2\pi X_0)$ goes into the unit automorphism of $\mathfrak{g}$. Consequently, half of the $\mathcal{L}(q,\tau)$ which exponentiate to global representations of $SL(2,\mathbb{R}) \simeq SU(1,1)$ yield representations of SO(2,1): those with τ integral. (If $\tau \in \mathbb{Z}/2$, one gets representations of $SL(2,\mathbb{R})$. For other real rational τ, one obtains representations of other Lie groups with finite center that cover $SL(2,\mathbb{R})$, while for irrational and non-real τ, only representations of $\tilde{G}$ are obtained. Since we wish to discuss perturbation-continuation in τ, we focus most of our attention on $\tilde{G}$ in what follows.)

The balanced $s\ell(2,\mathbb{R})$-modules $\mathfrak{L}(q,\tau)$ are then related to the Banach representation theory of the simply-connected group $\tilde{G}$ roughly as follows. If $V : \tilde{G} \to Aut(E)$ is a strongly continuous topologically completely irreducible (TCI) representation on a Banach space E [Wr], the parameters q and τ are determined by the images of two members of the enveloping algebra $\mathfrak{U} = \mathfrak{U}(s\ell(r,\mathbb{R}))$ under the associated infinitesimal representation dV : q is the scalar (by quasi-simplicity) value assumed by dV(Q) for $Q = X_0{}^2 - X_1{}^2 - X_2{}^2$ the Casimir element which generates the center $\mathfrak{z}(\mathfrak{U})$ of $\mathfrak{U}$, while $i\tau$ is most conveniently chosen to be the eigenvalue of $dV(X_0)$ with smallest $|Im(\tau)|$. Aside from certain minor technicalities (resulting from the fact that $\tilde{G}$ does not have finite center) the basis for this relationship between $\mathfrak{L}(q,\tau)$ and $dV(s\ell(2,\mathbb{R}))$ is a familiar combination of generalities from Chapter 4 of [Wr] and computations used by Bargmann [Bg] and Pukanszky [Pk]. We recall these, in the interests of self-containment and to advertise the choices leading to the "balanced" form of the Lie algebra modules. Our review proceeds in two steps. Assuming that $dV(X_0)$ has an eigenvector, we describe the construction of a "$\mathfrak{k}$-finite" basis $\{e_n\}$ and domain D_0 upon which $dV(s\ell(2,\mathbb{R}))$ acts as a module naturally isomorphic to (a direct summand of) the appropriate $\mathfrak{L}(q,\tau)$. Then the existence of such eigenvectors and the "largeness" properties of the domain D_0 are examined.

First, let $V: \tilde{G} \to Aut(E)$ be a TCI Banach representation, let $D_\omega = D_\omega(V)$ be the dense set of analytic vectors for V [Wr, Theorem 4.4.5.7, p. 279], and let $dV: \mathfrak{g} \to \mathfrak{A}(D_\omega)$ be the $\mathfrak{g}$ (and $\mathfrak{U}(\mathfrak{g})$) module induced on D_ω [Wr, p. 278] for $\mathfrak{g} = s\ell(2,\mathbb{R})$. Put $C_i = dV(X_i)$, $i = 0,1,2$, and $C_\pm = C_1 \pm iC_2$. Then, since the X_i (and C_i by the homomorphism property of dV) satisfy the commutation relations analogous to (2), it follows easily that $[C_0,C_\pm] = \pm iC_\pm$ and $[C_+,C_-] = 2iC_0$. (The first identities indicate as in Section 12D that $C_\pm$ should be "spectral shifts" for C_0...). In fact, if C_0 happens to have an eigenvector $u_0 \neq 0$ in D_ω with $C_0 u_0 = i\tau u_0$ for some $\tau \in \mathbb{C}$ (we recall later why it must), we have

$$C_0 C_\pm u_0 = C_\pm C_0 u_0 + [C_0, C_\pm] u_0 = i(\tau \pm 1) C_\pm u_0 \tag{6}$$

whence two nearby candidate-eigenvectors $C_\pm u_0 \in D$ exist if these are nontrivial. (Otherwise (6) is trivially true.) In fact, nontriviality turns out to depend precisely upon the nonvanishing of $\gamma_{-1}(q,\tau)$ and $\gamma_0(q,\tau)$ for

$$q = dV(Q) = dV(X_0^2 - X_1^2 - X_2^2) = C_0^2 - C_1^2 - C_2^2 = C_0^2 + \tfrac{1}{2}(C_+ C_- + C_- C_+),$$

the scalar value assigned to the central quadratic Casimir operator according to [Wr, 4.4.1.5, p. 257]. To see this, we have from the last form of $dV(Q)$ that

$$C_+ C_- + C_- C_+ = 2(q - C_0^2),$$

while the earlier identity

$$C_+ C_- - C_- C_+ = [C_+, C_-] = 2iC_0$$

leads by substitution to

$$\begin{aligned} C_+ C_- &= \tfrac{1}{2}[2(q - C_0^2) + 2iC_0] = q - C_0(C_0 - i), \\ C_- C_+ &= q - C_0(C_0 + i). \end{aligned} \tag{7}$$

Thus

$$\begin{aligned} C_+ C_- u_0 &= q + \tau(\tau - 1) u_0 = \gamma_{-1}^2(q,\tau) u_0 , \\ C_- C_+ u_0 &= q + \tau(\tau + 1) u_0 = \gamma_0^2(q,\tau) u_0 . \end{aligned} \tag{7'}$$

Consequently, if $\gamma_0(q,\tau) \neq 0$, $C_+ u_0 \neq 0$ and we may define a second "balancing" eigenvector by $u_1 = \gamma_0^{-1} u_0$, so that $C_+ u_0 = \gamma_0 u_1$ by definition and $C_- u_1 = \gamma_0^{-1} C_- C_+ u_0 = \gamma_0 u_0$ by (7'), in analogy to the comparable cases of (3) for $\mathcal{L}(q,\tau)$. Similarly, if $\gamma_{-1}(q,\tau) \neq 0$, $C_- u_0 \neq 0$ and $u_{-1} = \gamma_{-1}^{-1} C_- u_0$ has

$$C_+ u_{-1} = \gamma_{-1} u_0,\ C_- u_0 = \gamma_{-1} u_{-1}.$$

Replacing τ by $\tau\pm1$ (and, inductively, by $\tau\pm n$) in (6) and (7'), we can then recursively define an expanding bilateral sequence u_n of "balancing" eigenvectors in D_ω which satisfy the relations

$$C_0u_n = i(n+\tau)u_n,\ C_+u_n = \gamma_n u_{n+1},\ C_-u_n = \gamma_{n-1}u_{n-1}. \tag{3'}$$

This process can continue so long as $\gamma_n(q,\tau) \neq 0$ (to define u_{n+1} from u_n for $n > 0$) or $\gamma_{n-1}(q,\tau) \neq 0$ (to define u_{n-1} from u_n for $n < 0$). For "generic" q,τ, the quadratic function $\gamma^2_{q,\tau}(z) = q+(z+\tau)(z+\tau+1)$ has no integer roots at all, and in these cases (3') holds for all $n \in \mathbb{Z}$, setting up an isomorphism between $\mathfrak{L}(q,\tau)$ and a submodule of the one defined on D_ω by dV. (We defer further discussion of the density, core properties, etc., of $D_0 = \text{span}\{u_n: n \in \mathbb{Z}\}$ until later.) If $\gamma^2_{q,\tau}(z)$ has one integer root m_0, then the resulting sequence of eigenvectors is only one-sided infinite: if $m_0 \geq 0$, then it terminates at u_{m_0} and $C_+u_{m_0} = 0$, while if $m_0 < 0$, then it terminates at u_{m_0+1} with $C_-u_{m_0+1} = 0$. It is easy to see that, in this case, $\mathfrak{L}(q,\tau)$ splits as the direct sum of two submodules: $\mathfrak{L}_-(q,\tau)$ on $D_- = \text{span}\{e_n: n \leq m_0\}$ and $\mathfrak{L}_+(q,\tau)$ on $D_+ = \text{span}\{e_n: m_0 < n\}$; if $m_0 \geq 0$ our module on $D_0 = \text{span}\{u_n: n \leq m_0\}$ is isomorphic to $\mathfrak{L}_-(q,\tau)$ while if $m_0 < 0$, $D_0 = \text{span}\{u_n: n > m_0\}$ corresponds to D_+ and $dV(\mathfrak{g})$ to $\mathfrak{L}_+(q,\tau)$. Finally, if $\gamma_{q,\tau}(z) = [q-\frac{1}{4}+(\tau+z+\frac{1}{2})^2]^{\frac{1}{2}}$ has two integer zeros $m_1 < m_2$, then $\mathfrak{L}(q,\tau)$ splits into two semi-infinite summands $\mathfrak{L}_\pm(q,\tau)$ and a finite-dimensional $\mathfrak{L}_F(q,\tau)$ on $D_+ = \text{span}\{e_n: n > m_2\}$, $D_- = \text{span}\{e_n: n \leq m_1\}$ and $D_F = \text{span}\{e_n: m_1 < n \leq m_2\}$, respectively. Depending upon the signs of m_1 and m_2, the (densely-defined ... below) module in D_ω will be isomorphic to one of these three summands.

In fact, the submodule of the $\mathfrak{L}(q,\tau)$-module obtained above is necessarily algebraically irreducible: every algebraically irreducible $\mathfrak{L}(q,\tau)$-submodule is defined on a subspace $D_{k\ell} = \text{span}\{e_n: k < n \leq \ell\}$, where either $k = -\infty$ or

$\gamma_k(q,\tau) = 0$, and either $\ell = \infty$ or $\gamma_\ell(q,\tau) = 0$, and $\gamma_n(q,\tau)$ has no zeros between k and ℓ. This claim (which follows whether or not $D_{k\ell}$ is a priori the range of an isomorphism from the $\mathfrak{k}$-finite infinitesimal module of a TCI Banach representation ...) is easy to establish. Indeed, if $P : D \to D_0 = PD$ projects D onto a nontrivial $\mathfrak{L}(q,\tau)$-invariant subspace, then for some $n \in \mathbb{Z}$, we have $Pe_n \neq 0$. But since P commutes with $B_0(q,\tau)$, $Pe_n \in D_0$ is in the $i(n+\tau)$-eigenspace and is a multiple of e_n. (In fact, $Pe_n = e_n$.) Since both $B_\pm(q,\tau)$ commute with P as well, a repetition of the argument for the $C_\pm$ supplies an expanding bilateral sequence of multiples of $e_{n\pm k} \in D_0$ which can terminate only at an integer zero of $\gamma_{q,\tau}(z)$ as described above. Examination of the three cases ($\gamma_{q,\tau}(z)$ has two, one or no integer zeros) shows that non-overlapping invariant "irreducible" subspaces always result: the splitting at a root $\gamma_m(q,\tau) = 0$ yields $B_+(q,\tau)e_m = \gamma_m(q,\tau)e_{m+1} = 0$ $= \gamma_m(q,\tau)e_{m-1} = B_-(q,\tau)e_{m+1}$ so that both "upper" and "lower" subspaces are invariant. (Compare the discussion of subquotients in 12H.)

To obtain an eigenvector $u_0 \in D_\omega$, we proceed essentially as in [Pk]. That is, if $\tilde{K} = \{\exp(tX_0)\} \subset \tilde{G}$, the adjoint representation $\mathrm{Ad} : \tilde{G} \to \mathrm{Aut}(\mathfrak{g})$ sends $\tilde{K}$ onto the compact group $\{\exp(t\ \mathrm{ad}\ X_0): 0 \leq t \leq 2\pi\}$, with kernel $\{\exp(2\pi nX_0): n \in \mathbb{Z}\}$ contained in the center $Z(\tilde{G}) = \mathrm{kernel}(\mathrm{Ad})$. Hence, $V(\exp(2\pi nX_0))$ commutes with $V(G)$, and by Schur's Lemma for TCI Banach representation [Wr, Prop. 4.2.2.3] must be scalar; choose $\tau_0 \in \mathbb{C}$ so that $e^{i\tau_0} = V(\exp(2\pi X_0))$. Then $W(t) = e^{-it\tau_0}V(\exp(tX_0))$ is easily seen to be a 2π-periodic strongly-continuous one-parameter group of operators generated by $C_0 - i\tau_0$. For any $u \neq 0$ in dense D_ω and any $m \in \mathbb{Z}/2$ we then project out an im-eigenvector for $C_0 - i\tau_0$ by the usual "character formula":

$$u_m = \int_0^{2\pi} e^{-imt}W(t)u\,dt = \int_0^{2\pi} e^{-i(m+\tau_0)t}V(\exp(tX_0))u\,dt.$$

By "vector-valued Fourier analysis", we see that at least one of these $u_m \neq 0$, while [Wr, Lemma 4.4.5.15] ensures that $u_m \in D_\omega$, and

$$W(s)u_m = \int_0^{2\pi} e^{-imt}W(s+t)u\,dt = \int_0^{2\pi} e^{-im(r-s)}W(r)u\,dr$$
$$= e^{ims}u_m$$

ensures that u_m is an eigenvector for $C_0 - i\tau_0$ with eigenvalue im (i.e., $C_0u_m = i(m+\tau_0)u_m$). Take $\tau = \tau_0+m$, $u_0 = u_m$. Hence in all cases described above (relating 0 to the roots of $\gamma_{q,\tau}(z)$), the space $D_0 \subset D_\omega$ exists. Since it is invariant under $dV(\mathfrak{g})$, a standard analytic vectors result [Wr, Prop. 4.4.5.6] yields that $\bar{D}_0$ is invariant under V(G), whence by the TCI property, D_0 is dense in E.

In addition to its density, D_0 exhibits other "largeness" properties. For one, every irreducible (hence one-dimensional) representation of the Abelian group $\tilde{K}$ which occurs as a subrepresentation of $V|_{\tilde{K}}$ must have its representation space as one of the $\mathbb{C}u_m$: D_0 contains all "$\hat{\tilde{K}}$-isotypic components of V", and V is in this sense "$\tilde{K}$-finite". To check this remark, note that if $V(\exp(tX_0))u = e^{\lambda t}u$ for any $u \in E$, then $e^{2\pi\lambda}u = V(\exp(4\pi X_0))u = e^{2\pi i\tau_0}u$ by the above, or $(u \neq 0)$ $e^{2\pi(\lambda - i\tau_0)} = 1$ and $\lambda - i\tau_0 \in \mathbb{Z}$, so $\lambda = i(\tau_0+m)$. But then u is in the range of the bounded projection operator $P(m) = (\frac{1}{2}\pi)\int_0^{2\pi} e^{-imt}W(t)dt$, which is the closure of $P(m)D_0$ since D_0 is dense. If u_m occurs in D_0, then $P(m)D_0 = \mathbb{C}u_m$ agrees with its closure and contains u, while if u_m does not appear (i.e., the construction terminates before reaching u_m, owing to a zero of $\gamma_n(q,\tau)$), then $P(m)D_0 = \{0\}$ is closed and u = 0. Another largeness property of D_0 is its core property: every group-generator $\bar{C} = dV(X)^-$ for $X \in \mathfrak{g}$ is the closure of its restriction to D_0. (Similarly, the $\|\cdot\|_n$-closure of D_0 in the C^n-vectors $E^n(V)$ coincides with $E^n(V)$, and D_0 is dense in the Fréchet space $E^\infty(V)$ of C^∞-vectors.) In fact, if D_c is the closure of D_0 in the $\bar{C}$-graph-norm, it is easily seen to be invariant under $V(t,C) = V(\exp(tC))$, whence $D_c = D(\bar{C})$ by the pregenerator theorem. Indeed, if $v = \Sigma\{t^m/m!\, C^m u: 0 \leq m \leq n\}$, then $Cv = \Sigma\{t^m/m! C^{m+1}: 0 \leq m \leq n\}u = \Sigma\{t^m/m!\, C^m\}Cu$ for $u \in D_0$, and since $Cu \in D_0$ as well, there is a $\delta > 0$ such that $V_n(t,C)u =_{\text{Def}} \Sigma\{t^m/m! C^m u: 0 \leq m \leq n\} \to V(t,C)u$ and $V_n(t,C)Cu \to V(t,C)u$ as $n \to \infty$ for $|t| \leq \delta$. Thus, since $V_n(t,C)u \in D_0$, $V(t,C)u \in D_c$. Then the commutation relation $\bar{C}V(t,C) = V(t,C)\bar{C}$ extends this by limits to all $u \in D_c$. The arguments for the C^n-norms $\|\cdot\|_n$ are slightly more complicated, since they involve truncated commutation relations connecting $V_n(t,C)$ to the basis operators C_i, but they proceed in precisely the same way. (We omit details.)

The discussion above has been restricted to the description of a suitable module isomorphism relating the $\tilde{K}$-finite module $(dV(\mathfrak{g}),D_0)$ for a single TCI Banach representation $V: \tilde{G} \to \text{Aut}(E)$ to a suitable $\mathfrak{L}(q,\tau)$. Consequently, it does not make explicit the relationship between the series of exponentials of the $\{\mathfrak{L}(q,\tau): (q,\tau) \in \mathbb{C}^2\}$ (on $\ell^2(\mathbb{Z})$, say) and other familiar series of TCI Banach representations $\{V_{q,\tau}: (q,\tau) \in \Omega \subset \mathbb{C}^2\}$ considered in the literature. (By a series we mean, as usual, a parametrized family of representations of $\tilde{G}$ on a fixed Banach space E: a cross-section of some portion of the non-unitary dual space $(\tilde{G})^\wedge$ of Naimark-equivalence-classes of TCI Banach representations.) In order to understand the source of the extreme variations in (q,τ) dependence of these series (illustrated in detail in Section 12H), it is necessary to recast the construction in terms of (q,τ)-dependent intertwining operators. Briefly, we observe first that the module isomorphism is implemented by a linear bijection $W(q,\tau): D_0(q,\tau) \to D$ sending the dense $\tilde{K}$-finite domain $D_0(q,\tau)$ (for $V_{q,\tau}$ in E) into $D_0 \subset \ell^2(\mathbb{Z})$. Here, $D_0(q,\tau)$ may be dependent upon the continuation parameters (as it is in the "noncompact picture" of Kunze-Stein-Sally [Sℓ]-Section 12H). In other cases, $D_0(q,\tau) \subset E$ may not change as q and τ vary, but it is typically the case that the basis vectors $e_n = e_n(q,\tau)$ do vary with these parameters, at least via (q,τ)-dependent scalar multiples of fixed eigenvectors. (These are explicitly visible in the developments of Bargmann [Bg] and Pukanszky [Pk]. See, for example, our discussion of the "induced" principal series in the "compact picture": Section 12H.) It can be shown that $W(q,\tau)$ is necessarily closable and that it in most cases extends to a suitable $V_{q,\tau}$-invariant domain $D_\infty \subset E$ that is independent of (q,τ) and consists of C^∞-vectors for all $V_{q,\tau}$ and $dV_{q,\tau}(\mathfrak{g})$ as well. (For example, if $V_{q,\tau}$ runs through the unitary principal series, all $W(q,\tau)$ are then unitary, cf. Section 12H). Then we find that

$$dV_{q,\tau}(X_i) = W(q,\tau)^{-1}B_i(q,\tau)W(q,\tau), \quad i = 0,1,2. \tag{8}$$

Consequently, the (q,τ) dependence of the $C_i(q,\tau) = dV_{q,\tau}(X_i)$ is determined not only by the (q,τ)-dependence of the $B_i(q,\tau)$ but by that for the intertwining operators $W(q,\tau)$ as well. We believe that the "balanced" property of the $\mathfrak{C}(q,\tau)$ (further elucidated below) gives them best-possible perturbation behavior: the intertwining operators $W(q,\tau)$ can at best preserve the compact perturbation properties described in Section 12E, but they often destroy them. These remarks are elaborated in Sections 12E-H.

These intertwining operators from members of other series of representations lead to interesting norms on D beyond the $\ell^2(\mathbb{Z})$, E_p and $\ell^p(\mathbb{Z})$ norms: every such $W(q,\tau)$ can be used to transport the norm from $D_\infty \subset E$ back to D, and it appears to be fruitful to consider completing D with respect to this norm and seeking to exponentiate the $\mathfrak{C}(q,\tau)$ in the resulting space. It is interesting to note in this connection that all of the usual induced ("multiplier") non-unitary principal series representations lead to norms on D that are equivalent to the $\ell^2(\mathbb{Z})$ norm (unitarily so, for the unitary principal series), so that all of these lead to the same basic exponentiation problem that is solved here. It is also clear that the norms lifted back from the Kunze-Stein-Sally normalized non-unitary principal series do not in general share these equivalence properties (see Section 12H).

Finally, we elaborate upon earlier remarks concerning the extent to which the results of this chapter are known (or expected) to extend to other Lie groups. It is useful to separate the properties of the $\mathfrak{L}(q,\tau)$ and their exponentials into five different circles of ideas:

(A) compactly-perturbed analytic continuations of the unitary principal series, with respect to "characters" q but with <u>fixed</u> τ, which reach all other unitary series either as irreducibles or as orthogonal direct summands;

(B) unification of all TCI Banach-representations within a simple picture of analytic continuation and restriction to complemented subrepresentations;

(C) analytic continuation with respect to the τ parameter;

(D) analysis of the representation of interest as Phillips perturbations of (a) special "base-point" representation (s) whose operator-theoretic and exponentiation properties are especially transparent;

(E) existence of Lie algebra modules on spaces of ℓ^p-type for $p \neq 2$ which have "smeared exponentials" but not classical C_0 ones.

Briefly, a good deal of evidence is available to the effect that our results related to (A) can be generalized to all semisimple Lie groups, but the necessary results on "balanced representations of enveloping algebra modules" do not appear to be available in full generality, and we can only describe by example what they should look like. (Most of our discussion below will be focused upon this point, which appears to be the most important and applicable aspect of the theory.) As for (B), the picture already becomes much more complicated for $SL(2,\mathbb{C})$, where the "nonunitary dual" breaks into "integral" and "half-integral" series which cannot be joined by analytic continuations but all representations within either of these series can be linked by a suitable sequence of continuations and restriction to complemented subrepresentations (details below). For the de Sitter group $SO_e(4,1)$, the obvious analytic continuation of the unitary principal series described in Dixmier [Dx 2] can (easily) be seen to reach all other unitary series, the finite-dimensional series, and various nonunitary representations as direct summands in various complicated configurations, but it is not yet clear that every TCI Banach representation is equivalent to one of these direct summands. At the time of writing, our information concerning other groups is even less complete, but we are not aware of any evidence in the literature contrary to the conjecture that (B) is part of a general pattern. We shall not attempt to reproduce the complicated details behind these remarks, dropping the discussion of (B) at this point.

As suggested earlier, continuations of τ-type as in (C) are to be expected only for those Lie algebras $\mathfrak{g}$ associated with Hermitian symmetric spaces and "holomorphic discrete series" (See Helgason [Hg], Lüscher-Mack [L-M] and Vergne-Rossi [VR] for background.) Briefly, the continuous parameter τ arises from the fact that the compactly-embedded subalgebra $\mathfrak{k}$ has nontrivial center $\mathfrak{z}(\mathfrak{k})$ (here $\mathfrak{z}(\mathfrak{k}) = \mathfrak{k} = \mathbb{R}\,A_\infty$) whose exponential in $\widetilde{G}$ is a vector-group with a continuous series of representations. Hence, the allowable eigenvalues of these $A \in \mathfrak{z}(\mathfrak{k})$ form a (complex) continuum, while those for non-central $A \in \mathfrak{k}$ turn out to be confined to a discrete lattice of possibilities. In this sense, the representation theory of $SL(2,\mathbb{R})$ (and other finite-center groups with Lie algebra isomorphic to $s\ell(2,\mathbb{R})$) is more typical of what is to be expected for general $\mathfrak{g}$. Again, we shall not pursue details.

Discussion of the base-point phenomenon (D) is complicated by four quite different roles that $\mathfrak{L}_0$ plays in the present chapter: (i) as a convenient and easily-exponentiated reference point for Phillips perturbation in the Heisenberg matrix formalism (Sections 12D and E), (ii) as a representation easily identifiable via Fourier analysis with a "Schrödinger formalism" function-space representation that also appears as part of the analytic continuation series for the induced principal series in the "compact picture" (Section 12H), (iii) as a <u>non-unitary</u> representation illustrating nontrivial applications of our graph-density exponentiation theorem 9.2 via a direct computational check of the hypothesis GD, and finally (iv) as a Lie algebra of unbounded derivations on the convolution Banach algebra $\ell^1(\mathbb{Z})$ which can (easily) be shown <u>not</u> to possess a strongly-continuous exponential. As a practical matter, (i) is the only property of compelling interest for general $\mathfrak{g}$. As pointed out earlier, continuations in the Hilbert space theory can conveniently be treated by taking any balanced "infinitesimally unitary" $\mathfrak{g}$-module (with representing operators skew-symmetric) and exponentiating it by standard analytic vector methods. Then all other representations with the same $\mathfrak{k}$-isotypic components can be expected to arise as Phillips perturbations (modulo (A) and (B), of course). Since the purpose of (ii) is primarily expository, we have not investigated its generalizability. As for (iii), it is useful to observe that everything worked out for $\mathfrak{L}_0$ in Section 12B while checking the graph-density condition of Theorem 9.2 has a natural analog for the "infinitesimally unitary" balanced module $\mathfrak{L}(\frac{1}{4},-\frac{1}{2})$, which could reasonably be described as the simplest of the $\mathfrak{L}(q,\tau)$. That is, $\gamma_n(\frac{1}{4},-\frac{1}{2}) = n$ for all $n \in \mathbb{Z}$, whence the matrix-elements of the $B \in \mathfrak{g}$ (qua module-operators) are simple enough that certain recurrence relations involved in the check of graph-density can be solved explicitly, as they are in Section 12B for $\mathfrak{L}_0$. For other $\mathfrak{g}$, it turns out that careful choice of continuation parameters can yield substantial simplification of the form of the representing Heisenberg matrices, but the corresponding recurrences for the graph-density check are still sufficiently formidable that explicit solutions are an impractical means to this end. It appears likely that qualitative results can replace the present explicit computations, possibly leading to C^∞-vector exponentiation techniques for general Banach norms that rival the competing analytic vector methods in ease of application, but we have not yet identified these. Role (iv) of the base-point representation $\mathfrak{L}_0$ is really connected with (E), discussed next, and appears to be unique to $s\ell(S,\mathbb{R})$ (or direct sums of copies of it).

Finally, we point out that (E) is in one sense unique to $s\ell(2,\mathbb{R})$, and in another sense rather general. It is special in the sense that it relies upon the Abelian property of the maximal compactly-embedded subalgebra $\mathfrak{k} \subset \mathfrak{g}$ to apply "classical" Fourier analysis on $\mathbb{Z}$ and the circle T as a tool (Section 12C and D). But it can support a rather ad-hoc technique of greater generality, in order to produce modules of ℓ^p-type, for certain other $\mathfrak{g}$, which admit smeared exponentials but not strongly continuous ones when $p \neq 2$. That is, when copies of $s\ell(2,\mathbb{R})$ can be found in $\mathfrak{g}$ (they can always be found in $\mathfrak{g}_{\mathbb{C}}\ldots$), it can be shown that copies of the $\mathfrak{L}(q,\tau)$ modules on $\ell^p(\mathbb{Z})$ arise as direct summands of natural $\mathfrak{g}$ balanced modules on the ℓ^p direct sums of the ℓ^p-normed (finite-dimensional) $\mathfrak{k}$-isotypic components that are described, say, in Section 28 of Hewitt and Ross [HR]. One then concludes that if the $\mathfrak{g}$-modules admitted a strongly-continuous exponential, this would restrict to a (local) C_0 exponential for $\mathfrak{L}(q,\tau)$ on the $\ell^p(\mathbb{Z})$ subspace, contradicting our result in Section 12D. We suspect that a more direct attack upon this topic is possible, so we omit details of the bootstrap argument suggested above.

To conclude this section, we re-examine the "balanced"

property of the $s\ell(2,\mathbb{R})$-modules, $\mathfrak{L}(q,\tau)$, and the origin of their compact perturbation properties, within a broader conceptual framework which suggests why similar phenomena are to be expected in general. We assume throughout most of the discussion that the reader is conversant with standard terminology for semisimple Lie algebras and their representations (alias modules) as surveyed in Chapter 1 of [Wr]. As background, we observe from the defining equations (3) in the introductory paragraphs of the chapter that the module operators $B_i(q,\tau) \in \mathfrak{L}(q,\tau)$ for $i = 1,2$ are linear combinations of the operators $B_\pm(q,\tau)$, and conversely. Hence when τ is fixed and q changes to q', the B_i experience compact perturbations if and only if the $B_\pm$ do, and we focus upon compactness properties of

$$\Delta_\pm(q,q',\tau) = B_\pm(q,\tau)-B_\pm(q',\tau). \tag{9}$$

If we put $S_\pm e_n = e_{n\pm 1}$ and $\Gamma(q,\tau)e_n = \gamma_n(q,\tau)e_n$, extending these unitary shifts and diagonal operators to D by linearity, then (3) shows that

$$\begin{aligned} \Delta_+(q,q',\tau) &= S_+[\Gamma(q,\tau)-\Gamma(q',\tau)] = S_+\Delta(q,q',\tau), \\ \Delta_-(q,q',\tau) &= [\Gamma(q,\tau)-\Gamma(q',\tau)]S_- = \Delta(q,q',\tau)S_- \end{aligned} \tag{10}$$

where $\Delta(q,q',\tau) = \Gamma(q,\tau) - \Gamma(q',\tau)$. Hence, since the compact operators from an ideal and $S_\pm$ are invertible, compactness of perturbations is <u>equivalent</u> to compactness of $\Delta(q,q',\tau)$. (This reversibility of the reduction does not persist for other $\mathfrak{g}$.) A simple trick shows why $\Delta(q,q',\tau)$ is compact:

$$\begin{aligned} \Delta(q,q',\tau)e_n &= [\gamma_n(q,\tau)-\gamma_n(q',\tau)]e_n = \frac{\gamma_n^2(q,\tau)-\gamma_n^2(q',\tau)}{\gamma_n(q,\tau)+\gamma_n(q',\tau)}\, e_n \\ &= (q-q')\{\gamma_n(q,\tau)+\gamma_n(q',\tau)\}^{-1}e_n \end{aligned} \tag{11}$$

from (4), since $\gamma_n^2(q,\tau) = q + (n+\tau)(n+\tau+1)$ and the summands involving n cancel in the numerator. But $\gamma_n(q,\gamma) = \mathcal{O}(|n|)$, so $\Delta(q,q',\tau)$ is a diagonal operator with eigenvalues $\mathcal{O}(1/|n|)$ and is compact on $\ell^2(\mathbb{Z})$. Roughly speaking, the same mechanism seems to operate for general semisimple $\mathfrak{g}$, except that, in general, there are several pairs $B_\pm$ of operators, each of which is described as a linear combination of products of $S_+\Gamma$ or ΓS_-, form for a variety of "shifts" $S_\pm$ and diagonal multiplication operators Γ. Moreover, the general notion of "balance" seems to entail that character parameters (such as q) appear as summands in radicals in these Γ, while the "eigenvalue" (or "weight") parameters appear quadratically as other summands; whence computations like (11) lead to the same compactness conclusions in ℓ^2 spaces. We remark in passing that this additive decoupling of character and weight parameters within radicals in Γ seems to be an automatic consequence of the process of "balancing" the "unbalanced" module representations that naturally arise by inducing representations from characters of A on $L^2(G/AN)$ when G = KAN is Iwasawa-decomposed. (This will be illustrated in Section 12H for the usual induced principal series for $SL(2,\mathbb{R})$.)

For general $\mathfrak{g}$, one must start by replacing B_0 (more accurately, $\mathfrak{k} = \mathbb{R} B_0$) and its "eigenvalues" in two different ways: by the maximal compactly-embedded subalgebra $\mathfrak{k}$ that appears in a Cartan decomposition $\mathfrak{g} = \mathfrak{k} + \mathfrak{p}$, and by a "standard" (or "maximally toroidal") Cartan subalgebra $\mathfrak{h}$ for $\mathfrak{g}$ such that $\mathfrak{h}_\mathfrak{k} = \mathfrak{h} \cap \mathfrak{k}$ is a Cartan subalgebra of $\mathfrak{k}$. Having chosen a fundamental system of positive roots Φ_+ for $\mathfrak{h}$ (with an ordering), one considers all Cartan root-element pairs $X_{\pm\alpha} \in \mathfrak{p}_\mathbb{C} = \mathfrak{p} + i\mathfrak{p}$ for $\alpha \in \Phi_+$; the representatives $B_{\pm\alpha}$ of these as module operator play the role of $B_\pm$ for $s\ell(2,\mathbb{R})$. The eigenvectors in our $s\ell(2,\mathbb{R})$ modules are replaced in a general $\mathfrak{g}$-module in two ways: by "weight subspaces" (alias joint eigenspaces) for the Cartan subalgebra $\mathfrak{h}_\mathfrak{k}$ of $\mathfrak{k}$, and by the irreducible $\mathfrak{k}$-submodules, which are indexed up to equivalence by their "highest weights" ([Wr, Chapter 3]). In general, a given $\mathfrak{k}$-module can occur to multiplicity ≥ 1, and a given weight subspace can occur with dimensions > 1 within some of those $\mathfrak{k}$-submodules, but in most of the examples that we have studied, it is possible to specify a module basis by a highest weight (determining a unique $\mathfrak{k}$-submodule) and a weight (determining a unique one-dimensional weight subspace within that submodule); it will simplify discussion to suppose that this is possible. The $B_{\pm\alpha}$ can then be expressed as sums $B_{+\alpha} = \Sigma\, S^i_{+\alpha}\Gamma^i_+$, $B_{-\alpha} = \Sigma\, \Gamma^i_- S^i_{-\alpha}$ where the $S^i_{+\alpha}$ send weight vectors u with weight ω ($B_0 u = \omega(B_0)u$ for $B_0 \in \mathfrak{h}_{\mathfrak{k}}$) into ones with weight $\omega + \alpha$ in various $\mathfrak{k}$-submodules with highest weights ω_H that are Φ_+-next to that for the submodule where u originates ($\omega_H \pm \beta$ for $\beta \in \Phi_+$). The shifts in $B_{+\alpha}$ and $B_{-\alpha}$ occur in opposing pairs; the module basis can be described as "partially balanced" if the factors $\Gamma^i_{\pm}$ coincide when $S^i_{\pm}$ are opposing shifts. (Essentially, the matrix for $B_{-\alpha}$ is the transpose of that for B_α.) But more than this simple analogy with the $B_{\pm}(q,\tau)$ seems to be required for compactness of perturbations, and it available in known examples. "Full balancing" seems to involve (a) normalizing the basis within each $\mathfrak{k}$-submodule to put it in "infinitesimal unitary" form for $\mathfrak{k}$, and (b) choosing normalization between $\mathfrak{k}$-submodules so that pairs $S^i_+\Gamma^i$ and $\Gamma^i S^i_-$ admit factorization into "internal" shifts S_I and multipliers $\Gamma_{\pm}$ within $\mathfrak{k}$-submodules and other "external" shifts S_E and multipliers Γ_E between $\mathfrak{k}$-submodules in such a way that only Γ_E (not Γ_I) contains character parameters and the same Γ_E appears in all transitions connecting the submodules involved. (To clarify: the Γ_E are associated with roots $\pm\beta$ for $\beta \in \Phi_+$; they correspond to shifts linking submodules whose highest weights differ by $\pm\beta$ and they vanish on submodules for which no others with this weight-difference exist.) Detailed examination of Naimark's discussion for SL(2,$\mathbb{C}$) [Nk] and Dixmier's presentation for $SO_e(4,1)$ [Dx 2] may aid in making sense of these rather vague remarks. We regret that the lengthiness of details prevents reproduction of some of that material for self-contained illustration beyond $s\ell(2,\mathbb{R})$. When a clear, general formulation is obtained, it will be published separately.

12B. The graph-density condition and base-point exponentials

As indicated previously, this section serves two very different purposes: it illustrates the process of checking the hypotheses in our main infinitesimal exponentiation Theorem 9.2, and it supplies exponentiated "base-point representations" (on various Banach spaces) from which analytic series of group representations can be constructed via our perturbation-continuation Theorem 9.3.

The principal issue in illustrating Theorem 9.2 revolves about the verification of the graph-density condition (GD) for a generating set $S \subset \mathfrak{L}_0$: we choose $S = \{A_0, A_2\}$ and check that for $|\mathrm{Re}(\lambda)|$ large enough, the sets $(\lambda - A_0)D$ and $(\lambda - A_2)D$ are dense in D with respect to the C^1-norm

$$\|u\|_1 = \max\{\|A_i u\|, \|u\| : \quad i = 0,1,2\} \qquad (*)$$

defined by a "suitable" norm on D. For our applications here, the condition of "suitability" on $\|\cdot\|$ turns out to be a polynomial normalization condition for the basis $\{e_n\}$ used in the definition of A_0 and A_2 (cf. (1) in the Introduction) $\|e_n\| = \mathcal{O}(|n|^\nu)$ for ν a non-negative integer. It is easy to see that this condition is satisfied by all sequence space norms $\ell^p(\mathbb{Z})$, $1 \leq p \leq \infty$, and the $L^p(T)$ norms (identifying $e_n \simeq e^{inx}$). These constructions can be extended to the $\ell^p(\mathbb{Z}, \rho)$ norms for ρ any positive, polynomially bounded "weight function" (measure) on $\mathbb{Z}$. (For the $L^p(T)$ picture, one can generalize the measure and use the $L^p(T, \mu)$ graph-norm determined by any differential operator with bounded coefficients.) However, although A_0 is easily seen to be a pregenerator ($\bar{A}_0$ generates a group) for such examples, we check the pregenerator condition for A_2 in Theorem 9.2 only for $\ell^2(\mathbb{Z})$ and $L^p(T)$, using a variety of more technical methods (numerical range, vector field machinery from Chapter 11, analytic vector methods, etc.).

With respect to the construction of base-point representations for use in the analytic continuation program, we should point out immediately that the existence of exponentials for $\mathfrak{L}_0$ is hardly surprising from several perspectives: the exponential is nothing but the natural action of $G = SL(2, \mathbb{R})$ as a transformation group on $T \simeq G/AN$, acting upon appropriate function spaces (already visible in [Bg]), while various methods (some dating back to Harish-Chandra in [HCh]) show that D consists of analytic vectors for $\mathfrak{L}_0$ (hence is "big enough" for $\mathfrak{L}_0$ to determine a unique representation) in these spaces. Our methods have the advantages of unity with the emphasis of the rest of this monograph upon qualitatively different structural properties of the "normed module with basis" $(\mathfrak{L}_0, \{e_n\}, \|\cdot\|)$, and applicability in some cases not reached by the others.

As above, the vectors e_n are the canonical basis vectors, and D denotes the linear span of $\{e_n : n \in \mathbb{Z}\}$. In the following, we consider the completion of D with respect to a general norm on D.

12.1. Theorem

Let $\| \ \|$ denote a norm on D and suppose that there is a non-negative integer ν such that $\|e_n\| = \mathcal{O}(|n|^{\nu})$.

(a) Then the operators A_0 and A_2 both satisfy the τ_1-graph density condition.

(b) If A_0 and A_2 are sub-generators (i.e., restrictions to D of infinitesimal C_0 group generators), then the base-point Lie algebra $\mathfrak{L}_0$ exponentiates to a continuous group representation on the Banach space completion of D in the above norm.

Remark. The subgenerator condition in (b) cannot be relaxed, as is shown in Sections 12C-D below in the case of the $\ell^p(\mathbb{Z})$ norms for $p \neq 2$. There, only A_0 pregenerates a C_0 group in the classical sense and $\mathfrak{L}_0$ fails to exponentiate to a classical C_0 representation as we show in Section 12D. However, (a) above can be applied to show that (GD) holds in all of these cases. The example for $p = 1$ is especially interesting: there, $\ell^1(\mathbb{Z})$ is a convolution Banach algebra and $\mathfrak{L}_0$ consists of (unbounded) derivations with respect to the convolution product: $A(u*v) = (Au)*v + u*(Av)$ for all $A \in \mathfrak{L}_0$. If $\mathfrak{L}_0$ had an exponential on $\ell^1(\mathbb{Z})$, we show in Section 12D that this would yield an action of G as a group of algebra automorphisms of $\ell^1(\mathbb{Z})$; the idea of our proof that A_2 could not be a sub-generator is that the automorphism group of $\ell^1(\mathbb{Z})$ is too small to contain a copy of G (cf. Theorem 12.11).

Proof of Theorem 12.1. Let E denote the Banach completion $(D, \| \ \|)^{\sim}$. The given norm induces a corresponding C^1-norm $\| \ \|_1$ on D. For vectors u in D this norm is given by (*) above.

The completion $(D,\|\ \|_1)^{\sim}$ is denoted by D_1, and the corresponding dual spaces of continuous linear functionals on D_1 [resp. E] are denoted by D_1^* and E^*.

Let ΠE^* be the product of E^* with itself four times. Then by the remark following the statement of Theorem 9.2 the mapping of ΠE^* into D_1^* given by $(f,f_0,f_1,f_2) \to f^*$ with

$$f^*(u) = f(u)+f_0(A_0u)+f_1(A_1u)+f_2(A_2u) \quad \text{for } u \in D \tag{12}$$

is onto. Indeed, for every quadruple of elements in E^*, the linear functional f^* given by (12) is continuous on $(D,\|\cdot\|_1)$ with

$$|f^*(u)| \leq (\|f\|+\|f_0\|+\|f_1\|+\|f_2\|)\|u\|_1 \quad \text{for } u \in D, \tag{13}$$

where $\|f\| = \sup\{|f(u)|: u \in D, \|u\| = 1\}$ for $f \in E^*$. Conversely, it is shown in Chapter 9 that every $f^* \in D_1^*$ is given that way. The corresponding quadruple is not unique, since the quadruples (f,f_0,f_1,f_2) satisfying $f(u)+f_0(A_0u)+f_1(A_1u)+f_2(A_2u) = 0$ for all $u \in D$ represent $f^* = 0$.

We claim that for every $f^* \in D_1^*$ the sequence $f^*(e_n)$ grows at most like $\mathcal{O}(|n|^{\nu+1})$. By (13) it is enough to show that $\|e_n\|_1 = \mathcal{O}(|n|^{\nu+1})$. Now

$$\begin{aligned}\|e_n\|_1 &= \|e_n\|+\|ne_n\|+\|\tfrac{n}{2}(e_{n+1}+e_{n-1})\|+\|\tfrac{n}{2}(e_{n+1}-e_{n-1})\| \\ &\leq \|e_n\|+|n|(\|e_{n-1}\|+\|e_n\|+\|e_{n+1}\|) \quad \text{for } n \in \mathbb{Z}.\end{aligned} \tag{14}$$

Since $\|e_{n\pm1}\| = \mathcal{O}(|n|^{\nu})$ the claim follows.

Suppose, for given $\lambda \in \mathbb{C}$ and $f^* \in D_1^*$, that $f^*((\lambda-A_0)u)=0$ for all $u \in D$. Then the sequence of numbers $\xi_n = f^*(e_n)$ is a solution to $\lambda\xi_n = in\xi_n$ for all $n \in \mathbb{Z}$. It is clear that the sequence ξ_n must vanish when $\mathrm{Re}\lambda \neq 0$. The polynomial growth condition is not used.

Similarly, the condition, $f^*((\lambda-A_2)u) = 0$ for all $u \in D$, leads to the recursive identity

$$\lambda\xi_n = \frac{n}{2}(\xi_{n+1}-\xi_{n-1}) \tag{15}$$

for the sequence $\xi_n = f^*(e_n)$. Suppose, first, that λ is a positive integer. Then the difference equation has a non-zero solution given as follows. For λ odd, $\lambda = 2\tau+1$,

$$\xi_n = n^\lambda + a_2 n^{\lambda-2} + \ldots + a_{2\tau} n, \tag{16_O}$$

and for λ even, $\lambda = 2\tau$,

$$\xi_n = n^\lambda + a_2 n^{\lambda-2} + \ldots + a_{2\tau-2} n^2. \tag{16_E}$$

The coefficients a_{2k} are given by certain recursive formulas involving the binomial coefficients $\binom{m}{p} = \dfrac{m(m-1)\ldots(m-p+1)}{p!}$:

$$a_{2k} = \frac{1}{2k}\left[\binom{\lambda}{2k+1} + a_2\binom{\lambda-2}{2k-1} + \ldots + a_{2k-2}\binom{\lambda-2k+2}{3}\right] \tag{17}$$

for $k = 1,2,3,\ldots$. In particular, $a_2 = \frac{1}{2}\binom{\lambda}{3}$ for $\lambda \geq 3$, and for $\lambda = 2\tau+1$, $a_{2\tau} = (1/2\tau)[1+a_2+\ldots+a_{2\tau-2}]$.

Since the difference equation (15) is homogenous of order two, the solution space is two dimensional. All solutions $\{\eta_n\}$ have $\eta_0 = 0$. For all $a,b \in \mathbb{C}$ there is a unique solution $\{\eta_n\}$ with $\eta_{-1} = a$ and $\eta_1 = b$. Let λ be a non-negative integer, and let $\{\xi_n\}$ be the normalized solution described above. Let $\{\eta_n\}$ be an arbitrary solution. It is then clear that the sequences $\{\eta_n : n \in \mathbb{Z}_+\}$ and $\{\xi_n : n \in \mathbb{Z}_+\}$ are proportional. Similarly, the sequences $\{\eta_n : n \in \mathbb{Z}_-\}$ and $\{\xi_n : n \in \mathbb{Z}_-\}$ must be proportional. This is easy to check because all the solutions are given by the recursive identity

$$\xi_{n+1} = (2\lambda/n)\xi_n + \xi_{n-1} \quad \text{for } n > 0,$$

and by

$$\xi_{n-1} = -(2\lambda/n)\xi_n + \xi_{n+1} \text{ for } n < 0.$$

Hence, all the non-zero solutions grow like $|n|^\lambda$ when $n \to \pm\infty$.

The normalized solution is found as follows. First, substitute $\xi_n = n^\lambda + a_1 n^{\lambda-1} + a_2 n^{\lambda-2} + \ldots$ (with $a_1, a_2, \ldots$

unknown) into (15). Then equate the coefficients of $n^\lambda, n^{\lambda-1}, \ldots$. Using the binomial formula for $(n\pm1)^k$ one shows that a_ν is zero for ν odd, and that a_ν is given by (17) for ν even. The corresponding solutions $\{\xi_n\}$ are given by (16_0) and (16_E). It is clear that when λ and ν are integers and $\lambda > \nu \geq 0$ then the solution $\{\xi_n\}$ is not $O(|n|^\nu)$. Hence, for such λ sufficiently large the non-zero soluitons ξ_n cannot correspond to any $f^* \in D_1^*$ via $\xi_n = f^*(e_n)$. Hence, $f^*(e_n) = 0$ for all $n \in \mathbb{Z}$.

We finally observe that it is enough to consider positive values of λ, because of the following symmetry in (15). If $\{\xi_n\}$ is a solution to (15) for given λ, then the sequence $\{\xi_n'\}$ defined by

$$\xi_n' = \begin{cases} \xi_n & \text{for } n \text{ odd,} \\ -\xi_n & \text{for } n \text{ even,} \end{cases}$$

is a solution to the corresponding recursive identity with λ replaced by minus λ. Suppose $\|e_n\| = O(|n|^\nu)$; then for $\lambda = \pm(\nu+2), \pm(\nu+3), \ldots$ the range $(\lambda - A_2)D$ is τ_1 dense in D. This concludes the proof of part (a) of the theorem.

To obtain claim (b), we apply (a) and Theorem 9.2. In fact the subgenerator assumption in (b) combines with the τ_1-graph-density conclusion to establish that A_0 and A_2 are pregenerators (i.e., D is a core for the group-generators whose restrictions are A_0 and A_2). To see this, notice that $\|\cdot\|_1$-density of $D_\lambda = (\lambda - A_k)D$ in D certainly implies $\|\cdot\|$-density in D and E, so Lemma 5.6 implies that D is a core for A_k, $k = 0,2$. E.O.P.

The methods of Section 11A can be used (as indicated there) to prove a second abstract result, in the spirit of Theorem 12.1, which also suffices for the applications of this section. There, the condition $\|e_n\| = O(|n|^\nu)$ is replaced by the assumption that the shift operators $S_\pm e_n = e_{n\pm1}$ are bounded with respect to the norm. (It is possible to show that if $S_\pm$ are $\|\cdot\|$-bounded, then $\|e_n\| = O(e^{\alpha|n|})$ for some $\alpha > 0$ for the $\ell^2(\mathbb{Z},\rho)$ spaces with $\rho(n) = e^{\alpha|n|}$ it follows that $\|e_n\| = e^{\alpha|n|}$, $\|S_\pm\| \le e^\alpha$, and in fact both (GD) and the hypothesis of Theorem 12.1 fail. However, it is unlikely that, in general, either hypothesis contains the other.) The variant of Theorem 12.1 that can be obtained is as follows.

12.1'. Theorem

Suppose that $\|\cdot\|$ is a norm on D such that $S_\pm$ are bounded. Suppose further that A_0 and A_2 are pregenerators. Then these operators satisfy the graph density condition and $\mathcal{L}_0$ exponentiates to a group representation on the $\|.\|$-completion E of D.

Proof. Notice that $A_\pm = -iA_0S_\pm = -iS_\pm A_0 \pm S_\pm$. (Recall that $[A_0,S_\pm] = \pm iS_\pm$ from Section 12A.) Hence, if M = complex span $(\mathcal{L}_0) = \mathbb{C}A_+ \oplus \mathbb{C}A_- \oplus \mathbb{C}A_0$ it follows that $A_0 = C$ dominates M as in Theorem 11.4. Here, for any $\mu \in \rho(\bar{A}_0)$, it is easy to see that $R(\mu,A_0)e_n = (\mu-in)^{-1}e_n \in D$ for all n, whence $R(\mu,A_0)D \subset D$ and $(\mu-A_0)D = D$. Thus Theorem 11.4 applies to show that A_2 satisfies (GD). The fact that A_0 satisfied (GD) is a trivial consequnce of the observation in the proof of Theorem 12.1 (or Theorem 11.1) that $\|u\|' = \max\{\|u\|,\|A_0u\|\}$ is equivalent to $\|\cdot\|_1$. Hence this variant follows by Theorem 9.2.

Remark. A stronger result, assuming boundedness of shifts, can be obtained using analytic vector methods. As developed in Lemma 12.4 in Section 12C below, boundedness of $S_\pm$ implies by analytic dominance of A_0 that D consists of ("jointly") analytic vectors for $\mathcal{L}_0$. Consequently, it suffices, as in Theorem 12.1, to assume only that A_0 and A_2 are restrictions to D of group generators. (Then D must, in fact, be a core for these generators [Nℓ 1].) The exponential is then constructed by analytic vector methods.

The simplest and most important application of Theorem 12.1 (or Theorem 12.1') occurs for $\ell^2(\mathbb{Z})$, where $\mathcal{L}_0$ exponentiates to a non-unitary representation of G (in fact, of $SL(2,\mathbb{R})$ and $SO_e(2,1)$, cf. Section 12G). We use standard numerical range techniques to check the pregenerator claims, recalling that for an operator $A \in \mathbf{A}(D)$ in a Hilbert space E, the numerical range $W(A) = \{(Au,u): u \in D, \|u\| = 1\}$ has the property that for $u \in D$, $\lambda \notin W(A)^-$

$$\|(\lambda-A)u\| \geq \text{dist}(\lambda,W(A))\|u\|. \tag{18}$$

(For $\|u\| = 1$, $\|(\lambda-A)u\| \geq |((\lambda-A)u,u)| = |\lambda-(Au,u)| \geq \text{dist}(\lambda,W(A))$; in genreal, replace u by $\|u\|^{-1}u\dots$.)

12.2. Corollary

(a) In $\ell^2(\mathbb{Z})$, the numerical range $W(A_2)$ is contained in the strip $S(\frac{1}{2}) = \{z \in \mathbb{C} : |\text{Re}(z)| \leq \frac{1}{2}\}$, and A_2 pregenerates a C_0 group with

$$\|V(t,A_2)\| \leq e^{|t|/2}. \tag{19}$$

Also, A_0 pregenerates a unitary group.

(b) Consequently, $\mathcal{L}_0$ exponentiates to a strongly continuous representation of G on $\ell^2(\mathbb{Z})$.

Proof. (a) For $u = \Sigma u(n)e_n \in D$, we have

$$\begin{aligned}(A_2u,u) &= (\Sigma\, nu(n)/2(e_{n+1}-e_{n-1}),\Sigma u(m)e_m)) \\ &= \Sigma\frac{n}{2}(u(n)\bar{u}(n+1)-u(n)\bar{u}(n-1)) \\ &= \Sigma\frac{n}{2}u(n)\bar{u}(n+1)-\Sigma\frac{n+1}{2}\,u(n+1)\bar{u}(n) \\ &= i\Sigma n\,\text{Im}(u(n)\bar{u}(n+1))-\tfrac{1}{2}\Sigma u(n+1)\bar{u}(n).\end{aligned} \tag{20}$$

But $|\Sigma u(n+1)\bar{u}(n)| = |(u,S_+u)| \leq \|S_+\|\,\|u\|^2 \leq 1$ if $\|u\| = 1$, whence $|\text{Re}(A_2u,u)| \leq \frac{1}{2}$ when $\|u\| = 1$ and $W(A_2) \subset S(\frac{1}{2})$. But for $\lambda = n$ with $|n|$ large Theorem 12.1 ensures that $D_\lambda = (\lambda-A_2)D$ is $\|\cdot\|_1$-dense in D, hence $\|\cdot\|$-dense in $\ell^2(\mathbb{Z})$, and the estimate following (20) above ensures that $(\lambda-A_2)^{-1}: D_\lambda \to \ell^2(\mathbb{Z})$ satisfies $\|(\lambda-A_2)^{-1}\| \leq \text{dist}(\lambda,W(A_2))^{-1} \leq (|\text{Re}(\lambda)|-\frac{1}{2})^{-1}$, whence for all large $|n|$, $n \in \rho(\bar{A}_2)$ and $\|(\lambda-\bar{A}_2)^{-1}\| \leq (|\text{Re}(\lambda)|-\frac{1}{2})^{-1}$. Hence

A_2 is a pregenerator, and $V(t,A_2)$ satisfies the estimate (19) by the Hille-Yosida-Feller Theorem (of Appendix F). The claim for A_0 is trivial: the formula "$\exp(tA_0)e_n$" $= e^{int}e_n$ shows that A_0 is the restriction to D of the generator of a unitary group that leaves D invariant, whence by the pregenerator theroem, $\bar{A}_0$ generates that unitary group.

For (b), we now directly apply Theorem 12.1 or 12.1' to obtain the exponential.

<u>Remark</u>. If $\mathcal{L}_0$ is replaced by any one of the infinitesimally unitary $\{\mathcal{L}(q,\tau): q > 0, \tau \in \mathbb{R}\}$, then exponentials can be directly constructed by Nelson's "essentially self-adjoint Laplacian" method ([Nℓ 1, Theorem 5] or the Example of [Wr, p. 296]), which uses analytic vector methods related to our remark following Theorem 12.1'. Our proof above could be simplified slightly for one of these: $\mathcal{L}(\frac{1}{4},0)$. There $B_\pm(\frac{1}{4},0)e_n = (n\pm\frac{1}{2})e_{n\pm1}$ (as the reader may check) and the argument of Theorem 12.1 goes through without major change, while B_k, $k = 0,1,2$, are skew-symmetric, so the numerical range calculation may be bypassed in the proof above. It is interesting that $\mathcal{L}(\frac{1}{4},0)$ is, in a sense, the "average" of nonunitary $\mathcal{L}_0$ and its contragredient (or adjoint) Lie algebra $\mathcal{L}_0^*$: $A_\pm^*e_n = (n\pm1)e_{n\pm1}$. The arguments of Theorem 12.1 and Corollary 12.2 can be pushed through with equal ease for $\mathcal{L}_0^*$, and we will need this observation in the proof of Theorem 12.11 below. We omit the easy details.

Finally, we consider the norms on D obtained by lifting back the $L^p(T)$ norms from the circle (alias [0,1)) via Fourier series. For most of our purposes in other sections, it is most natural to work with the completed Banach spaces E_p as abstract completion of sequence spaces. Here, however, it is expedient to be more concrete: $E_p \simeq L^p(T)$ for $1 \le p < \infty$ and $E_\infty = C(T)$. It is easily verified in this function space setting A_0, A_1 and A_2 correspond to the vector fields d/dx, cos x d/dx and sin x d/dx, respectively, so that the machinery of Chapter 11 may be brought into play. More precisely, if we parametrize T by $[0,2\pi]$ via $x \to e^{ix}$ and if e_n corresponds to $(e^{ix})^n = e^{inx}$, then d/dx on $[0,2\pi]$ acts as A_0 on this basis. Moreover, since the shifts $S_\pm$ correspond to the multiplication operators determined by $e^{\pm ix}$, $A_\pm$ correspond to $e^{\pm ix}$ d/dx and $A_1 = (i/2)(A_+ + A_-)$ corresponds to cos x d/dx, etc. We compute the $L^p(T)$ norms with respect to normalized Lebesgue measure $dx/2\pi$.

12.3. Corollary.

(a) In E_p, A_2 has numerical range in the strip $S(1/p)$ and "sub-generates" a C_0 group $V^p(t,A_2)$ on E_p with $\|V^p(t,A_2)\| \le e^{|t|/p}$, $1 \le p \le \infty$. Also, A_0 subgenerates an isometry group there.
(b) Consequently, $\mathcal{L}_0$ has a strongly continuous exponential $V_0^p : G \to \mathrm{Aut}(E_p)$ for $1 \le p \le \infty$.

Proof: (a) Here, we apply Lemma 10.8 to the vector-field sin x d/dx, as defined on $C^\infty(T)$. By [Lg 1, Theorem IV.4], every vector field on the compact manifold T is complete. (In fact, sin x d/dx on $[0,2\pi]$ corresponds to the vector field on T which is tangent to the familiar conformal flow

$$F(t,e^{ix}) = (\cosh(t/2)e^{ix}-\sinh(t/2))(-\sinh(t/2)e^{ix}+\cosh(t/2))^{-1}.)$$

Moreover, $\mathrm{div}(A_2) = \mathrm{div}(d/dx \sin x) = \cos x$, so that $|\mathrm{div}(A_2)| \le 1$, whence Lemma 10.8 implies the existence of $V^p(t,A_2)$ with the properties described, and $W(A_2)$ (on this larger domain) is contained in $S(1/p)$. We note that the group-estimate can also be derived from the explicit form of the flow by a simple change-of-variables argument (cf. [Sℓ]). The complactly-embedded $A_0 \simeq d/dx$ sub-generates the "translation mod 2π" isometry group on $[0,2\pi]$ (alias the rotations of T).

For (b), we now use the full force of Theorem 12.1 to complete the proof, noting that $\|e_n\|_p \equiv 1$ by our normalization of the measure.

Remarks on the proof: (1) A proof more in the spirit of that for Corollary 12.2 can be given by means of the numerical range calculation in Proposition F.2, which yields the fact (cf. Appendix F) that $W(A_k) \subset \{z \in \mathbb{C} : |\mathrm{Re}(z)| \le 1/p\}$. (That argument actually reduces in essence to the one given in Corollary 12.2 when $p = 2$.) This alternative argument differs

from the one above in that it uses Theorems 12.1 and 9.2 in place of Lemma 10.8 (and, indirectly, Theorem 9.1).
(2) As pointed out previously, the exponentiated function-space representation that is obtained by any of these arguments is nothing but that induced by the natural transformation-group action of $SL(2,\mathbb{R})$ on T. That action can be regarded either as the (Palais-) integral of the vector-field Lie algebra [Pℓ] $\mathfrak{L}_0 = \mathbb{R}\, d/dx + \mathbb{R} \cos x\, d/dx + \mathbb{R} \sin x\, d/dx$, or as the natural action of the conformal group $SU(1,1) \simeq SL(2,\mathbb{R})$ on $T = \{z \in \mathbb{C} : |z| = 1\}$, as well as that of $SL(2,\mathbb{R})$ on its quotient modulo its solvable Iwasawa component
$AN = \left\{\begin{pmatrix} a & 0 \\ b & a^{-1}\end{pmatrix} : a,b \in \mathbb{R},\ a > 0\right\}$. From this perspective, the only real issue in the proof concerns the size of D (the trigometric polynomials) as a possible core for the vector-field operators, etc.: is D big enough to determine a <u>unique</u> exponential? Our use of the graph-density condition (Theorem 12.1 or 12.1') can be replaced by an easy application of the Weierstrass approximation theorem (for derivatives of C^∞ functions) in establishing this fact.

12C. <u>C^∞-integrals and smeared exponentials on ℓ^p</u>

Traditionally the representations of $G_1 = SL(2,\mathbb{R})$ have been studied in Hilbert spaces of square integrable functions on the circle, or in Hilbert spaces of analytic functions on the unit disc in the complex plane [Bg] (the "Schrödinger function space formalism"). In this section, the base-point Lie algebra $\mathfrak{L}_0$ is viewed as a Lie algebra of ∞ by ∞ matrices in the Banach spaces $\ell^p(\mathbb{Z})$ of sequences $u = \{u(n): n \in \mathbb{Z}\}$ with $\|u\|_p = (\Sigma\, |u(n)|^p)^{1/p}$ for $1 \le p < \infty$, and $\|u\|_\infty = \sup_n |u(n)|$. As before, D is the linear span of the canonical basis vectors (e_n). Then $\ell^p(\mathbb{Z})$ is the completion of D in the ℓ^p-norm. The completion of D in the ℓ^∞-norm is equal to the space c_0 of sequences $u = \{u(n)\}$ which vanish at $\pm\infty$, i.e. $u(n) \to 0$ as $|n| \to \infty$.

Let $\mathfrak{L}_0$ be the base-point Lie algebra spanned by the operators A_0, A_1, A_2 that were introduced in Section 12A. If E is one of the Banach spaces $\ell^p(\mathbb{Z})$ for $1 \le p < \infty$, or c_0, then $\mathfrak{L}_0$ is a Lie algebra of densely defined linear operators in E. It is quite easy to see that each of the operators A in

$\mathcal{L}_0$ have closed extension (viewed as operators in E). The corresponding least closed extensions are denoted by $\bar{A}^{(p)}$. (One way to check closability is to observe that the domain of the contragredient operator A*,

$$D(A^*) = \{f \in E^*: \exists g \in E^* \ni \langle Au,f\rangle = \langle u,g\rangle \text{ for } \forall\, u \in D\}$$

is weak*-dense in E* for all $A \in \mathcal{L}_0$. In the present case, D may be identified with a weak*-dense subspace of E* which easily can be shown to be contained in D(A*) for all A. In fact, D is mapped into itself by each A*. At several places in the remainder of this chapter, this duality $(\ell^p)^* = \ell^q$, $q = p(p-1)^{-1}$, simplifies the analysis considerably in a similar way.)

In considering the base-point Lie algebra $\mathcal{L}_0$ in each of the ℓ^p spaces it is convenient in this section to indicate the p-dependence as follows: $\mathcal{L}_0^{(p)}$ denotes the base-point Lie algebra viewed as a Lie algebra of operators in ℓ^p for each p. For each n and p we have corresponding spaces of C^n-vectors $D_n(\mathcal{L}_0^{(p)})$ defined as completions of D with respect to the natural C^n-topology defined by non-commutative monomials in A_0, A_1, A_2 of degree $\leq n$ and the ℓ^p-norm on D. (Cf. Sections 7A and 9B.) We denote by $D_\infty(\mathcal{L}_0^{(p)})$ the intersection of the C^n-spaces.

We first show informally that each $\mathcal{L}_0^{(p)}$ integrates to a differentiable representation on the appropriate space of C^∞-vectors $D_\infty(\mathcal{L}_0^{(p)})$. We later show that it also has a generalized "smeared" exponential on the original Banach Space ℓ^p.

Our principal tool is the observation that the generalized C^∞-spaces $D_\infty(\mathcal{L}_0^{(p)})$ are, in fact, independent of p. In fact, the common space D_∞ of C^∞-vectors is just the space of rapidly decreasing sequences (i.e., for all non-negative integer k, $\exists\, C_k < \infty$ such that $|u(n)| \leq C_k(1+|n|)^{-k}$ for all $n \in \mathbb{Z}$). This is a consequence of the fact that A_0 dominates the entire base-point Lie algebra $\mathcal{L}_0$, as demontrated in the following simple lemma (which does not seem to have appeared explicitly in the literature). The lemma states that C^1-dominance by a Lie algebra element implies C^∞ (and in fact analytic) dominance by the same element of the given Lie algebra of operators.

12.4. Lemma

Let $(D,\|\cdot\|)$ be a normed linear space and let $\mathcal{L}$ be a finite-dimensional Lie algebra of operators on D. We assume that there is a basis $B_0, B_1, \ldots, B_\ell$ for $\mathcal{L}$ and a constant C such that

$$\|B_j u\| \leq C(\|B_0 u\| + \|u\|) \text{ for all } u \in D \text{ and } 1 \leq j \leq \ell. \tag{21}$$

The following two sets of C^n-norms on D are then equivalent:

$$\|u\|_n = \max\{\|B_{i_1} \cdots B_{i_m} u\| : \ 0 \leq i_j \leq \ell, \ 0 \leq m \leq n\},$$

$$\|u\|_n' = \max\{\|B_0^m u\| : \ 0 \leq m \leq n\};$$

and, in fact, B_0 analytically dominates $\mathcal{L}$. (Here, as usual, the empty monomial (m = 0) is interpreted as the identity operator on D.)

Proof: The lemma is, in fact, a special case of a result of Nelson [Nℓ 1, Corollary 3.2]. To see this, note that our assumption (21) is precisely the first condition in Nelson's corollary. The remaining conditions of that corollary are also satisfied in our special case. In fact, our case is simpler in view of the estimates

$$\|(\text{ad } B_{i_1} \ldots \text{ad } B_{i_m})(B_0)u\| \leq k^m \|u\|_1' \tag{22}$$

for $u \in D$ and multi-indices with $0 \leq i_j \leq \ell$. Since $B_0 \in \mathcal{L}$, the verification of (22), in turn, is quite standard. For k we may take the maximum of the norms of the ad B_i, viewed as endomorphisms of $\mathcal{L}$ (finite-dimensional).

When the lemma is applied to $\mathcal{L}_0^{(p)}$, we get $D_\infty(\mathcal{L}_0^{(p)}) = D_\infty(A_0^{(p)})$. The fact that for each p, $D_\infty(A_0^{(p)})$ is equal to the space D_∞ of rapidly decresing sequences indexed by $\mathbb{Z}$, is an easy consequence of a "Sobolev estimate" for sequences. (Note first that $\ell^1 \subset \ell^p$ for all $p \geq 1$. Let k be a non-negative integer. Suppose a sequence $u = \{u(n)\}$ satisfies $\sup_n (1+|n|)^{k+2}|u(n)| = M < \infty$. Then

$$\Sigma(1+|n|)^k|u(n)| \leq M\Sigma(1+|n|)^{-2}. \tag{23}$$

Hence, D has the same closure with respect to all the ℓ^p-C^∞ topologies.)

By Corollary 12.2, $\mathcal{L}_0^{(2)}$ exponentiates to a C_0 (but non-unitary) representation on $\ell^2(\mathbb{Z})$. The restriction of this representation to the common space of C_∞-vectors D_∞ then induces a differentiable representation on D_∞. This is shown in [Ps 1, Proposition 1.2]. Since $D_\infty(\mathcal{L}_0^{(p)})$ is equal to D_∞ for all p, we conclude that each $\mathcal{L}_0^{(p)}$ integrates to a differentiable representation on D_∞ for each p.

This means that for each group element g in $SL(2,\mathbb{R})$, or the corresponding universal covering group G, we have continuous endomorphisms $V_\infty(g)$ on D_∞, and the map $g \to V_\infty(g)$ is differentiable in the strong operator topology of $L(D_\infty)$ with respect to the Fréchet space D_∞. Moreover

$$\frac{d}{dt} V_\infty(\exp tX_j)u\big|_{t=0} = A_j u \quad \text{for } u \in D \text{ and } 0 \leq j \leq 2.$$

We have thus in a sense obtained an exponential, namely V_∞, of each $\mathcal{L}_0^{(p)}$; but this exponential has certain serious disadvantages when $p \neq 2$. The problem is that $V_\infty(g)$ turns out not to extend to a bounded operator on ℓ^p, except for a few trivial values of the variable g.

For applications to Banach algebras, for instance, it would be very interesting if it were possible to obtain an exponential which gives bounded operators on $\ell^1(\mathbb{Z})$. Similarly, one wants to recover ℓ^p for other values of p. In order to obtain bounded operators on ℓ^p, we use a trick from quantum field theory: we "smear" the representation V_∞ with compactly supported C^∞-functions on the group G; or in mathematical language, regard V_∞ as a distribution. For every test function $\varphi \in \mathcal{D}(G)$ we consider the integral

$$\int_G \varphi(g)V_\infty(g)dg, \tag{24}$$

where dg denotes a fixed left-invariant (in fact bi-invariant) Haar measure on G. (Here $\mathcal{D}(G)$ means $C_0^\infty(G)$ with the usual inductive topology.)

The key observation is now that this integral (24) extends to a bounded operator on ℓ^p for each $\varphi \in \mathcal{D}(G)$.

Before the statement of our first theorem on the base-point representation in ℓ^p, we recall the following definition:

12.5. Definition

Let E be a Banach space and let G be a Lie group. A "smeared" or distribution representation V of G on E is a continuous homomorphism of the convolution algebra $\mathcal{D}(G)$ of compactly supported C^∞-functions on G into the algebra B(E) of bounded linear operators on E. (That is, V satisfies the algebraic condition $V(\varphi * \psi) = V(\varphi)V(\psi)$ for all $\varphi, \psi \in \mathcal{D}(G)$, and is continuous with respect to the inductive limit topology on $\mathcal{D}(G)$ [Sch, Sz] and the strong operator topology on B(E), respectively.)

In addition, the following two technical conditions are assumed:

(i) The linear span D(V) of the vectors $\{V(\varphi)u: \varphi \in \mathcal{D}(G), u \in E\}$ is dense in E.

(ii) The vector u = 0 is the only vector in E which satisfies the condition $V(\varphi)u = 0$ for all $\varphi \in \mathcal{D}(G)$.

Remark: It can be shown that continuity of V with respect to the strong- (or indeed weak-) operator topology on B(E) in fact implies continuity with respect to the uniform operator norm topology.

The elements in D(V) are called generalized Gårding vectors. If V is, in fact, defined by (24) from a continuous representation of G on E as in [Ps 1], then D(V) is precisely equal to the Gårding space [Gå] for this representation. Following a line of argument essentially introduced by L. Schwartz in his Tata Lectures [Sz] (and familiar from Lions [Ln]), we show that a smeared representation $V : \mathcal{D}(G) \to L(E)$ naturally induced a representation $dV : \mathcal{E}'(G) \to \mathbf{A}(D(V))$ of the convolution algebra of compactly-supported distributions into the endomorphism algebra of the generalized Gårding domain D(V). In fact, for every $T \in \mathcal{E}'(G)$, dV(T) is closable, and this remark applies in particular to each $T \in \mathfrak{U}(\mathfrak{g})$ in the enveloping algebra of the Lie algebra $\mathfrak{g}$ of G (viewed as a distribution supported at {e} - see below). On vectors $u = \Sigma_{k=1}^n V(\varphi_k)v_k \in D(V)$, dV(T) is defined by

$$dV(T)u = \sum_{k=1}^{n} V(T * \varphi_k)v_k. \tag{25}$$

(Here, one checks independence of (25) upon the representation $u = \Sigma\, V(\varphi_k)u_k = \Sigma\, V(\psi_k)v_k$ by noting that for all $\eta \in \mathcal{D}$

$$V(\eta)\ \{\Sigma V(T * \varphi_k)u_k - V(T * \psi_k)v_k\} = \Sigma\ V(\eta * T * \varphi_k)u_k - V(\eta * T * \psi_k)v_k$$

$$= V(\eta * T)\{(\Sigma\ V(\varphi_k)u_k - \Sigma V(\psi_k)v_k\} = 0$$

whence by (ii) $\Sigma\, V(T * \varphi_k)u_k = \Sigma\, V(T * \psi_k)v_k$.) Note that the convolution $T * \varphi_k$ is well defined. (In fact, $(T * \varphi_k)(g) = \int_G <T_h, \varphi_k(h^{-1}g)> dh$ is known [Sz] to be in $\mathcal{D}(G)$ so that $V(T * \varphi_k)$ is well defined.)

Let δ be the Dirac measure at the identity e in G, i.e., $\delta(\varphi) = \varphi(e)$ for $\varphi \in \mathcal{D}(G)$. An element X in the Lie algebra $\mathfrak{g}$ of right-invariant analytic vector fields on G is then identified with the element $X\delta \in \mathcal{E}(G)$. The distribution derivative $X\delta$ is given by $(X\delta)(\varphi) = \delta(-X\varphi)$ for $\varphi \in \mathcal{D}(G)$.

Let D be a dense linear subspace of E and $\mathfrak{L} \subset \mathbf{A}(D)$ a finite-dimensional real Lie algebra. Let G be a connected Lie group whose Lie algebra $\mathfrak{g}$ of right-invariant vector fields is isomorphic to $\mathfrak{L}$ via some isomorphism $\rho : \mathfrak{g} \to \mathfrak{L}$. Suppose every $A \in \mathfrak{L}$ is closable. If there is a smeared representation V of G in E such that the identity $\overline{\rho(X)} = V(X)$ holds for all $X \in \mathfrak{g}$, then $\mathfrak{L}$ is said to exponentiate to a smeared representation, and V is called a smeared exponential for ρ.

The main result of this section is the following.

12.6. Theorem

Each of the base-point Lie algebras $\mathfrak{L}_0^{(p)}$ exponentiates to a smeared representation $V^{(p)}$ on $\ell^p(\mathbb{Z})$ for $1 \le p < \infty$, and on c_0.

Remark: The representation $V^{(2)}$ is the integrated form of the basepoint representation on ℓ^2 which was constructed in Corollary 12.2.

It is shown, in the next section that, except for p = 2, the smeared representation $V^{(p)}$ is not the integrated form of a continuous "point representation".

The proof of the theorem is a special case of a general construction of smeared representations from differentiable representations on dense subspaces of C^{∞}-vectors. The general case is needed in the sequel, so we supply some considerations below.

Let D be a dense linear subspace of a Banach space E with norm $\|\cdot\|$. Let $\mathcal{L}$ be a finite-dimensional real operator Lie algebra with a closable basis $A_1,\ldots,A_d$. Let E_∞ be the intersection of the domains of all the operator monomials $\bar{A}_{i_1}\bar{A}_{i_2}\cdots\bar{A}_{i_n}$ where $1 \le i_j \le d$ and $n \ge 0$. Let τ_∞ be the weakest topology on E_∞ for which these monomials restrict to continuous endomorphisms of E_∞. As described in Chapter 7, every $A \in \mathcal{L}$ extends to an endomorphism A_∞ of E_∞, and $\mathcal{L}_\infty = \{A_\infty : A \in \mathcal{L}\}$ is a Lie algebra which is isomorphic to $\mathcal{L}$ via $A \to A_\infty$. The Lie algebra $\mathcal{L}_\infty \subset L(E_\infty,\tau_\infty)$ is said to be closed, and $\mathcal{L}_\infty$ is the closure of $\mathcal{L}$, cf. [Pw 1].

Let $E(\mathcal{L})$ be the associative enveloping algebra of $\mathcal{L}$. Suppose the τ_∞ topology is determined by some closable element $\Pi \in E(\mathcal{L})$ in the following sense. For all $\Gamma \in E(\mathcal{L})$ there is an integer n and a finite constant C such that

$$\|\Gamma u\| \le C \sum_{k=0}^{n} \|\Pi^k u\| \text{ for all } u \in D \tag{26}$$

($\Pi^0 = I$.) In that case, $\mathcal{L}$ is said to be C^{∞}-dominated by Π. Moreover, the τ_∞-topology is given by the seminorms $u \to \|(\bar{\Pi})^n u\|$ for $n \ge 0$, and hence $E_\infty = D_\infty(\bar{\Pi})$.

12.7 Lemma

Let $\mathcal{L} \subset \mathbf{A}(D)$ and Π be as above. Let G be a connected Lie group with Lie algebra isomorphic to $\mathcal{L}$. Suppose $\mathcal{L}$ integrates to a C^{∞}-representation of G in E_∞, and suppose that there is some real number in the resolvent set of $\bar{\Pi}$.

Then $\mathcal{L}$ has a smeared exponential which extends the C^{∞}-representation.

Remark: We give most details of proof here, and refer to [JM] for those omitted details.

Proof: Let V_∞ be the C^{∞}-representation which exists by assumption. Let da denote a fixed left-invariant Haar measure on G. Recall that $SL(2,\mathbb{R})$ and its covering group, being simply connected, are unimodular [Hℓ]. In order to limit the technicalities we assume that G is unimodular. Let $\mathfrak{g}$ be the Lie algebra of right-invariant vector fields on G, and let exp: $\mathfrak{g} \to G$ be the exponential mapping. We use the notation

$$(\tilde{X}\varphi)(a) = \frac{d}{dt}\varphi(a\exp(-tX))\Big|_{t=0} \text{ for } X \in \mathfrak{g},\ \varphi \in \mathcal{D}(G),$$

and $a \in G$, and put $\tilde{\mathfrak{g}} = \{\tilde{X} : X \in \mathfrak{g}\}$. The mapping $X \to \tilde{X}$ induces an anti-automorphism σ of the universal enveloping algebra of $\mathfrak{g}$ onto that of $\tilde{\mathfrak{g}}$. Let dV_∞ be the infinitesimal isomorphism of $\mathfrak{g}$ onto $\mathfrak{L}_\infty$, i. e. $dV_\infty(X) = \lim_{t\to 0} t^{-1}(V_\infty(\exp tX)-I)$ for $X \in \mathfrak{g}$. Let $P \in \mathfrak{U}(\mathfrak{g})$ be such that $dV(P) = \Pi_\infty$

For every $\varphi \in \mathcal{D}(G)$ the strongly convergent integral $\int\varphi(A)V_\infty(a)da$ defines a continuous endomorphism $V_\infty(\varphi)$ of E, since E_∞ is assumed to be a Fréchet space. The reader can verify the following $L(E_\infty)$ operator identity

$$V_\infty(\sigma(Q)\varphi) = V_\infty(\varphi)dV_\infty(Q) \tag{27}$$

for all $Q \in \mathfrak{U}(\mathfrak{g})$ and all $\varphi \in \mathcal{D}(G)$. For the verification of (27), suppose first that $Q = X$ is an element in $\mathfrak{g}$, and put $g(t) = \exp(tX)$. Then the following $L(E_\infty)$ identity holds

$$\begin{aligned} V(\tilde{X}\varphi) &= \int\frac{d}{dt}\varphi(a\ g(t)^{-1})\Big|_{t=0} V_\infty(a)\ da \\ &= \int\varphi(a)\frac{d}{dt} V_\infty(a)V_\infty(g(t))\Big|_{t=0} da \\ &= \int\varphi(a)V_\infty(a)dV_\infty(X)da. \end{aligned} \tag{28}$$

Hence $V(\tilde{X}\varphi) = V(\varphi)dV_\infty(X)$. For arbitrary Q, (27) follows from this and algebra.

Let λ be a point in the resolvent set of $\bar{\Pi}$. By local equicontinuity of V_∞ and the fact that $\lambda-\Pi$ dominates we have that for every compact subset K of G there is an integer n and a finite constant C such that

$$\|V_\infty(a)u\| \leq C\|(\lambda-\bar{\Pi})^n u\| \tag{29}$$

for all $a \in K$ and $u \in E_\infty$. For a fixed compact K, this means that the operator $V_\infty(a)R(\lambda,\Pi)^n$ extends uniquely to a bounded operator on E for all $a \in K$. We have denoted by $R(\lambda,\Pi)$ the operator $(\lambda-\bar{\Pi})^{-1}$ which is bounded on E, because λ belongs to the resolvent set of $\bar{\Pi}$. By assumption, $R(\lambda,\Pi)$ restricts to an endomorphism of E_∞. This is so because $D_\infty(=E_\infty) = D_\infty(\bar{\Pi}) = D_\infty(\lambda-\bar{\Pi})$. Hence, for all $\psi \in \mathcal{D}(G)$ with support in $K(\psi \in \mathcal{D}(K))$ the $L(E_\infty)$ integral

$$\int\psi(a)V_\infty(a)R(\lambda,\Pi)^n\, da \qquad (30)$$

extends to a bounded operator on E. Now, by (27) we have for $u \in E_\infty$ and $\varphi \in \mathcal{D}(K)$

$$V_\infty((\lambda-\sigma(P))^n\varphi)R(\lambda,\Pi)^n u = V_\infty(\varphi)(\lambda-\bar{\Pi})^n R(\lambda,\Pi)^n u = V_\infty(\varphi)u, \qquad (31)$$

recalling $R(\lambda,\Pi)^n u \in E_\infty$. With $\psi = (\lambda-\sigma(P))^n\varphi$ the left-hand side of this equation is given by the integral (30), and it follows that $V_\infty(\varphi)$ extends to a bounded operator $V(\varphi)$ on E. By (29) and (31),

$$\|V(\varphi)u\| \le C\int|(\lambda-\sigma(P))^n\varphi(a)|da\ \|u\| \qquad (32)$$

for all $u \in E$. Hence,

$$\|V(\varphi)\| \le C\int|(\lambda-\sigma(P))^n\varphi(a)|\ da \qquad (33)$$

As a consequence, the mapping $\varphi \to V(\varphi)$ is continuous from $\mathcal{D}(G)$ into the normed algebra L(E), in particular strongly continuous.

It is clear that V is a homomorphism from the convolution algebra $\mathcal{D}(G)$ into L(E), since $V_\infty(\varphi * \psi) = V_\infty(\varphi)V_\infty(\psi)$ for $\varphi, \psi \in \mathcal{D}(G)$, and E_∞ is dense in E.

The Gårding space of V_∞,

$$D(V_\infty) = \text{span}\{V_\infty(\varphi)u : \varphi \in \mathcal{D}(G),\ u \in E_\infty\}$$

is dense in E_∞, because V_∞ is by assumption a C^∞ representation of G in D_∞. By construction E_∞ is dense in E, and D(V) contains $D(V_\infty)$. Hence D(V) is dense in E.

In fact one can show a little more. Let K be a compact neighborhood of e in G, and let C and n be determined according to (33). Let D_n denote the domain of the operator $(\lambda-\bar{\Pi})^n$.

Let φ_ν be an approximate identity contained in $\mathcal{D}(K)$. Then for all $u \in D_n$,

$$\lim_\nu V(\varphi_\nu)u = u\ . \qquad (34)$$

This shows that the vectors in D_n are sufficiently well behaved. But we need (ii) for all vectors in E. The surprisingly difficult proof of this extension is given in [JM]. In the present case the following duality argument does the job. (The much easier case for one-parameter smeared representations is done in [Us].)

The duality argument uses the concept of a contragredient distribution representation, which we proceed to define.

Let G be a general Lie group and let E be a Banach space. Let V be a smeared representation of G on E. We now describe the contragredient representation $\tilde{V}$ as follows. E^* denotes the dual space, which is a Banach space with the dual norm. The map $\varphi \to V(\varphi)^*$ is an anti-representation because the transpose of a product is the product of the transpose in reverse order. If we define $\varphi^{\vee}$ by $\varphi^{\vee}(g) = \varphi(g^{-1})$ on $\mathcal{D}(G)$ and put $\tilde{V}(\varphi) = V(\varphi^{\vee})^*$, then $\tilde{V}$ is a homomorphism of $\mathcal{D}(G)$ into $B(E^*)$. It is continuous in the sense of Definition 12.5, and condition (ii) is satisfied for $\tilde{V}$. But the dual Gårding space $D(\tilde{V}) = \mathrm{span}\{\tilde{V}(\mathcal{D}(G))E^*\}$ need not be norm dense in E^*. It is if E is reflexive, but in general it is only $\sigma(E^*,E)$-dense.

Recall that in the actual verification of the defining conditions for a smeared representation, condition (ii) is the hardest to check. The following observation, which we state as a lemma, is useful.

12.8. Lemma

Let V be a continuous homomorphism of $\mathcal{D}(G)$ into B(E) such that (i) is satisfied, and

$$D(\tilde{V}) = \mathrm{span}\{V(\varphi)^*f : \varphi \in \mathcal{D}(G),\ f \in E^*\}$$

is $\sigma(E^*,E)$-dense in E^*. Then V is a smeared representation.

Proof: The conclusion of the Lemma is that (ii) is satisfied under the given assumptions on V and $\tilde{V}$. Suppose that $u \in E$ is such that $V(\varphi)u = 0$ for all $\varphi \in \mathcal{D}(G)$. Then we conclude that u is annihilated by $D(\tilde{V}) \subset E^*$. Since $D(\tilde{V})$ is assumed $\sigma(E^*,E)$-dense, $u = 0$. This concludes the proof.

For the present application to $SL(2,\mathbb{R})$ and the ℓ^p spaces, we must check that $D(\tilde{V})$ is indeed weak $*$-dense in E^*. To do this, we recall the construction of V_∞. Let $V^{(2)}$ be the base-point representation of Corollary 12.2 of Section 12B and let $\tilde{V}^{(2)}$ be the corresponding contragredient representation. As pointed out in Section 12B, $\tilde{V}^{(2)}$ has a common space of C^∞ vectors with $V^{(2)}$. Consequently, $\tilde{V}^{(2)}$ restricts to a C^∞ representation on $D_\infty = E_\infty$, by the same argument that we used for $V^{(2)}$. By the argument above applied to $\tilde{V}^{(p)}$ we conclude that for $1 < p$

$D(\widetilde{V}^{(p)})$ is in fact norm dense in $\ell^{p'} = (\ell^p)^*$, and in particular weak *-dense. Condition (ii) then follows from Lemma 12.8. The case $p = 1$ is a little different because c_0 is not the dual of ℓ^1. To apply Lemma 12.8 we must check that $D(\widetilde{V}^{(1)})$ is weak *-dense in $(\ell^1)^* = \ell^\infty$. But since $D(\widetilde{V}^{(1)})$ is norm dense in c_0, and c_0 is weak *-dense in ℓ^∞, the desired conclusion follows.

Note that this use of Lemma 12.8 in the present case is a quite economical simplification, since we avoid having to check the more difficult of the two defining properties [(i) and (ii)].

Remark: More will be said about Theorem 12.6 in the next section. As a consequence of the remark following Theorem 12.6 we note that the classical point-representation do not suffice for the representation theory on $\ell^1(\mathbb{Z})$. That may explain why the representations of G_1 have never been studied in the space $\ell^1(\mathbb{Z})$. Both the representations of G_1, and the Banach algebra $\ell^1(\mathbb{Z})$ are of importance in Fourier analysis.

Let G_0 denote the conformal group $SU(1,1)$ (conjugate to $G_1 = SL(2,\mathbb{R})$ by Section 12A), and let $(e^{ix},a) \to e^{ix}a : T \times G_0 \to T$ denote the natural action of G_0 on the circle. We then have:

12.9. Corollary

For every test function $\varphi \in \mathcal{D}(G_0)$ and every distribution f on the circle, the distribution h defined by $h(e^{ix}) = \int_{G_0} f(e^{ix}a)\varphi(a)da$ is in fact an infinitely differentiable function.

Proof: Since f is a distribution on T there is a non-negative integer n, and a function g with absolutely convergent Fourier series, such that $f(e^{ix}) = ((I-d/dx)^n g)(e^{ix})$. Note also that $d/dx = V_0^{(1)}(X_0)$ by the discussion preceding Corollary 12.3 above. Using the adjoint representation of G_0 on its own Lie algebra of right-invariant vector fields, we get the identity $V_0(a)(I-V_0^{(1)}(X_0))^n = P(a,X)V_0(a)$ for some variable coefficient differential operator $P(a,X)$ on G_0. This leads to the following equation for h,

$$h = \int_{G_0} \varphi(a)V_0(a)f\,da = \int_{G_0} \varphi(a)V_0(a)(I-V_0^{(1)}(X_0))^n g\,da$$

$$= \int_{G_0} \varphi(a)P(a,X)V_0(a)g\,da$$

$$= V_0^{(1)}(P^*(\cdot,X)\varphi)g. \tag{35}$$

We have used the symbol P^* for the transpose differential operator. Since g has absolutely convergent Fourier series and $V_0^{(1)}$ is a smeared representation on $\ell^1(\mathbb{Z})$, it follows that h has absolutely convergent Fourier series as well. But $(d/dx)^m h = V_0^{(1)}(X_0^m P^*(\cdot,X)\varphi)g$ for all $m = 1,2,\ldots$. Whence, all the distribution derivatives of h have absolutely convergent Fourier series, or equivalently $h \in C^\infty(T)$.

12D. The operators A_0, A_1 and A_2

In this section our attention is focussed upon the individual operators A_0, A_1, A_2. They are of independent interest in functional analysis, and they illustrate a number of counter-intuitive phenomena. In the Schrödinger formalism they correspond to the familiar vector fields d/dx, cos x d/dx, and sin x d/dx (cf. Corollary 12.3), respectively, so it is easy to appreciate the importance of a better understanding of these operators.

Since A_0, A_1 and A_2 are unbounded derivations in the convolution Banach *-algebra $\ell^1(\mathbb{Z})$, our observations provide usefull examples for the theory of unbounded derivations. See for example [PS, BR, Jo 4].

Let V be a smeared representation of a Lie group in a Banach space. An element X in the Lie algebra is said to have the <u>restriction property</u> if the operator V(X) exponentiates to a smeared representation of the additive group of the real line on the given Banach space. In [JM] we construct smeared representations with a dense set of Lie algebra elements which do <u>not</u> have the restriction property.

Let p belong to the union of the intervals $1 \le p < 2$ and $2 < p \le \infty$. It is shown below that no element in $\mathcal{L}_0$ different from A_0 pre-generates a C_0 one-parameter group on $\ell^p(\mathbb{Z})$. However, the following weaker positive result holds for all "timelike" $t(A_0+\varepsilon_1A_1+\varepsilon_2A_2) \in \mathcal{L}_0$: those with $\varepsilon_1^2+\varepsilon_2^2 < 1$ (cf. Section 12A).

12.10. Proposition

Every timelike element in $s\ell(2,\mathbb{R})$ has the restriction property. Consequently, there is always a basis of elements which have the restriction property.

Proof: We check the hypotheses of Lemma 12.7 for $\mathcal{L} = \mathbb{R}A$ $(G=\mathbb{R})$ to obtain the one-parameter distribution group. (The proof of condition (ii) in Definition 12.5 can be done by duality (cf. Lemma 12.8) or the argument for distribution semigroups in [Us, theorem 4.1] can be used.)

Now, let E be one of the Banach spaces $\ell^p(\mathbb{Z})$, $1 \le p < \infty$, or c_0. Using the shifts $S_\pm$ $(S_\pm e_n = e_{n\pm1})$ and the "symbolic" cosine and sine operators $W_1 = \frac{1}{2}(S_+ + S_-)$, $W_2 = (1/2i)(S_+-S_-)$ we get $A_k = W_kA_0$ for $k = 1,2$. With the notation $A = V(X)$ and $A_j = V(X_j)$, $0 \le j \le 2$ we have $A = t(I+\varepsilon_1W_1+\varepsilon_2W_2)A_0$. Here V denotes the smeared basepoint representation on E which exists by Theorem 12.6. We show now that the operator $\varepsilon_1W_1+\varepsilon_2W_2$ has norm less than one, such that A is a relatively bounded perturbation of A_0. A_0 itself, of course, is a C_0 group generator. It turns out that the mentioned similarity implies that A generates a smeared one-parameter group. We shall later see that unless $\varepsilon_1 = \varepsilon_2 = 0$ this distribution representation is not a point representation, let alone a C_0 group.

We have

$$\varepsilon_1W_1 + \varepsilon_2W_2 = \frac{\varepsilon_1-i\varepsilon_2}{2}\, S_+ + \frac{\varepsilon_1+i\varepsilon_2}{2}\, S_- \,,$$

and therefore

$$\|\varepsilon_1W_1 + \varepsilon_2W_2\| \le \tfrac{1}{2}|\varepsilon_1+i\varepsilon_2|\,(\|S_+\|+\|S_-\|) = (\varepsilon_1^2+\varepsilon_2^2)^{\frac{1}{2}} < 1 \qquad (36)$$

Hence, the operator $t(I+\varepsilon_1W_1+\varepsilon_2W_2)$ has a bounded inverse, and there is a finite constant C such that $\|A_0u\| \le C\|Au\|$ for all $u \in D$. Since $\|A_ku\| \le \|W_k\|\,\|A_0u\| = \|A_0u\|$ for $k = 1,2$ and $u \in D$, we get by Lemma 12.5 that the space of C^∞-vectors for the single operator A in E is the same as that for $\mathfrak{L}_0$. The latter space of C^∞-vectors is already known to be equal to $D_\infty(\bar{A}_0)$. Moreover $D_\infty(\bar{A}_0^{(p)}) = D_\infty$ independently of p.

Since $\mathfrak{L}_0$ has a continuous exponential $V^{(2)}$ on $\ell^2(\mathbb{Z})$, the one-parameter group $t \to V^{(2)}(\exp tX) = V^{(2)}(t,A)$ restricts to a C^∞-representation of the additive group of $\mathbb{R}$ on $D_\infty = D_\infty(\bar{A}^{(2)})$. Moreover, since A is a relatively bounded perturbation of the group-pregenerator A_0, it is easy to check by perturbation of resolvents (theorem IX 2.4 [Kt 2]) that A has all real λ with large $|\lambda|$ in its resolvent set. Hence Lemma 12.7 applies with $G = \mathbb{R}$ (or [Us, theorem 4.1]) and A has the restriction property.

As a corollary to the proof we get that for each p the operator $V^{(p)}(X) = \bar{A}^{(p)}$ integrates to a C^∞ one-parameter group on its own space of C^∞-vectors. Yet, when ε_1 and ε_2 are not both zero, A does not pregenerate a C_0 one-parameter group on ℓ^p for $1 \leq p \leq \infty$ and $p \neq 2$, as we show in the next theorem.

<u>12.11. Theorem</u>

The elements of $\mathfrak{L}_0$ that are different from A_0 do <u>not</u> pregenerate C_0 one-parameter groups on $\ell^p(\mathbb{Z})$ when $1 \leq p < 2$.

<u>Proof</u>: As observed in the Introduction to this chapter, the case $p = 1$ is of special interest, because $\ell^1(\mathbb{Z})$ is a Banach *-algebra (the group algebra of the additive group of the integers), when the product is convolution with respect to the counting measure on $\mathbb{Z}$, and the *-operation is given by $u^*(n) = \overline{u(-n)}$. The elements in $\ell^1(\mathbb{Z})$ are traditionally identified with functions, $f(e^{ix}) = \Sigma u(n)e^{inx}$ on the circle, with absolutely convergent Fourier series, and $\|f\| = \Sigma|u(n)|$. The proof is easiest for this case.

For ℓ^1, the conclusion of the theorem follows from a result of Liebenson [Li], as supplemented by an observation of Kanane [Ka].

We also point out that the result can be deduced from a certain theorem on mesures with bounded convolution powers. This was proved for the real line by Beurling and Helson [BH].

We shall need a modification to the circle group of the Beurling-Helson theorem. The latter result has advantage over the corresponding result of Liebenson, that it lends itself to generalization to ℓ^p for $p \neq 1$. In our proof of this generalization we shall in fact explicitly make use also of an estimate from [Li].

We first use the fact that the closures $\bar{A}$ of elements in $\mathcal{L}_0$ are closed derivations in ℓ^1. If we assume then that $\bar{A}$ is the infinitesimal generator of a one-parameter group $V(t,A)$ of bounded linear operators on ℓ^1, then it follows (as shown in detail below) from the derivation property of $\bar{A}$ that each $V(t,A)$ is in fact an automorphism of ℓ^1, i.e. $V(t,A)(uv) = (V(t,A)u)(V(t,A)v)$ for all $t \in \mathbb{R}$ and $u, v \in \ell^1$. (One first verifies the automorphism property for elements $u, v \in D(\bar{A})$ by differentiation, using the derivation property for $\bar{A}$. Introducing the notation $V_t = V(t,A)$ and recalling that $D(\bar{A})$ is invariant under V_t, we have for elements $u, v \in D(\bar{A})$ that

$$\frac{d}{dt} V_{-t}\{V_t u V_t v\} = V_{-t}[-\bar{A}\{V_t u V_t v\} + (\bar{A}V_t u)V_t v + V_t u(\bar{A}V_t v)] = 0 \qquad (37)$$

identically. The automorphism property follows for elements in $D(\bar{A})$. We then extend this property to all elements in ℓ^1, using the assumed ℓ^1-boundedness of $V(t,A)$. Note that the argument works when only weak $*$- (i.e., $\sigma(\ell^1,\ell^\infty)$-) continuity is assumed.

It turns out that the automorphisms of $\ell^1(\mathbb{Z})$ are quite well known. The maximal ideal space of the Banach algebra $\ell^1(\mathbb{Z})$ is equal to the dual group of $\mathbb{Z}$, which is the circle group T. Hence, every automorphism of $\ell^1(\mathbb{Z})$ induces in a natural way a transformation on T. Let α be an automorphism. For the moment the elements of $\ell^1(\mathbb{Z})$ are identified with functions on the circle, with absolutely convergent Fourier series. Then for $f \in \ell^1(\mathbb{Z})$,

$$\alpha(f)(e^{ix}) = f(e^{i\omega(x)}), \qquad (38)$$

where ω is a function from the line into itself. This function ω determines the transformation of the maximal ideal space, corresponding to α. Consider first the function f given by $f(e^{ix}) = e^{ix}$, i.e., e_1. Clearly $\alpha(f) = e^{i\omega(x)}$ belongs to

$\ell^1(\mathbb{Z})$ only if $e^{i\omega(0)} = e^{i\omega(2\pi)}$. That is, $\omega(2\pi)-\omega(0) = \nu 2\pi$ for some integer ν. A function ω which corresponds to an automorphism via (38) is called admissible. By the result of Liebenson [Li] and Kahane [Ka], a function ω is admissible if and only if it has the form

$$\omega(x) = \nu x + a \tag{39}$$

where ν is an integer and a is a real number. (Liebenson proved this result under the added assumption that ω be differentiable, and Kahane removed the assumption of differentiability of ω.)

Consequently the automorphisms of $\ell^1(\mathbb{Z})$ can be parameterized by ν and a. The automorphisms form a group $\mathbf{A}$ when multiplication is given by $(\nu',a')(\nu,a) = (\nu'\nu,\nu a'+a)$. But it is quite clear that there is only one continuous one-parameter subgroup of $\mathbf{A}$, namely the one given by $t \to (1,t)$. The corresponding group of automorphisms is just translation mod 2π : $V(t,A_0)f(e^{ix}) = f(e^{i(x+t)})$. One sees that the only $A \in \mathcal{L}_0$ which pregenerates a C_0 group is $A = A_0$.

The result of Liebenson is quite old, but it is not isolated in the literature. In fact, Beurling and Helson [BH] obtained slightly earlier a result which can be shown to contain the Liebenson-Kahane theorem. For the sake of completeness, we mention that [BH] was generalized by Helson [Heℓ] to Abelian locally compact groups with connected dual. It turns out that both of the (overlapping) results, [Li, Ka] and [BH], are crucial for our present development. It is only by a combination of methods from both works that we are able to generalize the ℓ^1-theorem to arbitrary ℓ^p for $p \neq 2$. But taken separately, neither of the works seem to generalize to ℓ^p.

Beurling and Helson's results for ℓ^1 state that if ω is real function, such that

$$\sup\{\|e^{in\omega(x)}\| : n \in \mathbb{Z}\} < \infty, \tag{40}$$

then ω must be of the form (39).

Strictly speaking, Beurling and Helson worked with the Banach algebra of functions on the real line which are Fourier-Stieltjes transforms of measures of bounded variation. But their method applies to our case (absolutely convergent Fourier series) and shows that the only functions ω for which the supremum in (40) is finite are the ones given by (39).

Note that the function f given by $f(e^{ix}) = e^{ix}$ has the property that $\|f^n\| = 1$ for all $n \in \mathbb{Z}$, where $\|\cdot\|$ denotes the $\ell^1(\mathbb{Z})$-norm. Hence, if an automorphism α is induced by some ω then

$$\sup\{\|e^{in\omega(x)}\| : n \in \mathbb{Z}\} = \sup_n \|\alpha(f^n)\| \leq \|\alpha\| < \infty. \tag{40'}$$

We now turn to the generalization of the ℓ^1 theorem to ℓ^p for arbitrary $p \neq 2$, $1 \leq p \leq \infty$. We first treat the interval $1 < p < 2$, using the Hausdorff-Young inequalities [Ti]. The same negative conclusion then follows for $2 < p \leq \infty$ by duality. For if we have a group generator on ℓ^p for some p with $2 < p \leq \infty$, then the contragredient group would act on ℓ^q and $1 \leq q < 2$. Using then the Phillips perturbation theorem [Ph, HP] we show that this is not possible.

Starting with the Banach spaces $\ell^p(\mathbb{Z})$ with $1 < p < 2$, we begin by showing that the situation here is analogous to the one for $\ell^1(\mathbb{Z})$. Let p, $1 < p < 2$, be fixed, and denote the ℓ^p norm by $\|\cdot\|$. The spaces D and D_∞ are norm dense in $\ell^p(\mathbb{Z})$. They are also algebras with the multiplication induced from $\ell^1(\mathbb{Z})$, but $\ell^p(\mathbb{Z})$ is not an algebra, of course. (To see that D_∞ is an algebra we may note that it is the space of C^∞ elements for a closed derivation $\bar{A}_0^{(1)}$ for example.) Since $D_\infty \subset \ell^1$, every element in D_∞ represents a function on the circle, with absolutely convergent Fourier series. Let $A \in \mathcal{L}_0$ be such that $\bar{A}$ (closure in $\ell^p \times \ell^p$) generates a C_0 one-parameter group $\{V(t,A): t \in \mathbb{R}\}$ on ℓ^p. We claim that $A = A_0$, and hence that the whole Lie algebra cannot have a C_0-exponential.

Note first that when restricted to D_∞, $\bar{A}$ is a derivation (which we denote by A_∞). Let $\{V_\infty(t,A): t \in \mathbb{R}\}$ be the C^∞ one-parameter group generated by A_∞. (Recall that $\mathcal{L}_0$ integrates to a C^∞ representation of G_1 on D_∞; in particular, the element A is integrable.) It is easy to see that $V_\infty(t,A)$ must coincide with the restriction of $V(t,A)$ to D_∞. Hence, $V(t,A)$ leaves D_∞ invariant, and since A_∞ is a derivation of D_∞, the endomorphisms $V_\infty(t,A)$ are automorphisms of D_∞.

Now let α be the automorphism $V_\infty(t,A)$ for some fixed real number t. Then α extends to a bounded linear operator $V(t,A)$ on $\ell^p(\mathbb{Z})$, as pointed out above. Furthermore, $\alpha(e_{\pm 1}) \in D_\infty$ and the convolution powers $\alpha(e_{\pm 1})^n \in D_\infty$ for $n = 2,3,\ldots$ since $\alpha(D_\infty) = D_\infty$ and α is an automorphism of the algebra D_∞. (Recall that D_∞ is closed under convolution and contained in ℓ^p for all p.)

As in the ℓ^1 case, we get the estimate

$$\sup\{\|\alpha(e_1)^n\| : n \in \mathbb{Z}\} \le \|V(t,A)\| < \infty \tag{41}$$

using $\|(e_{\pm 1})^n\| = 1$. So we are back at (40) with the only difference that the norm now is weaker $(\ell^1 \subset \ell^p)$. The Fourier transform of $\alpha(e_1)$ is a function h on the circle, since $\alpha(e_1) \in D_\infty \subset \ell^1$. The norm of h is equal to the ℓ^p norm of its Fourier coefficients. By (41),

$$\sup\{\|h^n(e^{ix})\| : n \in \mathbb{Z}\} < \infty. \tag{42}$$

We claim that h must assume its values on the circle $\{z \in \mathbb{C} : |z| = 1\}$.

The proof of the claim is relatively easy in the case $p = 1$, cf. [BH], because if $\sup\{\|h^n\| : n \in \mathbb{Z}\} < \infty$, the powers of h and 1/h must be uniformly bounded over the whole circle, so that $|h(z)| = 1$ for all $z \in T$. The proof in the present case, $1 < p < 2$ uses a deep (but by now, of course, quite standard) result in Fourier analysis, one of the Hausdorff-Young inequalities. Let $\|h^n\|_p$ denote the ℓ^p-norm of the Fourier series for h^n. Let $q = p/(p-1)$. Then by one of the Hausdorff-Young inequalities (see e.g. [Ti]),

$$\left(\int_T |h^n(z)|^q dz\right)^{1/q} \le \|h^n\|_p \tag{43}$$

for all $n \in \mathbb{Z}$. By the basic inequality (42), the right-hand side of (43) is uniformly bounded for $n \in \mathbb{Z}$. Suppose for the moment that there is a point $z_0 \in T$ such that $|h(z_0)| > 1$. Since h is continuous there must then be a neighborhood (of positive measure) J of z_0 and $\varepsilon > 0$ such that $|h(z)| \ge 1+\varepsilon$ for all $z \in J$. For $n = 1,2,3,\ldots$ we have by (43),

$$(\text{meas}(J))^{1/q}(1+\varepsilon)^n \le \left(\int_J |h^n(z)|^q dz\right)^{1/q}$$

$$\le \left(\int_T |h^n(z)|^q dz\right)^{1/q} \le \|h^n\|_p.$$

This leads to a contradiction, because

$\sup\{\|h^n\|_p : n \in \mathbb{Z}\} < \infty$ by (42) and $(1+\varepsilon)^n \to \infty$, whence $|h(z)| \le 1$ for all $z \in T$.

We show next that for all $\varepsilon > 0$, the set of points $H_\varepsilon = \{z \in T : \varepsilon < |h(z)| < 1-\varepsilon\}$ is empty. By continuity of h, this set is open for all $\varepsilon > 0$. Hence if $z_0 \in H_\varepsilon$ for some ε, then H_ε must contain a neighborhood J of z_0. Then for n = 1,2,3...

$$\text{meas}(J)^{1/q}(1-\varepsilon)^{-n} \le \left(\int_J |h^{-n}(z)|^q \, dz\right)^{1/q} \le \|h^{-n}\|_p.$$

This again leads to a contradiction by (42), since now $(1-\varepsilon)^{-n} \to \infty$. Hence $|h(z)|$ must be identically one, or identically zero. The last possibility is clearly inconsistent with (42), whence $|h(z)| \equiv 1$.

We now proceed with our study of h, and observe first that every branch of the complex logarithm function defines a continuous argument via $\omega(x) = -i \log h(e^{ix})$. Let $\omega(x)$ be such a continuous argument of $x \to h(e^{ix})$, i.e., $h(e^{ix}) = e^{i\omega(x)}$. Since h has ℓ^1-convergent Fourier series as well, $\omega(2\pi)-\omega(0) = \nu 2\pi$ for some integer ν. By integrating the vector field $U^{-1}AU$, where U denotes the Fourier transform, one can show that ω must be of class C^∞. Suppose that ω is <u>not</u> of the form (39). Let then for each n the Fourier series of $h^n(e^{ix})$ be denoted by $\{u^n(m) : m \in \mathbb{Z}\}$. It is shown in [Li] that there exist positive constants K and θ such that, for all $\varepsilon > 0$, there is an integer $n(\varepsilon) > 0$ with the property that

$$\sum_{\substack{m \\ |u^n(m)| < K\varepsilon}} |u^n(m)|^2 > \theta \tag{44}$$

for all n with $|n| \ge n(\varepsilon)$. In (44) we sum over those m for which $|u^n(m)| < K\varepsilon$. By assumption $1 \le p < 2$, so when $0 < |u^n(m)| < K\varepsilon$ we have $|u^n(m)|^{p-2} > (K\varepsilon)^{p-2}$. In the following estimates, the summation is over those indices m for which $0 < |u^n(m)| < K\varepsilon$. For every $|n| \ge n(\varepsilon)$,

$$\|h^n(e^{ix})\| \ge \left(\sum_m |u^n(m)|^p\right)^{1/p} = \left(\sum_m |u^n(m)|^2 |u^n(m)|^{p-2}\right)^{1/p}$$

$$\ge \left(\sum_m |u^n(m)|^2 (K\varepsilon)^{p-2}\right)^{1/p} \ge \theta^{1/p}(K\varepsilon)^{(p-2)/p} \tag{45}$$

Since the exponent $(p-2)/p$ is negative, the last number tends to ∞ when $\varepsilon \to 0$. Hence, $\lim_n \sup\|h^n(e^{ix})\| = \infty$ in contradiction with (42). The assumption that ω is not of the form (39) must then be false, and the proof is completed.

Returning now to the cases $2 < p < \infty$, suppose that $\bar{A}^{(p)}$ generates a C_0 group. Then, by reflexivity, the contragredient A^* generates a C_0 group on $\ell^q(\mathbb{Z})$ for $q = p(p-1)^{-1} \in (1,2)$ Now, A^* is not in general a derivation, since most $A \in \mathfrak{L}_0$ are not formally skew-symmetric, but we recall from the remark following Corollary 12.2 that $A_\pm^* e_n = (n\pm1)e_{n\pm1} = (A_\pm \pm S_\pm)e_n$, whence since $A_0^* = -A_0$ and $\mathfrak{L}_0 \subset$ complex span$(A_0, A_\pm)$ we get $A^* = -A+U$ where U is a (bounded) linear combination of shifts. Hence, if A^* generates a C_0 group on $\ell^q(\mathbb{Z})$, Phillips' perturbation theorem (Appendix B7) implies that A must, whence $A = A_0$ and we are done.

If $\bar{A}^{(\infty)}$ generates a C_0 group on c_0, then, by Phillips' perturbation in c_0, $\bar{A}^{(\infty)}-U$ generates a C_0 group as well. (U is as above but this time defined on c_0.) The corresponding contragredient one-parameter group on ℓ^1 will not be strongly continuous, but only weak*-continuous. The corresponding infinitesimal generator is $(A-U)^* = -A^{**}$. As indicated in the beginning of this proof, all the appropriate arguments work also for weak*-(i.e., $\sigma(\ell^1,c_0)$-) continuous groups. Hence, $\bar{A}^{(\infty)}$ cannot be the generator of a C_0 group.

Remarks: (a) In our proof that the elements in $\mathfrak{L}_0^{(p)}$, different from A_0, do not generate groups of bounded operators when $p \neq 2$, we make use of the derivation properties. Once the theorem is proved for the base-point Lie algebra, it extends by perturbation (9.3) to a wide class of other Lie algebras whose elements are not derivations. Thus our perturbation theorem enables us to draw quite general conclusions from the derivaiton property, although this property is in fact only satisfied for the elements in one single quite special Lie algebra of operators.
(b) The ℓ^p-result for $1 < p < 2$ is new, but our proof relies heavily on the detailed estimates (44) of [Li]. It is quite

surprising, for many reasons (cf. the introduction to this section), that the elements in $\mathfrak{L}_0$ different from A_0 do not pregenerate C_0 one-parameter groups on ℓ^p for $1 \leq p < 2$. In fact every such A has the form WA_0 where W is bounded and A_0 is a diagonal matrix with purely imaginary numbers in the diagonal. One would intuitively expect a multiplicative perturbation WA_0 to satisfy the Hille-Yosida condition. Such perturbations <u>on Hilbert spaces</u> are studied in [Jo 4]. The present examples show that the results of [Jo 4] fail outside Hilbert space. More surprisingly, the operators A_0, A_1, A_2 have a common dense set of analytic vectors, with respect to all the ℓ^p norms.

12.12. Proposition

The space D is a space of analytic vectors for each of the operators $\bar{A}_0^{(p)}$, $\bar{A}_1^{(p)}$, $\bar{A}_2^{(p)}$ when $1 \leq p \leq \infty$.

Proof: Every element in $\mathfrak{L}_0$ is on the form WA_0 where W is ℓ^p bounded. We conclude from Lemma 12.4 that A_0 dominates the entire Lie algebra $\mathfrak{L}_0^{(p)}$. Since every vector in D is clearly an analytic (even entire) vector for $A_0^{(p)}$, we then have that D is a common space of analytic vectors for $\mathfrak{L}_0^{(p)}$ for every $p \in [1,\infty]$. Using now the precise information of [Nℓ, theorem 1] we may conclude that there is a positive number δ such that the radii of analyticity for every $A \in \mathfrak{L}_0$ are $\geq \delta$. The last estimate is independent of p. E.O.P.

Open problems. We know from Proposition 12.10 that the elements in $\mathfrak{L}_0$ which lie inside the light cone have better generation properties than do the spacelike and lightlike elements. It would be interesting to know if, for every timelike A and every u in D, u is an entire vector for A. If this is true, we would in particular have a basis for $\mathfrak{L}_0$ with a common space of entire vectors. Unfortunately one does not seem to be able to show this with the present tools. It can be shown, however, that for every timelike A, there is a dense space D_A of entire vectors for $\bar{A}^{(p)}$ for all $p \in [1,\infty]$. It is not known whether every generator for a distribution one-parameter group has a dense set of entire vectors.

Conclusions: We have exhibited examples showing that the following two results do not have any obvious generalizations.

(i) A Lie algebra $\mathfrak{L}$ of skew-symmetric operators in a Hilbert space exponentiates to a unitary representation provided that there is a common dense set of analytic vectors for the operators belonging to some basis for $\mathfrak{L}$ [FSSS].

(ii) Let δ be an unbounded *-derivation in a C^*-algebra $\mathfrak{U}$, and let φ be a faithful state on $\mathfrak{U}$ such that $\varphi(\delta(a)) = 0$ for all a in the domain of δ. Suppose δ has a dense set of analytic elements. Then δ is the infinitesimal generator of an automorphism group [BR].

Finally, we contrast the numerical range properties of the operators $A \neq A_0$ in $\mathfrak{L}_0^{(p)}$ (for $p \neq 2$) with comparable properties in $\ell^2(\mathbb{Z})$ (Corollary 12.2) and $E_p \simeq L^p(T)$ (Corollary 12.3). We recall that in Corollary 12.2, we checked that the Hilbert space numerical range $W(A_2)$ is contained in the strip $S(\frac{1}{2}) = \{z \in \mathbb{C} : |\mathrm{Re}(z)| \leq \frac{1}{2}\}$. The corresponding claim for $W(A_1)$ is easily established by a similar calculation, and $W(A_0) \subset i\,\mathbb{R}$ since A_0 is skew-symmetric, so since numerical ranges are subadditive, $W(A) \subset S(\frac{1}{2})$ for any convec combination of $A_k, k = 0,1,2$. $(W(\Sigma\alpha_k A_k) \subset \Sigma\alpha_k W(A_k) \subset S(\frac{1}{2})$ if $\Sigma\alpha_k = 1)$. (Of course, each linear combination is a convex combination times a scalar.) A similar argument in the setting of Corollary 12.3 shows that in every E_p, if A is a convex combination of the A_k, then $W(A) \subset S(1/p)$ (with the convention $S(1/\infty) = i\mathbb{R}$. Here, $E_2 \simeq \ell^2(\mathbb{Z})$ nicely fits the patterns for other p.

One might expect comparable behavior for the $\ell^p(\mathbb{Z})$ norms, since they also provide a family of norms of L^p-type which naturally include $p = 2$, and they satisfy the useful Hausdorff-Young relations with respect to the $L^p(T)$ norms [Ti]. It is therefore rather startling that with the unique exceptions of the case $p = 2$ and the operators $A = tA_0$ $(t \in \mathbb{R})$ for general p, $W(A)$ exhausts the complex plane for all $A \in \mathfrak{L}_0$. In particular, this is true for A_1, A_2, and the timelike operators $t(A_0+\varepsilon_1 A_1+\varepsilon_2 A_2)$ $(t \in \mathbb{R}, 0 < \varepsilon_1^2 + \varepsilon_2^2 < 1)$ discussed in Proposition 12.10 above. (Recall that these operators pregenerate one-parameter distribution groups of bounded operators on each of the ℓ^p spaces.) These "pathologies" lend themselves to informal discussion, rather than formal proof, because the verification of $W(A) = \mathbb{C}$ for general $A \notin \mathbb{R} A_0$ and $p \neq 2$ involves lengthy computations whose qualitative characteristics can best be understood from examination of selected special cases.

We begin by recalling the definition of the numerical range for $A \in \mathbf{A}(D)$ in a general Banach space E: for $u \in D$ with $\|u\| = 1$, $\Delta_u = \{f_u \in E^* : f_u(u) = 1 = \|f_u\|\}$ denotes the functionals tangent to the unit ball in E at the point u, and $W(A) = \{f_u(Au) : u \in D(A), \|u\| = 1, f_u \in \Delta_u\}$. (In Section 10C, we compute Δ_u for $u \in L^p(M,u)$ and use this information to derive general bounds on numerical ranges of vector fields A on manifolds M; these bounds were used in Corollary 12.3. Here we need less detail.)

The simplest cases to treat involve A_2 (or A_1) and $p = \infty$ (or $p = 1$);since $p = \infty$ requires separate treatment from $1 \leq p < \infty$ anyway, we do it first. In fact, as u runs through the two-component vectors $u_m(s,\theta) = se_m + e^{i\theta}e_{m+1}$ in $D \subset c_0$, it is easy to see that the elements $f_{m\theta} = e^{i\theta}e_{m+1} \in \Delta_u$ (if they act via the natural conjugate-linear "inner product" action of $\ell^1(\mathbb{Z})$ on c_0), and that $\{(A_2u_m(s,\theta),f_{m\theta}): s \in [0,1], \theta \in [0,2\pi], m \in \mathbb{Z}\} = \mathbb{C}$ is contained in $W(A_2)$. That is, $\|e^{i\theta}e_{m+1}\|_1 = 1$ in $\ell^1(\mathbb{Z}) = c_0^*$, and $(u_m(s,\theta),f_{m\theta}) = 1$, while

$$A_2u_m(s,\theta) = s(m/2)(e_{m+1}-e_{m-1})+e^{i\theta}(m+1)/2(e_{m+2}-e_m) = (sm/2)e_{m+1}+v,$$

where

$v \perp e_{m+1}$ so that $(A_2u_m(s,\theta),f_{m\theta}) = e^{-i\theta}sm/2$ for all $m \in \mathbb{Z}$, $s \in [0,1]$ and $\theta \in [0,2\pi]$. As θ varies for fixed s and m, the circle of radius $s|m|/2$ is traced out in the plance $\mathbb{C}$, whence varying s through $[0,1]$ fills out the disc of radius $|m|/2$, and letting $m \to \infty$ fills the plane. Linear combinations of A_1 and A_2 exhibit the same sort of behavior, as the reader may check. (An extension of the calculation above, using

$A_1u_m(s,\theta) = ism/2(e_{m+1}+e_{m-1})+ie^{i\theta}(m+1)/2(e_{m+2}+e_m)$,

shows that for any $\alpha,\beta \in \mathbb{R}$,

$((\alpha A_1+\beta A_2)u_m(s,\theta),f_{m\theta}) = (\alpha+i\beta)e^{-i\theta}sm/2.$

Thus if α or β is nonzero, varying θ and s traces out a disc of radius $(\alpha^2+\beta^2)^{\frac{1}{2}}$ m/2 in $W(\alpha A_1+\beta A_2)$, exhausting the plane $m \to \infty$.)

For $1 \leq p < \infty$, we proceed as discussed in Appendix F: for $u \in \ell^p(\mathbb{Z})$ with $\|u\|_p = 1$, the function

$f_u(n) = u(n)|u(n)|^{p-2} \in \Delta_u \subset \ell^q(\mathbb{Z})$ $(1/p + 1/q = 1)$.

Hence if we apply this function (acting conjugate-linearly) to A_2u using a calculation analogous to that in the proof of Corollary 12.2 we get

$$(A_2u,f_u) = \Sigma\{\tfrac{n}{2}\, u(n)\bar{u}(n+1)|u|(n+1)^{p-2}\}-\Sigma\{\tfrac{n}{2}\, u(n)\bar{u}(n-1)|u|(n-1)^{p-2}\}$$

$$= \Sigma\{\tfrac{n}{2}\, u(n)\bar{u}(n+1)|u|(n+1)^{p-2}-\tfrac{(n+1)}{2}u(n+1)\bar{u}(n)|u|(n)^{p-2}\} \quad (46)$$

This contribution to $W(A_2)$ may be rearranged in a form which facilitates comparison with the special case $p = 2$:

$$(A_2u,f_u) = 2i\ \mathrm{Im}\,\Sigma\{\tfrac{n}{2}u(n)\bar{u}(n+1)|u|(n+1)^{p-2}\}-\tfrac{1}{2}\Sigma\{u(n+1)\bar{u}(n)|u|(n)^{p-2}$$

$$+ \Sigma\{\tfrac{n}{2}\, u(n+1)\bar{u}(n)[\,|u|(n+1)^{p-2}-|u|(n)^{p-2}]\} \quad (47)$$

Here, it is easy to see that the first two terms lie in the strip $S(\frac{1}{2})$ for all $1 \leq p < \infty$, just as in (20) (proof of Corollary 12.2) and that the third term vanishes exactly when $p = 2$. But otherwise this term can contribute large real part to $W(A_2)$ from summands with large n.

In order to prove that $W(A_2)$ actually exhausts the plane, we form $u_m(s,\theta) = se_m+te^{i\theta}e_{m+1} \in D$ for $s \in [0,1]$, $\theta \in [0,2\pi]$ and $s^p + t^p = 1$, noting that $f_{ms} = s^{p-1}e_m+t^{p-1}e^{i\theta}e_{m+1}$ and by the above

$$(A_2u_m(s,\theta),f_{ms})=\frac{m}{2}\ (st^{p-1}e^{-i\theta}-\frac{(m+1)}{2}\ s^{p-1}te^{i\theta})$$

$$=\frac{m}{2}\{(st^{p-1}-(1+\frac{1}{m})s^{p-1}t)\cos\theta-i(st^{p-1}+(1+\frac{1}{m})s^{p-1}t)\sin\theta\} \quad (48)$$

Here, if we choose fixed s so that

$$0 < st^{p-1}-2s^{p-1}t < st^{p-1}-(1+\frac{1}{m})s^{p-1}t,$$

then for all fixed $m > 0$, these number trace out an ellipse with minor axis longer than $(m/2)(st^{p-1}-2s^{p-1}t)$, and it is not difficult to check that these conditions persist as $s \to 0$, filling out a large elliptic disc in $\mathbb{C}$. As m varies, the ellipses become larger and exhust the plane. Linear combinations of A_1 and A_2 produce more complicated ellipses of the same kind, but the same basic argument can be used. If A_0 is included, it serves (as in the $p = \infty$ case) to translate the center of the ellipse. A much more complicated scheme involving suitably-chosen linear combinations of $e_{\pm m}$ and $e_{\pm m+1}$ can be used to move centers back near the origin, producing an argument qualitatively similar to the one indicated for $p = \infty$. We omit further details for this (essentially masochistic) exercise, noting only that for operators of the form $t(A_0+\varepsilon_1A_1 + \varepsilon_2A_2)$, the conclusion shows, by Proposition 12.10 above, that pregenerators A of distribution groups that arise by "restriction" from smeared representations of $SL(2,\mathbb{R})$ can still have $W(A) = \mathbb{C}$. That is, there is no necessary connection between such distribution-group properties and numerical range properties.

It would be interesting to find if there is a connection between C_0 group properties and the numerical range of the generator.

A general reference on numerical ranges is [BD], while [LP] treats the known connections between semigroups and numerical ranges.

12E. Compact and Phillips perturbations

We show here that when the balanced operator Lie algebras $\{\mathcal{L}(q,\tau):(q,\tau) \in \mathbb{C}^2\}$ in the Banach spaces E_p ($L^p(T)$ for $1 \leq p < \infty$ and $C(T)$ for $p = \infty$) are treated as perturbations of the base-point Lie algebra $\mathcal{L}_0$, the basis perturbations

$$U_k(q,\tau) = B_k(q,\tau)-A_k, \ k = 0,1,2, \tag{49}$$

satisfy the hypotheses of our analytic continuation theorem 9.3. Hence, we obtain an "analytic" series $\{V^p_{q,\tau}:(q,\tau) \in \mathbb{C}^2\}$ of representations of the simply-connected covering group G of SL(2,$\mathbb{R}$) on each of these Banach spaces.

In fact, the perturbations in (49) are better-behaved than is required in order to apply Theorem 9.3. It is trivially true that $U_0(q,\tau) = i\tau$, a scalar operator (cf. (1) and (3) above). A slightly more intricate computation using (1), (3) and (4) from the Introduction to this chapter (details below) reveals that the corresponding shift-operator perturbations are as well-behaved:

$$U_\pm(q,\tau) =_{\text{Def}} B_\pm(q,\tau)-A_\pm = [(\tau\pm\tfrac{1}{2})+M_\pm(q,\tau)]S_\pm \tag{50}$$

where $S_\pm e_n = e_{n\pm1}$ denote the unweighted shifts and $M_\pm(q,\tau)e_n = m_n^\pm(q,\tau)e_n$ is a diagonal operator which vanishes at $n = \pm\infty$ like $1/|n|$. Rapid decay of $m_n^\pm(q,\tau)$ turns out to imply that $M_\pm(q,\tau)$ is compact for all $(q,\tau) \in \mathbb{C}^2$, norm-continuous in (q,τ) on $\Omega_c = \{(q,\tau):\gamma_n^2(q,\tau) \notin (-\infty,0)i$ for all $n \in \mathbb{Z}\}$ and norm-analytic on $\Omega = \{(q,\tau): \gamma_n^2(q,\tau) \notin (-\infty,0]i\}$. This leads to unusually good behavior of the $V_{q,\tau}$ in the Casimir (or character) parameter q when τ is fixed: for $r \in \mathbb{C}$

$$U_k(q,r,\tau) =_{\text{Def}} B_k(q,\tau)-B_k(r,\tau), \ k = 1,2, \tag{51}$$

is compact and analytic in q, and for $x \in G$

$$V^p_{q,\tau}(x) = V^p_{r,\tau}(x) + K(q,r,\tau,x) \tag{52}$$

where K is compact and norm-continuous in q,r,τ and x on Ω_c (with analyticity in q and τ parameters on Ω). When p = 2, it follows that in terms of the quotient Calkin algebra $L(E_2)/K(E_2)$ (bounded operators on $E_2 \simeq \ell^2(\mathbb{Z})$ modulo compact operators), the q-dependence of the exponentials $V^2_{q,\tau}$ disappears: only the τ-dependence is "essential". (This manifests itself in the fact that the "essential spectra" of the $B(q,\tau) \in \mathcal{L}(q,\tau)$ are q-independent.)

These perturbation properties arise in large part from the "balanced" form of the $\mathfrak{L}(q,\tau)$ (cf. Section 12H): the radical within which q appers in $\gamma_n(q,\tau) = \mathrm{sgn}(n)(q+(\tau+n)(\tau+n+1))^{\frac{1}{2}}$ as in (4). Indeed, (1) and (3) yield

$$m_n^+(q,\tau) = \gamma_n(q,\tau)-n-(\tau+\tfrac{1}{2}); \; m_n^-(q,\tau) = \gamma_{n-1}(q,\tau)-n-(\tau-\tfrac{1}{2}) \quad (53)$$

whence whenever $\gamma_n+n \neq 0$ we get from (4)

$$m_n^+(q,\tau) = \frac{\gamma_n^2-n^2-(\tau+\frac{1}{2})(\gamma_n+n)}{\gamma_n+n} = (\gamma_n+n)^{-1}\{q+\tau(\tau+1)+(\tau+\tfrac{1}{2})(\gamma_n-n)\} \quad (54)$$

with a similar formula for $m_n^-(q,\tau)$. But

$$\gamma_n-n = (\gamma_n^2-n^2)(\gamma_n+n)^{-1} = [q+\tau(\tau+1)+(2\tau+1)n](\gamma_n+n)^{-1} \quad (55)$$

again, so one easily sees that γ_n-n is bounded in n and $m_n^+(q,\tau)$ goes to 0 at $\pm\infty$ like $(\gamma_n+n)^{-1} \simeq n^{-1}$. (More details are supplied in the proof of Lemma 12.13 below.)

Before proceeding, a few remarks are in order concerning the continuity domain Ω_c and the analyticity domain Ω: these avoid the branch-varieties in $\mathbb{C}^2$ introduced by our arbitrary choice of branch-cut for the square root in (4). (The "singularities" are therefore an artifact of our primitive parametrization of the $\mathfrak{L}(q,\tau)$ by $\mathbb{C}^2$ rather than by a more sophisticated complex variety that allows for the two-sheeted Riemann surface for the square root.) It is important that Ω_c not omit the "zeros" of $\gamma_n(q,\tau)$, since the discrete and finite-dimensional series live (as direct summands) at such points. Our choice is technically less convenient below than the natural branch cut along the negative reals (cf. the remark following Lemma 12.13), but it serves to include in Ω_c all points with $\tau \in \mathbb{R}$ and $q \in \mathbb{R}$, whence the natural "real-analytic continuation" of the unitary continuous (principal and complementary series) to "discrete triple" values $\tau = n/2$, is everywhere continuous.

The following result then establishes the infinitesimal perturbation-theoretic consequences for the special case p = 2, where E_2 may be identified with $\ell^2(\mathbb{Z})$ via the unitarity of the Fourier transform. The argument for this special case is simpler, and it is more readily generalized to $\ell^p(\mathbb{Z})$ and to other Lie algebras $\mathfrak{g}$ and their "balanced" series. We formulate the result in terms of $M_\pm(q,\tau)$, since the compact operators form a closed ideal in $L(E_2)$ which is invariant under products with the bounded shifts $S_\pm$.

12.13. Lemma

(a) For all (q,τ) and all $p \in [1,\infty)$ the operators $M_\pm(q,\tau)$ on D extend to compact operators on $\ell^p(\mathbb{Z})$ denoted by the same symbols.
(b) Moreover, if $\|\cdot\|_n$ denotes the C^n-norm on D induced by the base-point Lie algebra $\mathfrak{L}_0$, then the $M_\pm(q,\tau)$ are all $\|\cdot\|_n$-bounded for every n.
(c) The mappings $(q,\tau) \to M_\pm(q,\tau)$ are operator-norm-continuous on Ω_c and analytic on Ω, as maps into $L(D_n)$ (where D_n is the $\|\cdot\|_n$-completion of D).

Proof: As general background, we recall that diagonal operators $Me_n = m_n e_n$ on the Banach space $\ell^p(\mathbb{Z})$ are bounded if and only if $\|m\|_\infty = \sup\{|m_n| : n \in \mathbb{Z}\}$ is finite, and then $\|M\| = \|m\|_\infty$. (Recall that

$$\|Mu\|_p^p = \Sigma |m_n u(n)|^p \le \|m\|_\infty^p \,\Sigma |u(n)|^p \le \|m\|_\infty^p \|u\|_p^p$$

while $\|Me_n\| = |m_n|$ means that $\|M\|$ achieves its bound $\|M\|_\infty$.) This allows us to transfer operator-norm questions to matters involving the Banach algebra $\ell^\infty(\mathbb{Z})$.

As suggested above, the principal tool in checking the various claims of the lemma is boundedness of the functions $b_{q,\tau}(n) = n(\gamma_n(q,\tau)+n)^{-1} = (\gamma_n/n + 1)^{-1}$, where we agree to put $b_{q,\tau}(0) = 0$ even if $\gamma_0(q,\tau) = 0$. (This occurs only when $q = \frac{1}{4}$, $\tau = -\frac{1}{2}$, as the reader may check.) Notice that $\gamma_n(q,\tau) = -n$ cannot occur for any n by our conventions on the square root and signum. Also,

$\gamma_n(q,\tau)/n = (1+(2\tau+1)/n+(q+\tau(\tau+1))/n^2)^{\frac{1}{2}}$, whence $\lim\{\gamma_n(q,\tau)/n : n \to \pm\infty\} = 1$ and $\|b_{q,\tau}\|_\infty < \infty$. Moreover, if $K \subset \Omega_c$ is compact, we can view $b_{q,\tau}(n)$ as a continuous function on $K \times \{\mathbb{Z} \cup \{\infty\}\}$ by defining $b_{q,\tau}(\infty) = 1$ at the compactification point $\{\infty\}$ in the one-point compactification of $\mathbb{Z}$, whence there is a uniform bound $C_K \geq |b_{q,\tau}(n)|$ for all $(q,\tau,n) \in K \times (\mathbb{Z} \cup \{\infty\})$ and $\|b_{q,\tau}\|_\infty \leq C_K$ for $(q,\tau) \in K$.

For (a), then, the compactness claim is trivial, since we can write for $\gamma_n(q,\tau)+n \neq 0$

$$m_n^+(q,\tau) = n^{-1}b_{q,\tau}(n)\{q+\tau(\tau+1)+(\tau+\tfrac{1}{2})[q+\tau(\tau+1)+(2\tau+1)b_{q,\tau}(n)]\}, \tag{56}$$

so that m_n^+ vanishes at ∞ like $1/n$. Thus $m^+(q,\tau)$ is the $\|\cdot\|_\infty$-limit of its finite-cutoffs to $[-k,k]$, which corresponds to finite-rank operators in $\ell^p(\mathbb{Z})$ with range in $D_k = \text{span}\{e_n : |n| \leq k\}$. Consequently, $M_+(q,\tau)$ is an operator-norm limit of finite-rank operators and is compact. But these finite-rank operators are trivially operator-norm continuous on Ω_c and analytic on Ω, as linear combinations of the rank-1 operators which send $u = \Sigma\hat{u}(n)e_n$ into $m_n^+(q,\tau)\hat{u}(n)e_n = (\gamma_n(q,\tau)-n-(\tau+\frac{1}{2}))\hat{u}(n)e_n$, and our uniform estimates $\|b_{q,\tau}\|_\infty \leq C_K$ imply that the norm-convergence of these finite-rank operators to their limits $M_+(q,\tau)$ is uniform with respect to $(q,\tau) \in K$. Hence, for $n = 0$ in (c) (i.e., as maps into $L(E)$) the claimed continuity on Ω_c and analyticity on Ω follow by stability of these properties under uniform limits. Of course, the arguments for $M_-(q,\tau)$ are easy variants of those used above.

Then (b) and the $\|\cdot\|_n$-versions in (c) for $n \geq 1$ follow from the fact that diagonal operators have the same norms on all D_n when the latter are equivalently normed by the C^n-norms for the dominating (diagonal) operator A_0: $\|u\|_n' = \max\{\|A_0^k u\| : 0 \leq k \leq n\}$ (cf. Section 12C). That is, if $Me_n = m_n e_n$, then $[M,A_0] = 0$ so $\|A_0^k Mu\| = \|MA_0^k u\| \leq \|M\|\,\|A_0^k u\|$ and $\|Mu\|_n' \leq \|M\|\,\|u\|_n'$, while consideration of the $k = 0$ case shows that $\|M\|_n'$ achieves $\|M\|$. Hence the restriction mapping sends the algebra of diagonal operators in $L(E)$ isometrically into $L(D_n, \|\cdot\|_n')$, preserving boundedness and continuity/analyticity of the $M_\pm(q,\tau)$.

Remark on the proof: If the branch-cut for the square root in the definition (4) of $\gamma_n(q,\tau)$ is moved to the negative reals, then our non-constructive compactification trick for bounding $b_n(q,\tau)$ can be replaced by a simple explicit estimate, for then $\mathrm{Re}((q+(\tau+n)(\tau+n+1))^{\frac{1}{2}}) \geq 0$ for all $(q,\tau) \in \mathbb{C}^2$ and $\gamma_n(q,\tau)/n$ has non-negative real part, so that $|\gamma_n(q,\tau)/n+1| \geq 1$ for all n , q , τ and $|b_n(q,\tau)| \leq 1$ uniformly in n , q and τ. The advantage of the argument employed above is that it generalized easily to balanced modules for other Lie algebras $\mathfrak{g}$, using only qualitative information about shift-coefficients (that they "go to ∞ like the highest weight" on $\hat{K}$: roughly that the unboundedness of the operators in the diagonalized compact Cartan subalgebra is typical of that for all operators in the module. This last seems to be intimately connected with the fact that if $-\Delta_K$ is the positive "compact Casimir" or "compact Laplacian" operator, then $(1-\Delta_K)^{\frac{1}{2}}$ dominates the module in the unitary case).

For general E_p $(1 \leq p \leq \infty)$, a more delicate harmonic analysis argument is required to establish compactness of $M_\pm(q,\tau)$. The argument yields more when applied to $E_2 \simeq \ell^2(\mathbb{Z})$ than the one employed above: the operators $M_\pm(q,\tau)$ are actually Hilbert-Schmidt. However, neither the argument nor the conclusions appear to generalize beyond $s\ell(2,\mathbb{R})$, at least in any direct way. Consequently, we relegate the Hilbert-Schmidt conclusions to a remark and omit a few details in the proof.

12.14. Lemma

Let $E = E_p$ (isomorphic to the $\|\cdot\|_p$-completion of the trigonometric polynomials in $L^p(\mathbb{T})$ via the correspondence $e_n \leftrightarrow e^{inx}$). Then all conclusions in Lemma 12.13 hold.

Proof: Here, we take $\ell^2(\mathbb{Z})$ rather than $\ell^\infty(\mathbb{Z})$ as the basic reference space for analysis of $M_\pm(q,\tau)$, since $\mathcal{O}(1/|n|)$

sequences are square-summable. The scheme is to Fourier-transform the pointwise-multiplication operators $m^{\pm}(q,\tau)$ on $D \subset E_p$ into convolution operators $\mu^{\pm}(q,\tau)$ on $D \subset L^p(T)$, where $m^{\pm}(q,\tau) \in \ell^2(\mathbb{Z})$ implies that $\mu^{\pm}(q,\tau) \in L^2(T) \subset L^1(T)$, with $\|m^{\pm}(q,\tau)\|_2 = \|\mu^{\pm}(q,\tau)\|_2 \geq \|\mu^{\pm}(q,\tau)\|_1$ (normalized angular measure on T). The fact that the L^1 norm dominates the convolution-operator-norm is used to parlay $\ell^2(\mathbb{Z})$ information through $L^2(T)$ to $L^1(T)$ and to operator conclusions on E_p. (Hilbert-Schmidt conclusions, on the other hand, can be read off directly, since the ℓ^2 norm of a diagonal operator on $\ell^2(\mathbb{Z})$ is exactly the Hilbert-Schmidt norm. We omit details.)

With this background in mind, the argument runs in parallel to that for Lemma 12.13, using some of the computations. Here, the factorization (56) and the uniform estimate on the (q,τ)-dependent factor $b_{q,\tau}(n)\{\ \}$ there for $(q,\tau) \in K$ ensures that for some $C'_K < \infty$ we have

$$\|m^{+}(q,\tau)\|_2 \leq C'_K(1+\Sigma\{|n|^{-2} : n \in \mathbb{Z}\ ,\ n \neq 0\})^{\frac{1}{2}}. \tag{57}$$

(Notice that (56) does not hold at $n = 0$; the summand 1 and the value of C'_K take into account the $n = 0$ summand $|\gamma_0(q,\tau)-(\tau+\frac{1}{2})|^2$ in $\|m^{+}(q,\tau)\|_2^2$.) Thus $\|\mu^{+}_{q,\tau}\|_1$ is uniformly bounded for $(q,\tau) \in K$. Moreover, as before, $M^{+}(q,\tau)$ differs in $\ell^2(\mathbb{Z})$ from its finite-rank cutoff to $[-k,k]$ uniformly by $C'_K\ \Sigma\{|n|^{-2}\colon\ |n| > k\}$, which converges to 0 as $k \to \infty$, whence the corresponding finite-rank convolution operators converge in L^1(-alias-operator-) norm to the operator on L^p corresponding to $M_+(q,\tau)$, at a rate uniform in $(q,\tau) \in K$. Compactness, continuity, and analyticity then follow as before.

The C^m-norm-boundedness claim can also be obtained as in Lemma 12.13, but in this case we need to check that the A_0-norms agree topologically with the $\mathfrak{L}_0$-norms. The quickest argument is this: we checked in Corollary 12.3 that with respect to the usual $[0,2\pi)$ parametrication of T, A_0 corresponds to d/dx, A_1 to $\cos x\ d/dx$, and A_2 to $\sin x\ d/dx$. Consequently, boundedness of $|\sin x|$ and $|\cos x|$ below ensures that for all $u \in C^{\infty}(T) \subset L^p, \|A_k u\|_p \leq \|A_0 u\|_p$, $k = 1,2$. Consequently, A_0 dominates the basis for $\mathfrak{L}_0$, and the Nelson C^{∞}/analytic-domination argument in Section 12C. (Lemma 12.4) implies that the A_0 norms are equivalent to the $\mathfrak{L}_0$ ones. E.O.P.

The two versions of our basic lemma now permit the main exponentiation/perturbation/continuation theorem to be proved for all $1 \leq p \leq \infty$. It should be emphasized here that the case $p = 2$, where $E_p \cong \ell^2(\mathbb{Z})$, is of primary interest since it contains the unitary representation theory of the simply-connected group G and its quotients $SL(2,\mathbb{R}) \cong SU(1,1)$ and $SO_e(2,1)$ (cf. Section 12G).

12.15. Theorem

Let $\tilde{G}$ be the simply-connected Lie group with Lie algebra $\mathfrak{g} \cong s\ell(2,\mathbb{R})$, and let E_p be the completion of the sequence-space D with respect to $L^p(T)$-norm (via the Fourier correspondence) for $1 \leq p \leq \infty$.

(a) Every balanced operator Lie algebra $\mathfrak{L}(q,\tau)$, $(q,\tau) \in \mathbb{C}^2$, is exponentiable on E_p to a strongly continuous representation $V^p_{q,\tau}: G \to \mathrm{Aut}(E_p)$.

(b) For each integer $n \geq 0$, and fixed p, the operator Lie algebras $\{\mathfrak{L}(q,\tau): (q,\tau) \in \mathbb{C}^2\}$ all have a common (i.e., independent of q and τ) complete-normable space D^p_n of C^n vectors, and a common complete metrizable locally convex space D^p_{∞} of C^{∞} vectors (in fact, independent of p as well: $D^p_{\infty} \cong C^{\infty}(T)$.)

(c) The mapping $(q,\tau) \to V^p_{q,\tau}(x)$ is operator-norm continuous on Ω_c and analytic on Ω for each $x \in \tilde{G}$. Moreover, these properties extend to the restriction of $V^p_{q,\tau}(x)$ to the invariant (Banach) spaces D^p_n and (Fréchet) space D^p_{∞}, where either the bounded convergence operator topology from $L(D^p_{\infty})$ or the stronger locally m-convex projective limit topology from $L(D^p_n)$ spaces can be used on D^p_{∞}.

(d) For all fixed $(r,\tau) \in \mathbb{C}^2$, there exists a compact-operator-valued function $K^p_{r,\tau}(q,x)$ from $\mathbb{C} \times \tilde{G}$ to $L(E_p)$ such that for all $(q,x) \in \mathbb{C} \times \tilde{G}$

$$V^p_{q,\tau}(x) = V^p_{r,\tau}(x) + K^p_{r,\tau}(q,x); \tag{58}$$

$K^p_{r,\tau}$ is norm-continuous in $q \in \Omega_c$, norm analytic at $q \in \Omega$, and strongly continuous in $x \in \tilde{G}$.

(e) For $1 < p < \infty$, $K^p_{r,\tau}$ is operator-norm-continuous in (q,τ) on $\Omega_c \times \tilde{G}$.

Remarks: (a) We conjecture that (e) is true for $p = 1,\infty$ as well, but our present methods require reflexivity.
(b) Identity (58) has equivalent multiplicative formulations

$$V^p_{q,\tau}(x) = V^p_{r,\tau}(x)(1+R^p_{r,\tau}(q,x)) = (1+L^p_{r,\tau}(q,x))V^p_{r,\tau}(x) \tag{59}$$

with L and R compact. These functions share the properties of K described in (d) and (e). We leave details to the reader.

Proof: (a)-(b). In order to invoke Theorem 9.3 and Corollary 9.4, we must check that the $U_k(q,\tau)$ in (49) are bounded on all D^p_n, the C^n-vector domain for $\mathcal{L}_0$ (alias that for A_0). As remarked previously, the identities $U_1 = i/2(U_+ + U_-)$ and $U_2 = \frac{1}{2}(U_+ - U_-)$ in combination with (50) reduce this to boundedness of $M_\pm(q,\tau)$ (supplied by Lemmas 12.13 and 12.14) and of $S_\pm$. By equivalence of the $\mathcal{L}_0$ and A_0 norms on $D \subset E_p$, boundedness of $S_\pm$ reduces to a commutation problem for A_0 and $S_\pm$. But

$[A_0, S_\pm] = \pm S_\pm$ (and $(A_0 S_\pm - S_\pm A_0)e_n = ((n\pm 1)-n)e_{n\pm 1}$), so

$$\|A_0 S_\pm u\|'_n \le \|S_\pm A_0 u\|'_n + \|S_\pm u\|'_n \le \|S_\pm\|'_n \max\{\|A_0 u\|'_n, \|u\|'_n\} \le \|S_\pm\|'_n \|u\|'_{n+1}$$

inductively for the A_0 norms: boundedness of $S_\pm$ on E_p implies boundedness on all D^p_n. (Notice that $S_\pm$ correspond under the Fourier transform to multiplication by $e^{\pm ix}$ on $L^p(T)$, clearly isometries.)

For (c), continuity and analyticity of the maps $(q,\tau) \to M_{\pm}(q,\tau)$ into $L(D_n^p)$, supplied by claims (b) in Lemmas 12.13 and 12.14 leads to corresponding properties of $(q,\tau) \to U_k(q,\tau)$, $k = 1,2$, via analyticity of operator products and sums. The corresponding claim for $U_0(q,\tau) = i\tau$ is trivial, so the map $(q,\tau) \to (U_0(q,\tau),U_1(q,\tau),U_2(q,\tau))$ is continuous on Ω_c and analytic on Ω into $L(D_n^p) \times L(D_n^p) \times L(D_n^p)$. By composition, using the analyticity claims in Theorem 9.3 and Proposition 9.4, we obtain the asserted analyticity on Ω and continuity on Ω_c.

The proof of (d) is an easy application of canonical coordinates of the second kind, in combination with our strengthened version of Phillips' perturbation theorem (Appendix E). That is, for x near e $\in$ G, we write $x = \exp(t_0X_0)\exp(t_1X_1)\exp(t_2X_2)$ to obtain as in Chapters 8 and 9 that

$$\begin{aligned} V^p_{q,\tau}(x) &= V(t_0,B_0(q,\tau))V(t_1,B_1(q,\tau))V(t_2,B_2(q,\tau)) \\ &= V(t_0,[B_0(r,\tau)+U_0(q,r,\tau)])V(t_1,[B_1(r,\tau)+U_1(q,r,\tau)]) \\ &\qquad V(t_2,[B_2(r,\tau)+U_2(q,r,\tau)]) \end{aligned} \tag{60}$$

But $U_0(q,r,\tau) = 0$ since B_0 is independent of τ, and for $k = 1,2$ we get from Theorem E.1 that

$$V(t_k,[B_k(r,\tau)+U_k(q,r,\tau)]) = V(t_k,B_k(r,\tau))+K(t_k,U_k(q,r,\tau)) \tag{61}$$

where $K(\cdot)$ is compact. When (61) is substituted into (60) and products are distributed over sums, the left-most resulting summand becomes

$$V(t_0,B_0(r,\tau))V(t_1,B_1(r,\tau))V(t_2,B_2(r,\tau)) = V^p_{r,\tau}(x) \tag{62}$$

while the other three summands contain a compact factor and are compact. Since the U_k are norm-continuous or analytic in q, Theorem E.1 (or Phillips' Theorem) yields continuity or analyticity of the $K(t_k,U_k)$, whence all of the compact summands are products of factors independent of q and factors continuous or analytic in q, hence the sum of the compact terms exhibits the appropriate q-dependence. Strong continuity in x is immediate. By connectedness of $\widetilde{G}$, every $y \in \widetilde{G}$ is a product of the x near e to which the argument above applies, whence $V^p_{q,\tau}(y)$ can be factored into a product whose factors all have the compact additive perturbation property analogous to (61). Consequently, the substitution-distribution argument applies again to obtain $V^p_{q,\tau}(y)$ in terms of a left-most summand $V^p_{r,\tau}(y)$ (a product of factors $V^p_{r,\tau}(x)$) and a collection of compact summands with the appropriate q and x dependence. (We leave details to the reader.)

The proof in (e) is simply a refinement of that for (d), using the left-right smoothing property of compact factors in reflexive spaces (Lemma E.2, Proposition E.3): pointwise products of strongly-continuous B(E)-valued functions and norm-continuous compact-operator-valued functions are norm-continuous. Since the $K(t_k,U_k)$ in (61) are compact-valued and norm-continuous, when $1 < p < \infty$ the necessary reflexivity is available to infer that the compact summands obtained from substituting (61) into (60) are all norm-continuous in x, whence their sum $K^p_{r,\tau}(x,q)$ is. A similar argument then spreads this property from the original neighborhood of e to all $y \in \tilde{G}$. (Again, we leave details to the reader.) E.O.P.

12F. Perturbations and analytic continuation of smeared representations

In this section we study the analogue of Theorem 12.15 for the distribution (or smeared) representation in sequence spaces of the universal covering group $\tilde{G}$ of SL(2,ℝ): the results of that theorem carry over with only a few changes. That is, in Theorem 12.15 the objects under study are the C_0 (or continuous point-) representations of $\tilde{G}$ on $\ell^2(\mathbb{Z}) \approx L^2(T)$, while the objects under study in this section are the smeared representations (i.e., representations of the convolution algebra $\mathcal{D}(\tilde{G})$ of C_0^∞-functions on the group $\tilde{G}$) on the Banach spaces $\ell^p(\mathbb{Z})$.

As in Theorem 12.15 we obtain, for each balanced operator Lie algebra $\mathfrak{L}(q,\tau) \subset \mathbf{A}(D)$ indexed by $(q,\tau) \in \mathbb{C}^2$, a smeared exponential $V^{(p)}_{q,\tau} : \mathcal{D}(\tilde{G}) \to \mathrm{Aut}(\ell^p)$. The different smeared representations $V^{(p)}_{q,\tau}$ all have a common space D_∞ of C^∞-vectors. In fact, D_∞ is the space of Fourier series of functions in

$C^\infty(T)$, i.e., the common space of C^∞ vectors for the C_0 representations $V^p_{q,\tau}$ of Theorem 12.15.

As in the previous section, the regions Ω_c and Ω are domains of continuity and analyticity, respectively, of the appropriate representations. Each operator $V^{(p)}_{q,\tau}(\varphi)$, $(1 \le p \le \infty,\ (q,\tau) \in \mathbb{C}^2,\ \varphi \in \mathcal{D}(\widetilde{G}))$ is compact, and in fact nuclear [Sch]. Moreover, the mapping $q,\tau \to V^p_{q,\tau}(\varphi)$ is operator norm continuous on Ω_c [and analytic on Ω].

In the second part of this section, we show that none of the smeared representations $V^{(p)}_{q,\tau}$ are C_0 (point) representations. That is, for each $p \in [1,\infty]$ different from 2 and $(q,\tau) \in \mathbb{C}^2$ the operator $V^{(p)}_{q,\tau}(x)$, $x \in G$, is in general unbounded in ℓ^p.
For $p = 1$, we point out that as for the base-point representation the result follows from the Liebenson-Kahane-Beurling-Helson Theorem; but for $p > 1$ and $\neq 2$, the unboundedness depends on our generalization to ℓ^p of the L-K-B-H Theorem. The proof uses bounded perturbations to reduce the result to the non-C_0 nature of the basepoint representation (Theorems 12.6 and 12.11).

As usual, D denotes the linear span of the canonical basis vectors (e_n), $n \in \mathbb{Z}$. For given $(q,\tau) \in \mathbb{C}^2$, the balanced Lie algebras $\mathcal{L}(q,\tau)$ are given by sequences $\gamma_n = \gamma_n(q,\tau)$ of complex numbers satisfying the identity $\gamma_n(q,\tau)^2 = q + (\tau+n)(\tau+n+1)$ for $n \in \mathbb{Z}$. The operators $B_j = B_j(q,\tau)$ and $B_\pm = B_\pm(q,\tau)$ are given as in eqn. (3) of the Introduction to this chapter by the identities $B_1 = (i/2)(B_+ + B_-)$, $B_2 = \frac{1}{2}(B_+ - B_-)$, $B_+ e_n = \gamma_n(q,\tau)e_{n+1}$ and $B_- e_n = \gamma_{n-1}(q,\tau)e_{n-1}$. Similarly $\Omega_c = \{(q,\tau) : \gamma_n(q,\tau)^2 \notin (-\infty,0)i\}$ and $\Omega = \{(q,\tau) : \gamma_n(q,\tau)^2 \notin (-\infty,0]i\}$, cf. Section 13E.

For each positive integer k, $D_k(\mathcal{L}^{(p)}(q,\tau))$ denotes the usual D_k space, which by Lemma 12.13 is in fact independent of q, . We have

$D_k^{(p)} = D_k(\mathfrak{L}^{(p)}(q,\tau)) = \{\{u(n)\}: (1+|n|)^k u(n) \in \ell^p\}.$

We recall the following facts from Section 12E concerning the infinitesimal perturbation operators. Each of the perturbations

$$U_j(q,\tau;q_0,\tau_0) = B_j(q,\tau)-B_j(q_0,\tau_0) \tag{63}$$

and

$$U_j(q,\tau) = B_j(q,\tau) - A_j \tag{64}$$

are bounded on ℓ^p and on $D_k^{(p)}$ for all $j = 0,1,2$, q, $\tau \in \mathbb{C}^2$, $p \in [1,\infty]$ and $k = 1,2,\ldots$. For $j = 1$ and 2, the two sets of perturbation operators are constant operators plus compact weighted shifts with $\mathcal{O}(1/n)$ weights on the same series of Banach sequence spaces. For $j = 1,2$, the operators $U_j(q,\tau;q_0,\tau) = B_j(q,\tau)-B_j(q_0,\tau)$ are in fact compact weighted shifts with $\mathcal{O}(1/n)$ weights, cf. Lemma 12.13.

Before stating the main exponentiation/perturbation/continuation theorem for smeared representations on $\ell^p(\mathbb{Z})$, we note that the case $p = 2$ is covered by the essential part of Theorem 12.15, and that, for $p \neq 2$, we get genuine distribution (i.e., not C_0-) representations as pointed out below in Theorem 12.17.

<u>12.16. Theorem</u>

Let $\tilde{G}$ be the universal covering group of $SL(2,\mathbb{R})$.

(a) Every balanced operator Lie algebra $\mathfrak{L}(q,\tau)$, $(q,\tau) \in \mathbb{C}^2$, is exponentiable on $\ell^p(\mathbb{Z})$ for $1 \le p < \infty$, and on c_0, to a smeared representation $V_{q,\tau}^{(p)}: \mathcal{D}(\tilde{G}) \to L(\ell^p)$, or $L(c_0)$.

(b) For each integer $k \ge 1$ and fixed $p \in [1,\infty]$, the operator Lie algebras $\{\mathfrak{L}(q,\tau): (q,\tau) \in \mathbb{C}^2\}$ all have a common complete-normable space $D_k^{(p)}$ of C^k vectors, and a common Fréchet space D_∞ of C^∞ vectors. In fact

$D_\infty = s = \{\{u(n)\} : (1+|n|)^k u(n) \in \ell^\infty \text{ for all } k\}.$

(c) The mapping $(q,\tau) \to V_{q,\tau}^{(p)}(\varphi)$ is operator-norm continuous on Ω_c, and analytic on Ω, for each $\varphi \in \mathcal{D}(\widetilde{G})$. Moreover, each $V^{(p)}_{q,\tau}$ restricts to a smeared representation on the spaces $D^{(p)}_{k_\infty}$. The restriction of $V^{(p)}_{q,\tau}$ to D_∞ is the integrated form of a C^∞ representation. The continuity-and analyticity property in the parameters (q,τ) also hold for the restricted representations.

(d) Each of the operators $V^{(p)}_{q,\tau}(\varphi)$ on ℓ^p [resp. c_0, $p = \infty$] is compact, and in fact nuclear.

Proof: Several statements of the theorem are direct counterparts of statements in Theorem 12.15 with similar proofs. We will be sketchy at these parts of the present proof and omit details, thus placing the emphasis on the differences between Theorem 12.15 and the present Theorem 12.16. Whereas in Theorem 12.15 we derived the perturbation/continuation properties from our general theorem 9.3, we must in the present case of smeared exponentials give the proof in each special case without reference to a general result, since there is not yet an analogue of the perturbation Theorem 9.3 (even for one-parameter distrubution groups). Such a theorem would in fact be of independent interest.

By Theorem 9.3 or 12.15 all the Lie algebras $\mathcal{L}(q,\tau)$ exponentiate to continuous representations $V^{(2)}_{q,\tau}$ of G on $\ell^2(\mathbb{Z})$. Each of these representations restricts to a C^∞ representation $V^{(2)\infty}_{q,\tau}$ of G on D_∞ by [Ps 1, proposition 1.2]. Finally, the C^∞ representations may be extended to smeared representations $V^{(p)}_{q,\tau}$ on the ℓ^p-completion of D_∞ for all p with $1 \le p \le \infty$. By Lemma 12.7 we have

$$V^{(p)}_{q,\tau}(\varphi)u = \int \varphi(a) V^{(2)\infty}_{q,\tau}(a)u \, da \tag{65}$$

for $\varphi \in \mathcal{D}(\widetilde{G})$ and $u \in D_\infty$. The integral is convergent in the Fréchet space topology of D_∞.

Let Ω be the domain of analyticity. We have to show that $(q,\tau) \to V^{(p)}_{q,\tau}$ is analytic in Ω for all p with $1 \le p \le \infty$. (Recall that the representation space of V^∞ is the space c_0 of sequences which vanish at $\pm\infty$.) Let p be fixed, and let $\|\cdot\|$ denote the ℓ^p-norm. Then the operator $I-\bar{A}_0$ has a bounded inverse $R(1,A_0)$ on ℓ^p. We put $D_k = D((I-\bar{A}_0)^k)$ and $\|u\|_k = \|(I-\bar{A}_0)^k u\|$ for $u \in D_k$. Then D_k is a Banach space in the (graph) norm $\|\cdot\|_k$ for $k = 1,2,\dots$. Put $z = (q,\tau)$ to simplify the notation. By Theorem 12.15 the $L(\ell^2)$-operator valued functions $z \to V_z^{(2)}(a)$ are analytic for fixed $a \in \tilde{G}$. Similarly, the functions $z \to V_z^{(2)\infty}(a)$ are analytic from Ω into $L(D_\infty)$ equipped with the strong operator topology.

We must show for all $\varphi \in \mathcal{D}(\tilde{G})$ and all $u \in \ell^p$ that the ℓ^p-valued mapping $z \to V_z^{(p)}(\varphi)u$ is analytic on Ω. For given φ and u, it is clearly enough to show analyticity on relatively compact subsets of Ω. Let Z be a relatively compact subset of Ω, $(\bar{Z} \subset \Omega)$. Let φ be a given test function, and put $K = \mathrm{supp}(\varphi)$. We claim that there is a finite constant C and an integer k such that

$$\|V_z^{(2)}(a)u\| \le C\|u\|_k \tag{66}$$

for all $(z,a) \in Z \times K$ and $u \in D_\infty$. (In fact $k = 1$ works for all $p \in [1,\infty]$.) It is convenient to postpone the proof of the claim contained in (66). Assuming (66) we conclude that $V_z^{(2)\infty}(a)$ extends to a bounded linear mapping $V_z^{(2)k}(a)$ from $(D_k,\|\cdot\|_k)$ into ℓ^p for all $(z,a) \in Z \times K$. This in turn means $V_z^{(2)k}(a)R(1,A_0)^k$ is bounded on ℓ^p, and by (66)

$$\|V_z^{(2)k}(a)R(1,A_0)^k u\| \le C\|u\| \tag{67}$$

for all $u \in \ell^p$. Moreover, the $L(\ell^p)$-operator valued function $V_z^{(2)k}(a)R(1,A_0)^k$ is for fixed $a \in K$ analytic in the variable z; and for fixed $z \in Z$ continuous in the variable a. The last claim is contained in the proof of Lemma 12.7, and the first claim is proved in a similar way. (Analyticity and continuity are with respect to the uniform operator norm topology of $L(\ell^p)$.)

We shall need that the integral

$$\Psi(z) = \int_{\tilde{G}} \psi(a)V_z^{(2)k}(a)R(1,A_0)^k\, u\, da \tag{68}$$

depends analytically upon the parameter $z \in Z$ for every $u \in \ell^p$ and $\psi \in \mathcal{D}(K)$. Taking $\psi = (I-\tilde{X}_0)^k \varphi$ we then conclude that

$$V^{(p)}(z)(\varphi)u = \int_{\tilde{G}} ((I-\tilde{X}_0)^k \varphi)(a) V_z^{(2)k}(a) R(1,A_0)^k \, u \cdot da \quad (69)$$

depends analytically on z as claimed.

An application of Morera's theorem combined with Fubini's theroem implies analyticity of the integral $\Psi(z)$ given by (68). In order to apply those two theorems we must first verify that $z \to \Psi(z)$ is continuous. Once continuity of Ψ is known, it is clear that the integral of Ψ along the boundary of any closed triangle contained in Z must vanish. (We only have to note that the integrand is uniformly bounded, by (60), and analytic in the parameter z.)

Continuity of Ψ is a consequence of the Dominated Convergence Theorem (for vector value functions). If a sequence of points $\{z_j\} \subset Z$ converges to $z_0 \in Z$, then the corresponding sequence of integrable functions (all with support in K)

$$a \to \psi(a) V_{z_j}^{(2)k}(a) R(1,A_0)^k \, u$$

converges pointwise, and is ℓ^p-norm bounded by a function which is constant on K and vanishes off K. (The constant is $\sup|\psi(a)|C\|u\|$.) This concludes the proof of analyticity, modulo the claim contained in (66). It follows easily from Theorem 12.15(c) that the maps $(a,z) \to V_z^{(2)}(a)$ and $(a,z) \to V_z^{(2)1}(a)$ are jointly continuous, and hence that there is a constant C such that

$$\|V_z^{(2)}(a)u\|_{\ell^2} \le C\|u\|_{\ell^2} \quad \text{for all } u \in \ell^2, \quad (70)$$

and

$$\|V_z^{(2)}(a)u\|_{D_1^{(2)}} \le C\|u\|_{D_1^{(2)}} \quad \text{for all } u \in D_1^{(2)}, \quad (71)$$

and all $(z,a) \in Z \times K$ in the obvious notation. A simple exercise in Hölder's inequality for sequences (as in Section 12G) shows that there are finite constants C_p for $p \in [1,\infty]$ such that the Sobolev estimates

$$\|u\|_{\ell^2} \le \|u\|_{\ell^p} \le C_p\|u\|_{D_1^{(2)}} \quad (72)$$

hold for $p \in [1,2]$. Similarly, we have

$$\|u\|_{\ell^p} \le \|u\|_{\ell^2} \le C'_p\|u\|_{D_1^{(p)}} \tag{73}$$

for $p \in (2,\infty]$.

Combining (72) and (70) we get

$$\|V_z^{(2)}(a)u\|_{\ell^p} \le C_p\|V_z^{(2)}(a)u\|_{D_1^{(2)}} \le C_pC\|u\|_{D_1^{(2)}} \le C_pC\|u\|_{D_1^{(p)}} \tag{74}$$

for $p \in [1,2]$. Similarly, (71) and (72) lead to

$$\|V_z^{(2)}(a)u\|_{\ell^p} \le \|V_z^{(2)}(a)u\|_{\ell^2} \le C\|u\|_{\ell^2} \le CC'_p\|u\|_{D_1^{(p)}} ,$$

for $p \in (2,\infty]$. This concludes the proof of (66), and hence of those parts of (a) and (c) that concern the ℓ^p spaces.

The restriction properties of $V_{q,\tau}^{(p)}$ to the C^k spaces follow as in the previous section. (One can also construct the representations $V_{q,\tau}^{(p)k}$ from $V_{q,\tau}^{(2)\infty}$ in the same way as $V_{q,\tau}^{(p)}$ was constructed above from $V_{q,\tau}^{(2)\infty}$.) The proof of (b) is contained in the previous section and the introductory remarks to this section. (Cf. also Lemma 12.13 and Theorem 12.15.)

The remaining part of the proof concerns compactness (and in fact nuclearity) of $V_{q,\tau}^{(p)}(\varphi)$. Let m be a positive integer. Consider the identity

$$\begin{aligned} V_{q,\tau}^{(p)}(\varphi) &= (I-A_0^2)^{-m}(I-A_0^2)^m \, V_{q,\tau}^{(p)}(\varphi) \\ &= (I-A_0^2)^{-m} \, V_{q,\tau}^{(p)}((I-X_0^2)^m \, \varphi). \end{aligned} \tag{75}$$

The operator $V_{q,\tau}^{(p)}((I-X_0^2)^m\varphi)$ is bounded on ℓ^p since $(I-X_0^2)^m\varphi \in \mathcal{D}(\widetilde{G})$, and the operator $(I-A_0^2)^{-m}$ is diagonal with eigenvectors e_n : $(I-A_0^2)^{-m}\, e_n = (1+n^2)^{-m}\, e_n$. It is clear from this that $V_{q,\tau}^{(p)}(\varphi)$ is compact. Recall [Sch] that an operator

$T \in B(E)$ on a Banach space is <u>nuclear</u> if there exists elements $u_n \in E$, $f_n \in E^*$, and a sequence of complex number λ_n such that $\sup\{\|u_n\|,\|f_n\|\} < \infty$, $\Sigma\,|\lambda_n| < \infty$ and $T = \sum_n \lambda_n u_n \otimes f_n$. The tensor product notation $(u_n \otimes f_n)(u) = \langle u,f_n\rangle u_n = f_n(u)u_n$ is used.

We have $(I-A_0^2)^{-m} = \sum_n (1+n^2)^{-m}\, e_n \otimes e_n$ and

$V_{q,\tau}^{(p)}(\varphi) = \sum_n (1+n^2)^{-m}\, e_n \otimes (T^* e_n)$, where $T = V_{q,\tau}^{(p)}((1-X_0^2)^m \varphi)$.

Since $\|e_n\| = 1$ and $\|T^* e_n\|^* \le \|T\|$, $V_{q,\tau}^{(p)}(\varphi)$ is clearly nuclear. The same is true for the restricted smeared representations $V_{q,\tau}^{(p)k}$ on $D_k^{(p)}$, since $e_n \in D_k^{(p)}$ for all n and k. E.O.P.

<u>Remarks</u>: (1) We also note that $V_{q,\tau}^{\infty}(\varphi) = \int_G \varphi(x) V_{q,\tau}^{(2)\infty}(x)dx$ is nuclear as an operator on the Fréchet space D_∞, whereas the continuous operators $V_{q,\tau}^{(2)\infty}(x) \in L(D_\infty)$ are certainly not nuclear at $x = e$, hence at all x by the group property and the fact that the nuclear operators form an ideal in $L(E_2)$.

(2) The nuclearity of $V_{q,\tau}^{(p)}(\varphi)$ enables one to define the notion of a trace and a global character à la [H-Ch]. It turns out that this character coincides with the usual global character for the CCR representation $V_{q,\tau}^{(2)}$ of G on ℓ^2.

(3) We noted in the preamble to the theorem that the perturbation operators $B_j(q,\tau)-B_j(q_0,\tau)$ (with τ constant) are $\mathcal{O}(1/|n|)$-weighted shifts. A likely implication of this is that the smeared operators $\Delta_q(\varphi) = V_{q,\tau}^{(p)}(\varphi) - V_{q_0,\tau}^{(p)}(\varphi)$ are in fact integrals of well-behaved point-functions with values in the bounded operators. That is, $\Delta_q(\varphi) = \int \varphi(x)\{V_{q,\tau}^{(p)\infty}(x) - V_{q_0,\tau}^{(p)\infty}(x)\}dx$ on D_∞, where $V_{q,\tau}^{(p)\infty}(x) - V_{q_0,\tau}^{(p)\infty}(x)$ extends to a bounded operator on ℓ^p for all $x \in \widetilde{G}$.

A smeared representation V of a Lie group $\widetilde{G}$ in a Banach space E is said to be of <u>class C_0</u> if there is a C_0 representation W of $\widetilde{G}$ in E such that $V(\varphi) = \int \varphi(a)W(a)da$ for all $\varphi \in \mathcal{D}(\widetilde{G})$.

It remains to show that the smeared representations $V_{q,}^{(p)}$ for $1 \leq p \leq \infty$, $(q,\tau) \in \mathbb{C}^2$ are <u>not</u> of class C_0, except when $p = 2$. (The points of $\widetilde{G}$ are not represented by bounded operators.)

<u>12.17. Theorem</u>

The smeared representations $V_{q,\tau}^{(p)}$ are <u>not</u> of class C_0 when p belongs to the union of the intervals $1 \leq p < 2$, and $2 < p \leq \infty$.

<u>Proof</u>: Suppose for some $p \in [1,\infty]$ and $(q,\tau) \in \mathbb{C}^2$ that $V_{q,\tau}^{(p)}$ is of class C_0. Then by Lemma 12.13, the introductory remarks of this subsection and the perturbation theorem 9.3 we conclude then that the base-point Lie algebra $\mathfrak{L}_0$ exponentiates to a C_0 representation on ℓ^p. But this we know is not possible by Theorems 12.8 and 12.11. This contradiction concludes the proof.

12G. <u>Irreducibility, equivalences, unitarity, and single-valuedness</u>

In this section, we examine some of the standard representation-theoretic properties of the new series of C_0 exponentials for the $\{\mathfrak{L}(q,\tau): (q,\tau) \in \mathbb{C}^2\}$ that were constructed on the Banach spaces E_p, $1 \leq p \leq \infty$ in Section 12E; these matters are also considered informally in the context of the more exotic smeared representations on $\ell^p(\mathbb{Z})$ that were introduced in Section 12F (for $p \neq 2$). First, analytic vector methods are used to parlay algebraic irreducibility (or direct-sum reducibility) of the $\mathfrak{L}(q,\tau)$ into topological-complete-irreducibility (respectively, topological complemented reduction to TCI subrepresentations) of the global exponentials. Then the natural algebraic infinitesimal equivalences are shown to account for all possible Naimark-equivalence relations connecting (subrepresentations of) those exponentials. For $p = 2$ (i.e., on $\ell^2(\mathbb{Z})$) we show that every irreducible unitary representation of $\widetilde{G}$ is unitarily equivalent to a unitary direct summand of a suitable $V_{q,\tau}^2$. Finally, we verify that the exponentials have kernels containing those for the appropriate covering maps $\Pi : \widetilde{G} \to SL(2,\mathbb{R}) \simeq SU(1,1)$ and $\Pi^1: \widetilde{G} \to SO_e(2,1)$ (so that single-valued representations of these quotient groups are naturally produced) for exactly the appropriate values of τ.

Each of these matters is treated by a selected mixture of standard generalities, extensions of these generalities, and special calculations using explicit information concerning the representations in question. In this connection, we observe that the entire task can be carried out via generalities when τ is real rational, since then the C_0 exponentials can be showns to be representations of semisimple Lie groups with finite center (hence with "large compact subgroups" and K-finite representation theory). We make use here of special structure primarily when this turns out to be significantly cheaper than explaining how to fix up the standard generalities to fit the infinite-center situation for the simply-connected covering group $\tilde{G}$.

Our first result globalizes the reduction properties of the infinitesimal representations (described in Section 12A) to the strongly-continuous exponentials $V^p_{q,\tau} : \tilde{G} \to \mathrm{Aut}(E_p)$ from Section 12E, where $E_p \cong L^p(T)$ for $1 \le p < \infty$ and $E_\infty = C(T)$ as usual. In preparation for the single-valuedness discussion, we obtain more explicit information concerning the "finite-dimensional triple" situation.

<u>12.18. Theorem</u>

(a) If $(q,\tau) \in \mathbb{C}^2$ has $q+(n+\tau)(n+\tau+1) \neq 0$ for all $n \in \mathbb{Z}$, then the exponential $\{V^p_{q,\tau}(x): x \in \tilde{G}\}$ of $\mathcal{L}(q,\tau)$ on E_p is topologically completely irreducible (TCI).

(b) If $q+(n+\tau)(n+\tau+1)$ has a unique integer zero at $m \in \mathbb{Z}$, then $E_p = E^+_p \oplus E^-_p$ and $V^p_{q,\tau}$ restricts to a TCI representation on the invariant subspaces $E^\pm_p$. Here, if f_n is the functional on the dual E^*_p corresponding to e_n (integration against e^{-inx} in the $L^p(T)$-picture), then

$$E^+_p = \{u \in E_p: f_n(u) = 0 \quad \text{for } n \le m\}$$

and

$$E^-_p = \{u \in E_p: f_n(u) = 0 \quad \text{for } m < n\}.$$

(c) If $q+(n+\tau)(n+\tau+1)$ has two integer zeroes $\ell < m$, then $E_p = E^-_p \oplus F_p \oplus E^+_p$ and $V^p_{q,\tau}$ restricts to a TCI representation on each of the subspaces. Here $F_p = \text{span}\{e_n : \ell < n \leq m\}$ is finite-dimensional, E_p^+ is as in (b), and E_p^- is similar to (b) with ℓ replacing m. In this case, $2\tau \in \mathbb{Z}$ and $q \in \mathbb{R}$.

Proof: The invariance claims in (b) and (c) are easy consequences of generalities, once the spaces $E_p^{\pm}$ are identified as closures of the associated subspaces $D_+ = \text{span}\{e_n : n > m\}$ (etc.) that are invariant under $\mathcal{L}(q,\tau):D$ and these invariant subspaces of it consist of analytic vectors for $\mathcal{L}(q,\tau)$ (Section 12C), whence there closures are $V^p_{q,\tau}$-invariant by [Wr, Corollary 4.4.5.6, p. 279].

To check irreducibility, it is most natural to verify topological irreducibility and then invoke the easily established "FDS" properties of subrepresentations to infer TCI claims. In fact, suppose that E_0 is any closed invariant subspace of E. Then the bounded projections defined as in Section 12A by

$$P_n u = (2\pi)^{-1}\int_0^{2\pi} e^{-it(n+\tau)} V(t,B_0(q,\tau))u\, dt$$

map E_0 into itself since $V(t,B_0(q,\tau)) = V^p_{q,\tau}(\exp(tX_0))$, whence $D_0 = \text{span}\{e_n : P_n E_0 \neq \{0\}\} \subset D \cap E_0$. If $E_0 \subset E_p^+$, say, it is then easy to see from algebraic irreducibility of the action of $\mathcal{L}(q,\tau)$ on D_+ that $D_0 = D_+$ and $E_0 = E_p^+$, with similar arguments for E_p^- and F_p in (c) or D itself in (a). Hence the (sub-) representation spaces in (a)-(c) have no proper closed invariant subspaces and topological irreducibility follows. But the rank-1 projection operators P_n are the integrated forms of compactly-supported measures μ_n^τ on G defined by

$$\mu_n^\tau(f) = (2\pi)^{-1}\int_0^{2\pi} e^{-it(n+\tau)} f(\exp(tX_0))dt :$$

$$V^p_{q,\tau}(\mu_n^\tau)u = \int_G V^p_{q,\tau}(x)u\, d\mu_n^\tau(x) = P_n u \tag{76}$$

as the reader may check. Hence the span of the ranges of the finite-rank $V^p_{q,\tau}(\mu)$ for $\mu \in M_0(G)$ (compactly-supported measures) must be dense in E_p (or any closed invariant subspace E_0), whence by definition $V^p_{q,\tau}$ and its subrepresentations are finite-dimensionally spanned (FDS). By [Wr, 4.2.1.5, p. 231 and conventions on p. 242], it follows that topologically irreducible subrepresentations are TCI.

Finally, we turn to the numerology in (c). Suppose that $q+(\ell+\tau)(\ell+\tau+1) = 0 = q+(m+\tau)(m+\tau+1)$. Then $\ell^2+\ell(2\tau+1) = m^2+m(2\tau+1)$ by cancellation, and $2\tau+1 = (\ell^2-m^2)(m-\ell)^{-1} = -(\ell+m)$; whence $2\tau = -(\ell+m+1) \in \mathbb{Z} \subset \mathbb{R}$ and $q = -(\ell+\tau)(\ell+\tau+1) \in \mathbb{R}$.

E.O.P.

The corresponding conclusions for the smeared representations on $\ell^p(\mathbb{Z})$ (from Section 12F) can be established, once the appropriate definitions are agreed upon. It is more economical to postpone discussion of that situation until the C_0 representations have been completely treated on the E_p, so we shall proceed next to examine equivalences connecting the various $V^p_{q,\tau}$.

In preparation, we informally examine the algebraic equivalences that can connect (a submodule of) $\mathfrak{L}(r,\sigma)$ to (a submodule of) $\mathfrak{L}(q,\tau)$ for (r,σ), $(q,\tau) \in \mathbb{C}^2$. It turns out that these completely account for the Naimark-equivalences connecting pairs of the TCI Banach representations discussed in the theorem above. In fact, suppose that D_1, D_2 are subspaces of D, with D_1 invariant under $\mathfrak{L}(r,\sigma)$ and D_2 invariant under $\mathfrak{L}(q,\tau)$, while $S : D_1 \to D_2$ is a bijective intertwining operator for the restrictions (i.e., for all $u \in D_1$, $SB_k(r,\sigma)u = B_k(q,\tau)Su$, $k = 0,1,2$). Then $q = r$, $\tau-\sigma = m \in \mathbb{Z}$, and S is the restriction to D_1 of an algebraic equivalence between $\mathfrak{L}(r,\sigma)$ and $\mathfrak{L}(q,\tau)$ of the special form $Se_n = \alpha(n)e_{n+m}$, where for some $\beta \in \mathbb{C}$ we have $\alpha(n) \equiv \beta$ when $n > \max(0,m)$ and $\alpha(n+1)/\alpha(n) = \mathrm{sgn}(m+n)/\mathrm{sgn}(n)$ for all $n \in \mathbb{Z}$. Briefly, the Casimir values and spectra of compactly-embedded B_0 operators agree, and the bases are related by a trivial modification of a shift. To recall this, let $u \in D_1$. Then since S intertwines all elements in the associative algebras generated by $\mathfrak{L}(r,\sigma)$ and $\mathfrak{L}(q,\tau)$, $rSu = S(B_0^2-B_1^2-B_2^2)(r,\sigma)u = (B_0^2-B_1^2-B_2^2)(q,\tau)Su = qSu$ and $q = r$. But for each $e_n \in D_1$, we similarly obtain $i(n+\sigma)Se_n = SB_0(r,\sigma)e_n = B_0(q,\tau)Se_n$ whence Se_n is an eigenvector for $B_0(q,\tau)$: there must be an $m = m(n) \in \mathbb{Z}$ and $\alpha(n) \in \mathbb{C}$ such that $i(n+\sigma) = i(n+m(n)+\tau)$ (or $m(n) = \sigma-\tau$, independent of n) and $Se_n = \alpha(n)e_{m+n}$. But S also intertwines the shifts $B_\pm$ in the complexifications of the Lie algebras, whence by (3)

$$\gamma_n(r,\sigma)\alpha(n+1)e_{m+n+1} = SB_+(r,\sigma)e_n = B_+(q,\tau)Se_n$$

$$= \alpha(n)\gamma_{m+n}(q,\tau)e_{m+n+1} \qquad (77)$$

and since $q = r$, $\sigma = m+\tau$, we have by (4) in the Introduction to this chapter:

$$\gamma_{m+n}(q,\tau) = \operatorname{sgn}(m+n)(q+(m+n+\tau)(m+n+\tau+1))^{\frac{1}{2}}$$

$$= (\operatorname{sgn}(m+n)/\operatorname{sgn}(n))\operatorname{sgn}(n)(r+(n+\sigma)(n+\sigma+1))^{\frac{1}{2}}$$

$$= \operatorname{sgn}(m+n)/\operatorname{sgn}(n)\gamma_n(r,\sigma). \qquad (78)$$

Consequently, whenever $\gamma_{m+n}(q,\tau) \neq 0 \neq \gamma_n(r,\sigma)$, it follows that $\alpha(n+1)/\alpha(n) = \operatorname{sgn}(m+n)/\operatorname{sgn}(n)$. Given a value of α at any $n_0 \in \mathbb{Z}$, this relation determines $\alpha(n)$ for all n, with constancy outside the interval between m and 0 and alternating signs within that interval. (If either of $\gamma_{m+n}(q,\tau)$ and $\gamma_n(r,\sigma)$ vanishes, both do, and (77) does not restrict α; this occurs at the "boundaries" of the submodule.) But this formula can be used to choose $\alpha(n)$ for all n, extending S to all of D, intertwining $\mathcal{L}(r,\sigma)$ and $\mathcal{L}(q,\tau) = \mathcal{L}(r,\sigma+m)$.

This infinitesimal algebraic background then leads to the following global topological equivalence result. We recall in preparation that two representations $V_i : G \to \operatorname{Aut}(E_i)$ are <u>Naimark-related</u> if and only if there exists a closed, densely-defined injective intertwining operator $T : D(T) \to R(T)$ with $D(T) \subset E_1$, $R(T)$ dense in E_2, such that for all $\mu \in M_0(G)$ (compactly supported measures) $V_1(\mu)$ leaves $D(T)$ invariant, $V_2(\mu)$ leaves $R(T)$ invariant, and, for all $u \in D(T)$, $TV_1(\mu)u = V_2(\mu)Tu$. We observed in the proof of Theorem 12.18 above that all subrepresentations of the $V_{q,\tau}$ are FDS, whence by the discussion in Section 6 of Fell [Fℓ] (cf. also p. 232 [Wr]), Naimark-relatednes is an equivalence relation between these subrepresentations.

12.19. Theorem

Suppose that $1 \leq p(1), p(2) \leq \infty$ and $(r,\sigma),(q,\tau) \in \mathbb{C}^2$. Then $V^{p(1)}_{r,\sigma}$ has a subrepresentation in $E_{p(1)}$ that is Naimark-equivalent to a subrepresentation of $V^{p(2)}_{q,\tau}$ if and only if $\mathcal{L}(r,\sigma)$ and $\mathcal{L}(q,\tau)$ have corresponding algebraically equivalent submodules. Then $q = r$, $\sigma-\tau = m \in \mathbb{Z}$, and the Naimark equivalence can be implemented by the natural extension-by-limits of the algebraic equivalence described above. Moreover, either the intertwining operator or its inverse is bounded, and can be extended to a Naimark-equivalence of the representations as a whole.

Remark: The first claim above is the appropriate $\mathfrak{k}$-finite variant of a standard result of Godement [Wr, 4.5.5.2, p. 326] connecting Naimark and K-finite-algebraic equivalences. However, more is true here, and portions of the argument are simpler in the present setting, while a few steps in the argument of [Wr, 4.5.5.2] require modification in order to apply to this case, so we supply a few extra details below.

Proof: Without loss of generality, we may suppose that $p(1) \geq p(2)$ (otherwise, relabel).

If (a submodule of) $\mathcal{L}(r,\sigma)$ is equivalent to (a submodule of) $\mathcal{L}(q,\tau)$, then the argument given above ensures that $q = r$, $\sigma = \tau+m$, and S can be extended to all of D in the form $S = M_\alpha S_m$, where $S_m e_n = e_{m+n}$ is the m-shift and $M_\alpha e_n = \alpha(n)e_n$ is the α-multiplication operator. Clearly S_m is a $\|\cdot\|_{p(1)}$ isometry on D which extends to an isometry of $E_{p(1)}$. (It is a modulus-1 multiplication operator on $L^p(T)$ and $C(T)$.) But on D, M_α agrees with $\beta(P_+-P_-)$, where $P_\pm$ is the (bounded) projection onto the (closed) subspace of $E_{p(1)}$ generated by $\{e_n: \alpha(n) = \pm\beta\}$ whence M_α is a bounded bijection since $M_\alpha^2 = \beta^2 > 0$ (i.e., $M_\alpha^{-1} = \beta^{-2}M_\alpha$). Hence $M_\alpha S_m$ induces an autmorphism of $E_{p(1)}$ which intertwines $V^{p(1)}_{r,\sigma}$ and $V^{p(1)}_{q,\tau}$.

But then, since the natural injection $i_{12}: E_{p(1)} \to E_{p(2)}$ is bounded with dense range (i.e., $L^{p(1)}(T) \subset L^{p(2)}(T)$ for

$p(1) < \infty$ and $C(T) \subset L^{p(2)}(T)$ as well), the bounded operator $i_{12}M_\alpha S_m : E_{p(1)} \to E_{p(2)}$ extends $S : D \to D$ and is easily seen to intertwine $V^{p(1)}_{r,\sigma}$ and the restriction of $V^{p(2)}_{q,\tau}$ to $i_{12}E_{p(1)}$ (which obviously agrees with $V^{p(1)}_{q,\tau}$: these agree infinitesimally on the analytic vectors D).

Conversely, suppose that $S : D(S) \to R(S)$ sets up a Naimark-equivalence between a subrepresentation of $V^{p(1)}_{r,\sigma}$ in $E_{p(1)}$ and a subrepresentation of $V^{p(2)}_{q,\tau}$ in $E_{p(2)}$. Since D(S) is invariant under $V^{p(1)}_{r,\sigma}(\mu^\sigma_n) = P_n$, the projection onto $\mathbb{C}e_n$ (proof of Theorem 12.18 above), it contains the appropriate eigenvectors for $B_0(r,\sigma) \in \mathfrak{L}(r,\sigma)$; these determine a submodule of $\mathfrak{L}(r,\sigma)$. Similarly, R(S) contains the basis vectors for a submodule of $\mathfrak{L}(q,\tau)$. We argue (essentially as in the corresponding implication for the Theorem 4.5.5.2 in [Wr]) that S sets up an algebraic equivalence between these submodules. In fact, $V^{p(1)}_{r,\sigma}(\mu^\sigma_n)$ is easily shown to be the eigenprojection for $B_0(r,\sigma)$ onto the eigenspace with eigenvalue $i(n+\sigma)$, whence since closed S commutes with the differentiations involved in passing from the group representation to the infinitesimal representation, it follows that $SV^{p(1)}_{r,\sigma}(\mu^\sigma_n) = V^{p(2)}_{q,\tau}(\mu^\sigma_n)S$ projects onto an eigenspace of $B_0(q,\tau)$. One then argues as in our algebraic discussion that $Se_n = \alpha(n)e_{m+n}$ for $m = \sigma-\tau \in \mathbb{Z}$, so S sends a subspace of D into another subspace of D and the differentiation arguments in Theorem 4.5.5.2 of [Wr] then establish that it must supply an algebraic equivalence. (We refer the reader there for details.) This portion of the proof can also be carried out very concretely, with reference to the specific actions of $V(t,B_0(r,\sigma))$ on $u \in E_{p(1)}$ (qua $L^{p(1)}(T)$ or $C(T)$: $(V(t,B_0(r,\sigma))u)(e^{ix}) = e^{i\sigma t}u(e^{i(x+t)})$ and the Fourier nature of the eigenprojections. We omit details.) E.O.P.

The next result extends to "non-unitary" $(q,\tau) \in \mathbb{C}^2$ and E_p for $p \neq 2$ the result of Bargmann [Bg], Pukanszky [Pk] and Sally [Sℓ] concerning the values of τ for which our exponentials give rise to representations of the classical matrix groups $SL(2,\mathbb{R}) \simeq SU(1,1)$ and $SO_e(2,1)$.

12.20. Proposition

Let $1 \leq p \leq \infty$.
(a) The kernel $N_{q\tau}$ of $V^p_{q\tau}$ in $\tilde{G}$ contains the kernel of the covering map $\tilde{G} \to SL(2,\mathbb{R}) \simeq SU(1,1)$ if and only if $2\tau \in \mathbb{Z}$, whence this condition is necessary and sufficient for $\mathfrak{L}(q,\tau)$ to exponentiate to a representation of these isomorphic quotient groups.
(b) The kernel of the covering map $\tilde{G} \to SO_e(2,1)$ (allias the kernel of the adjoint representation of $\tilde{G}$ on $\mathfrak{g} \simeq s\ell(2,\mathbb{R})$) is contained in $N_{q\tau}$ if and only if $\tau \in \mathbb{Z}$, whence this is necessary and sufficient that $\mathfrak{L}(q,\tau)$ exponentiate to a representation of $SO_e(2,1)$.

Remarks: (1) In view of the equivalences catalogued in Theorem 12.19 above, this means that $SL(2,\mathbb{R})$ and $SU(1,1)$ have two inequivalent analytic (in q) series of representations on each E_p, for which the cases $\tau = 0$ and $\tau = 1/2$ may be taken as canonical representatives.
(2) Similarly, $SO_e(2,1)$ has only one q-analytic series on each E_p, which may be represented by the $\tau = 0$ case.
(3) Similar results and remarks apply to any other groups G' covered by $\tilde{G}$.

Proof: The key fact in the proof is that the kernels of all covering maps are contained in the subgroup $\tilde{K} = \{\exp(tX_0): t \in \mathbb{R}\}$ of the covering group that is generated by the compactly-embedded basis element $X_0 \in \mathfrak{g}$. This can be read off from the explicit parametrization of $\tilde{G}$ by $\Delta \times \mathbb{R}$ $(\Delta = \{z \in \mathbb{C} : |z| < 1\})$ given in [Sℓ] and mentioned in Section 12A above. Alternatively, it follows from generalities: the solvable Iwasawa component AN in $G' = SL(2,\mathbb{R})$ (the affine or "ax+b" group of the line) is already simply-connected by inspection or by Theorem VII.5.1 in [Hℓ], whence the product space $\tilde{K}AN$ is simply-connected and covers G'. By uniqueness of covering spaces/groups [Hc] $\tilde{G} = \tilde{K}AN$, whence the kernel of $\tilde{G} \to SL(2,\mathbb{R})$ is the kernel of $\tilde{K} \to K$. The $SO_e(2,1)$ case goes in the same way.

For $SL(2,\mathbb{R})$, we recall that X_0 can be identified with the matrix $\frac{1}{2}\begin{pmatrix} 0 & -1 \\ 1 & 0 \end{pmatrix}$ and $\exp(tX_0)$ is associated with the matrix $\begin{pmatrix} \cos t/2 & -\sin t/2 \\ \sin t/2 & \cos t/2 \end{pmatrix}$, which reduces to the unit matrix when $t = 4k\pi$ for $k \in \mathbb{Z}$: the kernel of the covering map is $\{\exp(4k\pi X_0): k \in \mathbb{Z}\}$ in $\tilde{G}$. The corresponding operator group has $V(t,B_0(q,\tau))e_n = e^{it(n+\tau)}e_n = e_n$ if and only if $t(n+\tau) = 2\ell\pi$ for some $\ell = \ell(n) \in \mathbb{Z}$ and all $n \in \mathbb{Z}$, whence $t\tau = 2\ell(0)\pi$. Here $\exp(4k\pi X_0) \in N_{q\tau}$ forces $4\pi\tau = 2\ell(0)\pi$ or $2\tau = \ell(0) \in \mathbb{Z}$, while then $4\pi(n+\tau) = 4\pi n + 2\ell(0)\pi = 2(2n+\ell(0))\pi$ follows for all $n \in \mathbb{Z}$ for τ of this form: $\exp(4k\pi X_0) \in N_{q\tau}$ if and only if $2\tau \in \mathbb{Z}$. This settles the $SL(2,\mathbb{R})$ and $SU(1,1)$ cases. For $SO_e(2,1)$ (alias the range of the adjoint representation) we know from the commutation relations that ad X_0 has eigenvalues $\pm i$ (acting upon $\mathfrak{g}_{\mathbb{C}}$), whence $\exp(t \text{ ad } X_0) = 1$ if and only if $t = 2k\pi$. The same reasoning as above then yields $\exp(tX_0) \in N_{q\tau}$ for $t = 2k\pi$ if and only if $\tau \in \mathbb{Z}$.

Next, we apply the results obtained above to an analysis of the relationship between the subrepresentations of $\{V^2_{q,\tau} :(q,\tau) \in \mathbb{C}^2\}$ (on $E_2 \simeq \ell^2(\mathbb{Z})$) and the classical unitary dual spaces of $\tilde{G}$, $SL(2,\mathbb{R})$, and $SO_e(2,1)$. Our first result surveys the general picture, while the second describes how the various classical series fit into our framework.

12.21. Proposition

Every topologically irreducible unitary representation $V : \tilde{G} \to \text{Aut}(H)$ on a Hilbert space H is unitarily equivalent to a unitary orthogonal direct summand of some $V^2_{q\tau}$ on $\ell^2(\mathbb{Z})$, where $\tau \in [0,1)$.

Proof: A topologically irreducible unitary representation V is necessarily TCI [Wr, 4.3.1.7, p. 247], so the discussion in Section 12A supplies an algebraic equivalence of the $\mathfrak{k}$-finite representation of $\mathfrak{g} \simeq s\ell(2,\mathbb{R})$ with a subrepresentation of $\mathfrak{L}(q,\tau)$. Moreover, since $dV(X_0) = C_0$ is skew-symmetric, it has imaginary spectrum so $\tau \in \mathbb{R}$, while the eigenvectors selected in Section 12A are orthogonal. But it is easy to see (as in [Bg] and [Pk]) that $C_1 = dV(X_1)$ and $C_2 = dV(X_2)$ are skew-symmetric if and only if $C_+^* = C_-$ and $C_-^* = C_+$ $(C_\pm = C_1 \pm iC_2)$, whence we have for all n such that $i(n+\tau)$ is an eigenvalue of C_0 with eigenvector u_n

$$\gamma_n(q,\tau)\|u_{n+1}\|^2 = (C_+u_n,u_{n+1}) = (u_n,C_-u_{n+1}) = \bar{\gamma}_n(q,\tau)\|u_n\|^2 .$$

Hence $\gamma_n(q,\tau) \in \mathbb{R}$ for all n, and by cancellation, $\|u_{n+1}\| = \|u_n\|$ whenever $\gamma_n(q,\tau) \neq 0$. It follows easily from reality of the $\gamma_n(q,\tau)$ that $B_+(q,\tau)^* = B_-(q,\tau)$ and vice versa in $E_2 = \ell^2(\mathbb{Z})$ (reverse the above argument for the unit vectors e_n and e_{n+1} in the range of the isomorphism, then apply bilinearity and orthogonality...). Consequently, $B_1(q,\tau)$ and $B_2(q,\tau)$ are skew-symmetric on D_0 and $V^2_{q\tau}$ restricts to a unitary representation on the closure of this (analytic vector) domain. Moreover, if some u_n is chosen normalized, all are, so that the algebraic equivalence preserves orthonormal bases and is unitary. Finally, to place τ in the fundamental interval [0,1], let $m = [\tau]$ (the greatest integer $\leq \tau$) and $\sigma = \tau - m$. By the argument in Theorem 12.18 above, it is then easy to see that $Se_n = \mathrm{sgn}(m+n)/\mathrm{sgn}(n)e_{m+n}$ defines a unitary equivalence on $\ell^2(\mathbb{Z})$ between $V^2_{q\tau}$ and $V^2_{q\sigma}$ which, when composed with the unitary equivalence above, puts V in unitary equivalence with an irreducible unitary subrepresentation of $V^2_{q\sigma}$ for $\sigma \in [0,1)$.

<u>12.22. Theorem</u>

Let $\tau \in [0,1)$.

(a) For all $\tau(1-\tau) < q < \infty$, $V^2_{q,\tau}: \tilde{G} \to \mathrm{Aut}(\ell^2(\mathbb{Z}))$ is irreducible and unitary: for $\frac{1}{2} < q < \infty$, representations in the <u>principal series</u> are obtained, while for $\tau(1-\tau) < q \leq \frac{1}{2}$, these lie in the <u>complementary series</u> for $\tilde{G}$. For $\tau = 0$ or $\frac{1}{2}$, these series lie in the unitary duals of $SL(2,\mathbb{R})$ and $SU(1,1)$, while for $\tau = 0$ they lie in the unitary dual of $SO_e(2,1)$.

(b) Let $0 \neq \tau \neq \frac{1}{2}$, put $\tau' = 1-\tau$, and suppose that $q < \tau(1-\tau)$

$=\tau'(1-\tau')$. Then there exists $\ell > 0$ such that $q = \ell(1-\ell)$ and the unitary discrete series representation D_ℓ^+ is contained as an (irreducible) direct summand of one of $V_{q\tau}^2$ and $V_{q\tau'}^2$, while D_ℓ^- is contained as a direct summand in the other. In both cases, the orthogonal complementary representations are nonunitary TCI representations.

(b') For $0 \neq \tau \neq \frac{1}{2}$, $\tau' = 1-\tau$ and $q = \tau(1-\tau) = \tau'(1-\tau')$, (the limiting case in (b) above), the first conclusion concerning containment of $D_\ell^\pm$ in (b) above holds, but the complementary representations for $D_\ell^\pm$ are the unitary discrete series representations $D_{\ell'}^\mp$, for $\ell' = 1-\ell$.

(c) For $\tau = \frac{1}{2}$ and $q < \tau(1-\tau)$, there exists a half-integer $\ell = k+\frac{1}{2} > 0$ such that $V_{q\tau}^2$ decomposes as the orthogonal direct sum of the two (irreducible) unitary discrete series representations $D_\ell^\pm$ for $SL(2,\mathbb{R})$ (hence $SU(1,1)$ and $\tilde{G}$) and the non-unitary finite-dimensional representation F_ℓ of dimension $2k+1$.

(c') For $\tau = \frac{1}{2}$ and $q = \tau(1-\tau) = \frac{1}{4}$, $V_{q\tau}^2$ decomposes as the orthogonal direct sum of the two "mock-discrete" unitary representations of $SL(2,\mathbb{R})$ (i.e., $V_{q\tau}^2$ is the "exceptional" or "Reducible principal series" representation).

(d) For $\tau = 0$ and $q = \ell(1-\ell)$ for $1 < \ell \in \mathbb{Z}$ $(q < 0)$ $V_{q\tau}^2$ decomposes as in (c) as the orthogonal direct sum of two unitary discrete series representations $D_\ell^\pm$ and the finite-dimensional non-unitary representation F_ℓ for $SO_e(2,1)$ (hence $SL(2,\mathbb{R})$, $SU(1,1)$ and $\tilde{G}$).

(d') For $\tau = 0 = q$, V_{00}^2 decomposes as the orthogonal direct sum of the two unitary discrete series representations $D_1^\pm$ and the trivial (hence unitary) representations on $C \approx \mathbb{C}e_0$ for $SO_e(2,1)$ (hence $SL(2,\mathbb{R})$, $SU(1,1)$ and $\tilde{G}$).

(e) The list given above exhausts the unitary subrepresentations of the $V_{q\tau}^2$ and accounts completely for the unitary duals of the group under discussion.

Remarks: (1) The exceptional point $(q,\tau) = (\frac{1}{4},\frac{1}{2})$ in (c') acts as a common limit point for the behaviors described in

(b') and in (c): the two integer zeroes for $\gamma_n(q,\tau)$ in (c) coalesce into a single zero of multiplicity 2, eliminating the non-unitary finite-dimensional component and exhibiting a limiting case of the unitary direct sum decomposition in (b'). We recall that the representations $D^{\pm}_{\frac{1}{2}}$ in (c') are not square-integrable ([Bg], [Lg]).
(2) The decomposition situations described in (b) and (b') extend to all τ with $\mathrm{Re}(\tau) \in [0,1)$, yielding non-unitary analytic continuations of the discrete series (and their complements) in the spirit of Sally's work [Sℓ] and recent generalizations due to Vergne-Rossi [VR] and others.
(3) Modulo the literature, (a)-(d') above constitute a second proof of our general result 12.21 above, since our list exhausts the unitaries described in [Pk].

Proof: As general background, we observe first that if $\tau \in [0,1)$ (or $\tau \in \mathbb{R}$ in general) then $\mathrm{Im}(q) \neq 0$ forces $\mathrm{Im}(q+(n+\tau)(n+\tau+1)) \neq 0$ for all $n \in \mathbb{Z}$, whence $\gamma_n(q,\tau) \notin \mathbb{R}$ for all $n \in \mathbb{Z}$, and $V^2_{q\tau}$ can have no unitary subrepresentations. Also, $V^2_{q\tau}$ as a whole is unitary if and only if

$$0 \leq \gamma_n^2(q,\tau) = q+(n+\tau)(n+\tau+1) = (q-\tfrac{1}{4})+(n+\tau+\tfrac{1}{2})^2 \text{ for all } n \in \mathbb{Z}.$$

Thus since $\tau \in [0,1)$, $(n+\tau+\frac{1}{2})^2$ has its minimum $(\tau-\frac{1}{2})^2$ at $n = -1$, and $q-\frac{1}{4} \geq -(\tau-\frac{1}{2})^2$ or $q > \tau(1-\tau)$ is equivalent to unitarity.

For (a), then, we observe that the conditions $q \geq \tau(1-\tau)$ and $\gamma_n(q,\tau) \neq 0$ for all $n \in \mathbb{Z}$ are necessary and sufficient for irreducible unitarity (Proposition 12.21), and $\gamma_n(q,\gamma) \neq 0$ for all $n \in \mathbb{Z}$ is equivalent to $q \neq \tau(1-\tau)$ in this case. (If $q = \tau(1-\tau)$, $\gamma_{-1}(q,\tau) = 0$, otherwise the minimum of $\gamma_n^2(q,\tau)$ is $q-\tau(1-\tau) > 0$.) The claims for the quotient groups follow by Proposition 12.20, while the nomenclature is standard ([Bg], [Pk], [Sℓ]).

The usual substitution $q = \ell(1-\ell)$ simplifies the discrete series analysis in (b)-(d'):

$$\gamma_n^2(q,\tau)=(q-\tfrac{1}{4})+(n+\tau+\tfrac{1}{2})^2=-(\ell-\tfrac{1}{2})^2+(n+\tau+\tfrac{1}{2})^2=(n+\tau+\ell)(n+\tau+1-\ell). \quad (79)$$

This product is non-negative wherever both factors have the same sign, and is negative where both are nonzero with differing signs. Since replacing ℓ by $1-\ell$ changes neither the factorization of γ_n^2 nor the expression for q, we may suppose without loss of generality that $\ell \geq 1-\ell$, so that $n+\tau+\ell \geq n+\tau+1-\ell$. Then both factors are ≤ 0 for $n \leq -(\tau+\ell)$, both are ≥ 0 for $n \geq -(\tau+1-\ell)$, and for any $-(\tau+\ell) < n < -(\tau+1-\ell)$, the signs differ and $\gamma_n^2(q,\tau) < 0$ or $\gamma_n(q,\tau) \notin \mathbb{R}$.

Consequently, since all irreducible unitary cases have been covered in (a) above, we can show that exactly the following possibilities occur:

(i) $V_{q\tau}^2$ is unitary on all of $\ell^2(\mathbb{Z})$ and one or both of $-(\tau+\ell)$ and $-(\tau+1-\ell)$ are integers, but there is no integer between (cases (b'), (c'), (d')),

(ii) only one of $-(\tau+\ell)$ and $-(\tau+1-\ell)$ is an integer, and an integer occurs between the two, so that $V_{q\tau}^2$ splits as the direct sum of an irreducible unitary and an irreducible nonunitary representation (case (b)), or

(iii) both $-(\tau+\ell)$ and $-(\tau+1-\ell)$ are integers, with integers between, so that $V_{q\tau}^2$ splits as the direct sum of three irreducibles: one finite-dimensional non-unitary and two unitary (cases (c), (d)).

For situation (i), which covers all reducible unitary cases, reducibility forces the vanishing of $\gamma_n(q,\tau)$, hence of at least one factor in $\gamma_n^2(q,\tau) = (n+\tau+\ell)(n+\tau+1-\ell)$, while unitarity precludes the existence of integers strictly between these vanishing points. If both factors vanish at the same integer $n = m$, it is easy to see that case (c') occurs: $\tau+\ell = \tau+1-\ell$ implies that $\ell = \frac{1}{2}$, whence $q = \frac{1}{4}$ and since $\tau \in [0,1)$ with $\tau+\frac{1}{2} \in \mathbb{Z}$, $\tau = \frac{1}{2}$ as well. (By Proposition 12.20, $V_{q\tau}^2$ then represents $SL(2,\mathbb{R})$, etc.) If both vanish, but at different integers, then $-(\tau+\ell)$ and $-(\tau+1-\ell)$ must be successive integers, leading quickly to case (d'): $-(\tau+\ell)+1 = -(\tau+1-\ell)$ implies that $\ell = 1$, so $q = 0$ and $\tau+\ell \in \mathbb{Z}$ with $\tau \in [0,1)$ forces $\tau = 0$. (Again, Proposition 12.20 establishes that $V_{q\tau}^2$ represents $SO_e(2,1)$, etc.) If $n+\tau+\ell$ vanishes and $n+\tau+1-\ell$ does not, then $m = -(\tau+\ell) \in \mathbb{Z}$ and unitarity entails that $-(\tau+1-\ell) < m+1 = -(\tau+\ell)+1$, whence $\ell < 1$. But $\ell \geq 1-\ell$ implies that $\ell \geq \frac{1}{2}$, so $\tau \in [0,1)$ shows that $\frac{1}{2} \leq \tau+\ell < 2$, and $\tau+\ell \in \mathbb{Z}$ yields $\tau+\ell = 1$, $m = -1$. In that case, $E_- =$ closed span$\{e_n : n \leq -1\}$ is invariant and $B_0(q,\tau)$ has eigenvalues $i(-1+\tau) = -i\ell$, $i(-1-1+\tau) = -i(\ell+1), \ldots, -i(\ell+k), \ldots$, so the restriction of $V^2_{q\tau}$ to E_- is the discrete series representation usually called D^+_ℓ (because the eigenvalues of $H_0(q,\tau) = iB_0(q,\tau)$ are the positive numbers $\ell, \ell+1 \ldots$). On the complementary invariant subspace E_+ = closed span $\{e_n : n > -1\}$, $B_0(q,\tau)$ has eigenvalues $i\tau = i(1-\ell), i(1+\tau) = i(2-\ell), \ldots, i(k+1-\ell), \ldots,$ corresponding to $D^-_{\ell'}$, for $\ell' = 1-\ell$. A corresponding analysis of the other case, where $n+\tau+\ell$ does not vanish on $\mathbb{Z}$ while $n+\tau+1-\ell = 0$ for $n=m$, yields the following conclusions: $-(\tau+1-\ell)-1 < -(\tau+\ell)$ so $\ell < 1$ again, whence $\tau+1-\ell \in \mathbb{Z}$ forces $\tau+1-\ell = 1$, $\ell = \tau$, E_- and E_+ are as before, but $V^2_{q\tau}$ restricts to $D^+_{\ell'}$ on E_- and to D^-_ℓ on E_+. We leave details to the reader, observing that this exhausts case (b'), and the discussion has exhausted (i) above. Replacement of τ by τ' is discussed below.

Turning to (ii), we observe that the discussion is quite similar to that above for the single-zero case, except that when $n+\tau+\ell$ has a zero at $m = -(\tau+\ell)$, $-(\tau+1-\ell) > m+1 = -(\tau+\ell)+1$ so that $\ell > 1$ (similar changes occur when $n+\tau+1-\ell$ vanishes on $\mathbb{Z}$). Here, the case $m = -(\tau+\ell) \in \mathbb{Z}$ yields the irreducible unitary discrete series representation D^+_ℓ on $E_- = \{e_n : n \leq m\}$, but the subrepresentation of $V^2_{q\tau}$ on $E_+ = \{e_n : n > m\}$ is nonunitary and TCI. When $n+\tau+1-\ell$ produces the zero, the subrepresentation on E_- is nonunitary TCI while that on E_+ is D^-_ℓ.

In order to sort out the single-zero cases as in (b) and (b'), we observe that when D^+_ℓ occurs on E_- for a given τ, we had $-(\tau+\ell) = m \in \mathbb{Z}$, whence $\tau'+1-\ell = 2+m \in \mathbb{Z}$ and $(n+\tau'+1-\ell)$ vanishes on $\mathbb{Z}$, so D^-_ℓ appears as the right-hand component of $V^2_{q\tau'}$. Similarly, when D^-_ℓ occurs in $V^2_{q\tau}$, D^+_ℓ occurs in $V^2_{q\tau'}$, as the reader may check. Notice also that the implicit restriction to $1 > \ell \geq \frac{1}{2}$ in (b') is now removed, since $0 < \ell < \frac{1}{2}$ now corresponds to $1-\ell = \ell' \geq \frac{1}{2}$.

Having exhausted (i) and (ii), we proceed to (iii), noting that the case of two integer zeroes for $\gamma^2_n(q,\tau)$ has already been covered in some detail in Proposition 12.20 above: $2\tau \in \mathbb{Z}$ and $\ell^2(\mathbb{Z})$ splits (orthogonally) into $E_- \oplus F \oplus E_+$

But our discussion above shows that the restrictions of $V^2_{q\tau}$ to $E_\pm$ must be unitary, and our assumption in (iii) ensures that $V^2_{q\tau}$ is nonunitary on F. Here, as above, $-(\tau+\ell) = m \in \mathbb{Z}$ leads to eigenvalues $i(m+\tau) = -i\ell, \ldots, -i(\ell+n)\ldots$ for B_0 on E_- and $-(\tau+1-\ell) = k \in \mathbb{Z}$ supplies eigenvalues $i(k+1+\tau) = i\ell, \ldots, i(\ell+n), \ldots$ on E_+, so that $D^\pm_\ell$ occur as the unitary subrepresentations. Also, when $\tau = 0$, then $\ell \in \mathbb{Z}$ and case (d) occurs by Proposition 12.20. But if $\tau = \frac{1}{2}$, then $\ell = 2m+1$ for $m \in \mathbb{Z}$ and case (c) occurs.

The analysis has now covered all unitary possibilities for subrepresentations of $V^2_{q\tau}$, establishing (c) by Proposition 12.20 (or by our remark.) E.O.P.

Turning finally to the question of possible variants of the results above for the smeared representations $V^{(p)}_{q\tau}$ on $\ell^p(\mathbb{Z})$ $(p \neq 2)$, it is clear that the unitarity results have no natural analogues in this setting. But our result on representations of $SL(2,\mathbb{R}) \simeq SU(1,1)$ and $SO_e(2,1)$ carries over easily and naturally: if $2\tau \in \mathbb{Z}$, $\mathcal{L}(q,\tau)$ exponentiates to a smeared representation of $SL(2,\mathbb{R})$ while if $\tau \in \mathbb{Z}$ a smeared representation of $SO_e(2,1)$ is obtained. To see this, notice that in these cases the representations $V^2_{q\tau} : G \to \mathrm{Aut}(\ell^2(\mathbb{Z}))$ restrict to differentiable representations (on the dense Fréchet space $D^{(2)}_\infty$ of C^∞ vectors) that have the same kernel, hence yield differentiable representations of $SL(2,\mathbb{R})$ (respectively $SO_e(2,1)$). Since this space is also the space of C^∞ vectors for $\mathcal{L}(q,\tau)$ in $\ell^p(\mathbb{Z})$, it follows as usual by Lemma 12.4 (which <u>did not</u> require simple-connectedness) that $\mathcal{L}(q,\tau)$ has a smeared exponential for $SL(2,\mathbb{R})$ (respectively $SO_e(2,1)$).

Proceeding backward through the section, we first observe that the natural notion of Naimark-relatedness for smeared representations simply replaces the convolution algebra $M_0(G)$ with $\mathcal{D}(G)$, retaining the other conventions: S must intertwine $V_1(\varphi)$ and $V_2(\varphi)$ for $\varphi \in \mathcal{D}(G)$. (The obvious exercise with closedness and smooth approximate identities $\varphi_n \to \delta_e$ in $\mathcal{D}(G)$ shows that if V_i is the integrated form of a

C_0 representation $i = 1,2$. Then S intertwines $V_i(\mathcal{D}(G))$ if and only if it intertwines $V_i(M_0(G))$, so our convention agrees with the usual one where the latter makes sense.) Theorem 12.19 then has an exact analogue in this setting, as does part of the discussion in Section 12A: the only Naimark-relations among subrepresentations of the smeared representations are the obvious (algebraic) ones, and every "suitable" TCI smeared representation of G (i.e., the $\mathcal{D}(G)$ algebra-representation is TCI) is Naimark-related to one of these. Indeed, the argument that algebraic equivalence of submodules of $\mathfrak{L}(r,\sigma)$ and $\mathfrak{L}(q,\tau)$ implies Naimark-relatedness of the entire smeared representations $V_{r\sigma}$ and $V_{q\tau}$ on $\ell^p(\mathbb{Z})$ goes just as in Theorem 12.19, except that the direction of bounded injections is reversed: $p(1) \leq p(2)$ implies that $i_{12}: \ell^{p(1)}(\mathbb{Z}) \to \ell^{p(2)}(\mathbb{Z})$ is bounded. The converse and Section 12A claims are more delicate: these smeared representations possess a sufficient vestige of FDS and $\mathfrak{k}$-finiteness to push the arguments through, if we confine attention to those smeared representations V such that $dV(X_0) = C_0$ is a group-pregenerator. (Then, although many $\mu \in M_0(G)$ will have $V(\mu)$ unbounded, the particular μ_n used to construct the eigenprojections $P_n = V(\mu_n)$ will go into bounded rank-1 projections...) We omit further details.

Concerning decomposition into irreducibles, it is easy to check that the smeared representations admit the same direct-sum reductions on the $\ell^p(\mathbb{Z})$(into pairs or finite-dimensional triples) that are described for the E_p in Theorem 12.18 for the various $(q,\tau) \in \mathbb{C}^2$. However, the irreducibility of the components is more delicate. By a (by now) standard argument ([Ps] or [Wr]) topological irreducibility of the components of $V^2_{q\tau}$ on $\ell^2(\mathbb{Z})$ is equivalent to topological irreducibility of the differentiable representation on D_∞ (common to all $\ell^p(\mathbb{Z})$). The obvious extension of that argument establishes equivalence of topological irreducibility for components of the smeared Banach representations in $\ell^p(\mathbb{Z})$ with topological irreducibility of their differentiable Fréchet component representations in D_∞. One then needs a "TCI if and only if topologically irreducible" lemma to obtain TCI properties of the smeared components. It turns out that the weakened FDS and $\mathfrak{k}$-finiteness properties of these smeared representations are quite adequate to push through the standard arguments.

We omit further details on these matters, pending further investigation toward finding a natural general framework for treatment of smeared representations for semisimple Lie groups without finite center. (Present arguments seem too ad hoc and dependent upon special structures.)

12H. Perturbation and reduction properties of other analytic series

This brief section is devoted to the infinitesimal study of four other analytic series of exponentials for $\tilde{G}$ and the groups that it cover. These are
(1) the "naive" analytic continuations of the principal p-series of induced (or multiplier) representations of SU(1,1) on $L^p(T)$, in the so-called "compact picture",
(2) the more delicate continuations for the multiplier representations of $SL(2,\mathbb{R})$ in $L^p(\mathbb{R})$, with continuation dependent p, in the "noncompact picture",
(3) Kunze-Stein's normalized principal series continuations on $L^2(\mathbb{R})$ for $SL(2,\mathbb{R})$ in the "noncompact picture", and
(4) Sally's normalized principal series for SU(1,1) lifted back (or "recompactified") to $L^2(T)$.

Our treatment is handled as two examples and several remarks which serve different purposes. First, perturbations are examined primarily in the "Schrödinger formalism" (viz., differential operators in function spaces) and contrasted with the (non-classical) pseudo-differential operators which arise in that formalism for the $V^p_{q\tau}$ of Section 12E. We obtain examples of interesting analytic perturbations of Lie algebras and group-generators where the perturbations are in some instances not bounded (let alone compact) but are relatively bounded. These lie beyond the scope of any analytic Lie algebraic perturbation theory known to the authors, hence they indicate a fruitful direction for further work in perturbation theory. Second, we compute the parameter-dependence of the intertwining operators between some of these series, in either the Schrödinger, or Heisenberg matrix-element formalisms. These calculations account in part for the major differences in analytic perturbation behavior. They also advertise the singularities in the intertwining operators that tend to arise at the values of (q,τ) where $\gamma_n(q,\tau)$ has zeroes in $n \in \mathbb{Z}$; these serve to account for strikingly different Jordan-Hölder reduction phenomena that arise at singular points. Third, the examples suggest a number of open problems whose solution is needed in order to evaluate the long-range possibilities for the use of our perturbation-continuation techniques in the study of the nonunitary duals of higher-dimensional groups.

Example 1: Continuation of Bargmann's principal series

In [Bg], Bargmann constructed on $L^2(T)$ two series of unitary representations for the Lie group SU(1,1), parameterized by $s = (\frac{1}{4} - q)^{\frac{1}{2}} \in \mathbb{R}$, which admit obvious analytic continuations to all $s \in \mathbb{C}$ and can be unified into a single two-parameter series in $(s,\tau) \in \mathbb{C}^2$ for the covering group $\widetilde{G}$ [Sℓ]. These constructions also extend routinely to series of representations on $L^p(T)$ for $1 \leq p < \infty$ and $C(T)$ (i.e., over all E_p, $1 \leq p \leq \infty$). We confine our discussion here to the simplest "integral series" [Bg] for SU(1,1) (and $SO_e(2,1)$ in order to simplify the calculations. Interested readers can easily extend our remarks to the general case using [Sℓ].

First, it is useful for our second example (and perhaps more transparent) to begin by reviewing a few generalities about "multiplier" representations of Lie groups G derived from actions of G on manifolds M ($G \times M \to M$, where for $x \in G$ and $m \in M$ we write mx for the image of m under the diffeomorphism of M induced by x). These have the form

$$[V_s(x)u](m) = \mu_s(x,m)u(mx) \tag{80}$$

where $\mu_s \in C^\infty(G \times M)$ is a suitable "multiplier" function that is determined by a "continuation parameter" $s \in \mathbb{C}$. (Recall that the homomorphism property $V_s(x)V_s(y) = V_s(xy)$ requires the "cocycle" identity $\mu_s(xy,m) = \mu_s(x,m)\mu_s(y,mx)$ while $V_s(e) = I$ requires $\mu_s(e,m) \equiv 1$.) In any L^p space on M, the infinitesimal representation for V can be computed for compactly-supported $u \in C^\infty(M)$ by uniform limits: for $X \in \mathfrak{g}$ (as a vector-field on G) corresponding to the vector-field $A(X)$ on M via the flow $(m,t) \to m\cdot\exp(tX)$ we get

$$\begin{aligned}[dV_s(X)u](m) &= \frac{d}{dt}[V_s(\exp(tX))u](m)\big|_{t=0} \\ &= (X\mu_s)(e,m)u(m) + \mu_s(e,m)[A(X)u](m) \qquad (81)\end{aligned}$$

thus

$$dV_s(X) = A(X) + X(\mu_s)(e,\cdot) \tag{82}$$

where A(X) is a continuation-independent vector-field (or differential operator) on M and $X(\mu_s)$ is an "infinitesimal multiplier" which additively perturbs A(X) and carries all of the continuation-dependence. It follows that, in any such situation, perturbations

$$dV_r(X) - dV_s(X) = [X\mu_r - X\mu_s](e,\cdot) \tag{83}$$

are operators of pointwise multiplication by smooth functions; these are bounded on L^p spaces if the functions are bounded, but can be shown to be compact only when the difference vanishes identically off sets of measure zero when the measure is Riemannian. (We omit details, since noncompactness is obvious from explicit calculations below.)

Example 2.

Here, $M = T = \{e^{i\theta}: 0 \le \theta \le 2\pi\}$ and $G = SU(1,1)$ acts on T in the usual way: if $x = \begin{pmatrix} \alpha & \bar\beta \\ \beta & \bar\alpha \end{pmatrix}$ then $e^{i\theta}x = (\alpha e^{i\theta}+\beta)(\bar\beta e^{i\theta}+\bar\alpha)^{-1}$. The multiplier $\mu_s(x,e^{i\theta}) = |\bar\beta e^{i\theta}+\bar\alpha|^{-1-2s}$ is a typical combination of the "modular function" $|\bar\beta e^{i\theta} + \bar\alpha|^{-1}$ for the action of the parabolic subgroup MAN (where

$M = \left\{\begin{pmatrix} 1 & 0 \\ 0 & 1 \end{pmatrix}, \begin{pmatrix} i & 0 \\ 0 & -i \end{pmatrix}\right\}$, $A = \left\{\begin{pmatrix} \cosh t & \sinh t \\ \sinh t & \cosh t \end{pmatrix}, t \in R\right\}$,

and $N = \left\{\begin{pmatrix} 1+it & -it \\ it & 1-it \end{pmatrix}, t \in \mathbb{R}\right\}$) on $T \simeq G/MAN$, and

the character (non-unitary if $Im(s) \neq 0$) of MAN (in fact, A) $|\bar\beta e^{i\theta} + \bar\alpha|^{-2s}$. If $\exp(tX) = \begin{pmatrix} \alpha(t) & \bar\beta(t) \\ \beta(t) & \bar\alpha(t) \end{pmatrix}$, then an easy calculation yields the formula

$$(X\mu_s)(e,e^{i\theta}) = -(1+2s)Re(\bar\beta'(0)e^{i\theta}) \tag{84}$$

while a more tedious exercise in the chain rule gives

$$A(X) = -2\{\alpha'(0) + i\,\mathrm{Re}(\beta(0))\sin\theta - \mathrm{Im}(\beta'(0))\cos\theta\}A_0 \tag{85}$$

where $(A_0 u(e^{i\theta}) = u'(e^{i\theta})$. We leave the details to the reader. Recalling that the "infinitesimal" matrix for X tangent to $\exp(tX)$ is $\begin{pmatrix}\alpha'(0) & \bar{\beta}'(0)\\ \beta'(0) & \bar{\alpha}'(0)\end{pmatrix}$ one gets for the representations of our standard basis-matrices $X_o = \frac{1}{2}\begin{pmatrix} i & 0\\ 0 & -i\end{pmatrix}$, $X_1 = \frac{1}{2}\begin{pmatrix}0 & -i\\ i & 0\end{pmatrix}$ and $X_2 = \frac{1}{2}\begin{pmatrix}0 & -1\\ -1 & 0\end{pmatrix}$ (cf. Section 12A) that

$$\begin{aligned} A_0(s) &=_{\mathrm{Def}} dV_s(X_0) = A_0 = \frac{d}{d\theta}\quad, \\ A_1(s) &=_{\mathrm{Def}} dV_s(X_1) = \cos\theta\; A_0 - (\tfrac{1}{2}-s)\sin\theta\quad, \\ A_2(s) &=_{\mathrm{Def}} dV_s(X_2) = \sin\theta\; A_0 + (\tfrac{1}{2}+s)\cos\theta\quad . \end{aligned} \tag{86}$$

At $s = -\frac{1}{2}$ (corresponding to $q = 0$) we get the "base-point" Lie algebra in its realization in $L^2(T)$. In fact, the standard trigonometric basis of eigenfunctions for $A_0 = A_0(s)$ is independent of s: $g_n(e^{i\theta}) = e^{in\theta}$. This basis yields the "Heisenberg matrix" form of (48):

$$\begin{aligned} A_0(s)g_n &= ing_n, \\ A_1(s)g_n &= \tfrac{1}{2}(e^{i\theta}+e^{-i\theta})ing_n-(\tfrac{1}{2}-s)\,\tfrac{i}{2}(e^{i\theta}-e^{-i\theta})g_n \\ &= i\tfrac{n}{2}(g_{n+1}+g_{n-1})-\tfrac{i}{2}(\tfrac{1}{2}-s)(g_{n+1}-g_{n-1}), \\ A_2(s)g_n &= \tfrac{n}{2}(g_{n+1}-g_{n-1})+\tfrac{1}{2}(\tfrac{1}{2}+s)(g_{n+1}+g_{n-1}). \end{aligned} \tag{49}$$

The corresponding shift operators then have the form

$$\begin{aligned} A_+(s)g_n &= i(n+\tfrac{1}{2}+s)g_{n+1} \\ A_-(s)g_n &= i(-n+\tfrac{1}{2}+s)g_{n-1} \end{aligned} \tag{50}$$

which is clearly not "balanced". Notice here that the shift perturbation

$$U_+(r,s)g_n = [A_+(r) - A_+(s)]g_n = i(r-s)g_{n+1}$$

is just a scalar multiple of the unitary shift operator, hence is clearly bounded and noncompact. A similar remark applies to the perturbation of A_- as s varies.

Associated with the unbalanced character of $\mathcal{L}_{s+\frac{1}{2}}$, we see that both $A_\pm(s)$ annihilate some basis vectors iff $\frac{1}{2} + s \in \mathbb{Z}$, and the "direct sum splitting" effect $A_+(s)g_n = 0 = A_-(s)g_{n+1}$ if $n+\frac{1}{2}+s = 0 = n+1-\frac{1}{2}-s = n+\frac{1}{2}-s$, or $n+\frac{1}{2} = \pm s = 0$.
Hence, for the integral series, there is no value of s where this direct-sum splitting occurs: every reducible module contains at least one uncomplemented submodule and further decompositions must be carried out in the quotient module. (This is the "subquotient" process of Harish-Chandra et al., illustrated for SL(2,R) in [Lg] and described for the general case in [Wr, Section 5.4].)

Now, as indicated in [Bg] (elaborated in [Pk] and [Sℓ]), for each "nonsingular" value of s we may renormalize the basis $\{g_n : n \in \mathbb{Z}\}$ to obtain a new (s-dependent) basis in $L^2(T)$ upon which $\mathcal{L}_{s+\frac{1}{2}}$ acts in a manner identical to the action of our balanced $\mathcal{L}(q,0)$ on its basis (for $q = \frac{1}{4} - s^2$). Specifically, we put $\omega_m(s) = \gamma_m(q,0)(n+\frac{1}{2}+s)^{-1}$, and $\rho_n(s) = \Pi\{\omega_m^{-1}(s)$: m inclusively between 0 and n-1}, with $e_n = \rho_n(s)g_n$. Then, for example, $A_+(s)e_n = \rho_n(s)A_+(s)g_n$ $= (n+\frac{1}{2}+s)\rho_n(s)g_{n+1} = (n+\frac{1}{2}+s)\omega_n(s)\rho_{n+1}(s)g_n = \gamma_n(q,0)e_{n+1}$. However, as s (equivalently q) approaches a "singular" value, where at least one of the $A_\pm(s)$ vanishes at some g_n (equivalently, the $B_\pm(q,0)$ vanish at e_n and e_{n+1} in a balanced way), the formulae for $\omega_n(s)$ diverge, and every $\rho_k(s)$ which contains ω_n^{-1} as a factor goes to zero. That is, no algebraic isomorphism exists at the singular values, and the isomorphisms connecting $\mathcal{L}_{s+\frac{1}{2}}$ to $\mathcal{L}(q,0)$ at nonsingular values have pathological limiting properties as these values are approached. To see this, notice that $\gamma_n(q,0)^2 = (q+n(n+1)) = (q-\frac{1}{4}+(n+\frac{1}{2})^2) = (n+\frac{1}{2})^2-s^2$ $= (n+\frac{1}{2}+s)(n+\frac{1}{2}-s)$ so that $\omega_n(s) = \mathrm{sgn}(n)(n+\frac{1}{2}-s)^{\frac{1}{2}}(n+\frac{1}{2}+s)^{-\frac{1}{2}}$.

Hence, as $s \to -(n+\frac{1}{2})$, $|\omega_n(s)| \to \infty$, and the other claims follow.

As indicated in Section 12A, the preceding discussion can usefully be formulated in terms of the intertwining operator $T(s)$ on $L^2(\mathbb{T})$ defined by $T(s)g_n = \rho_n(s)g_n = e_n$ for all $n \in \mathbb{Z}$ (extended by linearity). That is, we can use $T(s)$ to transport the balanced action of $A_\pm(s)$ on the s-dependent basis $\{e_n\}$ back to a corresponding balanced action of $B_\pm(s) = T(s)^{-1}A_\pm(s)T(s)$ on the fixed basis $\{g_n\}$: for example $B_+(s)g_n = T(s)^{-1}A_+(s)T(s)g_n = T(s)^{-1}A_+(s)e_n$ $= \gamma_n(q,0)T(s)^{-1}e_{n+1} = \gamma_n(q,0)g_{n+1}$. This realizes $\mathcal{L}(q,0)$ for $q = \frac{1}{4} - s^2$ as a module on the same basis as that for the infinitesimally induced representation, facilitating direct comparison. It is most instructive to write $A_\pm(s) = T(s)B_\pm(s)T(s)^{-1}$ and to note that, for $r \neq s$, the "raising and lowering" perturbations have the term

$$A_\pm(r)-A_\pm(s) = T(r)[B_\pm(r)-B_\pm(s)]T(r)^{-1}+\{T(r)B_\pm(s)T(r)^{-1} -T(s)B_\pm(s)T(s)^{-1}\}. \quad (87)$$

Now, one can show that $T(r)^{\pm 1}$ are bounded for all r (unitary for r real), whence since $B_\pm$ experiences a compact perturbation, the first term in the $A_\pm$ perturbation is compact, so the second must account for the noncompactness of the $A_\pm$ perturbation noted earlier. _A priori_, the { } factor could even be _unbounded_, since it contains unbounded factors $B_\pm(s)$, but explicit calculation shows that $T(r)B_+(s)T(r)^{-1}g_n$ $= \omega_{n+1}(r)B_+(s)g_n$ (with a related formula for B_-) and $T(s)B_+(s)T(s)^{-1}g_n = \omega_{n+1}(s)B_+(s)g_n$, whence { } reduces to $[W(r) - W(s)]B_+(s)$, with $W(r)g_n = \omega_n(r)g_n$. Formula (87) becomes $U_+(r,s) = T(r)(B_+(r) - B_+(s))T(r)^{-1}+(W(r)-W(s))B_+(s)$, and similarly for $U_-(r,s)$. A simple computation shows that $\omega_n(r) - \omega_n(s)$ is $O(1/|n|)$, whence that $[W(r) - W(s)]B_+(s)$ is bounded since the diagonal multipliers, $\omega_n(r) - \omega_n(s)$, and $\gamma_n(q,0)$, have a bounded product, { } is bounded. Indeed,

$$\omega_n(r)-\omega_n(s) = \gamma_n(p,0)(n+\tfrac{1}{2}+r)^{-1} - \gamma_n(q,0)(n+\tfrac{1}{2}+s)^{-1}$$
$$= [\gamma_n^2(p,0)(n+\tfrac{1}{2}+r)^{-2}-\gamma_n^2(q,0)(n+\tfrac{1}{2}+s)^{-2}][\gamma_n(p,0)(n+\tfrac{1}{2}+r)^{-1}$$
$$+ \gamma_n(q,0)(n+\tfrac{1}{2}+s)^{-1}]^{-1} \text{ for } p = \tfrac{1}{4} - r^2,\ q = \tfrac{1}{4} - s^2,$$

and the denominator expression is bounded. When the first factor is put over a common denominator $(n+\frac{1}{2}+r)^2(n+\frac{1}{2}+s)^2$ of degree 4 in n and $\gamma_n^2(p,0) = p - \frac{1}{4} - (n+\frac{1}{2})^2$ is substituted, one finds that the fourth-power terms in n cancel, leaving the numerator of degree exactly 3 in n. Thus $\omega_n(r) - \omega_n(s)$ behaves asymptotically exactly like $1/|n|$.

Finally, we note that the operators $B_\pm(s)$ discussed above are visibly not differential operators, but contain more complicated pseudodifferential operators for which $\gamma_n(q,0)$ is essentially the "symbol". Formally, $(q-\frac{1}{4}+(n+\frac{1}{2})^2)^{\frac{1}{2}}$, represents the square root of a second-order differential operator in terms of the symbol calculus. These remarks suggest that the balanced modules $\mathfrak{L}(q,\tau)$ can be expected to be rather intractable in the Schrödinger function-space formalism, involving operators like $(q - \frac{1}{4} + (-i\frac{d}{dx} + \frac{1}{2}I)^2)^{\frac{1}{2}}$ with technical complications arising from the fractional power. Our Heisenberg-formalism should be thought of as an effective tool for handling the algebra of symbols for the particular operator Lie algebras under consideration.

12I. A counter-theorem on group-invariant domains

We have already mentioned the fact that the K-finite vectors for the representations of $SL(2,\mathbb{R})$ are not group-invariant.

In view of the important Domain Invariance condition of Theorem 9.1 it seems relevant to state a general result on the structure of group-invariant domains. Suppose as in Theorem 9.1 that D is a dense linear subspace of a locally convex space E, and that $\mathfrak{L} \subset \mathfrak{U}(D)$ is a finite-dimensional real Lie algebra. Suppose $\mathfrak{L}$ exponentiates to a strongly continuous irreducible group representation, $V: G \to L(E)$. Then $D \subset D_\infty(V)$. We show that, unless E is finite-dimensional, a group-invariant domain D cannot be denumerably spanned. It is already known that, for compact G and irreducible V, the representation space E must be finite-dimensional ([Wr]).

12.23 Theorem

Suppose V is a strongly continuous irreducible representation of a connected Lie group G in a locally convex linear space E.

Let e_n be a denumerable (total) set of vectors contained in $D_\infty(V)$, and suppose that the space D of finite linear combinations of the vectors e_n is invariant under V(G).

Then E is finite-dimensional.

Proof: Define, for each n = 0,1,2,..., spaces

D_n = span $\{e_k: |k| \leq n\}$, and let u be a fixed non-zero vector in D. We show that the assumption,

$$V(G)u \subset D ,$$

implies that E is finite-dimensional.

Define, for each n, the following subsets of G:

$$G_n = \{g \in G: V(g)u \in D_n\} .$$

Now, each D_n is a finite-dimensional subspace of E. The locally convex space E is assumed (by definition) to be Hausdorff, so we conclude that D_n is a closed subspace of E for each n ([Sch]). Strong continuity of V then implies that the subsets G_n are closed. We have $G = \cup_{n=1}^{\infty} G_n$, in view of the assumption. The group G has the Baire property, which means that for some n, G_n contains a non-empty open set $\mathcal{O}$. We have $V(\mathcal{O})u \subset D_n$, and consequently,

$$F = \text{span } V(\mathcal{O})u ,$$

is finite-dimensional, and hence closed in E. Now $D_n \subset D_\infty(V)$, so we conclude that, for each $g \in \mathcal{O}$ and X in the Lie algebra $\mathfrak{g}$ of G, the vector,

$$dV(X)V(g)u = d/dt\ V(\exp tXg)u|_{t=0} ,$$

belongs to F, since $\mathcal{O}$ is open and F is closed. The infinitesimal operators dV(X) are linear, and it follows that F is invariant under dV(X) for all $X \in \mathfrak{g}$. In other words: F reduces the operator set $dV(\mathfrak{g})$. Restriction of $dV(\mathfrak{g})$ to F gives a matrix representation of $\mathfrak{g}$ in F, or equivalently a homomorphism, $X \to dV(X)|_F$ of $\mathfrak{g}$ into End(F).

Since F is finite-dimensional, and G is connected, it follows easily that F is then necessarily also invariant under V(G). Hence, $F \subset E$ reduces the representation V.

Let us show next that the initially given non-zero vector u is contained in F. If g belongs to $\mathcal{O}$ then

$$u = V(g^{-1})\, V(g)u \in V(G)F \subset F.$$

To summarize, the non-zero finite-dimensional (closed) subspace F is invariant under the group V(G), and thus F = E, since V is assumed irreducible. This concludes proof of the theorem.

This result clearly applies to the continuous representations of $SL(2,\mathbb{R})$ on ℓ^p and on the $L(\mathbb{T})$ Banach spaces: the domain D (K-finite vectors) cannot be group-invariant. It also applies less directly to the smeared representations on the ℓ^p spaces for p = 2; the underlying smooth representations V_∞ acting upon the Fréchet space D_∞ of C^∞-vectors cannot leave D invariant.

Indeed, comparable applications of this result can be made for irreducible representations of any connected semi-simple Lie group, if D is taken to be the domain of K-finite vectors.

Such phenomena are not confined to the semisimple setting. For example, the representation of the nilpotent 3-dimensional Heisenberg group implicit in the Weyl commutation relations cannot leave the span of the eigenstates of the harmonic oscillator invariant. (Alternatively put, the exponential of the Heisenberg "infinite-matrix" representation of the canonical commutation relations cannot leave the span of the standard unit vectors invariant.) Note, however, the connection with representations of $SL(2,\mathbb{R})$ via the discussion in Section 2C and Example 2 in Chapter 9.

Appendix to Part VII

THE TWO EXPONENTIATION THEOREMS

In the first theorem we shall need the generality of locally convex spaces (l.c.s.). The reader is refered to [Yo] for background. We recall that strongly continuous one-parameter groups, infinitesimal generators, exponentiability e.t.c., are defined, mutatis mutandis, for l.c.s. E as in the case of Banach spaces (real or complex). In addition to the usual strong continuity condition on the group representations a mild local equicontinuity assumption is imposed. We consider a given l.c.s. E and a fixed dense linear subspace $D \subset E$. An operator Lie algebra $\mathfrak{L}$ is a finite-dimensional real Lie subalgebra of End(D), the linear endomorphisms of D.

9.1. Theorem

Let S be a Lie generating subset of a given operator Lie algebra $\mathfrak{L} \subset \mathrm{End}(D)$ in a l.c.s. E. Suppose the one-parameter exponentials $\{V(t,A)\colon t \in \mathbb{R}\} \subset L(E)$ exist as C_0 groups (locally equicontinuous in t) for all elements A in S. If the following two conditions are satisfied then $\mathfrak{L}$ exponentiates to a C_0 Lie group representation in E:

(i) The domain D is invariant under the operator family $\{V(t,A)\colon t \in \mathbb{R}, A \in S\}$, i.e., $V(t,A)u \in D$ for all $u \in D$, $t \in \mathbb{R}$, and $A \in S$.

(ii) For each $A,B \in S$ there is a positive interval I (which may depend on u,A,B) such that the function $t \to B\,V(t,A)u$ is bounded in I.

The second theorem is stated only for the case when E is a Banach space. We shall need the C^1 topology τ_1 on D which is defined by the family of seminorms $u \to \|u\|$, and $u \to \|Bu\|$, for $B \in \mathfrak{L}$.

9.2. Theorem

Let $\mathfrak{L} \subset \mathrm{End}(D)$ be an operator Lie algebra in a Banach space E, and let $S \subset \mathfrak{L}$ be some Lie generating subset. Assume that the one-parameter exponentials $\{V(t,A)\colon t \in \mathbb{R}\} \subset L(E)$ exist as C_0 groups for all elements A in S. Then $\mathfrak{L}$ exponentiates if the following graph density condition (GD) is satisfied:

(GD). For $A \in S$ suppose $V(t,A)$ is of exponential type ω_A, and assume that there is a pair of complex numbers $\lambda_\pm$ such that $\mathrm{Re}\,\lambda_+ > \omega_A + |\mathrm{ad}\,A|$, $\mathrm{Re}\,\lambda_- < -\omega_A - |\mathrm{ad}\,A|$, and each of the operator ranges $R(\lambda_\pm I - A)$ is τ_1-dense in D.

Bounded perturbations of the operator Lie algebras (Section 9C)

A special case of Phillips' perturbation theorem concerns bounded perturbations A + U where A is the infinitesimal generator of a strongly continuous one-parameter group $\{V(t,A): t \in \mathbb{R}\} \subset L(E)$ in a Banach space E, and where $U \in L(E)$. The conclusion states ([Ph][HP][Kt 2][Yo]) that A + U is also an infinitesimal generator, and that $\{V(t,A + U): t \in \mathbb{R}\} \subset L(E)$ is given in terms of a norm convergent, time-ordered, integral of U against V(t,A). The integral formula is frequently credited to Dyson by physisists. Finally, Phillips showed that V(t,A + U) depends analytically on U.

Motivated by the applications to $s\ell(2,\mathbb{R})$ we recall here a generalization of Phillips' theorem (due to the authors) to the setting where the one-parameter group $\{V(t,A)\}_{t \in \mathbb{R}}$ is replaced by an arbitrary C_0 Lie group representation, and the single operator A by an Lie algebra $\mathfrak{L}_0$ of unbounded operators. If we think of $\mathfrak{L}_0$ as a "base point" Lie algebra we shall be concerned with exponentiability of $\mathfrak{L}_0$, and in particular the stability of this property under bounded perturbations. Once exponentiability has been verified for the perturbed Lie algebra, then continuous (resp., analytic) dependence on the perturbation parameters is quite easily verified from our Lie algebraic Dyson formulas. (The reader is referred to Chapter 9 for more details at this point.)

Let $\mathfrak{L}_0 \subset \mathrm{End}(D)$ be a finite-dimensional real operator Lie algebra on a normed space D, and let E be the norm completion. We fix a basis $\mathfrak{B}_0$ for $\mathfrak{L}_0$ and define the usual C^n-norms $\|\cdot\|_n$, n = 1,2,... . Here $\|\cdot\|_n$ is defined as a sum of seminorms $u \to \|B_1 \ldots B_r u\|$, $B_i \in \mathfrak{B}_0$, $i = 1,\ldots,r \leq n$. The completions $(D,\|\cdot\|_n)^\sim$ are denoted D_n. (In our $s\ell(2,\mathbb{R})$ application $\mathfrak{B}_0 = \{B_0, B_1, B_2\}$ with B_0 = "the periodic SO(2)-generator". We show in Lemma 12.3 that $\|\cdot\|_n$ is equivalent as a norm to $u \to \|u\| + \|B_0^n u\|$.) We consider $D_\infty = \cap D_n$, and maps $U \in \mathrm{End}(D_\infty)$, such that $\|Uu\|_n \leq \mathrm{const}_n \|u\|_n$ for all n. We say that U is <u>ultra-continuous</u> if it is $\|\cdot\|_n$-continuous for all n, — and we note that ultra-continuity is easily verified in the $s\ell(2,\mathbb{R})$ - module with singly generated $\|\cdot\|_n$-norms.

A strongly continuous representation V_o of a Lie group G_o is given at the outset. Let $\mathfrak{g}_o$ be the corresponding Lie algebra and define $\mathcal{L}_o = dV_o(\mathfrak{g}_o)$ on the domain D of C^∞-vectors for V_o:

$$D = C^\infty(V_o) = \{u \in E : \tilde{u}(g) = V_o(g)u \in C^\infty(G_o,E)\}.$$

Finally, let $S_o \subset \mathcal{L}_o$ be a Lie generating subset of the exact "base-point" operator Lie algebra $\mathcal{L}_o$.

<u>9.3. Theorem</u>

Let $\mathcal{L}_o$ be an exact "base-point" operator Lie algebra with Lie generating outset S_o as described above. We have $\mathcal{L}_o = dV_o(\mathfrak{g}_o)$, and $D = C^\infty(V_o)$. Let $f : S_o \to \mathrm{End}(D)$ be a function such that the subset $S = \{A + f(A) : A \in S_o\}$ Lie generates a finite-dimensional Lie subalgebra $\mathcal{L}$ of End(D), and finally assume that f(A) is ultra-continuous for all $A \in S_o$.

Then $\mathcal{L}$ exponentiates to a strongly continuous representation of the simply connected Lie group G with Lie algebra ismorphic to $\mathcal{L}$.

<u>9.5. Theorem</u>

Let $(\mathcal{L}_o,D)$ and $(\mathcal{L},D)$ be a pair of operator Lie algebras with common domain D, and let E be some norm completion of D. Consider a function $f: \mathcal{L}_o \to \mathrm{End}(D)$ satisfying the following two conditions:

(i) For all $A \in \mathcal{L}_o$, f(A) is continuous on D with respect to the original norm $\|\cdot\|$, as well as $\|\ \|_1$, where $\|\cdot\|_1$ is the D_1-norm defined from $\mathcal{L}_o$.

(ii) The mapping $A \to A + f(A)$ is a linear isomorphism of $\mathcal{L}_o$ onto $\mathcal{L}$.

Then each of the following two conditions implies exponentiability of $\mathcal{L}$ in E:

(a) Condition (GD) is satisfied for each operator in a linear basis $\mathcal{B}_o$ for $\mathcal{L}_o$.

(b) $\mathfrak{L}_o$ is exact, $\mathfrak{L}_o = dV_o(\mathfrak{g}_o)$ for some C_o representation V_o of G_o leaving D invariant, i.e., $V_o(g)D \subset D$ for all $g \in G_o$.

Our final theorem concerns analyticity of the perturbed exponential. We have omitted details. The theorem concerns admissible perturbations, $f: \Omega \times \mathfrak{L}_o \to \mathrm{End}(D)$.
The region Ω is contained in some ${}^o\mathbb{C}^\nu$, and $f(z,\cdot)$ is assumed to satisfy the conditions in Theorem 9.5 above. If $z \to f(z,A)$ is analytic (resp., continuous) on Ω for all A in $\mathfrak{L}_o$, then we show analyticity (resp., continuity) of the perturbed exponential $V_z = \exp(\mathfrak{L}_z)$ where $\mathfrak{L}_z$ is the Lie algebra generated by $\{A + f(z,A): A \in \mathfrak{L}_o\}$.

Vector-fields on C^∞ manifolds (Section 11A)

Definition A smooth vector-field X on a manifold M is said to be complete iff there is a flow $\gamma(t,\cdot)$, $-\infty < t < \infty$, such that

$$\frac{d}{dt}\gamma(t,\cdot)\Big|_{t=0} = X.$$

10.1. Lemma

Let D be the test function space $C_o^\infty(M)$ on a smooth manifold M, and let E be one of the following three locally convex algebras:

$C_\infty(M)$: continuous complex functions on M vanishing at ∞.
$\mathcal{E}(M)$: all smooth scalar functions on M.
$\mathcal{D}(M)$: $C_o^\infty(M)$, smooth and compact support.

Then a vector-field X on M is complete if and only if X is a pregenerator when regarded as an operator (a derivation, in fact) in the l.c.s. E.

10.2. Theorem

Let D and E be as in Lemma 10.1 and let $\mathfrak{L}$ be a finite-dimensional real Lie algebra of vector-fields on M. Then $\mathfrak{L}$ exponentiates in E if and only if there is a Lie generating subset $S \subset \mathfrak{L}$ of complete vectorfields.

Numerical ranges(Appendix F)

Definition. The numerical range $W(A)$ of an operator in a Banach space E is the set

$$W(A) = \{f(Au): u \in D(A),\ f \in E^*,\ 1 = \|f\| = \|u\| = f(u)\}.$$

Remark: Let E be the Banach space $L^1(X,\mu)$ for some measure space (X,μ). Let $u \in E$, and $S = \{x \in X: u(x) \neq 0\}$, $N = X \setminus S$. Let

$$v_o(x) = \begin{cases} u(x)\,|u(x)|^{-1}, & x \in S, \\ 0, & x \in N. \end{cases}$$

Then $f \in E^*$ satisfies $\|f\| = 1 = f(u)$ iff it can be Riesz represented $f = v_o + w \in L^\infty(X)$ where $\|w\|_\infty \leq 1$ and w vanishing on S.

GENERAL APPENDICES

"In Science - in fact, in most things - it is usually best *to begin at the beginning*. In *some* things, of course, it's better to begin at the *other* end. For instance, if you wanted to paint a dog green, it *might* be best to begin with the *tail*, as it doesn't bite at *that* end. And so -"

Sylvie and Bruno Concluded
The Professor's Lecture
LEWIS CARROLL

Appendix A

THE PRODUCT RULE FOR DIFFERENTIABLE OPERATOR VALUED MAPPINGS

An important tool in the analysis of semigroup commutation relations is the product rule for differentiable mappings taking values in topological vector spaces and in spaces of linear operators. In Chapter 3, and a number of other places, we used some version of an abstract product rule for such mappings.

Locally convex spaces are always assumed, by definition, to be Hausdorff, and the abbreviation l.c.s. is used. The topology on a given l.c.s. E is translation invariant and has a system of convex neighborhoods of the origin in E [Sch].

Let $I \subset \mathbb{R}$ be an open (possibly unbounded) interval. A mapping f of I into a l.c.s. E is said to be differentiable if the difference quotients of f converge at every point in I. The value of the derivative at a point $t \in I$ is denoted by $f'(t)$ and it is assumed that $f'(t)$ belongs to E. In particular, $f'(t_o)$ for a given $t_o \in I$ is defined as the limit of the difference quotients $(t - t_o)^{-1}(f(t) - f(t_o))$ for $t \to t_o$. The higher-order derivatives (if they exist) are denoted by $f^{(n)}(t)$ for $n = 2, 3, \ldots$ and $t \in I$. If the derivatives up to order n exist then f is said to be of class C^n.

Let E and F be locally convex spaces. The vector space $L(E,F)$ of continuous linear mappings of E into F is equipped with the strong operator topology. Equipped with this topology, $L(E,F)$ itself becomes a l.c.s. which we denote by $L_s(E,F)$. A generic neighborhood of the origin in $L_s(E,F)$ is given by a finite subset S of E and a neighborhood Ω of the origin in F as follows:

$$\{A \in L(E,F);\ Au \in \Omega \quad \text{for all } u \in S\}.$$

We also study differentiable locally equicontinuous mappings K of the interval I into $L_s(E,F)$. A mapping $K: I \to L_s(E,F)$ is said to be locally equicontinuous if for all compact subintervals $I_o \subset I$ the image $K(I_o)$ is an equicontinuous subset of $L(E,F)$. If K is continuous and E is barreled then the local equicontinuity is automatic by the Banach-Steinhaus Theorem [Sch]. Furthermore, it is easy to see that if K is differentiable and the derivative K' is locally equicontinuous, then K itself must be locally equicontinuous.

For mappings $f: I \to E$ and $K: I \to L_s(E,F)$ the product $H(t) = K(t)f(t)$ for $t \in I$ is defined as follows. The operator $K(t)$ is applied to the vector $f(t)$. The resulting vector is denoted by $H(t)$ and belongs to F. Hence, H maps I into F.

Ultimately, we are (mainly) interested in Banach spaces but the generality of this appendix is required for the applications in Chapter 3 (and elsewhere) to Lie algebras $\mathfrak{L} \subset \mathbf{A}(D)$ of unbounded operators on a dense domain D in a Banach space. The domain D is equipped with some l.c.s. topology and the product rule is applied. On one occasion the algebra $\mathbf{A}(D)$ of linear endomorphisms in D is equipped with a l.c.s. topology and a finite-dimensional ad-orbit $\mathcal{O}_A(B) \subset \mathbf{A}(D)$ is studied together with a differentiable mapping

$$K : \mathbb{R}_+ \to L_s(\mathcal{O}_A(B), \mathbf{A}(D)).$$

A.1. Theorem (The Product Rule)

Let $I \subset \mathbb{R}$ be an open interval. Let $K: I \to L_s(E,F)$ be a differentiable locally equicontinuous mapping, and let $f: I \to E$ be differentiable. Then the product mapping $H: I \to F$ defined by $H(t) = K(t)f(t)$ for $t \in I$ is differentiable.

The first-order derivative is given by

$$H'(t) = H(t)f'(t) + K'(t)f(t) \quad \text{for } t \in I. \qquad \text{(A.1)}$$

Moreover, suppose f and K are both of class C^n and $K^{(n-1)}$ is locally equicontinuous. Then H is of class C^n and the Leibnitz rule for the nth derivative of a product is valid.

Proof: Let $t_o \in I$ and let Ω_1 be a closed convex neighborhood of the origin in F. Then by local equicontinuity there is a neighborhood Ω_2 of the origin in E such that

$$K(t)\Omega_2 \subset \Omega_1 \tag{A.2}$$

for all $t \in I$ with $|t-t_o| \leq 1$.

We can pick a positive δ $(0 < \delta \leq 1)$ such that

$$(t-t_o)^{-1}\ (f(t) - f(t_o)) \in \Omega_2 + f'(t_o) \tag{A.3}$$

and

$$(t-t_o)^{-1}\ (K(t)-K(t_o))f(t_o) \in \Omega_1 + K'(t_o)f(t_o) \tag{A.4}$$

for all $t \in I$ with $0 < |t-t_o| < \delta$. This is possible since f and K are both assumed differentiable with respect to the l.c.s. topologies on E and $L_s(E,F)$ respectively.

For such t we have

$$\begin{aligned}(t-t_o)^{-1}\ (H(t)-H(t_o)) &= K(t)(t-t_o)^{-1}(f(t)-f(t_o))+(t-t_o)^{-1}(K(t)-K(t_o))f(t_o)\\ &\in K(t)(\Omega_2+f'(t_o))+\Omega_1+K'(t_o)f(t_o)\\ &\subset \Omega_1+K(t)f'(t_o)+\Omega_1+K'(t_o)f(t_o)\end{aligned}$$

where (A.3), (A.4) and (A.2) have been used in this order. Letting $t \to t_o$ one sees that the difference quotient $(t-t_o)^{-1}(H(t)-H(t_o))$ converges. For the limit $H'(t_o)$ we have

$$H'(t_o) \in 2\Omega_1 + K(t_o)f'(t_o) + K'(t_o)f(t_o).$$

The neighborhood Ω_1 was arbitrary from the beginning and F is assumed Hausdorff. Hence the formula (1) for $t = t_o$ has now been proved.

If each of the two terms on the right-hand side of (A.1) are differentiable, then one obtains a comparable formula for $H^{(2)}(t)$. It is now clear how to derive the last part of Theorem A.1 by induction. Under the assumption that f and K are of class C^n and $K^{(n-1)}$ is locally equicontinuous, one gets that H is of class C^n with nth order derivative given by

$$H^{(n)}(t) = \sum_{m=0}^{n} \binom{n}{m} K^{(m)}(t)f^{(n-m)}(t) \text{ for } n = 2,3,\dots \text{ and } t \in I.$$

E.O.P.

Remark: Suppose K is locally equicontinuous and of class C^{∞}. Suppose in addition that for some $A \in L(E,E)$ the formula $K'(t) = K(t)A$ holds for all $t \in I$. Then all the derivatives $K^{(n)}$ are locally equicontinuous. This remark covers all the applications given.

Corollary: Let E_1, E_2, and E_3 be locally convex spaces. Let $I \subset \mathbb{R}$ be an open interval, and let K and M be differentiable operator valued mappings $K : I \to L_s(E_2,E_3)$ and $M : I \to L_s(E_1,E_2)$.

Suppose K is locally equicontinuous. Then the product mapping defined by $H(t) = K(t)M(t)$ for $t \in I$ is differentiable from I into $L_s(E_1,E_3)$.

Proof: We have to show differentiability of $H: I \to L_s(E_1,E_3)$ with respect to the strong operator topology, or equivalently that for each vector $u \in E_1$ the mapping $t \to H(t)u$ of I into E_3 is differentiable. So for given $u \in E_1$ we define a mapping $f(t) = M(t)u$ $(t \in I)$. Then $H(t)u = K(t)f(t)$, and the theorem applies.

Appendix B

A REVIEW OF SEMIGROUP FOLKLORE, AND INTEGRATION IN LOCALLY CONVEX SPACES

The motivation for the generality of locally convex spaces is the same in this appendix as in Appendix A. Ultimately, we are interested in operator theory in Banach spaces, but in the proofs we often have occasion to use some auxiliary locally convex spaces.

Our treatment follows the first parts of [Ko, Ps 1] and [Aa 2] to some extent. A number of results from representation theory in Banach spaces ([Nℓ 1, and Ps 1]) generalize to locally convex spaces [Mr 4], [Jo 1].

Let E be a l.c.s., and let (S,μ) be a measure space, where S is a locally compact Hausdorff space, and μ is a complex regular Borel measure on S, of finite total variation. Let $\varphi : S \to E$ be a continuous function with bounded range, that is $\varphi(S)$ is a bounded subset of E. The integral

$$I = \int_S \varphi(s)d\mu(s) \tag{B.1}$$

is defined in the weak sense. For all $u^* \in E^*$

$$\langle u^*, I\rangle = \int_S u^*(\varphi(s))d\mu(s).$$

The dual space E^* of continuous linear functionals on E is equipped with the topology of uniform convergence on bounded subsets of E, denoted by E^*_β . The estimate

$$|\langle u^*, I\rangle| \le \|\mu\| \sup|u^*\varphi(s)|$$

(where $\|\mu\|$ is the total variation of μ) shows that the integral (B.1) belongs to the bi-dual $(E^*_\beta)^*$.

In special cases where E is reflexive, quasi-complete, or sequentially complete the integral actually belongs to E. If E is not known to have any completeness properties the integral may not belong to E but only to the sequential completion $\tilde{E}$ of E.

B.1. Lemma

Let E be a l.c.s. Suppose either that E is sequentially complete, or else that E is quasi-complete. Then there is a vector $u \in E$ such that

$$u^*u = \int_S u^*\varphi(s)\, d\mu(s) \tag{B.2}$$

for all $u^* \in E^*$.

Proof: Let the sequential completion of E be denoted by $\tilde{E}$, where E is an arbitrary l.c.s. The topology on E is given by a calibration $\Gamma = \{p\}$ of (continuous) semi-norm [Sch]. Every $p \in \Gamma$ extends uniquely to a continuous semi-norm $\tilde{p}$ on $\tilde{E}$, and the topology on $\tilde{E}$ is given by the semi-norms $\tilde{\Gamma} = \{\tilde{p};\ p \in \Gamma\}$. It is well known ([Bk], [As]) that for all compacts $K \subset S$ there is a vector $u_K \in \tilde{E}$ such that

$$u^*u_K = \int_K u^*\varphi(s)\, d\mu(s)$$

for all $u^* \in E^*$. Since the total variation of μ is finite ($\|\mu\| < \infty$) there are compacts $K_n \subset K_{n+1}$ such that $|u|(S\setminus K_n) \to 0$. For every $\tilde{p} \in \tilde{\Gamma}$ and $m < n$ we have

$$\tilde{p}(u_n - u_m) \leq \sup\{p(\varphi(s));s \in S\}\,|\mu|(S\setminus K_m).$$

Hence, the sequence u_n is convergent in $\tilde{E}$, i.e. the limit u belongs to $\tilde{E}$, and

$$u^*u = \lim_n \int_{K_n} u^*\varphi(s)d\mu(s) = \int_S u^*\varphi(s)d\mu(s)$$

for all $u^* \in E^*$. In particular, if E is sequentially complete the value of the integral u belongs to E itself, and (B.2) holds.

A similar argument applies to the case where E is quasi-complete. The details are well known [Bk, As] and left to the interested reader. E.O.P.

The lemma applies to the construction of regular vectors for semigroups of operators and for group representations..

Let E be a l.c.s. with sequential completion $\tilde{E}$. Let $\{V(t);\ 0 \leq t < \infty\} \subset L(E)$ be a continuous locally equicontinuous (c.l.e.) semigroup. The infinitesimal generator A is defined on its domain D(A) of vectors $u \in E$ such that the limit $\lim_{t \to 0_+} t^{-1}[V(t)u-u]$ exists. The limit is denoted by Au. A standard argument (identical to the one given in ([Ko, Ps 1]) shows that a vector u belongs to D(A) if and only if there is a vector $v \in E$ such that

$$V(t)u - u = \int_0^t V(s)v\, ds \quad \text{for } t > 0. \tag{B.3}$$

In that case $Au = v$. A simple consequence of this formula is that the generator A is closed.

The graph of A is clearly contained in $E \times E$. Let $\widetilde{A}$ denote the closure of A viewed as an operator in $\widetilde{E}$. The semigroup V extends uniquely to a c.l.e. semigroup $\{\widetilde{V}(t);\ 0 \leq t < \infty\} \subseteq L(\widetilde{E})$ such that $\widetilde{V}(t)$ is the unique extension of V(t) to $\widetilde{E}$ for all $t \in [0,\infty)$.

B.2. Lemma

The infinitesimal generator of $\widetilde{V}$ is equal to $\widetilde{A}$.

Proof: Apply formula (B.3) to $\widetilde{V}$ and the infinitesimal generator of $\widetilde{V}$.

B.3. Lemma

Let $D_\infty(\widetilde{A})$ be equal to $\bigcap_1^\infty D(\widetilde{A}^n)$ equipped with the C^∞ topology given by the semi-norma $u \to \widetilde{p}(\widetilde{A}^n u)$, for $\widetilde{p} \in \widetilde{\Gamma}$ and $n = 1,2,\ldots$.

Then for each $u \in \widetilde{E}$ and $\varphi \in C_0^\infty(0,\infty)$ the integral

$$\int_0^\infty \varphi(t)\, \widetilde{V}(t)u\, dt \tag{B.4}$$

belongs to $\widetilde{E}$. There is a linear operator $V(\varphi)$ on $\widetilde{E}$ with the property that $V(\varphi)$ maps $\widetilde{E}$ continuously into $D_\infty(\widetilde{A})$ for each $\varphi \in C_0^\infty(0,\infty)$. Moreover

$$\widetilde{A}^n V(\varphi)u = \int_0^\infty (-1)^n \varphi^n(t)\, \widetilde{V}(t)u\, dt \tag{B.5}$$

for $n = 1,2,\ldots,$ and $u \in \widetilde{E}$.

Proof: Lemma B.1 guarantees existence of the integrals. Repeated application of formula (B.3) to $\widetilde{V}$ shows that the domain of $\widetilde{A}^n$ is equal to the space of vectors $u \in \widetilde{E}$ such that the mapping $t \to \tilde{V}(t)u$ is of class C^n from $(0,\infty)$ to $\tilde{E}$, for each $n = 1,2,\ldots$. If u_1 is the vector defined by the integral (4) for some $u \in \tilde{E}$, then it is clear that the mapping $t \to \tilde{V}(t)u_1$ is of class C^∞, and that the formula (B.5) holds. For $\tilde{p} \in \tilde{\Gamma}$ we then get the estimate

$$\tilde{p}(\tilde{A}^n V(\varphi)u) = \tilde{p}(V((-1)^n \varphi^{(n)})u)$$

$$\leq \sup \{|\varphi^{(n)}(t)| \cdot \tilde{p}(\tilde{V}(t)u) : t \in \operatorname{supp}\varphi\}.$$

The right-hand side is finite because the support of φ is compact and $\tilde{V}$ is locally equicontinuous. It follows that $V(\varphi)$ is continuous from $\tilde{E}$ to $D_\infty(\tilde{A})$ for each $\varphi \in C_0^\infty(0,\infty)$.

B.4. Corollary (Poulsen [Ps 1, Corollary 1.3])

Let D be a dense linear subspace of E (and therefore also of $\tilde{E}$). Suppose that D is invariant under V(t) for all $t \in [0,\infty)$ and contained in $D(A^n)$ for some positive integer n. Then D is a core for $\tilde{A}^n$ for all n, that is $\tilde{A}^n = \overline{A^n|_D}$ where the closure is taken in $\tilde{E}$.

Proof and remark: The corollary as stated is slightly more general than Corollary 1.3 in [Ps 1], but the proof is the same. In the case where E is a Banach space and n = 1 the conclusion of the corollary has been known (with a different proof) for some time.

Poulsen obtained a comparable result for group representations. It is stated in [Ps 1, Theorem 1.3] for the case where E is a Banach space, but the proof generalizes word for word to locally convex spaces.

B.5. Theorem (Poulsen [Ps 1, Theorem 1.3]).

Let V be a continuous locally equicontinuous representation of a Lie group G in a sequentially complete (or quasi-complete) l.c.s. E. For $n = 1,2,\ldots,\infty$, let $D_n(V)$ be the spaces of C^n vectors for V. Let D be a dense linear subspace of E which is contained in $D_n(V)$ and invariant under V(G). Then D is dense in $D_n(V)$.

Again, let V be a c.l.e. semigroup in a l.c.s. E, and let $\tilde{E}$ be the sequential completion of E. The infinitesimal generator A of V is closed, but not necessarily densely defined in E. This is not a serious problem (for our applications) because the existence of a dense linear subspace D of E contained in D(A) and invariant under A is always assumed at the outset.

The transposed A^* of A is defined on the domain $D(A^*)$ of vectors $u^* \in E^*$ such that for some $v^* \in E^*$ we have $\langle Au,u^*\rangle = \langle u,v^*\rangle$ for all $u \in D(A)$. The smeared operators $V(\varphi)$ belong to $L(\widetilde{E})$, and the transposed $V(\varphi)^*$ belongs to $L(E^*_\beta)$ for all $\varphi \in C_0^\infty(0,\infty)$.

Recall that the dual of $\widetilde{E}$ coincides with E^*, i.e., $(\widetilde{E})^* = E^*$, because every $u^* \in E^*$ extends uniquely to $\widetilde{E}$. The following two pointwise topologies on E^* are used, $\sigma(E^*,E)$ and $\sigma(E^*,\widetilde{E})$, the topology of pointwise convergence on E and $\widetilde{E}$, respectively. The latter is stronger than the former. So a subset of E^* which is $\sigma(E^*,\widetilde{E})$ dense is in particular $\sigma(E^*,E)$ dense.

For a c.l.e. semigroup V, the linear span of the vectors

$$\{V(\varphi)^*u^* \ ; \ \varphi \in C_0^\infty(0,\infty),\ u^* \in E^*\}$$

is called the dual Gårding domain, and is denoted by $G^*(V)$. A simple Hahn-Banach argument shows that $G^*(V)$ is a $\sigma(E^*,\widetilde{E})$ dense linear subspace of E^*.

B.6. Theorem

Let D be a dense linear subspace of a l.c.s. E, and let the algebra of linear endomorphisms in D be denoted by $\mathbf{A}(D)$. Let V be a c.l.e. semigroup with infinitesimal generator A.

(a) Suppose $D \subset D(A)$. Then the dual Gårding domain $G^*(V)$ is contained in $D(A^*)$ and invariant under both A^* and $V(\varphi)^*$ for $\varphi \in C_0^\infty(0,\infty)$.

(b) The duality $\langle D,G^*(V)\rangle$ turns $\mathbf{A}(D)$ into a (Hausdorff) l.c.s. for which the continuous seminorms are given by elements $u \in D$ and $u^* \in G^*(V)$:

$$p_{u,u^*}(C) = |\langle Cu,u^*\rangle| \qquad \text{for all } C \in \mathbf{A}(D).$$

(c) Suppose D is invariant under A. (With an abuse of notation we write $A \in \mathbf{A}(D)$.) Suppose for some $B \in \mathbf{A}(D)$ that the ad-orbit $\mathcal{O}_A(B)$ is finite-dimensional. Let $t > 0$ be given. Then the mapping $K\colon (0,t) \to L_s(\mathcal{O}_A(B), \mathbf{A}(D))$ defined by

$$K(s)C = V(t-s)\,C\,V(s) \quad \text{for all } C \in \mathcal{O}_A(B)$$

is locally equicontinuous and of class C^∞.

Proof: (a) Transposition of the two well-known identities

$$V(\varphi)Au = V(-\varphi')u ,$$

and

$$V(\psi)V(\varphi) = V(\psi*\varphi) \text{ for } \varphi,\psi \in C_0^\infty(0,\infty) \text{ and } u \in E$$

leads to

$$A^*V(\varphi)^*u^* = V(-\varphi')^*u^* \quad \text{for } u^* \in E^*$$

and

$$V(\varphi)^*\,V(\psi)^* = V(\psi*\varphi)^* ,$$

respectively. The conclusion in (a) is an immediate consequence.
(b) The topology described in (B.6) is Hausdorff because $G^*(V)$ is $\sigma(E^*,E)$-dense in E^*.
(c) The mapping described in (c) is relevant to the proof of Theorem 3.2. The finite-dimensional $\mathcal{O}_A(B)$ is equipped with the relative topology from $\mathbf{A}(D)$. But there is only one Hausdorff topology which turns $\mathcal{O}_A(B)$ into a topological vector space [Sch]. Differentiability of K is verified in the proof of Theorem 3.2. Local equicontinuity of K is automatic since $\mathcal{O}_A(B)$ is barreled.

In Chapter 12, we use bounded Phillips perturbations to describe the analytic continuations of the representations of $SL(2,\mathbb{R})$. References for that part of semigroup theory are [HP, p. 389] and [Kt 2, p. 495]. For the sake of completeness we describe here the analytic dependence of the perturbed one-parameter (semi-) group upon the bounded perturbation. In principle, Chapter 12 generalizes bounded Phillips-perturbations from the case of a single unbounded group generator to the one of a Lie algebra of unbounded operators, (Theorems 9.3, and 12.15).

For a given Banach space E we denote by L(E) the Banach algebra of all bounded linear operators on E. The norms on E and L(E) are both denoted by $\|\cdot\|$. For $n = 0,1,2,\ldots$ the product of L(E) with itself n times is designated by $L_n(E)$, i.e., $L_n(E) = L(E) \times \ldots \times L(E)$ with the convention $L_0(E) = \mathbb{C}$. A mapping of $L_n(E)$ into L(E) which is linear separately in each variable is called n-linear. A function f mapping L(E) into itself is said to be entire (analytic in L(E)) if there exists n-linear functions $f_n : L_n(E) \to L(E)$ and constants $c_n \geq 0$ for n = 0,1,2,... such that

(i) $f(K) = \Sigma_{n=0}^{\infty} f_n(K,K,\ldots,K)$ for all $K \in L(E)$,

(ii) $\|f_n(K,\ldots,K)\| \leq c_n\|K\|^n$ for all $n \geq 0$ and $K \in L(E)$,

and

(iii) the function φ defined by $\varphi(z) = \sum_n c_n z^n$ for $z \in \mathbb{C}$ is entire.

Let M and ω be non-negative constants. We denote by $C_0(M,\omega)$ the class of infinitesimal generators of C_0 one-parameter groups $\{V(t): -\infty < t < \infty\}$ on E such that $\|V(t)\| \leq M \exp(\omega|t|)$ for all $t \in \mathbb{R}$. If A is an unbounded operator in E and $K \in L(E)$, then the domain of A + K is by definition equal to the domain D(A) of A. The following result is proved in [HP,Kt, loc.cit.] for semigroups. Extension to one-parameter groups is trivial.

B.7. Theorem

For every $A \in C_0(M,\omega)$ and $K \in L(E)$, the operator A + K belongs to $C_0(M,\omega+\|K\|M)$. Moreover the one-parameter group V(t,A+K) generated by A + K is for fixed t an entire function of K.

Remark on proof: For t > 0, V(t,A+K) is constructed byy successive approximation as follows.

$$V(t,A+K) = \sum_{n=0}^{\infty} V_n(t)$$

where $V_0(t) = V(t,A) = V(t)$, and

$$V_n(t) = \int_0^t \int_0^{s_{n-1}} \cdots \int_0^{s_1} V(t-s_{n-1})K\ldots V(s_1-s_0)KV(s_0)ds_0\ldots ds_{n-1} \qquad (B.6)$$

Hence $\|V_n(t)\| \leq M^{n+1}\|K\|^n\, t^n/n!\, \exp(\omega t)$ for n = 0,1,2,... .

The expression on the right-hand side of (6) is clearly of the form $f_n(K,\ldots,K)$ for some n-linear f_n. Also the function $\varphi(z) = \Sigma(zMt)^n/n! = \exp(zMt)$ is entire, so by construction V(t,A+K) is entire in K for fixed t > 0. Moreover, the estimate $\|V(t,A+K)\| \leq M \exp(\omega t+\|K\|Mt)$ is an immediate consequence. The same arguments apply to $t < 0$.

In the applications of Chapter 12 one is given an analytic function $\lambda \to K(\lambda)$ defined on a domain Ω in $\mathbb{C}^2$ and taking values in $L(E)$. We leave to the reader to show, in that case, that the function $\lambda \to f(K(\lambda))$ is analytic in Ω whenever $f : L(E) \to L(E)$ is an entire function.

Appendix C

THE SQUARE OF AN INFINITESIMAL GROUP GENERATOR

In Chapter 4, we discussed the (semi-) group generation properties of certain quadratic operators $M = \Sigma\, a_i A_i^2$ in terms of comparable properties of their (commuting) component squares A_i^2. If A is the generator of a strongly continuous group on a Banach space, it is well-known that A^2 generates a strongly continuous semigroup $\{V(t,A^2)\colon t \in [0,\infty)\}$ that can be represented in terms of the Gauss kernel and the group generated by A (the idea dates back to Gelfand [Gf] in 1939). The qualitative fact that these semigroups admit analytic continuations into the open right half-plane is also familiar, and the case where A generates an isometry group has received detailed discussion in the literature (Yosida [Yo], Moore [Mr 8]), but we require here some folklore about detailed behavior near the imaginary axis that seems never to have been recorded in the appropriate form. Consequently, we present here for reference an elementary treatment of the required facts, using standard facts about analytic continuation of the Gauss kernel.

For $x \in \mathbb{R}$ and $\zeta \in \mathbb{C}$ (with $\mathrm{Re}(\zeta) > 0$) the Gauss kernel $p_\zeta(x)$ is given explicitly by

$$p_\zeta(x) = (4\pi\zeta)^{-\frac{1}{2}} \exp(-x^2/4\zeta), \tag{C.1}$$

where $\zeta^{-\frac{1}{2}}$ has the determination which is positive for positive ζ. For real numbers a we have by inverse Fourier transform

$$\exp(-\zeta a^2) = \int_{-\infty}^{\infty} p_\zeta(x)\exp(ixa)\,dx.$$

It is shown below that this formula makes sense when the variable ia is replaced by an operator A which is the infinitesimal generator of a strongly continuous one-parameter group.

It is well-known that the Gauss kernel $p_\zeta(x)$ decays fast at infinity and acts like an approximate identity. The reader can verify that for $\omega \geq 0$ the integral $\int_{-\infty}^{\infty} |p_\zeta(x)|e^{\omega|x|}dx$ is convergent. In fact for $\zeta \in S$ (the open right half-plane)

$$\int_{-\infty}^{\infty} |p_\zeta(x)| e^{\omega|x|} dx \le 2 \left(\frac{|\zeta|}{\text{Re } \zeta}\right)^{\frac{1}{2}} \exp\left(\frac{\omega^2 |\zeta|^2}{\text{Re } \zeta}\right) . \tag{C.2}$$

Given $\delta > 0$ and $c > \omega \geqq 0$ we have the following estimate

$$\int_{|x| > \delta} |p_\zeta(x)| e^{\omega|x|} dx \leqq (4\pi|\zeta|)^{-\frac{1}{2}} \exp\left[-\frac{\delta^2 \text{Re } \zeta}{4|\zeta|^2} + c\delta\right] \times \int_{|x| > \delta} e^{(\omega - c)|x|} dx \tag{C.3}$$

for all $\zeta \in S$ with $2c|\zeta|^2 \le \delta \text{ Re } \zeta$. These estimates can easily be verified by the reader. The complex kernels form a semigroup of functions, i.e.,

$$p_\zeta * p_{\zeta'} = p_{\zeta+\zeta'} \quad \text{for } \zeta, \zeta' \in S. \tag{C.4}$$

C.1. Theorem

Let A be the infinitesimal generator of a strongly continuous one-parameter-group V on a Banach space E. Then for every $\zeta \in S$ and $u \in E$ the integral

$$T(\zeta)u = \int_{-\infty}^{\infty} p_\zeta(x) V(x) u \, dx \tag{C.5}$$

is convergent, and defines a holomorphic semigroup $T(\zeta)$ in the open right half-plane in the following sense:

(i) The operators $T(\zeta)$ are bounded and satisfy

$$T(\zeta)T(\zeta') = T(\zeta+\zeta') \quad \text{for all } \zeta, \zeta' \in S. \tag{C.6}$$

(ii) For every $u \in E$, the mapping $\zeta \to T(\zeta)u$ from S to E is holomorphic.

(iii) For every sequence $\{\zeta_n\} \subseteq S$ such that $\zeta_n \to 0$, and $\sup_n |\zeta_n|/\text{Re } \zeta_n < \infty$ we have $\lim_n T(\zeta_n)u = u$. (Non-tangential limits at zero exist.)

(iv) The infinitesimal generator of $T(\zeta)$ is equal to A^2, in the extended sense that $u \in D(A^2)$ if and only if for every sequence ζ_n as in (iii) we have

$$A^2 u = \lim_{n\to\infty} \zeta_n^{-1}(T(\zeta_n)u - u). \tag{C.7}$$

Remark: In particular, (iv) implies that for any $\pi/2 > \theta > -\pi/2$ it is true that when $u \in D(A^2)$

$$d/dt\ T(e^{i\theta}t)u_{|t=0} = e^{i\theta}A^2u.$$

Proof: (i) If $\rho(x) = \|V(x)\|$ denotes the operator norm of $V(x) \in L(E)$ for $x \in \mathbb{R}$, then $\rho(x+y) \leq \rho(x)\rho(y)$. Hence, there are finite constants M and ω such that $\rho(x) \leq Me^{\omega|x|}$ for all $x \in \mathbb{R}$. Combining this with (C.2) and Lemma B.1 we get that the integral (C.5) defines a bounded operator $T(\zeta)$ for each $\zeta \in S$. The right-hand side of (C.2) gives an upper bound on the operator norm of $T(\zeta)$. Integration of (C.4) against $V(x)$ with respect to Lebesgue measure on the line leads to the semigroup property (6), by an application of Fubini's theorem combined with Lemma B.1.

(ii) By a standard result on holomorphic vector valued mappings, it is enough to show that for given $u \in E$ and $u^* \in E^*$ the mapping $\varphi(\zeta) = \langle T(\zeta)u,u^*\rangle$ is holomorphic in S.

We show first that $\varphi(\cdot)$ is continuous, and then apply Morera's theorem. Continuity in S is easiest obtained from a standard application of Lebesgues' theorem on dominated convergence (of vector valued functions). Consider a sequence $\{\zeta_n\} \subseteq S$ converging to a point $\zeta_0 \in S$. We may assume that there are finite positive numbers α and β such that $\mathrm{Re}\ \zeta_n \geq \alpha$ and $|\zeta_n| \leq \beta$ for all n. Then in view of (C.2) the sequence of functions $x \to p_{\zeta_n}(x) < V(x)u,u^* >$ is dominated by an integrable function, with integral less than or equal to a constant times

$$(\beta/\alpha)^{\frac{1}{2}} \exp(\omega^2\beta^2/\alpha).$$ Hence, $\lim_n \varphi(\zeta_n) = \varphi(\zeta_0)$, by Lebesgues' theorem.

By Morera's theorem it is now enough to show that the integral $\int_B \varphi(\zeta)d\zeta$ vanishes whenever B is the boundary of a closed triangle contained in S. For a given triangle we have by Fubini's theorem

$$\int_B \varphi(\zeta)d\zeta = \int_{-\infty}^{\infty}\int_B p_\zeta(x)d\zeta < V(x)u,u^* > dx,$$

where the right-hand side vanishes because $p_\zeta(x)$ is holomorphic in ζ for fixed x.

(iii) Let $\{\zeta_n\} \subseteq S$ be a sequence of points such that

$\zeta_n \to 0$, and $\tau = \sup_n |\zeta_n|/\mathrm{Re}\ \zeta_n < \infty$. We show that $\lim_n T(\zeta_n)u = u$ for fixed $u \in E$. Let $\varepsilon > 0$ be given. Pick $\delta > 0$ such that $\|V(x)u-u\| \leq \varepsilon$ for all $|x| \leq \delta$. Pick some $c > \omega$. Since $\zeta_n \to 0$ and τ is finite, we may assume that $2c|\zeta_n|^2 \leq \delta\ \mathrm{Re}\ \zeta_n$ for all n, and the estimate (3) holds for all ζ_n. As usual the region of integration is divided into two parts:

$$\|T(\zeta_n)u-u\| \leq \left(\int_{|x| \leq \delta} + \int_{|x| > \delta}\right) |p_{\zeta_n}(x)| \, \|V(x)u-u\| dx.$$

The first integral is dominated by $\sqrt{\tau}\ \varepsilon$, and the second integral by a constant times $|\zeta_n|^{-\frac{1}{2}}\exp(-\delta^2/\tau 4|\zeta_n|)$, in view of (C.3). Since the last sequence of numbers converges to zero for $n \to \infty$, the desired conclusion follows.
(iv) Let C be the operator whose domain D(C) is equal to the set of vectors u such that the limits in (C.7) exist. The value Cu for $u \in D(C)$ is equal to that limit.

We show first that C is closed. For every complex number z with Re $z > 0$ ($z \in S$), let $\gamma(z)$ be the line segment $\{tz: 0 \leq t \leq 1\}$. The following claim holds: A vector u belongs to D(C) and Cu = v if and only if

$$T(z)u - u = \int_{\gamma(z)} T(\zeta)v\, d\zeta \quad \text{for all } z \in S. \tag{C.8}$$

Suppose that (C.8) is satisfied for some u and v. Let $\{z_n\} \subseteq S$ be a sequence as in (iii). Suppose $\{z_n\}$ is contained in $\Gamma_\varphi = \{\zeta : -\pi/2 + \varphi < \mathrm{Arg}\ \zeta < \pi/2 - \varphi\}$ for some $\varphi > 0$. By (iii) there is for given $\varepsilon > 0$ a $\delta > 0$ such that $\|T(\zeta)v-v\| < \varepsilon$ for all $\zeta \in \Gamma_\varphi$ with $|\zeta| < \delta$. For sufficiently large n we have $\|z_n^{-1}[T(z_n)u-u]-v\| \leq \varepsilon$ by (C.8). Hence $u \in D(C)$ and Cu = v.

Conversely, suppose that $u \in D(C)$. To verify the formula (C.8) with v = Cu for given $z \in S$ we note that by the remark at the end of (iv) the function $t \to T(zt)u$ is differentiable at t = 0 with $d/dt\ T(z\cdot t)u\big|_{t=0} = z\,Cu$. It is easy to show differentiability for $t > 0$, as well as $d/dt\ T(z\cdot t)u = zT(z\cdot t)Cu$. Hence by the fundamental theorem of calculus (for vector valued functions)

$$T(z)u - u = \int_0^1 z\, T(z\cdot t)Cu\, dt.$$

Since this integral is equal to the right-hand side of (C.8), the second part of the claim follows.

By a double application of (C.8) it is now quite easy to show that C is closed, (cf. [Ko, p. 262]).

The operator A^2 is closed as well by a result of Taylor (cf. [DS, p. 602]). Conclusion (iv) states precisely that the operators C and A^2 are equal.

Put $D_+ = \text{span}\{T(\zeta)u : u \in E, \text{Re}\,\zeta > 0\}$. Now D_+ is clearly invariant under all $T(\zeta)$ for Re $\zeta > 0$, (also under $T(0) = I$) and is dense by (iii). The easily verified identity $V(x)T(\zeta) = T(\zeta)V(x)$ for $x \in \mathbb{R}$ and $\zeta \in S$ implies that D_+ is also invariant under $V(x)$ for $x \in \mathbb{R}$.

By (ii) the mapping $\zeta \to T(\zeta)u$ is holomorphic (hence differentiable) for all $u \in D_+$. Therefore D_+ is contained in $D(C)$. We show next that D_+ is contained in $D(A^2)$ as well and that $Cu = A^2u$ for all $u \in D_+$.

Indeed, given $u = T(\zeta)v \in D_+$ ($\zeta \in S$, $v \in E$) there is a sequence $\{v_n\}$ from $D(A^2)$ such that $v_n \to v$. Then

$$CT(\zeta)v_n = d/dz\, T(z)v_n\big|_{z=\zeta} = \int_{-\infty}^{\infty} p''(x)V(x)v_n\, dx$$

for all n. (The notation $p''_\zeta(x)$ is used for $(d/dx)^2 p_\zeta(x)$.) Therefore $T(\zeta)v_n \in D(A^2)$ and $CT(\zeta)v_n = A^2T(\zeta)v_n$. Now $v_n \to v$. Hence $T(\zeta)v_n \to u$, and $A^2T(\zeta)v_n \to Cu$. Since A^2 is closed one gets that $u \in D(A^2)$ and $Cu = A^2u$ as claimed.

Now D_+ is a core for A^2 by the pregenerator theorem B.5. Suppose that D_+ is also a core for C. Then, since C and A^2 coincide on D_+, we have the desired conclusion

$$C = \overline{C_{|D_+}} = \overline{A^2_{\ |D_+}} = A^2.$$

Let C_1 be the operator defined on the vectors u for which the limit (C.7) exists when the sequence $\{\zeta_n\}$ is restricted to the positive real axis, and C_1u is the value of the limit. Then clearly $C \subseteq C_1$ and C_1 is equal to the infinitesimal generator of the restricted semigroup $T(\zeta)$ for $\zeta \in [0,\infty)$. In particular, C_1 is closed and by a second application of the pre-generator theorem $C_1 = \overline{C_{1|D_+}}$. Since C is contained in C_1 and closed, D_+ is a core for C as well. This concludes the proof of (iv). The identity $C = C_1$ follows as a corollary to the proof.

Remarks: (1) There is an alternative approach to the problem. To show that A^2 generates a (holomorphic) semigroup one needs estimates on the resolvent of A^2 together with the Hille-Yosida theorem. The identity $R(\lambda^2,A^2) = -R(\lambda,A)R(-\lambda,A)$ enables one to get these estimates from the corresponding estimates on the resolvents of A.
(2) The theorem has a non-commutative generalization. Given a continuous representation V of a Lie group on a Banach space, and elements $X_1,\dots,X_r$ in the Lie algebra, then the closure of the operator $dV(\Sigma X_k^2)$ is the infinitesimal generator (in the extended sense) of a holomorphic semigroup in the open right half-plane. This semigroup is given as in (C.5) by an integral with respect to a generalized (non-elliptic) Gauss kernel on the Lie group. The reader is referred to [Jo 2] for details. The theorem and one of its non-commutative generalizations may be viewed as analytic continuation of results due to Poulsen [Ps 1].

Appendix D

AN ALGEBRAIC CHARACTERIZATION OF $\mathcal{O}_{A^2}(B)$

This appendix supplies several results on the adjoint action of squares in associative algebras which extend and clarify the discussion of ad-orbits $\mathcal{O}_C(B)$ for $C = A^2$ that appears in Chapters 1, 4 and 11.

When A and B belong to a finite-dimensional Lie algebra, $\mathcal{O}_C(B)$ for $C = A^2$ is in general infinite-dimensional. Here, A and B belong to $dV(\mathbf{g})$ for some group representation V and the ad-orbits depend on V. There is, however, a purely algebraic condition which does not use any Lie algebra structure: A certain polynomial equation in ad A and B is equivalent to finite-dimensionality of $\mathcal{O}_{A^2}(B)$.

Let $\mathbf{A}$ be an algebra with unit over a field F of characteristic zero. We denote by F[t] or F[s,t] the ring of polynomials in one variable t, or two variables s and t, respectively. Given a linear endomorphisms H in $\mathbf{A}$ ($H(aA+B) = aH(A)+H(B)$ for $a \in F$ and $A, B \in \mathbf{A}$) and a polynomial $p(t) = a_0+a_1t+\ldots+a_nt^n \in F[t]$, the endomorphism p(H) is given by the usual functional calculus $p(H) = a_0+a_1H+\ldots+a_nH^n$. A similar remark applies to polynomials in two variables.

For given $A \in \mathbf{A}$ the endomorphisms of left-multiplication and right-multiplication by A are denoted by L_A and R_A respectively, and $\text{ad } A = L_A - R_A$ by definition. With the above notation we have

$$\text{ad}(A^2) = 2\ \text{ad}(A)R_A + (\text{ad } A)^2. \tag{D.1}$$

Indeed, L_A, R_A and ad A all commute by associativity of $\mathbf{A}$, and $\text{ad}(A^2) = L_A^2 - R_A^2 = (L_A-R_A)(L_A+R_A) = \text{ad } A(2R_A+\text{ad } A)$.

Concerning the orbit $\mathcal{O}_A(B)$ the following easy equivalences hold:

(i) $\mathcal{O}_A(B)$ is finite-dimensional.

(ii) For some non-zero $p(t) \in F[t]$

$$p(\text{ad } A)B = 0 \tag{D.2}$$

(iii) The elements $\{(\text{ad } A)^k B: 0 \leq k < \infty\}$ are linearly dependent over F.
In fact, if p(t) is a polynomial of minimal degree which satisfies (D2), then the dimension of $\mathcal{O}_A(B)$ is equal to the degree of p(t).

One more result about $\mathcal{O}_A(B)$ is used without mentioning:

D.1. Proposition

(a) For each $A \in \mathbf{A}$, the set $\mathcal{F}_A(\mathbf{A}) = \{B \in \mathbf{A} : \mathcal{O}_A(B)$ is finite-dimensional$\}$ is an associative (and hence Lie) subalgebra of $\mathbf{A}$.
(b) Thus if $\mathcal{S} \subset \mathbf{A}$,

$$\mathcal{F}_{\mathcal{S}}(\mathbf{A}) = \cap\{\mathcal{F}_A(\mathbf{A}): A \in \mathcal{S}\}$$

is an associative and Lie subalgebra.

Proof: (b) follows from (a). For given $A \in \mathbf{A}$ we denote by δ the derivation ad A. Let $B_1, B_2 \in \mathcal{F}_A(\mathbf{A})$ and let $p_1, p_2 \in F[t]$ be such that $p_i(\delta)B_i = 0$ for $i = 1,2$. Then $p = p_1p_2$ satisfies $p(\delta)(B_1+B_2) = 0$; but to find some $p_3 \in F[t]$ such that $p_3(\delta)(B_1B_2) = 0$ seems not so easy. By the Leibnitz formula,

$$\delta^n(B_1B_2) = \Sigma\{\tbinom{n}{k}\delta^k(B_1)\delta^{n-k}(B_2): 0 \leq k \leq n\} ,$$

we have

$$\mathcal{O}_A(B_1B_2) = \text{span}\{\delta^n(B_1B_2): 0 \leq n\} \subseteq \text{span}[\mathcal{O}_A(B_1)\mathcal{O}_A(B_2)].$$

Hence, $\dim \mathcal{O}_A(B_1B_2) \leq \dim \mathcal{O}_A(B_1)\dim \mathcal{O}_A(B_2)$. E.O.P.

It is a bit more complicated to give conditions on A and B, similar to the ones above, which are equivalent to finite-dimensionality of $\mathcal{O}_{A^2}(B)$:

Consider $X = X(s,t) \in F[s,t]$ given by $X(s,t) = 2st + s^2$. The symbol F[X] denotes the ring of polynomials

$$p(s,t) = \sum_{k=0}^{n} a_k(2st+s^2)^k$$

with $a_k \in F$ and $n = 0,1,2,\dots$. ($n = 0$ is interpreted in the usual way.) If $f(X) = \Sigma_0^n a_k X^k$ then we write

$$p(s,t) = f(2st + s^2) \tag{D.3}$$

On polynomials the partial derivatives D_s and D_t are defined purely algebraically, and it is easy to check that a given $p(s,t) \in F[s,t]$ belongs to $F[X]$ if and only if it satisfies the differential equation $sD_s p - (s+t)D_t p = 0$.

D.2. Proposition

For given elements $A, B \in \mathbf{A}$ the ad-orbit $\mathcal{O}_{A^2}(B)$ is finite-dimensional if and only if there is a non-zero $p(s,t) \in F[X]$ such that

$$p(\text{ad } A, R_A)B = 0. \tag{D.4}$$

Proof: Suppose $p(s,t) \in F[X]\setminus\{0\}$ satisfies (D.4). We have $p(s,t) = f(2st+s^2)$ for some non-zero polynomial f in one variable. Whence, in view of (D.1), the polynomial f satisfies

$$f(\text{ad}(A^2))B = 0 \tag{D.5}$$

and it follows that $\mathcal{O}_{A^2}(B)$ is finite-dimensional as a result of the above remark, (ii) ⇒ (i).

Conversely, the relation (D.5) is clearly satisfied for some non-zero f whenever $\mathcal{O}_{A^2}(B)$ is finite-dimensional. The corresponding polynomial $p(s,t)$ defined by (D.3) then satisfies (D.4); again by (D.1). This completes the proof.

Remarks: (1) It is clear that there is a similar condition of the form $p(\text{ad } A, L_A)B = 0$.
(2) It may be expected at first thought that $\mathcal{O}_{A^2}(B)$ is finite-dimensional whenever (D.4) is satisfied for some non-zero $p(s,t)$. This is not the case, of course, because $F[s,t]$ contains elements which do not belong to $F[X]$.
(3) The formula (D.1) is a special case of the identity

$$\operatorname{ad}(A_1A_2) = \operatorname{ad} A_1 \operatorname{ad} A_2 + R_{A_1} \operatorname{ad} A_2 + R_{A_2} \operatorname{ad} A_1 \tag{D.6}$$

for A_1, $A_2 \in \mathbf{A}$.

The latter formula has for commuting A_1 and A_2 the following consequence.

D.3. Proposition

Let $\mathbf{A}_0$ be an Abelian sub-algebra of $\mathbf{A}$, and let $B \in \mathbf{A}$ be given. Then the set of elements

$$N = \{A \in \mathbf{A}_0 : (\operatorname{ad} A)^n B = 0 \text{ for some } n\}$$

is a sub-algebra of $\mathbf{A}_0$.

Proof: We leave to the reader to check that the sum of two elements from N belongs to N. Consider elements A_1, $A_2 \in N$ and denote by δ_1 and δ_2 the respective derivations ad A_1 and ad A_2. The right-hand side of (D.6) is a sum of three commuting operators. Using the notation $(i^n\ j\ k)$ for the coefficients $n!/i!j!k!$ in the tri-nomial formula, we get

$$\operatorname{ad}(A_1A_2)^n = \sum_{\substack{i+j+k=n \\ i,j,k \geq 0}} (i^n\ j\ k)(\delta_1\delta_2)^i (R_{A_1}\delta_2)^j (R_{A_2}\delta_1)^k$$

whence

$$\operatorname{ad}(A_1A_2)^n B = \Sigma(i^n\ j\ k)\delta_1^{i+k}\ \delta_2^{i+j}\ (B)A_1^jA_2^k \tag{D.7}$$

where the summation is as above. Suppose specifically that $\delta_1^{n_1}(B) = 0$ and $\delta_2^{n_2}(B) = 0$. Let n be an integer, $n \geq n_1+n_2$. For the summation indices $i, j, k \geq 0$ with $i + j + k = n$ we have $i + k \geq n_1$ or else $i + j \geq n_2$. Whence the right-hand side of (D.7) vanishes. E.O.P.

Appendix E

COMPACT PERTURBATIONS OF SEMIGROUPS

Here, we give a short proof of a result, generalizing a theorem of Vidav [Vd], concerning compactness of the additive perturbation of a C_0 semigroup that results from a compact perturbation of its generator. We are endebted to R.S. Phillips for the suggestion upon which our proof is based.

E.1. Theorem

Suppose that A generates a C_0 semigroup $\{V(t,A): t \in [0,\infty)\}$ on a Banach space E, and that $U \in L(E)$ is compact. Let the perturbed semigroup generated by $B = A + U$, $\{V(t,B): t \in [0,\infty)\}$ be written in the standard form

$$V(t,B) = V(t,A) + K(t,U) \tag{E.1}$$

for $K(t,U) = \int_0^t V(t-s,A)\, U\, V(s,B)\, ds$ (strongly convergent).

(a) Then $K(t,U)$ is compact for all $t \in [0,\infty)$ and operator-norm continuous in t.

(b) If A generates a group, then $K(t,U) = V(t,A)R(t,U)$ $= L(t,U)V(t,A)$ where both $R(t,U)$ and $L(t,U)$ are compact and $R(t,U)$ is norm-Lipschitz in t.

(c) If in (b), $\{V(t,A)^* : t \in \mathbb{R}\}$ is C_0 on E^*, then $L(t,U)$ and $R(t,U)$ are norm-C^1 in t. For example, if E is reflexive, this is true.

Proof: Recall first that B generates a C_0 semigroup by Phillips perturbation theorem [HP] and that (E.1) is derived as follows: Let $u \in D(A)$ and use $V(s,B)D(B) \subset D(B) = D(A)$ to conclude that $V(t-s,A)V(s,B)u$ is differentiable in s by the product rule, so that

$$V(t,B)u - V(t,A)u = \int_0^t d/ds\, V(t-s,A)V(s,B)u\, ds$$

$$= \int_0^t V(t-s,A)(B-A)V(s,B)u\, ds = K(t,U)u.$$

Since D(A) is dense, this identity extends by limits to all $u \in E$, yielding (E.1).

To obtain (a), put $G(s,t) = V(t-s,A)U$. Then by a standard result (Lemma E2 below), G is operator-norm continuous in (s,t) for $s \leq t$, whence for any $T < \infty$, $K_T = \{G(s,t): 0 \leq s \leq t \leq T\}$ is operator-norm compact; it also consists of compact operators since the latter form a two-sided ideal in L(E). But then if S is a bounded set in E and $S_T = \{V(s,B)u: u \in S, s \in [0,T]\}$, an easy Ascoli argument shows that $K_T S_T$ is precompact in E, so that

$$t^{-1}K(t,U)S = t^{-1} \int_0^t G(s,t)V(s,B)S\, ds \in \overline{co}(K_T S_T)$$

is precompact and K(t,U) is compact. Norm-continuity of G(s,t) and uniform norm-boundedness of G(s,t)V(s,B) then yields the norm-continuity claim.

To obtain precompactness of $K_T S_T$, think of S_T as a bounded (hence equicontinuous) subset of L(L(E),E): $u \in S_T$ sends $G \in K_T$ into $Gu \in X$. Then since K_T has compact closure and each $G \in K_T$ is compact (so GS_T has compact closure), equicontinuous S_T has compact closure for the topology of uniform convergence on K_T. But when $\overline{S_T}$ carries that topology, the map $\overline{S_T} \times K_T \to K_T S_T$ is jointly continuous and sends this compact product space onto a compact set, so $K_T S_T$ is precompact.

In (b), clearly $R(t,U) = \int_0^t V(-s,A)UV(s,B)ds$ is an integral of a uniformly norm-bounded function on any [0,T], hence is norm-Lipschitz, while $R(t,U) = V(-t,A)K(t,U)$ and $L(t,U) = K(t,U)V(-t,A)$ are both compact.

For (c), we revert to Vidav's proof-scheme, checking for R(t,U) that $F(s) = V(-s,A)\,U\,V(s,B)$ is norm-continuous by Lemma E.2 below, hence that its integral is norm-C^1. For L(t,U), it is quickest to verify that

$$K(t,U) = \int_0^t V(s,B)UV(t-s,A)ds = \int_0^t V(s,B)UV(-s,A)ds\; V(t,A)$$

$$= L(s,U)V(t,A),$$

so that the same argument will apply to norm-continuous V(s,B)UV(s,A). The Lemma that we need (both here and in Section 12E) is Vidav's [Vd].

E.2. Lemma

(a) If U is compact and $V(r)$ is a strongly continuous function on some locally compact topological space T into $L(E)$ then $V(r)U$ is norm-continuous into $L(E)$.
(b) If $V^*(r)$ is strongly continuous into $L(E^*)$ then $UV(r)$ is also norm-continuous into $L(E)$.
(c) If $V_1(r)$ is norm-continuous and compact-valued while $V_2(r)^*$ is strongly continuous ($V_2(r) \in L(E)$) then $(V_1V_2)(r)$ is norm-continuous into $L(E)$.

Proof: (a) If S is the unit ball of E, $\overline{US}$ is compact, and the topology of pointwise convergence (strong operator) on $\overline{US}$ agrees with the topology of uniform convergence there: $r \to r_o$ implies that $V(r) \to V(r_o)$ uniformly on $\overline{US}$, whence $V(r)U \to V(r_o)U$ uniformly on S: i.e. in operator norm.
(b) Duality: U^* is compact and $[UV(r)]^* = V^*(r)U^*$ is then norm-continuous into $L(E^*)$ by (a), so $UV(r)$ is norm-continuous into $L(E)$.
(c) As $r \to r_o$, $V_1(r_o)V_2(r) \to V_1(r_o)V_2(r_o)$ by (b) with $U = V_1(r_o)$, while $\|V_1(r)-V_1(r_o)\| \to 0$ implies that $V_1(r)V_2(r) - V_1(r_o)V_2(r_o) = (V_1(r) - V_1(r_o))V_2(r) + V_1(r_o)(V_2(r) - V_2(r_o)) \to 0$, since $\|V_2(r)\|$ is uniformly bounded on any compact neighborhood of r_o. E.O.P.

Applying the Lemma, we need only observe that if $V(t,A)^*$ is stronly continuous on E^*, then its Phillips perturbation by $U^* \in L(E^*)$ is also strongly continuous; this clearly agrees with $V(t,B)^*$. Hence we can take either $V(-s,A)$ or $V(s,B)$ as $V_2(s)$ and $V(-s,A)U$ or $V(s,B)U$ as V_1, using Lemma E.2(a) to obtain norm-continuity for the latter in (a) and (c) of Theorem E.1. This completes the proof of the Theorem.

Remarks: (1) Vidav [Vd] obtains only (a), for reflexive E, by observing as above that $V(t-s,A)UV(s,B)$ is a norm-continuous compact-valued function whose norm-convergent Riemann integral must be compact. This argument fails in general: for rank-1 operators of the form $U = v \otimes f$ for $v \in E$, $f \in E^*$ (acting via $(v \otimes f)(u) = f(u)v$) it is easy to check that $V(t-s,A)UV(s,B) = (V(t-s,A)v \otimes V^*(s,B)f$, and that if $V(t-s,A)v \neq 0$ then norm-

continuity of the integrand for $K(t,U)_*$ (in S) holds iff $s \to V^*(s,B)f$ is norm-continuous into E^*. Hence if $V^*(s,B)$ (equivalently, $V^*(s,A)$) fails to be strongly continuous on all of E^*, there exist U for which the integrand is <u>not</u> norm-continuous in s, so Vidav's argument does not apply.
(2) For reflexive E, another argument applies that gives more: for rank-1 $U = v \otimes f$ as above, $V(t-s,A)UV(s,B) = (V(t-s,A)v) \otimes (V^*(s,B)f)$ is continuous into $E \otimes E^*$ with any "cross-norm" in the sense of Schatten (cf. Rickart [Rk]), as the reader may check. Extending by linearity and cross-norm limits, one obtains for any $U \in E \tilde{\otimes} E^*$ in the cross-norm completion that $V(t-s,A)UV(s,B)$ is a continuous $E \tilde{\otimes} E^*$-valued function whose integral $K(t,U)$ is in $E \tilde{\otimes} E^*$. In particular, if U is Hilbert-Schmidt or trace-class, for E a Hilbert space, so is $K(t,U)$, and the latter is continuous in t for the Hilbert-Schmidt-(respectively, trace-)norm. We omit details.

Appendix F

NUMERICAL RANGES AND SEMIGROUPS ON L^p SPACES

In Chapters 10 and 12, we have found it useful to discuss the group-generation properties of unbounded Banach space operators A in terms of their numerical ranges. Although the relationship between contraction semigroups and dissipative operators is well known, neither the mild generalizations that we require nor their applications in L^p spaces appear to be discussed in the literature. For the reader's convenience, we review the main ideas and sketch the more important proofs.

Our approach to numerical ranges here is based upon Nelson's notion of dissipativity for an operator A with domain D(A) dense in a Banach space E([Nℓ3], see also [RS II, p. 235]): A is dissipative iff $\mathrm{Re}(f(Au)) \le 0$ for all $u \in D(A)$ and $f \in E^*$ with $\|f\| = \|u\|$ and $f(u) = \|u\|^2$. That is, we take the numerical range W(A) of A to be

$$W(A) = \{f(Au)\colon u \in D(A),\ f \in E^*,\ 1 = \|f\| = \|u\| = f(u)\} \quad (\mathrm{F.1})$$

and observe that A is dissipative iff W(A) is contained in the left half-plane. (Notice that a nontrivial pair $(0,0) \neq (u,f) \in D(A) \times E^*$ with $\|f\| = \|u\|$ and $f(u) = \|u\|^2$ can be replaced by the pair $(\|u\|^{-1}u,\ \|f\|^{-1}f) = (u_1,\ f_1)$ with $1 = \|u_1\| = \|f_1\| = f_1(u_1)$; $\mathrm{Re}(f(Au)) \le 0$ iff $\mathrm{Re}(f_1(Au_1)) \le 0$.) Readers familiar with the technically different Lumer-Phillips approach ([LP], [Yo]) will recall that there, one selects for each $u \in E$ a functional $f_u \in E^*$ with $f_u(u) = \|u\|^2$, puts $[u,v] = f_v(u)$ for the semi-inner-product of u and v, and A is called dissipative iff $\mathrm{Re}([Au,u]) \le 0$ for all $u \in D(A)$. For many spaces, such as the L^p spaces for $1 < p < \infty$, f_u is uniquely determined by u, so that $W(A) = \{[Au,u]\colon u \in D(A)$ and $\|u\| = 1\}$ follows. In general, more than one choice of f_u is possible, yielding more than one semi-inner product and a non-canonical notion of dissipativity, but it is shown in [LP] that a semigroup-pregenerator is dissipative for one choice of semi-inner product iff it is dissipative for all. A similar remark holds for the slightly more general group-pregeneration result stated below, so no generality turns out to be lost by using the numerical range defined in (F.1) rather than the potentially smaller "Lumer numerical range" $W(A,[\,,\,]) = \{[Au,u]: u \in D(A), \|u\| = 1\}$. (It has recently been proved that the different dissipativity notions coincide in general.)

F.1. Theorem (Lumer-Nelson-Phillips)

Let A be closable and densely-defined, with domain D. Then A is a pregenerator of a C_0 group $\{V(t,A): t \in \mathbb{R}\}$ of pure exponential type ω ($\|V(t,A)\| \le e^{\omega|t|}$ for all $t \in \mathbb{R}$) iff it satisfies the following two conditions.
(a) The numerical range $W(A)$ is contained in the ω-strip $S(\omega) = \{\lambda \in \mathbb{C}: |\mathrm{Re}(\lambda)| \le \omega\}$.
(b) There exist λ_+,λ_-, with $\mathrm{Re}(\lambda_+) > \omega$, $\mathrm{Re}(\lambda_-) < -\omega$, such that $D_+ = (\lambda_+-A)D$ and $D_- = (\lambda_--A)D$ are dense in E.

Proof Sketch: Condition (a) is easily seen to be equivalent to the assumption that $A_+ = A - \omega$ and $A_- = -A-\omega$ are both dissipative. ($f(A_\pm u) = \pm f(Au) - \omega f(u) = \pm f(Au) - \omega$, whence $f(Au) \in S(\omega)$ iff $\mathrm{Re}(f(A_\pm u)) \le 0$.) Similarly, $V(t,A)$ is of pure exponential type ω iff $\|V(t,A_\pm)\| \le 1$ for $t \ge 0$: $A_\pm$ pregenerate C_0 contraction semigroups. The theorems of Lumer-Phillips [LP] and Nelson [Nℓ] then apply to establish that in the presence of (b), $A_\pm$ are dissipative iff they pregenerate such C_0 contraction semigroups.

In showing that (a) and (b) are sufficient, here are the two main ideas in reducing the argument to the Hille-Yosida-Feller (HYF) theorem in Chapter 6. First, one shows that for any A and $\lambda \notin W(A)$

$$\|(\lambda-A)u\| \ge \text{distance } (\lambda,W(A))\,\|u\| \tag{F.2}$$

for all $u \in D(A)$. Indeed, if $\|u_1\| = 1 = f_1(u_1) = \|f_1\|$.

$$\|(\lambda-A)u_1\| \ge |f_1((\lambda-A)u_1)| = |\lambda f_1(u_1)-f_1(Au_1)| = |\lambda-f_1(Au_1)|$$
$$\ge \text{distance } (\lambda, W(A)); \tag{F.3}$$

the general case follows upon replacing (nonzero) $u \in D(A)$ by $u_1 = \|u\|^{-1}u$ and then multiplying both sides of (F.3) by $\|u\|$ to obtain (F.2). If $(\lambda-A)D(A)$ is dense in E and $\operatorname{dist}(\lambda,W(A) > 0$, then $(\lambda-A)^{-1}$ exists and extends to a bounded inverse $R(\lambda,A) = (\lambda-\bar{A})^{-1}$ for $\bar{A}$ with $\|R(\lambda,A)\| \le \operatorname{dist}(\lambda,W(A))^{-1}$. The argument mentioned in Lemma 5.7'(b) then shows that such a λ is contained in a subset of $\rho(\bar{A}) \sim \overline{W(A)}$ which is open and relatively closed in $\mathbb{C} \sim \overline{W(A)}$ and the resolvent estimate extends to the entire component of λ in $\mathbb{C} \sim \overline{W(A)}$. In the cases under discussion above, this yields the HYF estimates for $R(\lambda,A_+)$ in the right half-plane. We omit further discussion of the necessity of (a) and (b).

Remark: An argument in [LP] is easily adapted to show that if A satisfies (a), it is necessarily closable ($A_\pm$ are dissipative, hence closable ...). In the applications discussed here, the domain $D(A^*)$ of the adjoint is always trivially weak-* dense in E^*, yielding the same information by more familiar methods.

The next two results then give a concrete interpretation of the calculations involved in checking hypothesis (a) in Theorem F.1 when the Banach space E in which A acts is $L^p(X,\mu)$ for $1 \le p < \infty$ and μ is a measure on a space X. That is, we describe the functionals $f \in E^*$ associated with $u \in E$ via $1 = \|u\| = \|f\| = f(u)$ in terms of their Riesz representatives $v \in L^q(X,\mu)$ ($q = p/(p-1)$ for $p > 1$, $q = \infty$ if $p = 1$), so that the numerical range consists of numbers of the form $f(Au) = \int_X Au\,\bar{v}\,d\mu$.

F.2. Proposition

Suppose $1 < p < \infty$ and u is a measurable function in $L^p(X,\mu)$ with $\|u\|_p = 1$. Then there is a unique functional f with $\|f\| = f(u) = 1$, whose Riesz representative may be chosen to be the $v \in L^q(X,\mu)$ such that $v(x) = u(x)|u(x)|^{p-2}$ when $u(x) \ne 0$, $v(x) = 0$ when $u(x) = 0$.

Proof: If $f \in L^p(X,\mu)^*$ is represented by v then $\left|\int_X u\bar{v}d\mu\right| = 1 = \|f\|\ \|u\|_p = \|v\|_q\|u\|_p$, so Hölder's inequality becomes an equality, and a well-known result ensures that for a suitable $c \in \mathbb{C}$, $v(x) = cu(x)|u(x)|^{p-2}$ a.e. where $u(x) \neq 0$, [Yo, p. 34] whence $1 = \int_X u\bar{v}d\mu$ implies that $c = 1$. It is equally well-known that every v of this form defines a functional f with $\|f\| = 1 = f(u)$. (The verification is routine).

E.O.P.

The situation for $p = 1$ is slightly more complicated, since functions u which vanish on sets of positive measure are associated with more than one dual "tangent functional"

F.3. Proposition

Let $u \in L^1(X,\mu)$ be a measurable function such that $\|u\|_1 = 1$. Let $S = \{x \in X: u(x) \neq 0\}$ and $N = X \sim S$, and let $v_o = u|u|^{-1}$ on S, $v_o = 0$ on N. Then a functional $f \in L^1(X,\mu)^*$ has $\|f\| = 1 = f(u)$ iff it can be Riesz-represented by $v = v_o + v_N \in L^\infty(X,\mu)$, where $\|v_N\|_\infty \leq 1$ and v_N vanishes on S.

Proof: Since v_o and v_N live on disjoint sets and $\|v_o\|_\infty = 1$ by construction, any such $v = v_o + v_N$ has $\|v\|_\infty = \max\{\|v_o\|_\infty, \|v_N\|_\infty\} = 1$ while $\int_X u\bar{v}d\mu = \int_X u\bar{u}|u|^{-1}d\mu + \int_X u\bar{v}_N d\mu = \|u\|_1 + 0 = 1$, so every such v represents a normalized $f \in (L^1)^*$ with $f(u) = 1$. Conversely, suppose that v represents such an f, and put $v_N = v - v_o$. It suffices to verify that v_N vanishes a.e. on S, for then $\|v\|_\infty = \max\{\|v_o\|_\infty, \|v_N\|_\infty\}$ will follow and ensure that $\|v_N\|_\infty \leq 1$; v_N may then be modified on a set of measure 0 to vanish off N.

To check that v_N must vanish a.e. on S, put $w_N(x) = -|u(x)|u(x)^{-1}v_N(x)$ for $x \in S$, so that $v(x) = u(x)|u(x)|^{-1}(1 - w_N(x))$ on S. Then $\|v\|_\infty = 1$ forces

$0 \le w_N(x) \le 2$ a.e. on S. But

$$0 = 1 - 1 = \int_X u\bar{v}_0 d\mu - \int_X u\bar{v} d\mu = \int_X u(-\bar{v}_N) d\mu$$

$$= \int_S u\, \bar{u}|u|^{-1}\, w_N d\mu = \int_S |u| w_N d\mu,$$

so the non-negative function $|u|w_N$ must vanish a.e. on S, whence w_N and v_N vanish a.e. there since u is nonzero on S.

E.O.P.

Finally, we consider the case $p = \infty$, in the following guise: X is a locally compact Hausdorff space and $E = C_\infty(X)$, the continuous complex-valued functions vanishing at ∞ on X, with the sup-norm $\|\cdot\|_\infty$

F.4. Proposition

Let $u \in C_\infty(X)$, with $\|u\|_\infty = 1$. Then a functional $f \in C_\infty(X)^*$ satisfies $\|f\| = 1 = f(u)$ iff it is Riesz-represented by a measure of the form $\bar{u}\pi$, where π is a probability measure supported in the compact maximum set $M = \{x \in X: |u(x)| = 1\}$.

Proof: If f is represented by $\bar{u}\pi$, then for any $v \in C_\infty(X)$, $|f(v)| \le \int_M |\bar{u}|\, |v| d\pi \le \|v\|_\infty$, while $f(u) = \int_M |u|^2 d\pi = 1$, so $\|f\| = 1 = f(u)$. Conversely, suppose that μ represents a functional f with $\|f\| = 1 = f(u)$, so that the total variation $|\mu|(X) = \|f\|$. Then for every $\varepsilon > 0$, μ is supported in $M(\varepsilon) = \{x \in X: |u(x)| \ge 1-\varepsilon\}$, since otherwise

$$1 = |f(u)| \le \int_{M(\varepsilon)} |u| d|\mu| + \int_{X-M(\varepsilon)} |u| d|\mu|$$

$$\le |\mu|(M(\varepsilon)) + (1-\varepsilon)|\mu|(X-M(\varepsilon)) < 1,$$

a contradiction. Hence μ is supported in $M = \cap\{M(\varepsilon): \varepsilon > 0\}$. Let $\pi = u\mu$, necessarily supported in M. Then $\pi(X) = \int_X u d\mu = 1$, while $|\pi|(X) \le \|u\|_\infty |\mu|(X) \le 1$, whence $|\pi|(X) = 1$ and π must be a probability measure. But then $\bar{u}\pi = |u|^2\mu = \mu$ since $|u| \equiv 1$ on the support M of μ. E.O.P.

Appendix G

BOUNDED ELEMENTS IN OPERATOR LIE ALGEBRAS

For a variety of reasons, it is important to know just where in an operator Lie algebra $\mathcal{L} \subset \mathbf{A}(D)$ bounded operators can occur. This issue arises, for example, in our discussion of the possible structural changes in an operator Lie algebra that can occur as a result of certain special types of Phillips perturbations (Section 9E). According to Doebner and Melsheimer [DM] this matter also has important implications in the application of Lie Theory ("symmetry and degeneracy groups") in mathematical physics. The prototype for our discussion is an early result of Singer [Sr 2], which asserts in one form that a Lie algebra $\mathcal{L}$ of bounded skew adjoint operators on a Hilbert space must have the special Levi decomposition $\mathcal{L} = \mathfrak{k} \oplus \mathfrak{z}$, where the radical $\mathfrak{z}$ is central (hence Abelian) and the semisimple part $\mathfrak{k}$ is compact. Hence ([DM]) any Lie algebra of skew-symmetric operators which does not have this special form must contain at least one unbounded operator. This is in particular true if $\mathcal{L}$ is noncompact semisimple; Doebner-Melsheimer exhibit examples such as SO(2,1) where two elements are unbounded, and conjecture without proof that all must be unbounded. We prove two stronger results below for $\mathcal{L}$ an exponentiable operator Lie algebra in a Banach space E.

(1) If $\mathcal{L}$ is simple, either it consists entirely of bounded operators, or it has <u>property U</u> : every nonzero $A \in \mathcal{L}$ is unbounded.
(2) If $\mathcal{L}$ is <u>totally noncompact semisimple</u> (a direct sum of noncompact simple Lie ideals) and the exponential of $\mathcal{L}$ is uniformly norm-bounded on G (equicontinuous), then $\mathcal{L}$ necessarily has property U.

The first of these results follows from essentially easy consideration, while the second is a corollary of the considerably deeper generalization of Singer's theorem that forms the main result of the section.

Turning to specifics, let $\mathcal{L} \subset \mathbf{A}(D)$ be an operator Lie algebra on a dense C^∞ domain D in a Banach space E. Then, as in Chapter 9, we say that $A \in \mathcal{L}$ is bounded iff it is bounded on the normed subspace $(D, \|\cdot\|)$ of $(E, \|\cdot\|)$, hence extends to a bounded $\bar{A} \in L(E)$. It is trivial to see that the subset $\mathfrak{B}(\mathfrak{L}) = \mathfrak{B}$ of bounded operators in $\mathfrak{L}$ is a Lie subalgebra, and examples at the end of the section show that no more than this can be concluded without assuming that $\mathfrak{L}$ is exponentiable, even if $\mathfrak{L}$ consists of skew-symmetric operators in a Hilbert space. (Of course, $\mathfrak{B}(\mathfrak{L})$ is automatically exponentiable...). Given exponentiability of $\mathfrak{L}$, we can draw the following more interesting conclusion:

G.1. Proposition

If $\mathfrak{L}$ is exponentiable, then $\mathfrak{B}(\mathfrak{L})$ is a Lie ideal in $\mathfrak{L}$.

Remark: Our argument uses a bit less than exponentiability: it suffices that $\mathfrak{L}$ be "sub-exponentiable" in the sense that there exists a representation $V: G \to \mathrm{Aut}(E)$ of the simply connected G with Lie algebra $\mathfrak{g} \cong \mathfrak{L}$ such that $D \subset C_\infty(V)$ and $\mathfrak{L}$ is the restriction to D of $dV(\mathfrak{g})$. That is, it need not be the case that D is a core for any of the unbounded generators $dV(X)$ of one-parameter groups $V(\exp(tX))$, $X \in \mathfrak{g}$.

Proof: For any $A \in \mathfrak{L}$ and $B \in \mathfrak{B}(\mathfrak{L})$, $V(t,A)\bar{B}\,V(-t,A) \in L(E)$. But if $\tilde{A}$ and $\tilde{B}$ denote the restriction of $\bar{A}$ and $\bar{B}$ to $D_\infty = C_\infty(V)$, then $V(t,A)\tilde{B}\,V(-t,A) = \exp(-t \text{ ad } \tilde{A})(\tilde{B}) = \exp(t \text{ ad } A)(B)^\sim$. Hence, $\exp(t \text{ ad } A)(B)^\sim$ and $\exp(t \text{ ad } A)(B)$ are restrictions to D_∞ and D respectively of the bounded operator $V(t,A)\bar{B}\ V(-t,A)$. Hence $\mathfrak{B}(\mathfrak{L})$ is invariant under $\exp(\text{ad } \mathfrak{L})$ and under ad $\mathfrak{L}$, so it is a Lie ideal. (Notice that the operator norm gives $\mathfrak{B}(\mathfrak{L})$ its unique linear finite-dimensional topology, whence ad $A(B)$ is the operator-norm-limit of $t^{-1}(\exp(t \text{ ad } A)(B) - B)$ as $t \to 0$, and is bounded).

G.2. Corollary

If $\mathfrak{L}$ is exponentiable and semisimple, then $\mathfrak{L} = \mathcal{U} \oplus \mathfrak{B}(\mathfrak{L})$, as the direct sum of two (commuting) ideals, where $\mathcal{U}$ has the property U and $\mathfrak{B}(\mathfrak{L})$ is entirely bounded. In particular, if $\mathfrak{L}$ is simple, either $\mathfrak{L}$ has property U or $\mathfrak{L} = \mathfrak{B}(\mathfrak{L})$.

Proof: $\mathfrak{L} = \mathfrak{L}_1 \oplus \ldots \oplus \mathfrak{L}_k$, as the direct sum of ideals, where each of the $\mathfrak{L}_i$ is simple (has no proper ideals), by the structure theory for semisimple Lie algebras [Hℓ]. Hence for each i, $\mathfrak{B}_i = \mathfrak{L}_i \cap \mathfrak{B}(\mathfrak{L})$ is an ideal in $\mathfrak{L}$, hence in $\mathfrak{L}_i$, so by

Proposition G.1, either $\mathbf{B}_i = \mathcal{L}_i$ or $\mathbf{B}_i = \{0\}$. Consequently, $\mathbf{B}(\mathcal{L}) = \oplus \{\mathcal{L}_i : \mathbf{B}_i = \mathcal{L}_i\}$ and $U = \oplus \{\mathcal{L}_i : \mathbf{B}_i = \{0\}\}$ yields the claimed decomposition of $\mathcal{L}$.

Again, examples given at the end of the section indicate that we cannot expect definitive results like Corollary G.2 when $\mathcal{L}$ has a more general structure, unless further constraints are imposed upon the exponential. The appropriate constraint is uniform boundedness, which of course includes unitarity when E is a Hilbert space.

G.3. Theorem (Generalized Singer Theorem)

Suppose that $\mathcal{L}$ is exponentiable, and that its exponential is uniformly norm-bounded on G (for some $M < \infty$, $\|V(x)\| \leq M$ for all $x \in G$). Let $\mathcal{L} = S +)R$ be a (semidirect) Levi decomposition of $\mathcal{L}$ into a semisimple subalgebra S, and a solvable radical ideal R containing the nil-radical N.

(a) Then $\mathbf{B} = \mathbf{B}(\mathcal{L}) = \mathbf{k}_{\mathbf{B}} \oplus \mathfrak{z}_{\mathbf{B}}$ as the direct sum of two commuting ideals $\mathbf{k}_{\mathbf{B}} = S \cap \mathbf{B}$ and $\mathfrak{z}_{\mathbf{B}} = R \cap \mathbf{B}$, where $\mathbf{k}_{\mathbf{B}}$ is compact semisimple and $\mathfrak{z}_{\mathbf{B}}$ is central in $\mathbf{B}$.

(b) If $E \subset S$ is the direct sum of all noncompact simple ideals in S, then E has property U (hence meets $\mathbf{B}$ trivially) and commutes with $\mathbf{B}$: $E \cap \mathbf{B} = [E,\mathbf{B}] = \{0\}$.

(c) The nil-radical $N \supset [R,R]$ commutes with $\mathbf{B}$..

(d) $S = E \oplus \mathbf{k}_U \oplus \mathbf{k}_{\mathbf{B}}$, where $\mathbf{k}_U$ is a (possibly trivial) sum of compact simple ideals $\mathbf{k}_i$ with the property U. Each such $\mathbf{k}_i$ either commutes with all of $\mathbf{B}$ or has a direct sum of irreducible sub-modules contained in $\mathfrak{z}_{\mathbf{B}}$ (while necessarily commuting with $\mathbf{k}_{\mathbf{B}}$).

Remarks: (1) If $R \neq N$, then it appears that there may be (necessary unbounded) operators A in R such that ad A acts nontrivially on $\mathbf{B}$. The set of these seems to possess no necessary linear structure, and (c) implies that it is not closed under formation of commutators. Likewise, the set R_U of unbounded elements in R has no apparent structure of interest.

(2) It appears that $\mathbf{k}_U$ in (d) above can fail to commute with $\mathbf{B}$, although the condition that every $\mathbf{k}_i \subset \mathbf{k}_U$, that does not, must have nontrivial irreducible submodules in $\mathfrak{z}_{\mathbf{B}}$, strongly constrains the relation between the possible structure of these $\mathbf{k}_i$ and the possible dimension of $\mathfrak{z}_{\mathbf{B}}$.

Proof: The result turns upon the observation that the natural "adjoint representation" of $\mathbf{L}$ on $\mathbf{B}_{\mathbb{C}}$ has its range in operators (matrices) with pure imaginary eigenvalues of ascent zero. We first check that (a), (b) and (c) will follow from this. Let $S_{\mathbf{B}} = \mathbf{B} \cap S$ and $R_{\mathbf{B}} = \mathbf{B} \cap R$; it is clear since $\mathbf{B}$ is an ideal in $\mathbf{L}$ that $S_{\mathbf{B}}$ is an ideal in the subalgebra S and that $R_{\mathbf{B}}$ is an ideal in $\mathbf{B}$. Moreover, $S_{\mathbf{B}}$ is then the direct sum of a subset of the simple ideals in S, hence semisimple, and $R_{\mathbf{B}}$ is solvable, hence $\mathbf{B} = S_{\mathbf{B}} +) R_{\mathbf{B}}$ is a Levi decomposition for $\mathbf{B}$.

Now, let $N \subset R$ be the nilpotent nil-radical for $\mathbf{L}$, and recall that $[\mathbf{L}, R] = N$. Hence for any $B \in \mathbf{B}$ and $A \in N \subset R$, $\operatorname{ad} A\,(B) \in N_{\mathbf{B}} \subset N$, whence $(\operatorname{ad} A)^{s+1}(B) = 0$ for s the degree of nilpotency of N, so ad A has zero as its only eigenvalue on $\mathbf{B}$ and the ascent 0 condition on $\mathbf{B}_{\mathbb{C}}$ then implies that $\operatorname{ad} A = 0$ on $\mathbf{B}$: , and hence $N_{\mathbf{B}} \subset N$ commutes with $\mathbf{B}$ (so (c) is proved) and $N_{\mathbf{B}}$ is central there: $N_{\mathbf{B}} = \mathfrak{z}_{\mathbf{B}}$. But now if $A \in R_{\mathbf{B}} = \mathbf{B} \cap R$ and $B \in \mathbf{B}$, $\operatorname{ad} A(B) \in N_{\mathbf{B}} = \mathfrak{z}_{\mathbf{B}}$ so $(\operatorname{ad} A)^2(B) = 0$ and the argument above shows that $\operatorname{ad} A = 0$ on $\mathbf{B}$ again, so $R_{\mathbf{B}} = \mathfrak{z}_{\mathbf{B}}$ is central. Clearly the radical of the ideal $\mathbf{B}$ is $R \cap \mathbf{B}$, so we conclude that the radical of $\mathbf{B}$ is central.

Next, the semisimple part S is the direct sum of a family S_i of simple ideals (in S), and since $\mathbf{B} \cap S_i$ is an ideal in S_i for each i, either $S_i \subset \mathbf{B}$, or $S_i \cap \mathbf{B} = \{0\}$, for each i, so $S \cap \mathbf{B} = \oplus\{S_i : S_i \subset \mathbf{B}\}$. Now, for each $S_i \subset \mathbf{B}$, $\operatorname{ad} S_i : (S_i)_{\mathbb{C}} \to (S_i)_{\mathbb{C}}$ has pure imaginary eigenvalues of ascent 0, so for each $A \in S_i$, the Killing form, trace $((\operatorname{ad} A)^2)$, is (by conjugation-invariance of the trace) just the sum of the squares of these eigenvalues, hence negative. (Note that the adjoint representation of S_i on itself is an isomorphism, so every $A \in S_i$ must have non-zero eigenvalues for ad A.) Thus the Killing form of S_i is negative-definite, and S_i is compact. This completes the proof of (a).

Turning to (b) and (d), we observe that each simple $S_i \subset S$ with $S_i \cap \mathbf{B} = \{0\}$ either commutes with $\mathbf{B}$ or has the property that $A \to \mathrm{ad}\, A|_{\mathbf{B}_{\mathbb{C}}} = \tilde{A}$ is a Lie algebra isomorphism. We check that in the latter case, $\tilde{S}_i$ (hence S_i) must be compact, whence the sum E of all noncompact $S_i \subset S$ must commute with $\mathbf{B}$. To prove compactness, we note that $\tilde{S}_i$ is a matrix Lie algebra such that the bilinear form $(\tilde{A},\tilde{B})_\sim = -\mathrm{tr}(\tilde{A}\tilde{B})$ is a real inner product since (as above) $\mathrm{tr}(\tilde{A}^2)$ is a sum of squares of imaginary eigenvalues, not all 0. But since

$$\begin{aligned}(\exp(t\,\mathrm{ad}\,\tilde{C})(\tilde{A}),\ \exp(t\,\mathrm{ad}\,\tilde{C})(\tilde{B}))_\sim &= -\mathrm{tr}(\exp(t\tilde{C})\,\tilde{A}\exp(-t\tilde{C})\exp(t\tilde{C})\tilde{B}\exp(-t\tilde{C}))\\ &= -\mathrm{tr}(\exp(t\tilde{C})(\widetilde{AB})\ \exp(t\tilde{C})^{-1})\\ &= -\mathrm{tr}(\widetilde{AB})\\ &= (\tilde{A},\tilde{B})_\sim .\end{aligned}$$

The group G generated by $\exp(\mathrm{ad}\,\tilde{S}_i)$ consists of $(\,,)_\sim$ orthogonal transformations on $\tilde{S}_i$. This proves that $\mathrm{int}(\tilde{S}_i)$ is compactly-embedded in $\mathrm{Aut}(\tilde{S}_i)$, so $\tilde{S}_i$ and S_i must be compact as claimed.

Finally, we turn to the check of the eigenvalue claims. Since $\mathfrak{L}$ is exponentiable, we may replace D by $D_\infty = C_\infty(V) \supset D$, and $\mathfrak{L}$ by $\mathfrak{L}_\infty \cong \mathfrak{L}$ on D_∞. We omit subscripts ∞ for brevity, noting that the new $D = D_\infty$ is invariant under $V(t,A)$ for all $A \in \mathfrak{L} = \mathfrak{L}_\infty$. Then by Chapter 3 (or standard results on infinitesimal representations derived from group representations) we obtain for all $A \in \mathfrak{L}$ and $B \in \mathbf{B} = \mathbf{B}(\mathfrak{L})$

$$V(t,A)\,B\,V(-t,A) = \exp(t\ \mathrm{ad}\ A)(B).$$

Now, this identity extends as usual to $B \in \mathbf{B}_{\mathbb{C}}$, so as in Chapters 5 - 6 we may choose B to be one of the members of a basis for $\mathbf{B}_{\mathbb{C}}$ consisting of generalized eigenvectors for ad A with eigenvalue α and ascent s, so that

$$\begin{aligned}V(t,A)\,B\,V(-t,A) &= \exp(t((\mathrm{ad}\ A-\alpha) + \alpha))(B)\\ &= e^{\alpha t}\,\Sigma\{t^k/k!\ (\mathrm{ad}\ A-\alpha)^k(B)\ :\ 0 \le k \le s\}.\end{aligned}$$

But since every $B \in \mathbf{B}_{\mathbb{C}}$ is by definition a bounded operator, and

$$\|V(t,A)\| = \|V(\exp t A)\| \leq M < \infty \text{ for all } t,$$

we have

$$M^2 \|B\| \geq \|V(t,A)\, B\, V(-t,A)\|$$
$$= e^{\mathrm{Re}(\alpha)t} \| \Sigma \frac{t^k}{k!} (\mathrm{ad}\ A - \alpha I)^k (B)\| \quad \text{for all } t \in \mathbb{R}.$$

This is possible only if $\mathrm{Re}(\alpha) = 0$ and the coefficients of t^k for $k > 0$ in the operator polynomial

$$\Sigma\{t^k/k!\ (\mathrm{ad}\ A - \alpha)^k\ (B) : 0 \leq k \leq s\}$$

are zero: Hence $\sigma(\mathrm{ad}\ A|_{\mathbf{B}_{\mathbb{C}}})$ consists of pure imaginary eigen-values of ascent 0 as claimed. This completes the proof of the theorem.

REFERENCES

[Aa 1] Aarnes, J.F., Fröbenius reciprocity of differentiable representations, Bull.Amer.Math.Soc., 80 (1974) 337-340.

[Aa 2] Aarnes, J.F., Differentiable representations I: Induced representations and Fröbenius reciprocity, (Trondheim preprint 1973).

[AM] Abraham, R., and Marsden, J.E., Foundation of Mechanics, 2nd edn., The Benjamin Cummings Publ. Co., Reading, Mass., 1978.

[As] Arveson, W., On groups of automorphisms of operator algebras, J.Funct.Anal., 15 (1974) 217-243.

[Bb] Babalola, V.A., Semigroups of operators on locally convex spaces, Trans.Amer.Math.Soc., 199 (1974) 163-179.

[BB] Barut, A.O., and Brittin, W., Lectures in theoretical physics 10 : de Sitter and conformal groups and their applications, Colorado Associated University Press, 1971.

[Bg] Bargmann, V., Irreducible unitary representations of the Lorentz group, Ann.Math., 48 (1947) 568-640.

[BD] Bonsal, F.F., and Duncan, J., Complete Normed Algebras, Springer, New York, 1973.

[BES] Brezin, J., Ellis, R., and Shapiro, L., Recognizing G - induced flows, Israel J.Math., 17 (1974) 56-65.

[BH] Beurling, A., and Helson, H., Fourier-Stieltjes transforms with bounded powers, Math.Scand., 1 (1953) 120-126.

[Bk 1] Bourbaki, N., Integration, (rev.edn.) Hermann, Paris, 1965.

[Bk 2] Bourbaki, N., Groupes et Algèbres de Lie, Hermann, Paris, 1975.

[BR] Bratteli, O., and Robinson, D.W., Unbounded derivations in C*-algebras I and II, Comm.Math.Phys., 42 (1975) 253-268, and 46 (1976) 11-30.

[B Ra] Barut, A.O., and Raczka, R., Theory of group representations and applications, Polish Scientific Publishers, Warsaw, 1977.

[By] Bony, J.M., Principe du maximum, inégalité de Harnack et unicité du problème de Cauchy pour les opérateurs elliptiques dégénérés, Ann.Inst.Fourier, Grenoble, 19 (1969) 277-304.

[Ch] Chilana, A.H., Relatively continuous linear operators and some perturbation results, J.London Math.Soc. (2) (1970) 225-231.

[CM] Chernoff, P., and Marsden, J.E., On continuity and smoothness of group actions, Bull.A.M.S., 76 (1970) 1044.

[Co] Connes, A., Sur la théorie non-commutative de l'integration., Lect. Notes in Math. 725, 19-143, Springer Verlag, Berlin, 1979.
See also, C*-Algèbres et géométrie differentielle, C.R. Acad.Sci., Paris, 290, Ser A, 599-604, and: A survey of foliations and operator algebras, in: Operator Algebras and Applications, Proc.Symp.Pure Math., 38, vol. 1, Amer.Math.Soc., Providence R.I. 1980, 521-632.

[Cz] Chazarain, J., Problèmes de Cauchy abstraits et applications à quelques problèmes mixtes, J.Funct. Anal. 7 (1971) 386-446.

[Db] Dembart, B., On the theory of semigroups of operators on locally convex spaces, J.Funct.Anal., 16 (1974) 123-160.

[Dc] Dirac, P.A.N., The Principles of Quantum Mechanics, 3rd edn., Clarendon Press, Oxford, 1947.

[DM] Doebner, H.D., and Melsheimer, O., On representations of Lie algebras with unbounded generators: I Physical consequences. Il Nuvo Cimento, v. ILA, N.1 (1967), 73-97.

[DS 1-3] Dunford, N., and Schwartz, J.T., Linear Operators, Vols. I, II and III, Interscience, New York, 1958.

[Dx 1] Dixmier, J., Sur la relation i(PQ-QP) = I, Compositio Math., 13 (1958) 263-270.

[Dx 2] Dixmier, J., Représentations integrables du groupe de De Sitter, Bull.Soc.Math.France, 89 (1961) 9-41.

[Dx 3] Dixmier, J., Algèbres Envellopantes, Gathier-Villars, Paris, 1974.

[F] Fell, J.M.G., Non-unitary dual spaces of groups, Acta Math., 114 (1965) 267-310.

[FC] Foias, C., and Colojoară, I., Theory of Generalized Spectral Operators, Gordon and Breach, New York, 1968.

[Fd] Folland, G.B., Sub elliptic estimates and functional spaces on nilpotent Lie groups, Arkiv för Mat., 13 (1975) 161-208.

[FdeV] Freudenthal, H., and de Vries, H., Linear Lie Groups, Academic Press, New York, 1969.

[Fℓ] Feller, W., On the generation of unbounded semigroups of bounded linear operators, Ann.Math. (2), 58 (1953) 166-174.

[Frö] Fröhlich, J., Application of commutator theorems to the integration of representations of Lie algebras and commutation relations, Comm.Math.Phys., 54 (1977) 135-150.

[FS] Flato, M., and Simon, J., Separate and joint analyticity in Lie group representations, J.Funct.Anal., 13 (1973) 268-276.

[FSSS] Flato, M., Simon, J., Snellman, H., and Sternheimer, D., Simple facts about analytic vectors and integrability, Ann.Ecol.Norm.Sup. Paris, 5 (1972) 423-434.

[FSt 1] Flato, M., and Sternheimer, D., Sur l'integrabilité des représentations anti-symmetriques des algèbres de Lie compacts, C.R.Acad.Sci. Paris, Ser.A 227 (1973) 939-942.

[FSt 2] Flato, M., and Sternheimer, D., Remarks on the connection between external and internal symmetries, Phys. Rev.Letters, 15 (1965) 934-935.

[Ft] Freudenthal, H., Lecture notes on matrix groups, Washington University (St. Louis).

[Fu 1] Fuglede, B., On the relation PQ - QP = -iI, Math.Scand., 20 (1967) 79-88.

[Fu 2] Fuglede, B., Commuting and non-commuting self-adjoint operators, Preprint, University of Copenhagen, 1980.

[Fu 3] Fuglede, B., Conditions for two selfadjoint operators to commute or to satisfy the Weyl relation, Mathematica Scandinavica, 51 (1982) 163-178.

[Gå 1] Gårding, L., Note on continuous representations of Lie groups, Proc.Nat.Acad.Sci. USA 33 (1947) 331-332.

[Gå 2] Gårding, L., Vecteurs analytiques dans les representations des groupes de Lie, Bull.Soc.Math. France, 88 (1960) 73-93.

[Gd 1] Goodman, R., Analytic and entire vectors for representations of Lie groups, Trans.Amer.Math.Soc., 143 (1969) 55-76.

[Gd 2] Goodman, R., Analytic domination by fractional powers of a positive operator, J.Funct.Anal., 3 (1969) 246-264.

[Gd 3] Goodman, R., One-parameter groups generated by operators in an enveloping algebra, J.Funct.Anal., 6 (1970) 218-236.

[Gd 4] Goodman, R., Differential operators of infinite order on a Lie group, J.Math.Mech., 19 (1970) 879-894.

[Gd 5] Goodman, R., Some regularity theorems for operators in an enveloping algebra, J.Differential Equations, 10 (1971) 448-470.

[Gf] Gelfand, I.M., On one-parameter groups of operators in a normed space, Dokl.Akad.Nauk. SSSR, NS.25 (1939) 713-718.

[GJ] Goodman, F.M., and Jorgensen, P.E.T., Lie algebras of unbounded derivations, J.Funct.Anal., (in press).

[GJKS] Giles, J.R., Joseph, G., Koehler, D.O., and Sims, B., On numerical ranges of operators on locally convex spaces, J.Austral.Math.Soc., (to appear).

[GK] Giles, J.R., and Koehler, D.O., On numerical ranges of elements of locally convex algebras, Pacific J.Math., 49 (1973) 79-91.

[Hc] Hochschild, G., The Structure of Lie Groups, Holden Day, San Francisco, 1965.

[H-Ch] Harish-Chandra, Representations of a semisimple Lie group on a Banach space I, Trans.Amer.Math.Soc., 75 (1953) 185-243.
(This is the first paper in a series. The later papers appeared in Trans.Amer.Math.Soc., 76 (1954) 26-65 and 234-253: Amer.J.Math., 77 (1955) 743-777; 78 (1956) 1-41 and 564-628.)

[Heℓ] Helson, H., Isomorphisms of Abelian group algebras, Ark.Mat., 2 (1953) 475-487.

[Hℓ] Helgason, S., Differential Geometry and Symmetric Spaces, Academic Press, New York, 1962.

[Hö] Hörmander, L., Hypoelliptic second order differential equations, Acta Math., 119 (1968) 147-171.

[HP] Hille, E., and Phillips, R.S., Functional analysis and Semigroups, AMS Colloquium Publications, Providence, R.I., 1957.

[HR] Hewitt, E., and Ross, K.A., Abstract Harmonic Analysis I, Academic Press, Ne York 1963.

[HS] Hausner, M., and Schwartz, J.T., Lie groups and Lie Algebras, Gordon and Breach, New York, 1968.

[HW] Hazewinkel, M., and Willems, J.C., (ed.), Stochastic Systems: The Mathematics of Filtering and Identification and Applications, D. Reidel Pub. Co., Dordrecht, 1980.

[Hz] Hazewinkel, M., Formal groups and applications, Academic Press, New York, 1978.

[Hz 1] Hazewinkel, M., On deformation, approximations, and non-linear filtering, Systems and Control Letters, 1 (1981), 32-36.

[Hz M] Hazewinkel, M., and Marcus, S.I., On Lie algebras and finite-dimensional filtering, Stochastics, 7 (1982) 29-62.

[JM] Jorgensen, P.T., and Moore, R.T., Distribution represenations of Lie groups II (in preparation); Distribution representations ... I, (by P.T. Jorgensen), J.Math.Anal.Appl., 65 (1978) 1-19.

[J-Mu] Jorgensen P.E.T., and Muhly, P.S., Selfadjoint extensions satisfying the Weyl operator commutation relations, J. d'Analyse Math., 37 (1980) 46-99.

[Jo 1] Jorgensen, P.T., Some criteria for integrability of Lie algebras of unbounded operators, (Thesis) Aarhus, 1973.

[Jo 2] Jorgensen, P.T., Representations of differential operators on a Lie group, J.Funct.Anal., 20 (1975) 105-135.

[Jo 3] Jorgensen, P.T., Perturbation and analytic continuation of group representations, Bull.Amer.Math.Soc., 82 (1976) 921-924.

[Jo 4] Jorgensen, P.T., Approximately reducing subspaces for unbounded linear operators, J.Funct.Anal., 23 (1976) 392-414.

[Jo 5] Jorgensen, P.T., Approximately invariant subspaces for unbounded linear operators II, Math.Ann., 227 (1977) 177-182.

[Jo 6] Jorgensen, P.T., Selfadjoint extension operators commuting with an algebra, Math. Zeit., 169 (1979) 41-62.

[Jo 7] Jorgensen, P.T., Partial differential operators and discrete subgroups of a Lie group, Math.Ann., 247 (1980) 101-110.

[Jo 8] Jorgensen, P.E.T., The integrability problem for infinite-dimensional representations of finite-dimensional Lie algebras, Expositiones Mathematicae, Vol. 1, No. 3 (July 1983).

[Jt] Jost, R., Eine Bemerkung zu einem Letter von L. O'Raifeartaigh und einer Entgegnung von M. Flato und D. Sternheimer, Helv.Phys.Acta, 39 (1966) 369-375.

[Ka] Kahane, J.-P., Sur les fonctions sommes de series trigonometriques absolument convergentes, C.R.Acad. Sci. Paris, Ser. A 240 (1955) 36-37.

[Ki] Kisyński, J., An integration of Lie algebra representations in Banach space, International Atomic Energy Agency, Trieste, Internal Report No. 130 (1974).

[Km] Kallman, R.R., Unitary groups and automorphisms of operator algebras, Amer.J.Math., 91 (1969) 785-806.
[Ko] Komura, T., Semigroups of operators on locally convex spaces, J.Funct.Anal., 2 (1968) 258-296.
[Kr] Krein, S.G., Linejnye differentialnye uravnenija v Banachovom prostranstve, 1967 (Preprint).
[KS] Kunze, R.A., and Stein, E., Uniformly bounded representations and harmonic analysis of the real 2 × 2 unimodular group, Amer.J.Math., 82 (1960) 1-62.
[Kt 1] Kato, T., On the commutation relation AB - BA = c, Arch.Rat.Mech.Anal., 10 (1962) 273-275.
[Kt 2] Kato, T., Perturbation Theory for Linear Operators, Springer, New York, 1966.
[Lg] Lang, S., "SL(2,ℝ)", Addison-Wesley, Reading, Mass., 1975.
[Lg 1] Lang, S., Introduction to Differentiable Manifolds, Interscience, New York, 1962.
[Li] Liebenson, Z.L., On the ring of functions with absolutely convergent Fourier series, Uspehi Matem. Nauk, N.S. 9 (1954) 157-162.
[Ln] Lions, J.L., Les semi-groupes distributions, Port. Math., 19 (1960) 141-164.
[LP] Lumer, G., and Phillips, R.S., Dissipative operators in Banach space, Pacific J.Math., 11 (1961) 679-698.
[Lr] Luscher, M., Analytic representations of semisimple Lie groups and their continuation to contractive representations of holomorphic Lie semigroups, Deutsches Elektronen Synchroton, 75/51 (1975).
[Lu] Lumer, G., Spectral operators, Hermitian operators, and bounded groups, Acta Sci.Math., 25 (1964) 75-85.
[Mc] Michael, E.A., Locally multiplicatively-convex topological algebras, Mem.Amer.Math.Soc., 11 (1952) 1-82.
[Mr 1] Moore, R.T., Lie algebras of operators and group representations on Banach spaces, Thesis, Princeton University, 1964, 1-165 + ix.
[Mr 2] Moore, R.T., Exponentiation of operator Lie algebras on Banach spaces, Bull.Amer.Math.Soc., 71 (1965) 903-908.
[Mr 3] Moore, R.T., Commutation relations among unbounded operators and bounded semigroups, Notices Amer.Math. Soc., 12 (1965) 557.
[Mr 4] Moore, R.T., Differential harmonic analysis and locally convex representations of Lie groups, (Lecture notes, Univ. of California, Berkeley, 1966) 1-290.

[Mr 5] Moore, R.T., Measurable, continuous and smooth vectors for semigroups and group representations, Amer.Math. Soc., 19 (1972) 1-80.

[Mr 6] Moore, R.T., Banach algebras of operators on locally convex spaces, Bull.Amer.Math.Soc., 75 (1969) 68-73.

[Mr 7] Moore, R.T., Adjoint, numerical ranges, and spectra of operators on locally convex spaces, Bull.Amer. Math.Soc., 75 (1969) 85-90.

[Mr 8] Moore, R.T., Generation of equi-continuous semigroups by Hermitian and sectorial operators I, II, Bull. Amer.Math.Soc., 77 (1971) 224-229 and 368-373.

[Mr 9] Moore, R.T., Core Fréchet spaces of C^∞-vectors for families of unbounded operators, (in preparation).

[MZ] Montgomery, D., and Zippin, L., Topological Transformation Groups, Interscience, New York, 1955.

[Nk] Naimark, M.A., Linear Representation of the Lorentz Group, Pergamon, London, 1964.

[Nℓ 1] Nelson, E., Analytic vectors, Ann.Math., 70 (1959) 572-615.

[Nℓ 2] Nelson, E., Notes on non-commutative integration, J.Funct.Anal., 15 (1974) 103-116.

[Nℓ 3] Nelson, E., Feynman integrals and the Schrödinger equation, J.Math.Phys., 5 (1964) 332-343.

[Nℓ 4] Nelson, E., Topics in Dynamics, Princeton Lectures Notes, Princeton University Press, New Jersey, 1969.

[NS] Nelson, E., and Stinespring, W., Representation of elliptic operators in an enveloping algebra, Amer.J. Math., 81 (1959) 547-560.

[O'R 1] O'Raifeartaigh, L., Mass differences and Lie algebras of finite order, Phys.Rev.Lett., 14 (1965) 575-577.

[Ors] Ørsted, B., A model for interacting quantum fields, J.Funct.Anal., 36 (1980) 53-77. (See also, Composition series for analytic continuation of holomorphic discrete series, Trans.Amer.Math.Soc., 260 (1980) 563-573.)

[Ou] Ouchi, S., Semigroups of operators in locally convex spaces, J.Math.Soc. Japan, 25 (1973) 265-276.

[Ph] Phillips, R.S., Perturbation theory for semi-groups of linear operators, Trans.Amer.Math.Soc., 74 (1954) 199-221.

[Pk] Pukanszky, L., The Plancherel theorem for the universal covering group of SL(2,ℝ), Math.Ann., 156 (1964) 96-143.

[Pℓ] Palais, R., A global formulation of the Lie theory of transformation groups, Mem.Amer.Math.Soc., 22 (1957).

[Ps 0] Poulsen, N.S., Lecture notes on representations of Lie groups, Aarhus, 1972, (unpublished).

[Ps 1] Poulsen, N.S., On C^∞-vectors and intertwining bilinear forms for representations of Lie groups, J.Funct.Anal., 9 (1972) 87-120.

[Ps 2] Poulsen, N.S., On the canonical commutation relations, Math.Scand., 32 (1973) 112-122.

[Pw 1] Powers, R.T., Self-adjoint algebras of unbounded operators, Commun.Math.Phys., 21 (1971) 85-124.

[Pw 2] Powers, R.T., Self-adjoint algebras of unbounded operators II, Trans.Amer.Math.Soc., 187 (1974) 261-293.

[PS] Powers, R.T., and Sakai, S., Unbounded derivations in operator algebras, J.Funct.Anal., 19 (1975) 81-95.

[Ra] Rao, R.R., Unitary representations defined by boundary conditions - the case of $s\ell(2,\mathbb{R})$, Acta Math., 139 (1977) 185-216.

[Rd] Rudin, W., Functional Analysis, McGraw Hill, New York, 1973.

[Rd 1] Rudin, W., The automorphisms and the endomorphisms of the group algebra of the unit circle, Acta Math., 95 (1956) 39-55.

[Rk] Rickart, C.E., General Theory of Banach Algebras, Van Nostrand, Princeton N.J., 1960.

[RS] Reed, M., and Simon, B., Methods of Modern Mathematical Physics, Vols. I and II, Academic Press, New York, 1973.

[Ru 1] Rusinek, J., The integrability of a Lie algebra representation, Letters Math.Phys., 2 (1978) 367-371.

[Ru 2] Rusinek, J., On the integrability of representations of real Lie algebras in a Banach space, Preprint (9), Warsaw University, 1981.

[Sch] Schaefer, H.H., Topological Vector Spaces, (3rd printing), Springer, New York, 1971.

[Se] Serre, J.P., Lie algebras and Lie groups, Lecture Notes, Harvard University, 1964.

[Sg 1] Segal, I.E., A class of operator algebras which are determined by groups, Duke Math.J., 26 (1959) 549-552.

[Sg 2] Segal, I.E., An extension of a theorem of L.O'Raifeartaigh, J.Funct.Anal., 1 (1967) 1-21.

[Sg 3] Segal, I.E., Mathematical Cosmology and Extragalactic Astronomy, Academic Press, New York, 1976.

[Sg 4] Segal, I.E., Positive-energy particle modles with mass splitting, Proc.Nat.Acad.Sci. USA, 57 (1967) 194-197.

[Sh] Shale, D., Linear symmetrics of free Boson fields, Trans.Amer.Math.Soc., 103 (1962) 149-167.

[Si] Simon, J., On the integrability of finite-dimensional real Lie algebras, Commun.Math.Phys., 28 (1972) 39-46.

[Sℓ] Sally, P., Analytic continuation of the irreducible unitary representations of the universal covering group of SL(2,ℝ), Mem.Amer.Math.Soc., 69 (1962).

[Sm] Simms, D.J., Lie groups and quantum mechanics, Lecture Notes in Mathematics 52, Springer, New York, 1968.

[Sr 0] Singer, I.M., Uniformly continuous representations of Lie groups, Ann.Math. (2), 56 (1952) 242-247.

[Sr 1] Singer, I.M., Lie algebras of unbounded operators, Thesis, University of Chicago, 1950.

[St 1] Sternheimer, D., Extensions et unifications de algèbres de Lie, J.Math.Pures et Appl., 47 (1968) 249-289.

[St 2] Sternheimer, D., Propriétes spectrales dans les représentations de groupes de Lie, J.Math.Pures et Appl., 47 (1968) 289-319.

[SW] Singer, I.M., and Wermer, J., Derivations on commutative normed algebras, Math.Ann., 129 (1955) 260-264.

[Sz] Schwartz, L., Lectures on mixed problems in partial differential equations and the representation of semigroups, Tata Inst.Found.Research, 1958.

[Ta] Taylor, A.E., Linear operators which depend analytically on a parameter, Ann.Math. (2), 39 (1938) 574-593.

[Ti] Titchmarsh, E.C., Introduction to the Theory of Fourier Integrals, Oxford, 1937.

[TW] Tits, J., and Waelbroech, L., The integration of Lie algebra representations, Pacific J.Math., 26 (1968) 595-600.

[Us] Ushijima, T., On the generation and smoothness of semigroups of linear operators, J.Fac.Soc. Tokyo, 19 (1972) 65-127.

[Vd] Vidav, Ivan, Spectra of perturbed semigroups with applications to transport theory, J.Math.Anal.Appl., 30 (1970) 264-279. (See also, Perturbations of strongly continuous semigroups, Preprint, 1975, University of Ljubljana, Yugoslavia).

[VR] Vergne, M., and Rossi, H., Analytic continuation of the holomorphic discrete series of a semi-simple Lie group, Acta Math., 136 (1976) 1-59.

[Wk 1] Waelbroeck, L., Etude spectrales des algèbres complètes, Mémoires de l'Académie Royale de Belgique Cℓ. des Sc. 1962.

[Wk 2] Waelbroeck, L., Les semi-groupes differentiables, Deuxième colloque d'Analyse Functionelle CBRM, (1964) 97-103.

[Wk 3] Waelbroeck, L., Differentiable mappings into b-spaces, J.Funct.Anal., 1 (1967) 409-418.

[Wk 4] Waelbroeck, L., Differentiability of Hölder - continuous semigroups, Proc.Amer.Math.Soc., 21 (1969) 451-454.

[Wm] Wightman, A.S., The problem of existence of solutions in quantum field theory. In: D. Feldman (ed.), Proc. Fifth Annual Eastern Theoretical Phys.Conf., Benjamin, 1967.

[Wr] Warner, G., Harmonic Analysis on Semi-Simple Lie Groups I, Springer-Verlag, Berlin-Heidelberg-New York, 1971.

[Ya] Yao, T., Unitary irreducible representations of SU(2,2) I, II and III, J.Math.Phys., I : 8 (1967) 1931-1954. II : 9 (1968) 1615-1626. III : 12 (1971) 315-342.

[Yo] Yosida, K., Functional Analysis, (3rd edn.) Springer, New York 1971.

References to Quotations

Bohr, Niels, Atomic theory and mechanics, Nature, 116 (1925) 845-852.

Born, Max, and Wiener, Norbert, A new formulation of the laws of quantization of periodic and aperiodic phenomena, J. Math. Phys., MIT, 5 (1926) 84-98.

Dirac, P.A.M., The Development of Quantum Theory, Gordon and Breach Publishers, New York, 1971.

Einstein, Albert, Autobiographisches Skizze, in C. Seelig (ed.), Helle Zeit-Dunkle Zeit, Zurich-Stuttgart-Vienna, 1956.

Eucken, Arnold (ed.), Die Theorie der Strahlung und der Quanten, Verhandlungen auf einer von E. Solvay einberufenen Zusammenkunft, Halle an der Saale, Wilhelm Knapp, 1914.

Heaviside, Oliver, On operators in physical mathematics, Part II, Proc. Roy. Soc. (London), 54 (1893) 105-143.

Marlow, A.R., Mathematical Foundations of Quantum Theory, Academic Press, New York-San Francisco-London, 1978.

Mehra, J., and Rechenberg, H., The Historical Development of Quantum Theory, Vols. 1-4, Springer-Verlag, New York-Heidelberg-Berlin, 1982.

Pauli, Wolfgang, Uber den Zusammenhang des Abschlusses der Elektronengruppen im Atom mit Komplexstruktur der Spektren, Zeit. Phys., 31 (1925) 765-783.

Wiener, Norbert, I am a Mathematician: The Later Life of a Prodigy, The MIT Press, Cambridge, Massachusetts, 1956.

INDEX

LIST OF SYMBOLS